Additive Manufacturing

Second Edition

Additive Manufacturing

Second Edition

Edited by

Amit Bandyopadhyay

Susmita Bose

CRC Press is an imprint of the
Taylor & Francis Group, an **informa** business

CRC Press
Taylor & Francis Group
6000 Broken Sound Parkway NW, Suite 300
Boca Raton, FL 33487-2742

First issued in paperback 2021

ISBN 13: 978-1-03-223859-3 (pbk)
ISBN 13: 978-1-138-60925-9 (hbk)

Publisher's Note
The publisher has gone to great lengths to ensure the quality of this reprint but points out that some imperfections in the original copies may be apparent.

Library of Congress Control Number: 2019949036

Visit the Taylor & Francis Web site at
http://www.taylorandfrancis.com

and the CRC Press Web site at
http://www.crcpress.com

Contents

Preface

The field of additive manufacturing is seeing an explosive growth in recent years due to renewed interest in manufacturing in the United States and other parts of the world. The experience of drawing something in a computer and then see that part being printed in a 3D printer, which can be touched or felt, is still fascinating to many of us. Now we are seeing the same in our kids who are only in their middle school or high school and experience the revolution of additive manufacturing/3D printing through their own creation. Such transformative change in our society is only possible due to significant reduction in the price of various 3D printers. Even a few years back, a good 3D printer was more than $50,000 in the United States. Due to the high cost of 3D printers, the majority of our population was only able to see a picture or a video of different 3D printers. As the cost of the printer came down below $1000, most businesses, universities and schools now have some kind of 3D printers, which allows many people to explore and innovate with this fascinating technology. Therefore, we felt that our book will be quite timely where we have tried to capture some of these fascinating developments using "3D Printing" or "Additive Manufacturing" technologies.

We understand that there are a few other books in the market that deal with additive manufacturing in some form. When we reviewed the landscape, we realized that majority of those books are developed with greater emphasis on mechanical engineering aspects. However, at present, most of the printing technology is quite mature and majority of the innovation is coming in the areas of materials development, related critical issues and their applications. Therefore, we have focused our work more in the applications of additive manufacturing than the core 3D printing technologies. Our hope is that the reader will be able to see how these technologies are currently being used and then contribute to the field with their own innovation. We have designed the book in a way that can also be used in a class room setting. The first few chapters are focused on different additive manufacturing technologies based on various materials that are typically used. Then we discuss some of the application areas where additive manufacturing is making a significant impact. Finally, some discussions on educational aspects and regulatory issues have also been added since those are becoming important as the additive manufacturing is becoming a more mature technological platform in many industries. For this Second Edition, we have more emphasis on design issues in additive manufacturing and new information related to first-generation additive manufacturing processes for polymers. We have also added suggested further reading materials and some questions for educators if used as a textbook.

We like to thank all the authors for their contributions in this book. Without their help our project would have never been completed. Also, without some sacrifice at home from both our boys, Shohom and Aditya, we could not have completed this work.

Even after working in this area for more than 20 years, we still learn new things regularly related to applications of additive manufacturing. We hope that this book will be useful to many veteran researchers as well as those who are entering this field now, and help them understand the subject better to contribute toward their work, which can make a positive difference in our society.

Amit Bandyopadhya and Susmita Bose
Pullman, WA, USA
September 2019

Editors

Amit Bandyopadhyay, the Herman and Brita Lindholm Endowed Chair Professor in the School of Mechanical and Materials Engineering at Washington State University (WSU), received his BS in metallurgical engineering from Jadavpur University (Kolkata, India) in 1989, MS in metallurgy from the Indian Institute of Science (Bangalore, India) in 1992, and PhD in materials science and engineering from the University of Texas at Arlington (Arlington, TX) in 1995. In 1995, he joined the Center for Ceramic Research at Rutgers University for his postdoctoral training. In 1997, he joined WSU as an assistant professor, promoted to an associate level in 2001, and to the full professor level in 2006. His research expertise is focused on additive manufacturing of hard materials toward structural and biomedical applications. He has communicated over 300 technical articles, inventor of 19 issued patents, edited 10 books, and has supervised over 40 graduate students for their degrees in physics, mechanical engineering, and materials science and engineering. Among others, Professor Bandyopadhyay received the CAREER award from the US National Science Foundation, and the Young Investigator Program award from the US Office of Naval Research. Professor Bandyopadhyay is a fellow of the US National Academy of Inventors, the American Ceramic Society, the American Society for Materials, the American Institute for Medical and Biological Engineering, the American Association for the Advancement of Science, Society for Manufacturing Engineers, and the Washington State Academy of Sciences. His work has been cited over 17,000 times and the current "H" index is >69 (Google Scholar, 09/2019).

Susmita Bose is Herman and Brita Lindholm Endowed Chair Professor in the School of Mechanical and Materials Engineering (MME) at Washington State University. Professor Bose received her BS from Kalyani University (India), MS from the Indian Institute of Technology (IIT)—Kanpur, and PhD from Rutgers University (NJ, USA) in 1998 in Physical Organic Chemistry. Since 1998, she is working with 3D printing of bone tissue engineering scaffolds with controlled chemistry, especially with calcium phosphates, surface modification of metallic implants, and drug delivery. In 2001, she started as an assistant professor in MME, promoted to associate professor in 2006, and to full professor in 2010. Her awards include the Presidential Early Career Award for Scientist and Engineers (PECASE, the highest honor given to a young scientist by the US President at the White House) award from the National Science Foundation, the Schwartzwalder-Professional Achievement in Ceramic Engineering award, and Richard M. Fulrath award from the American Ceramic Society (ACerS). Prof. Bose was named as Life Science Innovation Northwest Women to Watch Honoree by the Washington Biotechnology and Biomedical Association. She has supervised over 40 graduate students, published over 250 technical papers, edited 9 books, and inventor of 11 issued patents. Her research papers have been cited over 16,000 times ("H" index 68, Google Scholar, 09/2019). Dr. Bose is a fellow of the Materials Research Society, US National Academy of Inventors, the American Society for Materials, the American

Association for the Advancement of Science, the Royal Society for Chemistry, the Washington State Academy of Sciences, the American Institute for Medical, and Biological Engineering and the ACerS. Dr. Bose's group research on 3D printed bone tissue engineering scaffolds has been featured by the AP, BBC, NPR, CBS News, MSNBC, ABC News, and many other TV, radio stations, magazines, and news sites all over the world.

Contributors

Vamsi Krishna Balla
Bioceramics and Coating Division
CSIR – Central Glass & Ceramic Research Institute
Kolkata, India

Amit Bandyopadhyay
W. M. Keck Biomedical Materials Research Lab
School of Mechanical and Materials Engineering
Washington State University
Pullman, Washington

F. S. L. Bobbert
Department of Biomechanical Engineering
Faculty of Mechanical, Maritime, and Materials Engineering
Delft University of Technology (TU Delft)
Delft, the Netherlands

Susmita Bose
W. M. Keck Biomedical Materials Research Lab
School of Mechanical and Materials Engineering
Washington State University
Pullman, Washington

Christian Carpenter
Apogee Boost, LLC
Monroe, Washington

Denis Cormier
Department of Industrial and Systems Engineering
Rochester Institute of Technology
Rochester, New York

Mitun Das
Bioceramics and Coating Division
CSIR – Central Glass & Ceramic Research Institute
Kolkata, India

Michael D. Dickey
Department of Chemical and Biomolecular Engineering
North Carolina State University
Raleigh, North Carolina

Thomas Gualtieri
W. M. Keck Biomedical Materials Research Lab
School of Mechanical and Materials Engineering
Washington State University
Pullman, Washington

Forough Hafezi
Department of Manufacturing and Industrial Engineering
Sabanci University
Istanbul, Turkey

Bryan Heer
W. M. Keck Biomedical Materials Research Lab
School of Mechanical and Materials Engineering
Washington State University
Pullman, Washington

Edward D. Herderick
Rapid prototype+manufacturing (RP+M)
Avon Lake, Ohio

Dongxu Ke
W. M. Keck Biomedical Materials Research Lab
School of Mechanical and Materials Engineering
Washington State University
Pullman, Washington

Bahattin Koc
Faculty of Engineering and Natural Sciences
Department of Manufacturing and Industrial Engineering
Sabanci University
Istanbul, Turkey

Caitlin Koski
School of Mechanical and Materials Engineering
Washington State University
Pullman, Washington

Can Kucukgul
Department of Manufacturing and Industrial Engineering
Sabanci University
Istanbul, Turkey

Mukesh Kumar
Advanced Process Technology Group
Biomet Inc.
Warsaw, Indiana

Y. Li
Department of Biomechanical Engineering
Faculty of Mechanical, Maritime, and Materials Engineering
Delft University of Technology (TU Delft)
Delft, the Netherlands

M. J. Mirzaali
Department of Biomechanical Engineering
Faculty of Mechanical, Maritime, and Materials Engineering
Delft University of Technology (TU Delft)
Delft, the Netherlands

Bryan Morrison
One Patient Solutions
Biomet Inc.
Warsaw, Indiana

S. Burce Ozler
Department of Manufacturing and Industrial Engineering
Sabanci University
Istanbul, Turkey

Dishit Paresh Parekh
Department of Chemical and Biomolecular Engineering
North Carolina State University
Raleigh, North Carolina

Clark Patterson
Rapid prototype+manufacturing (RP+M)
Avon Lake, Ohio

Naboneeta Sarkar
W. M. Keck Biomedical Materials Research Lab
School of Mechanical and Materials Engineering
Washington State University
Pullman, Washington

Timothy W. Simpson
Paul Morrow Professor of Engineering Design and Manufacturing
Director, Additive Manufacturing & Design Graduate Program
Co-Director, Penn State CIMP-3D
The Pennsylvania State University
University Park, Pennsylvania

Kellen D. Traxel
School of Mechanical and Materials Engineering
Washington State University
Pullman, Washington

Sahar Vahabzadeh
W. M. Keck Biomedical Materials Research Lab
School of Mechanical and Materials Engineering
Washington State University
Pullman, Washington

Ranji Vaidyanathan
Department of Materials Science and Engineering
School of Materials Science and Engineering
Oklahoma State University
Tulsa, Oklahoma

A. A. Zadpoor
Department of Biomechanical Engineering
Faculty of Mechanical, Maritime, and Materials Engineering
Delft University of Technology (TU Delft)
Delft, the Netherlands

Yanning Zhang
School of Mechanical and Materials Engineering
Washington State University
Pullman, Washington

1 Introduction to Additive Manufacturing

Amit Bandyopadhyay, Thomas Gualtieri, Bryan Heer, and Susmita Bose

CONTENTS

1.1 INTRODUCTION

Additive manufacturing (AM) is a technology that is rapidly developing and is being integrated into manufacturing and also our day-to-day lives. Its emergence into the commercial world has been labeled by a variety of names, such as three-dimensional (3D) printing, rapid prototyping (RP), layered manufacturing (LM), or solid free-form fabrication (SFF). Conceptually, AM is an approach where 3D designs can be built directly from a computer-aided design (CAD) file without any part-specific tools or dies. In this freeform layer-wise fabrication, multiple layers are built in the X-Y direction on top of one another to generate the Z or 3rd dimension. Once the part is built, it can be used for touch and feel concept models, tested for functional prototypes, or used in practice. The everyday consumer should realize that AM can be a way to connect with manufacturers on a new level. AM is much more than a process that can be used to make personalized novel items or prototypes. With new developments in AM, we live in an age that is on the cusp of industrialized rapid manufacturing taking over as a process to mass produce products and make it economically feasible to design and create new ones in a timely fashion. As a result, the manufacturing process of sectors across the globe will adapt to these developments while incorporating a new style of customer–manufacturer interaction. AM allows people to contribute to the design process from almost any location at all and will break the barriers of localized engineering and emerge on a global scale. Just as the Internet has given us the ability to spread and access information from any location at all, digital designing and CAD have given people the ability to make, change, and critique designs from anywhere. With AM, those designs can be made and tested from almost any location at all with very little lead time. The capabilities of AM machines have become sophisticated to a point where thinking of the design and drafting the model in CAD is the limitation instead of the manufacturability of the product.[1] As a new generation grows up with CAD technology and the abilities and availability of AM machines grow, the process of designing a product will mature from being created by a select group of engineers to being created by the consumer and company together, with the final product being able to be manufactured anywhere in the world in a timely manner.

1.2 HISTORY OF AM

1.2.1 Start of 3D Printing

AM was first developed in the 1980s when a man named Charles "Chuck" Hull invented the first form of 3D printing called stereolithography (SLA). It was the advancements in laser technology, along with Mr. Hull's innovation on the materials and process that he used, that first made this conceptual method a reality.[2] Stereolithography is a system where an ultra-violet (UV) light source is focused down into an UV photo-curable liquid polymer bath, where, upon contact, the polymer hardens. Patterns can be drawn using the ultra-violet source to semi-cure the polymer layer. Uncured polymer stays in the bath and provides support to the part that is being built. After a layer of printing is done, the hardened polymer layer moves down on a build plate in the liquid medium, and the next layer of polymer is available on top for the following layer. This process continues until the part is finished based on the CAD design and is removed from the liquid medium. In most cases, further curing is needed before the part can be touched. It was in 1983 when Chuck Hull invented this new technology, and, subsequently, in 1986, he formed the very first company to develop and manufacture 3D printers called 3D Systems.[2] This was the first step in the history of making an RP machine outside of science fiction movies or books. Mr. Hull was also the first to devise a way to allow CAD files to communicate with the RP system in order to build computer modeled parts. Such an endeavor was not trivial. To accomplish this, 3D CAD models had to be sliced in a virtual world, and then each slice could be used to build a layer using the 3D printer. In the first-generation CAD files for 3D printers, only the surface files mattered, which were termed. *stl* files from the stereolithography process. After developing this technology, the patent application was filed in August 1984 and was approved in 1986 by the United States Patent and Trademark Office (USPTO), making it the first patent of an RP system.[3] Though Chuck Hull patented this technology in 1986, it took several years for 3D Systems to launch the first solid state stereolithography system.[2]

1.2.2 Development of Other RP Technologies

While 3D Systems was developing and patenting this technology, other innovators started to develop new types of AM machines that used different methods and materials. At the University of Texas at Austin, an undergraduate student named Carl Deckard and an assistant professor Dr. Joe Beaman started working on a new technology known as selective laser sintering (SLS). SLS worked by first spreading powdered material on a build plate where a laser selectively sintered the powder in certain areas of the plate. Another layer of powder was then distributed over the previous layer and the process was repeated. In the end, the powder from each layer was sintered together in overlapping regions to produce the 3D part. Deckard and Dr. Beaman started working on this technology in 1984 and made the first SLS machine in 1986. They then commercialized the technology, creating the first SLS company called Nova Automation, which later became DTM Corp. In 1989, they made the first commercial machines which were called Mod A and Mod B, and they continued advancing and making more SLS machines until the company was sold to 3D Systems in 2001.[4]

Around the same time, Scott Crump and his wife Lisa, both graduates of Washington State University, were developing another AM technology in their garage. Scott wanted to make a toy for his daughter, so he invented the technology referred to as fused deposition modeling (FDM).[5] This technology involved the heating of a thermoplastic to a semi-liquid state, which was then deposited onto a substrate where it built the part layer by layer.[6] Scott and Lisa went on to start the company Stratasys, Inc. in 1989, selling this technology, in addition to having it patented in 1992.[7,8] Stratasys, Inc. has continued to grow, and the company now is a leader in polymer 3D printing with many printers in markets ranging from commercial industry to educational.

At the same time, another man named Roy Sanders was developing a new RP method. His company, formerly known as Sanders Prototype Inc., now named

Solidscape®, released its first 3D printer called the ModelMaker™ 6 Pro in 1994.[9] This machine used an inkjet approach to build a part.[10] This method essentially acts the same as stereolithography, but instead of a laser being focused into a liquid medium, hot thermoplastic wax liquid is sprayed onto a plate to build each layer of a part. This machine could make high resolution wax models which were very popular for businesses that perform complex investment casting such as the jewelry industry.[11] The company experienced commercial success and was later bought by Stratasys, Inc. in May of 2011.[12]

The above mentioned are just some of the original RP systems that were being developed at that time, yet these founders were not the only people who saw the significance of these technologies. Once 3D Systems patented stereolithography, companies in other countries started to develop this technology as well. In Japan, two companies called NTT Data CMET and Sony/D-MEC started to develop stereolithography systems in 1988 and 1989, respectively.[13] In addition, companies in Europe such as Electro Optical Systems (EOS) and Quadrax developed stereolithography systems in 1990.[13] Many companies around the globe were starting to develop their own 3D printing devices and coming up with unique ways to improve the process. It was apparent 3D printing had sparked worldwide interest at its inception, and since the early 1990s, rapid development of the process has been present across the globe.

1.2.3 Moving from RP to AM

At this point, most of the technologies were developed to produce polymeric objects and had not been able to process other materials such as metals or ceramics. Inherently, these machines were limited to the prototyping stage and not at the full-on manufacturing stage where the finished parts are made to be used. It took further development in RP technology in order to produce parts out of metals and ceramics. In order to push the boundaries of RP and bring the technology to a manufacturing level, several companies were trying to develop a metal AM machine. One of the first was a company mentioned before called EOS, which was started in 1989 by Dr. Hans J. Langer and Dr. Hans Steinbichler.[14] They began working on printing plastic parts using stereolithography systems and then SLS. In the early 1990s, they started to research using SLS to make metal parts and presented their first prototype of a direct metal laser sintering (DMLS) machine in 1994. Subsequently, they launched the first DMLS system in 1995.[14] This process worked similarly to SLS, but could sinter metal powders. Initially, the metals that could be used in this process were many general engineering materials such as aluminum, cobalt, nickel, stainless steel, and titanium alloys.[14] In 1997, EOS sold its stereolithography product line to 3D Systems and took over the global patent rights for laser sintering technology.[14] Since then, it has significantly advanced SLS and DMLS and has made it one of the most popular AM processes in manufacturing. As a result, this has made EOS one of the most successful and competitive AM companies in the world.

Around this same time, another AM technology that could produce metal parts called laser engineered net shaping (LENS®) was being developed in Albuquerque, New Mexico. It was developed by Sandia National Laboratories in 1997 and was eventually commercialized by Optomec.[15] The first machine was sold in 1998.[16,17]

The LENS® system worked by depositing powder under a high-power laser where it was melted and solidified on a substrate. The base and head were both mobile which made it so the metal could be deposited in selected areas of the substrate. The metal was then deposited layer by layer until the desired part was built.[18] Optomec has continued to advance LENS® technology and is constantly updating its product line to fit the ever-changing consumer interests.

Another popular type of AM process also being developed in the late 1990s was electron beam melting (EBM). A company called Arcam AB started in 1997 creating EBM technology.[19] EBM worked by shooting an electron beam at a powder bed in selective areas. Once a layer of powder had been melted in selected areas, another layer of powder was laid on top of the previous one, and the process was continued until the part was complete.[20] Working with Chalmers University of Technology, Arcam AB released its first EBM machines and sold them to two clients in 2002.[19] In 2007, a manufacturer of orthopedic implants made a Fixa Ti-Por hip implant that was Conformité Européenne (CE)-certified using EBM technology.[19] Since then, more implants have been made using EBM. EBM is also being used in the aerospace industry, and as it continues to develop, so do the number of applications that are found that benefit from its use.

1.2.4 Impact of AM

Since the emergence of these technologies and companies, the AM industry has constantly been expanding, growing, and advancing with much enthusiasm. With many industries seeing the lucrative value and abilities of AM, the global market has been expanding very quickly, as shown in Figure 1.1. Many new types of RP and AM methods have been created since these original pioneers first started developing the technology. Some new technology has been novel and some were just variations of the established methods. There has also been a lot of development in the materials that can be used, as well as research into making their properties optimal for end use. These original technologies all started as RP, layered manufacturing, or solid free-form fabrication methods, where they were designed to only make quick prototypes

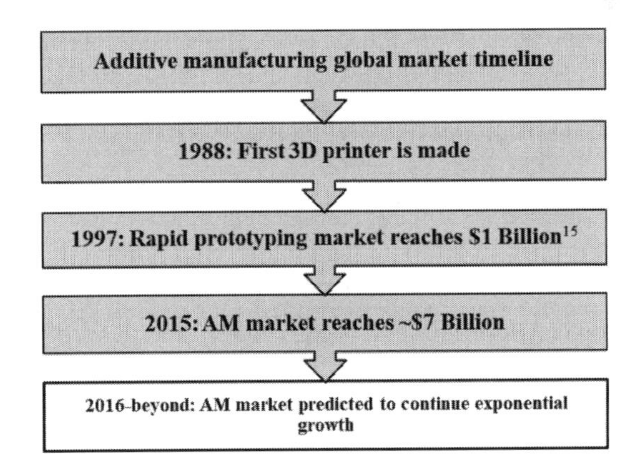

FIGURE 1.1 Growth of additive manufacturing industry.

or "show and tell parts" using polymeric materials. Over the years it has moved into AM, where functional prototypes and parts can be made to perform in a variety of environments. As AM has been incorporated into industry, its global market has steadily increased since the first 3D printer was made. Due to this continuous exponential growth in interest and adoption of the technology, it has been predicted that >10% of products and components produced will be influenced by AM.[21] AM will continue to become more integrated into industry and our personal lives as the technology and availability continue to grow.

1.3 CURRENT MANUFACTURING CHALLENGES

1.3.1 Part-versus Systems-Level Manufacturing

In a broad sense, manufacturing can be broken down into part-level and systems-level manufacturing, each with its own inherent number of challenges and issues. Part-level manufacturing is where manufacturing methods are focused on creating certain parts to be shipped to consumers. Systems-level manufacturing is taking these parts and combining them with others to create the entire system. Let's take the example of manufacturing an airliner. Millions of parts working seamlessly together require intensive control and monitoring for part integration, and a streamlined assembly line is essential for successful systems-level manufacturing. Similarly, intensive manufacturing control is essential for part-level manufacturing in order to ensure parts are reliable, reproducible, and accurate to the design in order to be properly integrated into the airplane. As seen in Figure 1.2, these part-level manufacturing controls can be broken down into four subcategories (design, materials, processes, and quality control) that each have their own challenges. During conventional manufacturing, part design was the largest bottleneck for designers and manufacturers, as the design dictated the complexity of tooling costs, the amount of processes required to reach the finished product, and of course, the lead time to produce the part. Typically, complicated designs are burdened with rising manufacturing costs, as more processes and high tooling costs are involved in order to realize the final design. Also, the type of material and quality control can dramatically increase the manufacturing cost. If engineering specifications require, say, an exotic metal of high material cost or quality control for strict material homogeneity, defect control, or microstructural control, the cost increases. Many factors actually go into part-level manufacturing to ensure that future systems-level manufacturing is feasible and reliable.

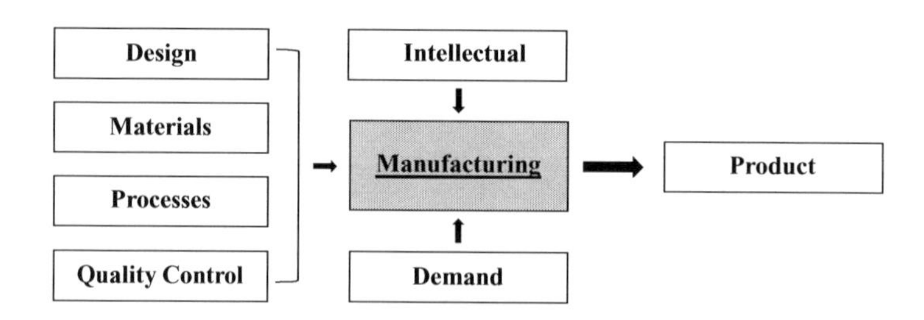

FIGURE 1.2 Challenges toward part-level manufacturing.

Much more goes into manufacturing in addition to these four subcategories, such as intellectual property and demand. Through conventional manufacturing, availability of a skilled workforce and manufacturing facilities are factors for creating a product in high demand. For example, processes like forging, pressing, and extrusion require heavy machinery, large floor space, and a skilled workforce in order to ensure efficiency during and between manufacturing steps. Then, with demand determining part volume requirements, efficiency is crucial to reduce lengthy lead times for large volumes of parts. With a limited production volume, lead times can remain long for high volume production. On the other hand, small volume production is economically inefficient due to the long lead time, as well as a high tooling cost/part ratio. With AM, however, lead times can be reduced, as the entire part is printed in a single machine with no tooling requirements, which eliminates the need for multiple machines in order to create the final product. This can be viewed in Figure 1.3, where AM can reduce the time it takes to go from the first generation design to the final design in the part manufacturing flowchart compared to conventional processes, which require more intensive manufacturing steps. Also, with the only tooling cost being the AM machine itself, and no specialized training required, the cost per part remains relatively constant over any production volume range. The shorter lead time to produce 3D concept models and less machine requirements are some of the key advantages over conventional manufacturing and were the driving forces for the first generation of additive manufacturing technologies.

1.3.2 Centralized and Projection-Based Manufacturing Issues

Presently, the standard distribution of goods and products is done by large-scale production. Apart from this system's many advantages such as the low cost of standard goods and high rates of production, it also has disadvantages and issues. Projection-based manufacturing is a method where past sales and statistical forecasts for future estimated sales contribute to the number of parts manufactured for a certain product. While this method works well, there can be a waste of products when demand changes due to a variety of reasons such as social-, political-, or

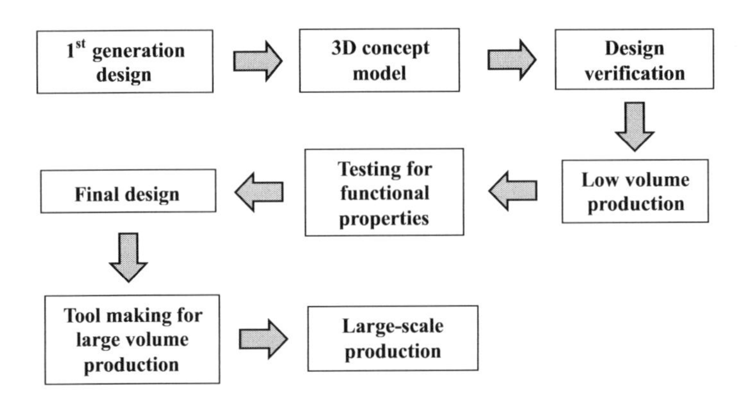

FIGURE 1.3 Part manufacturing flowchart.

economic-based decisions by the consumer. Waste of product and sometimes loss of jobs can be associated with an unpredicted change of demand, and if products were made on-demand, these issues could be alleviated. However, if mass production is based in other countries then there is the danger of dependence on foreign markets. If the foreign market is impacted by events such as a natural disaster or war, the production does as well.[22] This happened in 2011 when Japan was hit by a tsunami. Many automotive manufacturing plants were damaged, which impacted not only Japanese automotive companies, but American companies as well, and consequently significantly disrupted vehicle production.[23] There are also issues such as the distributors curtailing product availability. The retailers do not always display the new product because they would rather devote shelf space to a proven seller.[24] This could cause a scenario where the newest version of a product could fit the need/want of the consumer, but is not available because the store is not buying the product in order to preserve shelf space. Of course, these are general issues that tend to arise in large-scale production and could be reduced if the products were made using a small-scale on-demand manufacturing method.

1.3.3 Generalized Designs: Consumer Settling for Only Adequate Products

Another issue that large-scale production often leads to is a product that suits most people's needs, but does not cater to the individual consumer's taste. Parts are designed by engineers to fix an issue or fulfill a need of a consumer. Yet, for many things, the general item that is made does not fit the exact needs of the individual buyer because it is built for the general population instead. Likewise, the designers that create the part are sometimes not experiencing the issue firsthand. They only make the product to fit the specifications they receive however they receive them. This trail of information is not always reliable or effective to make an optimal product. Also, on a mass-scale production line, it can be hard if not impossible to make custom products based on the tooling and methods that have been established. This can limit the ability of manufacturers to make custom products.[25] Therefore, the large-scale production of many goods does not accommodate the individual likes and tastes of consumers. As a consequence, this way of manufacturing results in the consumer settling for something that is just adequate. Though this is OK for many circumstances, if we have the resources to make things exactly how we want them, then why wouldn't we?

Let's take a scenario where someone wants a desk and desires certain shelves, drawers, and size of the desk to fit their office. Possibly, they want some custom designs built into the desk to make it personal. A manufacturer has built a series of desks with a certain configuration of storage space and has made it so the desk can fit through a standard door. Yet maybe none of the available desks are designed exactly as this particular person wishes. The person could try and contact the manufacturer and see if they could custom build a desk for him. Since this would disrupt the production line, it would likely cost a substantial amount more, if it could even be done at all. Also, there usually tends to be an issue with communication from two different locations, and describing exactly what the person wants could be difficult. In the end, this person will most likely buy the desk that fits their needs the closest and will settle for something that is not quite exactly what they wanted. Now back to the question from before, what if making a perfect desk is possible, could be done easily, and was cost effective; do you think the consumer would spend a little more money and effort to buy that? Of course, they would, and with the current state and development of AM, as well as the availability and improvement of CAD, this could soon be a reality across a variety of consumer products.

1.4 AM: UNPARALLELED MANUFACTURING PARADIGM

1.4.1 Current State of AM and How It Generally Works

AM has now reached a point where it is ready to be implemented for industrial use. Its advantages over traditional manufacturing methods have caught the interest of most industries. The advantages stem from the machine's ability to create complex geometries using a layer by layer build system in contrast to conventional methods. Though there are many types of methods and machines, they all generally work using the same principal process. First, a CAD model of a part or object is made. The CAD file is then converted to an STL file. An STL file is the standard file type for nearly all AM machines which were created by Chuck Hull.[2] The system then cuts the item apart into layers on the computer in the easiest direction to build. Then, by various methods, it deposits or binds material layer by layer, stacking each one on the next, until the part is built. This system allows for incredibly complex geometries to be built relatively easily out of a variety of materials. Parts that could not be made using any other manufacturing method now can be made using this technology. Due to its capabilities of making such shapes, the amount of applications it can be used for is unparalleled by any other manufacturing method.[1] Industries such as art, aerospace, and medical have applications where AM could be used. This makes it a promising new method that likely will be incorporated into the industrial mass production of products. It will change the way parts are manufactured, designed, and distributed, as well as the customer-manufacturer relationship.

1.4.2 Advantages of AM: No Restriction on Design

AM will have a profound effect on the manufacturing process of many goods in many different industries and has the ability to make parts that could not be made before. AM is a start-to-finish process that can make an entire part without multiple machines or processes and can effectively build complex geometries that are very difficult, costly, or impossible using conventional methods. This gives the designer added freedom when making a part. Many times, the optimal design is not feasible with the types of manufacturing processes available presently. With AM, there is no restriction besides the size of the part, which has to fit in the machine. Now, the designer only has to design their part so that it can fit the specifications for its own application instead of largely considering manufacturing restrictions into their design. Another benefit to this process is the only tooling involved is a single AM machine, so a constant tooling cost is eliminated. Though some parts must be machined afterwards to have the right surface finish, there is much less tooling required as compared to conventional methods. This reduced tooling eliminates a huge cost of production. The only other cost is maintenance of the machine. AM also saves material because it is an additive technique as opposed to reduction technique. An example of a reduction manufacturing method is milling, where the product starts as a block with dimensions larger than the final product. The excess material is then removed until the final dimensions are achieved. The waste material is then either disposed of or recycled, for which the manufacturer usually has to pay. With AM, material is added until the product is made so little to no material is lost, which can equate to a 75% reduction in material use and can lower the production time and cost by 50%.[26] These huge savings are one of the reasons that AM has sparked so much interest with manufacturers.

Figure 1.4 demonstrates how the cost per part generally changes for part volume and complexity through additive manufacturing and conventional manufacturing processing. At low part volume and high complexity, the cost per part is the highest. In contrast, AM processes are generally a horizontal line in both graphs, which begins to show the regimes where AM can outperform conventional processing. It should also be noted that with the technological revolutions in the AM community, as well as the widespread adoption of additive technology throughout multiple industries, the AM cost per part vs. part volume trend is decreasing over time. In the future, this trend will likely continually decrease with increased implementation of the technology.

1.4.3 Advantages of AM: Versatility in Manufacturing

Another key aspect that makes AM so lucrative is its versatility in the parts that it builds. If it is found that the design that is being produced has a flaw, or there is something that can be changed that would optimize its use, it can be changed instantly.

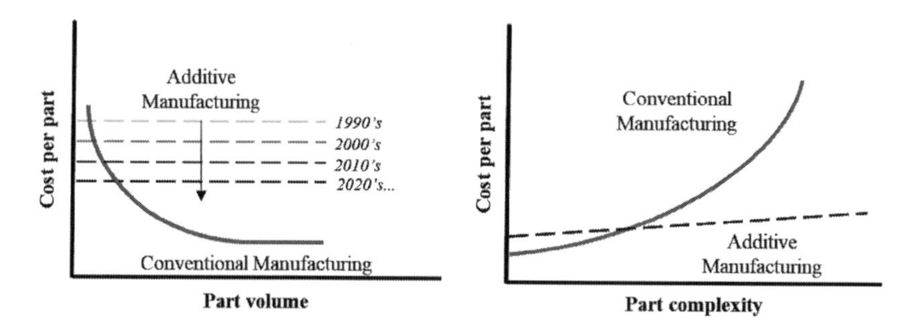

FIGURE 1.4 Comparison of cost per part vs. part volume and complexity for conventional manufacturing and additive manufacturing.

In many traditional manufacturing methods this can be very difficult. For example, in a casting process, once an expensive die is made, it cannot be changed that day to accommodate an alteration in the design. This is why AM started as a process that could be used to make a new part fast and cheap to test to see if it would work for the application. AM still has this capability, which is an undoubtedly powerful tool, but now has moved into being able to produce ready-to-use parts. It also makes it so on-demand building is much easier and costs less. If a designer wants to try something new, or a customer wants a custom part, it can be built easily without disrupting normal production.

1.4.4 Advantages of AM: Altering Materials for Enhanced Performance

AM can now also use many different materials, such as various plastics, metals, composites, and ceramics. The type of material depends on the type of AM process. The most popular materials are plastics because they have been studied the most and have been around the longest.[1] It has been found that not only can AM use different materials, but researchers are finding ways that these processes can be used to alter materials and change their properties. Some of the new freeform fabrication techniques can bond materials like a ceramic and metal to create a composite that has increased wear properties.[27] Or, these techniques can be used to deposit a ceramic coating on a metal substrate to increase the material's thermal and wear resistance. Another way AM processes are used is for the repair of broken parts and structures. When a material is broken or has experienced material loss, instead of replacing the part, an AM machine simply adds material back or bonds the two parts together. This process is known as laser cladding and can lower the cost of many industries that must replace parts or structures frequently.[28] This shows that AM processes do not just provide advantages for making complex geometries, but can also optimize material properties to make the final part even more effective, as well as be able to fix damaged parts.

1.4.5 AM Already Incorporated into Modern Manufacturing

AM is on the horizon for being a standard part of manufacturing, and companies have already created facilities that are dedicated to additive manufacturing of parts. GE Aviation opened a facility in December 2013 that accommodated up to 60 electron beam melting and direct laser sintering machines, and more and more companies are making space for AM expansion.[29] The aerospace industry is already using AM machines to build parts that will go into engines. AM makes it so aerospace manufacturers can optimize parts, lower weight, reduce material loss, and increase the buy-to-fly ratio.[26] Buy-to-fly refers to the time it takes between purchasing the material (generally expensive metals) to the time it is flying and making money. Along with GE, many other aerospace manufacturers are starting to use AM or already are implementing it into their production line. This just shows that AM is already being adopted into mainstream production in multiple competitive industries. Even though the aerospace industry is one that can typically afford these expensive manufacturing processes, it is a precursor to manufacturing other goods. Issues such as surface finish and material properties still pose some issues, but overall, AM now has the ability to make parts ready for use.[26]

1.4.6 Evolution of CAD to AM and Its Influence on Manufacturing

The immense capabilities of these different machines cause the real constraint in the manufacturing process to be the design.[1] One of the key factors that makes AM so groundbreaking is it can build a ready-to-use part from a CAD file. As the advancements in CAD have made it so almost anything can be designed, essentially any part that can be theorized can now be made in a digital format. This CAD file can then be transferred to an STL file and made on an AM machine. This capability has given engineers and designers the ability to design more complex and efficient prototypes and parts than when it was done on paper. Yet, even when CAD was developed, designers still had to design parts based on inherent constraints in available manufacturing technologies. Now, with the advent of additive manufacturing technologies such manufacturing constraints do not exist, which gives engineers design flexibility that was not available before.

Figure 1.5 summarizes advantages and challenges of additive manufacturing as discussed before. Though there are many challenges, usage of additive manufacturing will become more widespread due to ease of operation, ability to explore creativity, and various other reasons previously mentioned. It is anticipated that with current and future technological innovation, the young generation will lead the development and applications of AM technologies worldwide.

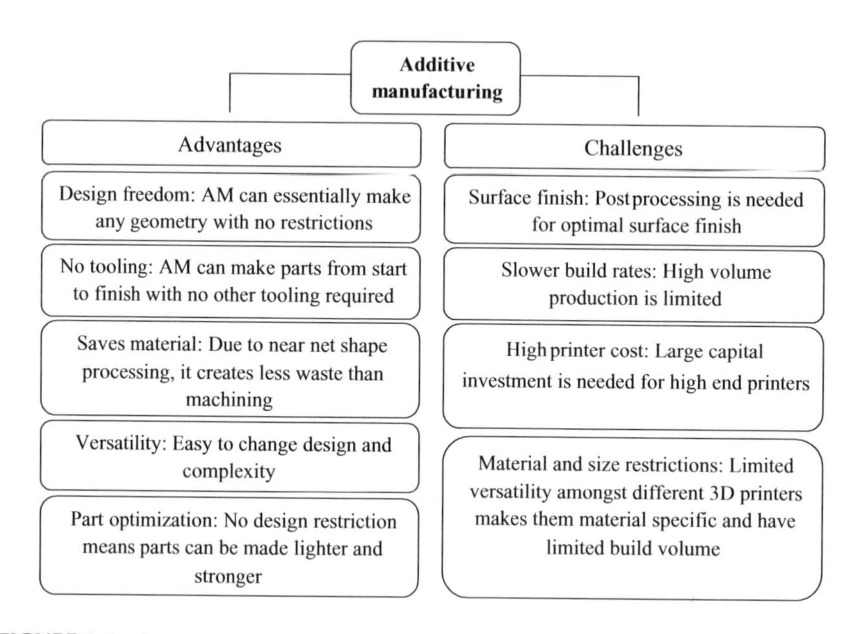

FIGURE 1.5 Summary of advantages and challenges of additive manufacturing.

1.5 GLOBAL ENGINEERING AND AM

1.5.1 Moving from Localized to Globalized Engineering

AM's ability to build parts from a CAD file will make it so companies and people exist in a world where the ability to communicate information will have no geo-political boundaries. Companies will be able to effectively and efficiently communicate designs and concepts anywhere in the world; this means anyone who has the ability to operate CAD will have the ability to create a part or alter a design. All designs created will be able to be made into real working parts from a physical distance of 10,000 kilometers or more without any problem. Except it will not just only be able to spread ideas, but physical objects as well. Creation of a part can be accomplished from anywhere on earth, or beyond earth, which will make it so we now have the ability to break away from localized engineering and move into global engineering. Figure 1.6 gives a few examples of different design concepts that are made simply manufacturable through AM. Anything from simply engineered parts to large architectural designs can be conceptualized and realized from any part of this world. Such freedom in design and manufacturing will transform the next generation of products due to inputs from the local population who will eventually use them.

FIGURE 1.6 Visuals of parts with different features that are simply manufacturable with AM. (From Roschli, A. et al., *Addit. Manuf.*, 25, 275–285, 2019.)

1.5.2 Engineer from Anywhere in the World Efficiently and Effectively

AM and CAD will make commercial designing of parts much more efficient and able to be done from anywhere. The main advantage this technology gives is in making communication of ideas and designs much easier and straightforward. Many companies have multiple branches in different countries and locations and clear communication between them is imperative to operation. Having key people in the right location to work on a project is not always an option, and sometimes miscommunication between facilities in different locations working on different parts of a project can slow production. In the past, schematics or drawings could be sent from one designer to another to try and interpret what has been done. In current times, CAD files can be sent over the Internet, and the part can be viewed in a 3D, interactive space. This has a much greater advantage over trying to interpret 2D images. Yet, still testing and seeing something on the computer is not the same as seeing the real thing and knowing how it will act in real life.

This can lead to issues with pairing parts. Boeing experienced this problem with the production of the 787. They had many parts built in many different areas of the globe, and when they were all brought back to Everett, Washington to be built, some of the parts did not integrate as originally planned.[31] AM makes it so these issues are less of a problem, as it allows for design teams in different areas to relay information across the globe by giving them a literal model of the part. Designs can be sent to one location, altered, and sent to another location, etc. This is a cost-effective process because they do not need to make new tooling, send people to different sites, or wait very long to have the part built. This capability will have huge effects that will change the way design can be done. AM is already being implemented for these uses in companies like Lockheed Martin. They originally had five different business sectors that worked independent of each other and qualification of different parts was a slow and difficult process. Now they have implemented 3D printers in different areas of production to try and speed this up. This makes it so each location is working with the same systems and machines in order to speed up qualification.[26] Now, with this integrated AM design process, it is efficient and feasible to have a design team anywhere in the world to quickly create a properly integrated product.

1.5.3 Manufacturing in Space: No Longer a Dream

This capability can even go beyond earth. NASA currently is working on trying to develop a robust technology so settlements can be made on the moon and are developing AM machines in order to utilize in-situ resources to build structures or parts.[33] If, or more likely when, we make settlements outside of earth on the moon or mars, communication between earth and astronauts can be very effective and clear using AM machines. If there is a problem and some complex part or device must be made at a satellite location, engineers on earth will be able to send up CAD files to be printed at the location. Figure 1.7 gives a visual image of this ability AM will provide. Just like a company designing from multiple locations, AM allows the same thing to be done over any space where a signal can transfer data.

FIGURE 1.7 Examples of demonstration parts fabricated out of Martian simulant. (From Goulas, A. et al., *Addit. Manuf.*, 10, 36–42, 2016.)

1.6 FUTURE TRENDS

1.6.1 On-Demand Manufacturing of Custom Products

This idea of global engineering does not stop at companies and engineers; it will incorporate the everyday consumer as well. CAD is now a standard tool that most all people are familiar with. Today, kids are learning how to use some sort of 3D modeling software starting from a young age in schools. It is no longer only a drafting tool that is taught in colleges, but has trickled down to being learned as early as middle school.[34] It is almost at the point where it is like typing; it's assumed that if you were born after a certain year, you know how to do it. This makes it so essentially anyone can design something that they want, as long as it is not beyond their CAD abilities. Now AM also makes it so they can build anything they design.

AM will enable on-demand manufacturing of custom goods to be a reality. As mentioned earlier, the mass manufacturing method of producing goods has its flaws. In many cases, it would be more economical if the goods could be made closer to the location of sale and made on-demand to the customer's exact needs. This system of course would be very costly for most goods and would cause their price to increase dramatically. AM makes it possible to be economic to manufacture volumes of one.[1] Goods like tables and chairs and other moderately priced home items could now be made using AM technologies and be completely customizable to the consumer. Many industries which are characterized by increasing demand for individual customization, such as furniture, are incorporating AM.[35] Figure 1.8 shows a very unique and intricate chair that has been 3D printed and shows the complexity and detail that AM can achieve. Also, if another customer wanted to change the design or alter it, they could with ease. AM centers could be in many areas to ensure the items would not have to be shipped as far. This makes the customization of those goods very easy as well. In the beginning of this chapter, the example of the person wanting a

FIGURE 1.8 3D printed cellular loop chair. (From Andersen, G., Valve Handle by Geir, MakerBot, Thingiverse, Published on June 19, 2011, http://www.thingiverse.com/thing:9450.)

specific table was brought up. If that person knew how to use CAD, and modeled the exact table they wanted, they could just send it to an AM manufacturer and have it built exactly how they want it. The person could now get the table they need, as well as add on any custom parts they want, such as their name engraved on it. Assuming these AM machines were built to make these kinds of structures, it should cost no more than what a standard table would cost when made there. Even if the person did not have CAD experience, in a modern AM company, a consumer ideally would be able to sit down with a designer and make the ideal chair or table for themselves.

1.6.2 Allowing People's Creativity to Become a Reality

This practice allows consumers to easily have almost everything custom made to their liking with little to no extra charge. With these capabilities, people have the aptitude to be creative and come up with new things that were never thought of before. The design team has now moved from the small group of engineers to the collective brains of everybody. Now every person can come up with an idea and have it become a reality with ease. How many times in people's lives have they thought of an invention that could help them in their day-to-day lives or just be something unique they want in their house. Then they let the idea pass by because they do not have the time, resources, or skills to make this idea come to life. By having readily available AM machines nearby and ready to make parts, all a person has to do is create their idea in a CAD file, and it can be made.

FIGURE 1.9 3D printed hose nozzle handle. MakerBot® Replicator Desktop 3D Printer was used to build this. (Courtesy of MakerBot®; From Folkway University of the Arts, Bionic Manufacturing Program, Photo by Nathalie Richter, Design by Anke Bernotat, Partners: Authentics, Plant Biomechanics Group Freiburg, Folkwang University of the Arts, Fraunhofer IWM, Fraunhofer UMSICHT, Fruth, KIT, RPM. Funded by the BIONA funding program of the Federal Ministry of Education and Research, March 1, 2013.)

1.6.3 Personal AM Machines as a Standard Household Application

There is also a growing industry of personal AM machines or 3D printers. People do not even have to send their CAD file out to be built. This gives the ability to make custom items for people's own homes, as well as day-to-day items. For instance, if your hose handle breaks, instead of buying a new one, you can just design and 3D print one. Figure 1.9 is an image of a hose valve handle that was 3D printed to replace a broken one. As stated before, the boundaries on what can be made are what the user can imagine. Many companies such as MakerBot® are making printers that are made for home use and are becoming more affordable. Figure 1.10 is an image of the MakerBot® Replicator Desktop 3D Printer, which is priced and designed to target at-home usage.[38] As these companies and the technology develop, the price of household 3D printers are continually dropping, and it won't be long before 3D printers become a standard household item. This just adds to the global engineering by making it easier for people to spread ideas and design new things. Now if somebody wants a custom part such as a unique lamp shade, all they have to do is design it and print it from home.

FIGURE 1.10 MakerBot® Replicator Desktop 3D Printer. (Courtesy of MakerBot®; From MakerBot® Replicator Desktop 3D Printer, Makerbot.com, 2009–2014. http://store.makerbot.com/replicator.)

1.6.4 AM Advancing Medical Technology and Helping Lives

This globalized engineering does not only help with commercial goods; it can be life-changing in the medical industry.[39–45] When it comes to an implant or tissue replacement, nobody wants to settle for something that most closely fits their needs. The patient wants the product to be perfect and is generally willing to spend more money to make that happen. Currently in the medical field there are different sizes of implants to fit different patients. Though there is a lot of versatility in different types of implants, it is sometimes necessary to have custom implants or patient-matched implants.[45,46] A custom implant could also better ensure the implant will be successful particularly for trauma patients with complex fractures. Making a custom implant can be difficult using traditional manufacturing methods and also tends to involve a long demanding adaptation phase before an optimum result is achieved.[47] Whether the implant is going to replace a bone or act as a scaffold to be placed in a damaged bone or tissue, it will more than likely be a hard or impossible part to machine using traditional methods. AM now has the ability to make implants that fit optimally into a patient, as well as create new implants that could not be made before. Figure 1.11 shows an image of some spinal implants made by EOS using AM. The parts are very complex and would not be able to be made using conventional forms of manufacturing. These implants could be made quickly with lower lead and healing time, as well as make it possible to fix problems that we could not before. It also is more cost effective for the hospital, which can lower the price of the procedure and increase patient care.[47] Things such as tissue-engineered cranial implants and porous bone scaffolds can now be made using AM.[45,48] Just like the example of furniture, these implants can be made on-demand when the patient comes in. From an X-ray, a CAD model can be made of the injury.[45,49] That model can then

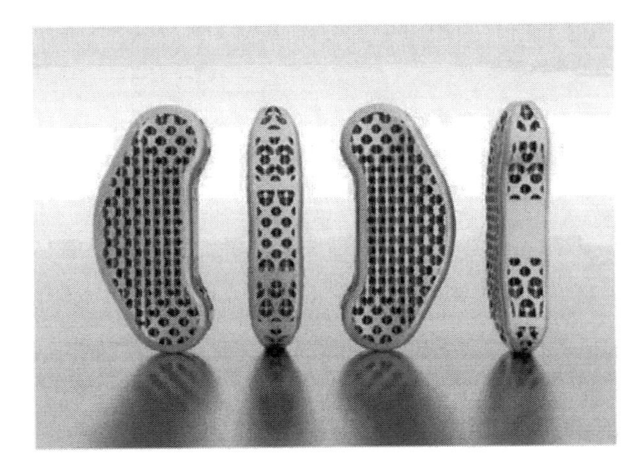

FIGURE 1.11 Spinal implants made of EOS Titanium Ti6Al4V. (Courtesy of EOS; From MakerBot® Replicator Desktop 3D Printer, Makerbot.com, 2009–2014. http://store.makerbot.com/replicator.)

be sent to the AM machine, whether it be down the hall or in another country, and an implant can be printed exactly how the physician thinks is best. This will help to treat patients where there was no treatment before, as well as decrease the time it takes for an injury to heal. This is just another area where AM has already made a difference and will continue to in the future.

1.6.5 AM of Bi-Metallic and Multi-Material Structures

Another area of additive manufacturing that is rapidly growing is bi-metallic and multi-material structures.[50–58] In this approach, multiple materials of different compositions can be deposited in the same operation allowing AM processed parts to be designed with various functionalities. For example, if a ceramic coating can improve wear resistance in a certain area of a part, a part can be designed with a metal base and a ceramic outer layer, which can then be manufactured in one AM operation without the need for a separate coating-based manufacturing operation.[53,56,58] Similarly, if the part needs to be magnetic in one area and non-magnetic in another area, a compositionally gradient part can be designed and manufactured using AM.[54] Finally, some of the bi-metallic parts with two different alloys can also be produced by AM-based manufacturing processes.[52,55,57] Most of these bi-metallic or multi-material structures are being manufactured using directed energy deposition (DED)-based AM processes.

1.7 SUMMARY

Overall, AM will have a profound effect on the manufacturing of many goods, as well as create a world of global engineering where ideas and designs can be spread in the most effective way. The current state of large-scale manufacturing leaves consumers settling for products that are not quite exactly what they want. Also, it does not provide the consumer with customization of products. AM is a method that has the ability to make complex geometries, alter material properties, and allow for versatility in the production of parts. The applications it can be used for are unparalleled by any other manufacturing method. By incorporating it into companies, it will give them the ability to manufacture what was once not possible. One of the most powerful things it will do is make the communication of designs and parts uncomplicated and effective. This will make it so companies, consumers, and anybody with access to CAD can design, critique, or customize a part that can be printed on-location and physically tested. AM will make it so companies and design teams in multiple locations can work together in the most effective way possible. AM makes the engineering of a part have no borders or barriers and changes the world from localized to globalized engineering.

PROBLEMS

1. State the advantages and disadvantages of additive manufacturing compared to traditional manufacturing approaches.
2. State advantages and disadvantages of on-demand manufacturing over projection-based manufacturing.
3. What is the difference between rapid prototyping and additive manufacturing?
4. Discuss various factors that need to be considered before making any decision related to manufacturing any part.
5. Discuss how AM is influencing global and concurrent engineering involving multiple sites.
6. What are the advantages and disadvantages of multi-material additive manufacturing?
7. What are the advantages and disadvantages of patient-matched implants?

REFERENCES

1. N. Hopkinson, R.J.M. Hague, and P.M. Dickens. *Rapid Manufacturing. An Industrial Revolution for the Digital Age*. John Wiley & Sons, Chichester, UK, 2006.
2. 3D Systems. The Journey of a Lifetime. www.3dsystems.com. 3D Systems, Inc., 2014.
3. C.W. Hull. Apparatus for production of three-dimensional objects by stereolithography. Uvp, assignee. Patent US 4575330 A. March 11, 1986.
4. Selective Laser Sintering, Birth of an industry. Department of Mechanical Engineering and Faculty Innovation Center, Cockrell School of Engineering. The University of Texas at Austin, Austin, TX, 2013. http://www.me.utexas.edu/news/2012/0612_selective_laser_sintering.php.
5. Inventor of 3D Printing Scott Crump: "My dreams started in a garage." 3D Printing.com. September 17, 2013. http://on3dprinting.com/2013/09/17/ inventor-of-3d-printing-scott-crump-my-dreams-started-in-a-garage/.
6. FDM Technology. Stratasys for a 3D world. Stratasys, 2014. http://www.stratasys.com/3d-printers/technologies/fdm-technology/faqs.
7. S.S. Crump. Apparatus and method for creating three-dimensional objects. Stratasys, assignee. Patent 5121329. June 9, 1992.

8. J.P. Pederson. *International Directory of Company Histories*, Vol. 67. St. James Press, 2005. Funding Universe.com.
9. Solidscape. About Us. Solidscape: A Stratasys Company. Solid-scape.com, 2013.
10. S. Crawford. How 3-D printing works? March 1, 2011. HowStuffWorks.com http://computer.howstuffworks.com/3-d-printing.htm.
11. Solidscape. News Release: Sanders Prototype, Inc. changes name to Solidscape, Inc. Solid-scape.com. Solidscape, Inc.: A Stratasys Company, 2013. https://www.solidscape.com/news/sanders-prototype-inc-changes-name-to-solidscape-inc/.
12. Solidscape. News Release: Solidscape sells to 3D printer maker, Stratasys. Solidscape, 2011. https://www.solidscape.com/news/solidscape-inc-sells-to-3d-printer-maker-stratasys/.
13. T. Wohlers and T. Gornet. History of additive manufacturing. State of the industry. Wohlers Report 2011. Wohlers Associates, Fort Collins, CO, 2011.
14. EOS. history. EOS e-Manufacturing Solutions. http://www.eos.info/about_eos/history.
15. Sandia National Laboratories. News Releases: Creating a complex metal part in a day is a goal of commercial consortium. Sandia National Laboratories, December 4, 1997. http://www.sandia.gov/media/lens.htm.
16. Optomec. Company milestones. Optomec. Production grade 3D printers with a material difference, 2014. http://www.optomec.com/company/milestones/.
17. Optomec. Company overview. Optomec. Production grade 3D printers with a material difference, 2014. http://www.optomec.com/company/.
18. Optomec. LENS technology. Optomec. Production grade 3D printers with a material difference, 2014. http://www.optomec.com/printed-metals/lens-technology/.
19. Arcam. Arcam history. Arcam AB® CAD to Metal®. http://www.arcam.com/company/about-arcam/history/.
20. Arcam. EBM® Electron beam melting—In the forefront of additive manufacturing. Arcam AB® CAD to Metal®. http://www.arcam.com/technology/electron-beam-melting/.
21. R. Jiang, R. Kleer, and F.T. Piller. Predicting the future of additive manufacturing: A Delphi study on economic and societal implications of 3D printing for 2030. *Technological Forecasting and Social Change* 117, 84–97, 2017.
22. S. Chand. The advantages and disadvantages of large scale production. YourArticleLibrary.com: The Next Generation Library, 2014. http://www.yourarticlelibrary.com/economics/the-advantages-and-disadvantages-of-large-scale-production/10901/.
23. B. Canis. The motor vehicle supply chain: Effects of the Japanese earthquake and tsunami. Congressional Research Service. CRS Report for Congress. 7-5700. R41831, May 23, 2011. www.crs.gov.
24. L. Perner. Distribution: Channels and logistics. Department of Marketing. Marshall School of Business. USC Marshall. Lars Perner 1999–2008. http://www.consumerpsychologist.com/intro_Distribution.html.
25. G. Hamel. The definition of high volume manufacturing. Small Business by Demand Media. Houston Chronicle: Houston, TX, 2014.
26. G. Warwick. Print to build. *Aviation Week & Space Technology* 176(11), 40–43, 2014.
27. M. Roy, V.K. Balla, A. Bandyopadhyay, and S. Bose. Compositionally graded hydroxyapatite/tricalcium phosphate coating on Ti by laser and induction plasma. *Acta Biomaterialia* 7, 866–873, 2011.
28. Laser Cladding Services. Services we offer. www.lasercladding.com, 2014.
29. G. Warwick. Adding power. *Aviation Week & Space Technology* 176(11), 43–44, 2014.
30. A. Roschli, K.T. Gaul, A.M. Boulger, B.K. Post, P.C. Chesser, L.J. Love, F. Blue, and M. Borish. Designing for big area manufacturing. *Additive Manufacturing* 25, 275–285. 2019.

31. D. Gates. Boeing 787's problems blamed on outsourcing, lack of oversight. *The Seattle Times Company*, Published on February 2, 2013, Seattle, WA, 2014.
32. A. Goulas, R.A. Harris, and R.J. Friel. Additive manufacturing of physical assets by using ceramic multicomponent extra-terrestrial materials. *Additive Manufacturing* 10, 36–42. 2016.
33. M. Green, and T. Talbert. Lunar settlement: Piecing together a full moon picture. Office of the Chief Technologist. NASA, Washington, DC. October 31, 2012.
34. H. Livingston. Get 'em while they're young. *Cadalyst*, Winter, 2012. http://www.cadalyst.com/cad/get-039em-while-they039re-young-14242.
35. J. Gausemeier, N. Echterhoff, M. Kokoschka, and M. Wall. Thinking ahead the future of additive manufacturing—Analysis of promising industries. Direct Manufacturing Research Center. Heinz Nixdorf Institute. University of Paderborn, Paderborn, Germany, 2011.
36. G. Andersen. Valve handle by Geir. MakerBot. Thingiverse. Published on June 19, 2011. http://www.thingiverse.com/thing:9450.
37. Folkway University of the Arts. Bionic Manufacturing Program. Photo by Nathalie Richter, Design by Anke Bernotat, Partners: Authentics, Plant Biomechanics Group Freiburg, Folkwang University of the Arts, Fraunhofer IWM, Fraunhofer UMSICHT, Fruth, KIT, RPM. Funded by the BIONA funding program of the Federal Ministry of Education and Research. March 1, 2013.
38. MakerBot® Replicator Desktop 3D Printer. Makerbot.com, 2009–2014. http://store.makerbot.com/replicator.
39. S. Bose, D. Ke, H. Sahasrabudhe, and A. Bandyopadhyay. Additive manufacturing of biomaterials. *Progress in Materials Science* 93, 45–111, 2018.
40. S. Bose, M. Roy, and A. Bandyopadhyay. Recent advances in bone tissue engineering scaffolds. *Trends in Biotechnology* 30(10), 546–554, 2012.
41. S. Bose, S. Vahabzadeh, and A. Bandyopadhyay. Bone tissue engineering using 3D printing. *Materials Today* 16(12), 496–504, 2013.
42. S.A.M. Tofail, E.P. Koumoulos, A. Bandyopadhyay, S. Bose, L. O'Donoghue, and C. Charitidis. Additive manufacturing: Scientific and technological challenges, market uptake and opportunities. *Materials Today* 21(1), 22–37, 2018.
43. S. Bose, S.F. Robertson, and A. Bandyopadhyay. Surface modification of biomaterials and biomedical devices using additive manufacturing. *Acta Biomaterialia* 66, 6–22, 2018.
44. S. Bose, S.F. Robertson, and A. Bandyopadhyay. 3D printing of bone implants and replacements. *American Scientist* 106, 112–119, 2018.
45. A. Bandyopadhyay, S. Bose, and S. Das. 3D printing of biomaterials. *MRS Bulletin* 40(2), 108–115, 2015.
46. Specialized and custom fitted hip implant options. BoneSmart, 2014. http://bonesmart.org/hip/hip-implants-specialized-and-custom-fitted-options/.
47. EOS. Additive manufacturing in the medical field. EOS. E-Manufacturing Solutions, 2013. http://ip-saas-eos-cms.s3.amazonaws.com/public/b674141e654eb94c/c5240ec3f487106801eb6963b578f75e/medicalbrochure.pdf.
48. EnvisionTEC. Tissue engineering bone implants using EnvisionTEC 3D printer. EnvisionTEC, Dearborn, MI. The benchmark in 3D printing, 2014.
49. M. Veselinovic, D. Stevanovic, M. Trajanovic, M. Manic, S. Arsic, M. Trifunovic, and D. Misic. Method for creating 3D surface model of the human tibia. In *34th International Conference on Production Engineering*. September 2011.
50. A. Bandyopadhyay, and B. Heer. Additive manufacturing of multi-material structures. *Materials Science and Engineering R: Reports* 129, 1–16, 2018.

51. A. Bandyopadhyay, and K. Traxel. Invited review article: Metal-additive manufacturing—Modeling strategies for application-optimized designs. *Additive Manufacturing* 22, 758–774, 2018.
52. H. Sahasrabudhe, R. Harrison, C. Carpenter, and A. Bandyopadhyay. Stainless steel to titanium bimetallic structure using LENS™. *Additive Manufacturing* 5, 1–8, 2015.
53. T. Gualtieri and A. Bandyopadhyay. Additive manufacturing of compositionally gradient metal-ceramic structures: Stainless steel to vanadium carbide. *Materials & Design* 139, 419–428, 2018.
54. B.T. Heer, and A. Bandyopadhyay. Compositionally graded magnetic-nonmagnetic bimetallic structure using laser engineered net shaping. *Materials Letters* 216, 16–19, 2018.
55. B. Onuike, B. Heer, and A. Bandyopadhyay. Additive manufacturing of Inconel 718—copper alloy bimetallic structure using laser engineered net shaping. *Additive Manufacturing* 21, 133–140, 2018.
56. Y. Zhang, and A. Bandyopadhyay. Direct fabrication of compositionally graded Ti-Al_2O_3 multi-material structures using laser engineered net shaping. *Additive Manufacturing* 21, 104–111, 2018.
57. B. Onuike, and A. Bandyopadhyay. Additive manufacturing of Inconel 718—Ti6Al4V bimetallic structures. *Additive Manufacturing* 22, 844–851, 2018.
58. K. Traxel, and A. Bandyopadhyay. Reactive-deposition-based additive manufacturing of Ti-Zr-BN composites. *Additive Manufacturing* 24, 353–363, 2018.

2 Additive Manufacturing of Polymers

Amit Bandyopadhyay, Kellen D. Traxel, Caitlin Koski, and Susmita Bose

CONTENTS

2.1 INTRODUCTION

Polymer-based additive manufacturing (AM) was originally developed in the 1980s for research and development as well as touch-and-feel purposes. Currently, polymer AM has developed into a diverse technology through advances in computer-aided design (CAD) and widespread use of the technology among manufacturers, hobbyists, and researchers. AM, in general, allows for the creation of application-specific parts without the need for conventional machining or the utilization of molds and can therefore create complex features not possible using traditional manufacturing techniques. The first commercial use of polymer AM emerged in 1987 through the use of stereolithography (SLA) from the company 3D Systems. Three AM technologies were commercialized by 1991 including fused deposition modeling (FDM) from Stratasys, solid ground curing from Buital, and laminated object manufacturing (LOM) from Helisys in addition to SLA. In 1992, selective laser sintering (SLS) from DTM was commercially available, quickly followed by the development of the Sanders Prototype using rapid ink jet prototyping [1]. These techniques quickly matured with widespread interest and demand for more reliable processes that allow for the creation of components that are easily visualized, as well as the development of functional and complex parts.

TABLE 2.1
Classification of Additive Manufacturing Techniques with Regard to Polymeric Materials

Classification	Relevant Process	Description
Binder Jetting	Inkjet and aerosol jet	Liquid bonding agent is selectively deposited to fuse powder materials
Material Extrusion	FDM Fused filament fabrication (FFF) 3D dispensing 3D bioplotting	Material is selectively dispensed through a nozzle
Material Jetting	Inkjet-printing	Droplets of build material (i.e., photopolymer or thermoplastic materials) are selectively deposited
Powder Bed Fusion	SLS	Thermal energy (e.g., laser or an electron beam) selectively fuses regions of a powder bed
Sheet Lamination	LOM	Sheets of material are bonded together to form an object
Vat Polymerization	Stereolithography	Liquid photopolymer in a vat is selectively cured by light activation polymerization

There are many individual AM processes currently available which vary in their method of layered manufacturing of polymers. In 2008, the ASTM International Committee F42 first defined the different additive manufacturing categories to establish the rapidly growing manufacturing field. The specific categories that apply to polymer 3D printing processes are included in Table 2.1 [2], and each is explained in detail within this chapter.

2.1.1 Polymeric Materials for AM

A range of polymers are utilized in AM including thermoplastics, thermosets, and elastomers, among others. Thermoplastic polymers are formed through addition polymerization and are made up of linear or branched molecular chains, which interact by weak intermolecular forces. Additionally, thermoplastic polymers can be reshaped by repeated heating and cooling. Thermoset polymers are pre-polymers in a soft-solid or viscous state formed by condensation polymerization that permanently hardens after curing. Thermoset polymers have cross-linked or 3D-network structures that are made up of strong covalent bonds. Thermoset polymers are hard, strong, and more brittle compared to thermoplastic polymers and cannot be reshaped once cured. Thermoplastics such as nylon, acrylonitrile butadiene styrene, and polyether ether ketone are commonly implemented in polymer additive manufacturing processes including material jetting and material extrusion [3]. Thermoset polymers that cure at ambient temperature including urea formaldehyde resins, unsaturated polyester, and epoxy resins are typically used in binder jetting-based systems and vat

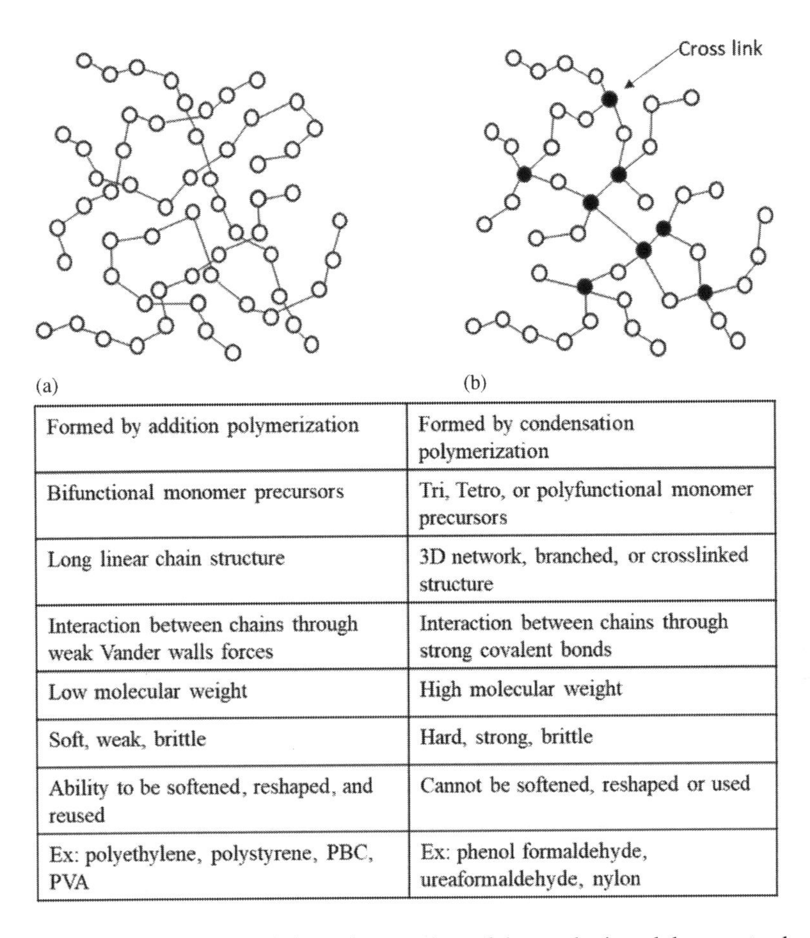

Formed by addition polymerization	Formed by condensation polymerization
Bifunctional monomer precursors	Tri, Tetro, or polyfunctional monomer precursors
Long linear chain structure	3D network, branched, or crosslinked structure
Interaction between chains through weak Vander walls forces	Interaction between chains through strong covalent bonds
Low molecular weight	High molecular weight
Soft, weak, brittle	Hard, strong, brittle
Ability to be softened, reshaped, and reused	Cannot be softened, reshaped or used
Ex: polyethylene, polystyrene, PBC, PVA	Ex: phenol formaldehyde, ureaformaldehyde, nylon

FIGURE 2.1 General characteristics and comparison of thermoplastic and thermoset polymers for additive manufacturing applications. (a) Thermoplastic polymer and (b) Thermoset polymer.

polymerization [4]. Common thermoplastic polymers utilized for powder bed fusion include polystyrene, polymer ester, polyamides, polyurethane, and polyether ether ketone. Material jetting and vat photopolymerization processes typically utilize curable polymers including acrylates, acrylics, polylactic, and epoxies. Material extrusion utilizes thermoplastic filaments, typically acrylonitrile butadiene styrene, polylactic, acrylics, polycarbonate, polyetherimide, or high impact polystyrene. Many of these polymers have been commercialized for SLS and SLA implementation (Figure 2.1).

2.1.2 Vat Polymerization (Stereolithography)

SLA is the original AM process initially patented by Chuck Hull in 1984 for helping engineers visualize and fit-test their designs [5]. This process is governed by the controlled movement of a UV light across the top surface of a monomer bath, or vat, to outline a layer-pattern that is determined from the slice file. Dependent on the vat material and the processing parameters, a depth of material is cured, the

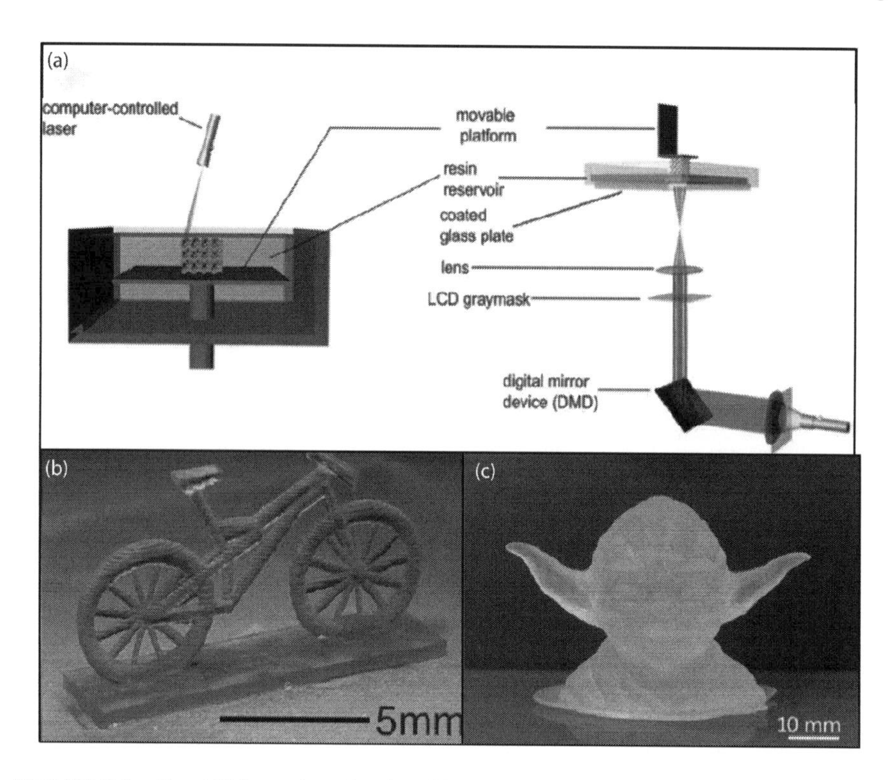

FIGURE 2.2 Capabilities and mechanics of SLA. (a) Schematic of two different SLA setups involving a top-down approach (left) and bottom-up approach (right). Both involve a movable platform and light source for the curing process. (From Melchels, F.P.W. et al., *Biomaterials*, 31, 6121–6130, 2010.) (b) Bicycle printed with high-resolution spokes on small length scale. (Adapted from Lee, M.P. et al., *Sci. Rep.*, 5, 2015.) (c) SLA-processed figurine with glass reinforcement. (From Sano, Y. et al. *Addit. Manuf.*, 2018.)

build platform lowered, and the process repeated (see Figure 2.2a). Another variation (shown in the same image) involves laser curing starting from the bottom of the vat, with the build platform moving upwards to create additional layers. After many of these controlled layers are completed, a component is produced, and the part is removed from the build-plate. Complex internal features and functionality can be incorporated into the component by tight control of the thickness of the features within each layer. More specifically, the designer can include thin sections in each layer that form a strut (or support structure) that are cured and provide a support for other features such as overhangs. Because these supports are designed to be thin relative to the functional features, they are typically removed after printing by simple surface finishing. These capabilities make SLA a common technique for creating functional parts as well as "fit-check" prototypes in mechanical systems for different end applications.

Because of the desire to decrease part build times, modern SLA systems use a "flood" light which enables multiple areas to be scanned at once, significantly increasing the throughput for production applications. In addition, the tight controls on the selectively cured areas enable components with microscale features and resolution (Figure 2.2b). The main challenge with this technique, however, is multi-material processing. The vat must be completely emptied and refilled with a different material to build on top of the previous material. Despite these challenges, variations have been proposed that utilize a ceramic phase (continuous or discontinuous) mixed in with the monomer bath, resulting in a polymer-ceramic composite with higher strength than the single polymer component alone (see Figure 2.2c) [6]. As an example, fiberglass reinforcement to an epoxy resin can increase the strength of a final component by a factor of 7.2 [6]. This is an advantageous variation as it enables a part strength-increase without significant modification to the existing setup. Elsewhere, multi-step SLA methods have been developed to realize multi-material components through the use of a rotating spindle and a novel layering-reset technique that allows a user to print a specific section of a part in one vat and a different section of the part in a separate vat (and subsequently different materials and/or color) [6].

This foundational technology initially led to the formation of the company *3D Systems* (South Carolina, USA), which is still known today for manufacturing versatile SLA machines that produce polymer structures with fine features and complex geometries. Other emerging machine manufacturers are *Formlabs* (Massachusetts, USA), and *XYZ* (San Diego, CA, USA). Because SLA relies on a layer-by-layer photopolymerization process to produce bulk structures, it is limited to the implementation of thermoset polymers such as acrylics, acrylates, and epoxies. These precursor materials can be toxic and challenging to work with, which is the main drawback to using this technology. A process that was initially meant to form prototypes has turned into a scientific platform for manufacturing bioresorbable scaffolds [7], micro-optical instruments [8], surgical planning procedures [9], among many other applications. This process also laid the foundation for other techniques discussed in subsequent chapters.

2.2 MATERIAL EXTRUSION (FUSED DEPOSITION MODELING)

FDM is one of the most widely utilized polymer additive manufacturing types owing to its ease of use and wide material selection (schematic and example parts are shown in Figures 2.3 and 2.4). Originally patented by S. Scott Crump (*Stratasys*, Minnesota, USA) in 1989 [13], many new machine manufacturers have become successful by manufacturing low-cost, easy-to-operate FDM machines for hobbyists and engineers. Some well-known manufacturers of these machines are *Formlabs* (Massachusetts, USA), *Makerbot* (NYC), *Ultimaker* (Netherlands), *WanHao* (China), *XYZ printing*, *LulzBot* (Colorado, USA), among others. In this process, a spool of thermoplastic wire is extruded through a nozzle assembly and heated to its softening temperature (anywhere from 100°C to 200°C). After reaching the nozzle exit, it is selectively deposited onto a substrate (or previously deposited layer) to form the current layer. Because of the cyclic softening-hardening nature of this process, thermoplastics are the only materials suitable for use. For highly complex designs, support material is extruded through a separate nozzle before the print-material is deposited and provides a structural support for the next layer to form sharp overhangs,

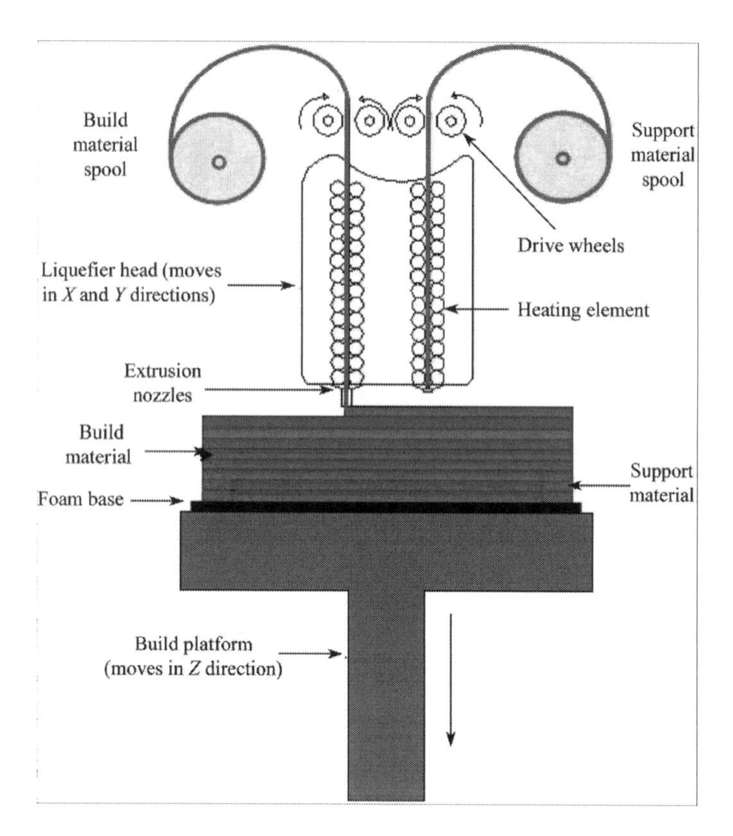

FIGURE 2.3 Schematic of a standard FDM-based printer with both feedstock and support material extruders. (From Mohamed, O.A. et al., *Adv. Manuf.*, 2015.)

closed-off sections, or complex internal features. The support material deposition is often governed by the build software, which determines which areas require more support in comparison to others. The support material is typically a dissolvable plastic which can be set in a detergent bath after removing the part from the build-plate, post-printing. Depending on the build material and the specific support material used, different times are required for dissolving the supports. When the support structures are easily accessible, however, a simple grinding or breaking process is more efficient for removal.

Critical parameters that govern the properties and characteristics of these parts are the layer-by-layer deposition orientation relative to the build plate, layer thickness, air gap distance between raster-paths, width of deposition, and extrusion head temperature, among others [14]. In general, parts with larger layer thicknesses (300–400 μm) lead to lower tolerancing capability than with lower layer thickness (10–50 μm), particularly in places of changing cross section from layer to layer [14]. In addition, for acrylonitrile butadiene styrene (a common FDM thermoplastic), lower reciprocating wear performance has been reported for larger layer thicknesses due to the debris, cracks, and fracture that can be caused in the microstructure [15]. Elsewhere, polylactic acid fatigue properties have been shown to be strongly affected with the build

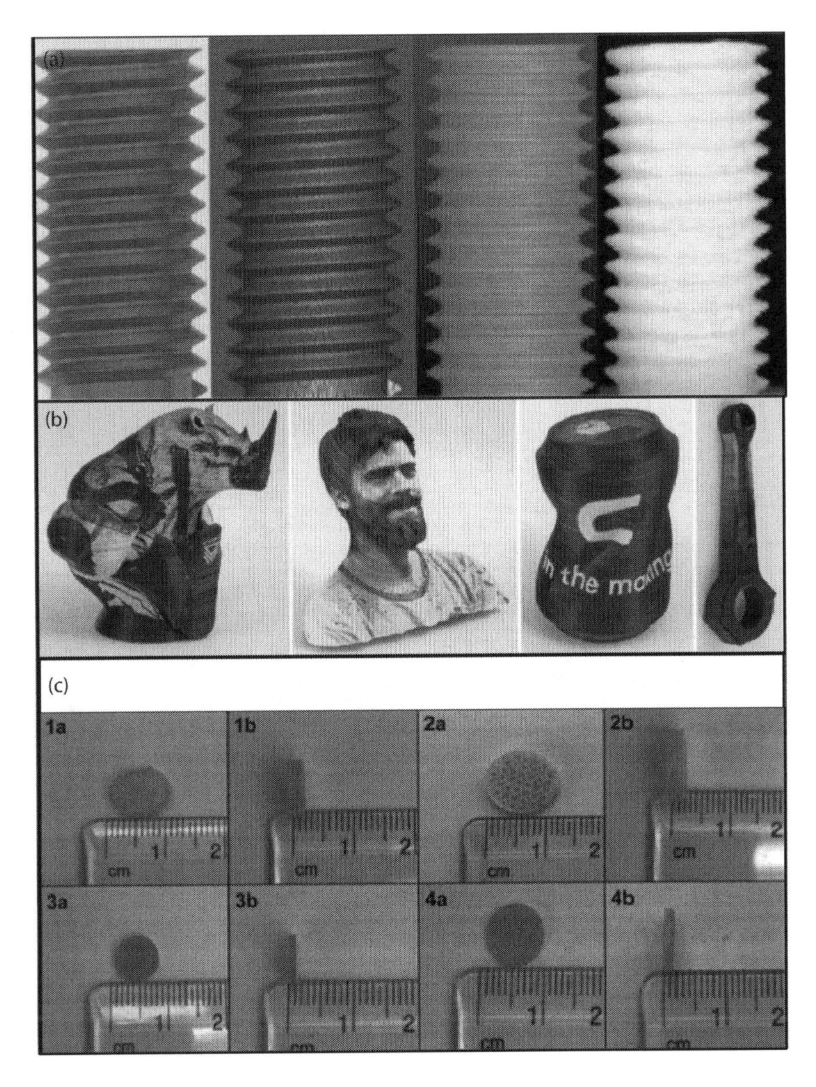

FIGURE 2.4 Capabilities of commercial FDM machines. (a) Threading printed via FDM process. From right to left: CAD design, STL representation (M10 × 1.5 screw design), STL file, sliced toolpaths, and final printed screw. (From Tronvoll, S.A. et al., *Procedia Manuf.*, 26, 763–773, 2018.) (b) Examples of structures (~10–20 cm tall) fabricated using FDM technique. (From Kuipers, T. et al., *Comput. Graph.*, 2018.) (c) Examples of drug-loaded tablets for sustained drug release via FDM. (From Solanki, N.G. et al., *J. Pharm. Sci.*, 2018.)

direction, namely, alternating 45° layers showed the strongest fatigue life, indicating that anisotropy is a challenge with this process [16]. Because of the heat input using this process, warping can be another by-product of large layer thickness and must be accounted for with the optimization of baseplate heating and extrusion temperature to minimize thermal gradients.

The ease of use of this process has led to its implementation in many different "maker-space" communities across the globe. Students can prototype and test different designs rapidly in comparison to conventional metal-working techniques, and because the thermoplastic feedstock is safe and economic, learning how to use the machine is very simple. Because of these aspects, this process is also commonly employed in different applications for manufacturing non-structural aerospace components as well as medical surgery preparation and parts for functional prototyping. Variations of this process include the use of multiple nozzles for multi-material and/or color capability, as well as the extrusion of slurry-based materials for biomedical applications, which have become a standard process in the emerging field of bioprinting. Elsewhere, the concept of 4D printing (developed at Massachusetts Institute of Technology) has been developed with this technology, which includes the integration of not only complex geometries and features, but pre-designed shape-memory or smart-material aspects within printed components [17].

2.3 POWDER BED FUSION

Powder bed fusion, also known as SLS, was developed at the University of Texas at Austin in 1986. While processes such as SLA and FDM use either monomer or spool-based feedstock, SLS was intended to fabricate components from plastic in powder form. The setup involves the use of two different powder beds, one for powder spreading and the other for powder solidification and subsequent part formation (Figure 2.5a). A laser heating source traces the top layer of the central powder bed according to the outline of the slice file, softening the polymer particles near the surface, and forcing them to sinter together and densify. After a layer is traced out, the main powder bed drops one-layer depth and a re-coater blade (or roller) moves additional powder (from the other bed) over the top surface of the previously solidified layer, and the process repeats until a part is realized. For complex geometries and overhangs, the powder bed itself can be used as a support structure, or thin walled struts can be selectively sintered to support larger-scale features as is similar with other methods. After the machine has finished tracing its last layer, extensive cleanup is required due to the large relative amount of unused powder, which is one of the main disadvantages of this technique. These parts typically have rough, grainy surface finish and require some post processing (Figure 2.5b). Variations include the use of premixed powder compositions and composites, as well as multi-laser setups for decreased layer-time and higher throughput.

SLS was commercialized by Nova Automation, which became DTM Corporation in 1992 [21], but is now marketed by *3D Systems*, with parallel development of laser sintering from EOS (Germany). A wide range of thermoplastic materials can be used such as polyacrylate and polyether ether ketone, leading to high performance polymer parts with application-specific mechanical properties. Applications include functional prototypes, support parts such as jigs and fixtures, orthopedic and neurological implants, as well as pharmaceuticals with specific and customizable surface topography as shown in Figure 2.5c.

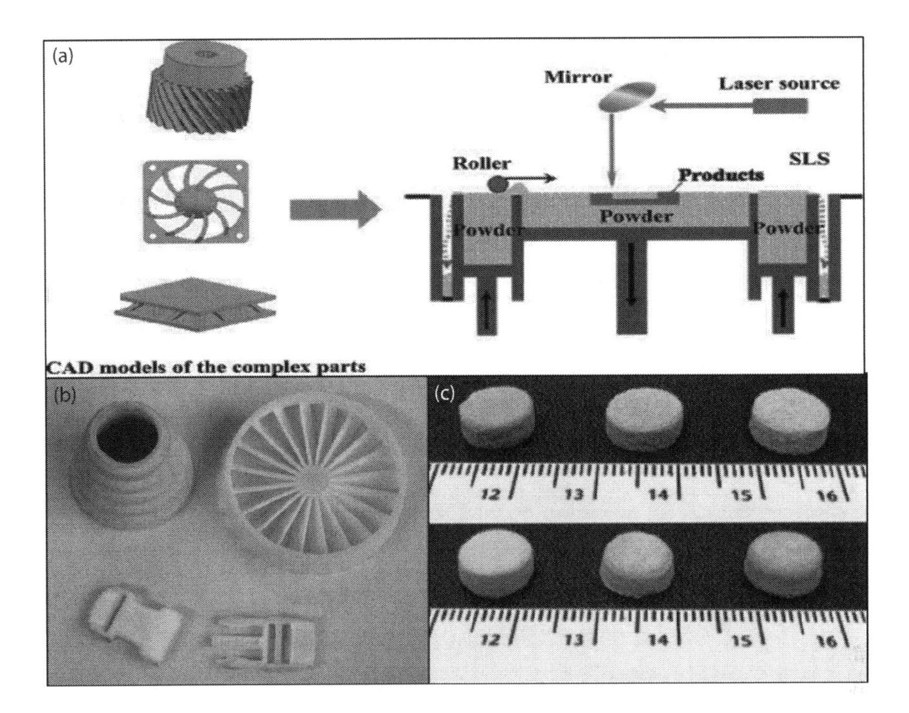

FIGURE 2.5 Selective laser sintering. (a) Process schematic process. (From Jin, L. et al., *Ceram. Int.*, 44, 20992–20999, 2018.) (b) Example SLS parts utilizing polymers. (From Kruth, J.P. et al., *Assem. Autom.*, 2003.) (c) Example of SLS utilized to fabricate oral drug polymer-based products using Kollicoat IR and Eudragit L 100-55. (From Fina, F. et al., *Int. J. Pharm.*, 2017.)

2.3.1 Binder Jetting

Binder jetting was first developed at Massachusetts Institute of Technology in the late 1980s and commercialized by *ExOne* (Pennsylvania, USA) and *Z Corporation* (South Carolina, USA) [25]. Similar to SLS, binder jetting setup includes two different powder beds, one for powder spreading and the other for powder solidification and subsequent part formation. To start the process, a liquid polymer binder (typically a thermosetting resin) is selectively deposited onto the main powder bed, forming an agglomerate that outlines the contour of the layer as defined by the slice file [25]. After the binder is deposited, the powder bed is externally heated through a radiation heat source to partially cure the binder in order to improve mechanical strength to build the next layer. After this occurs, another layer of powder is swept over the top of the part, and the process repeated until the final green part has been printed. Binder-particle interactions are critical for this process, and depending on quality control of the precursor materials, variability will exist in the final dimensions and quality of the components. This aspect can also lead to very fine features if fine powders are used in the bed.

Application areas for binder jetting include prototypes particularly due to binder jetting's full color and visual form, green parts for use in other manufacturing

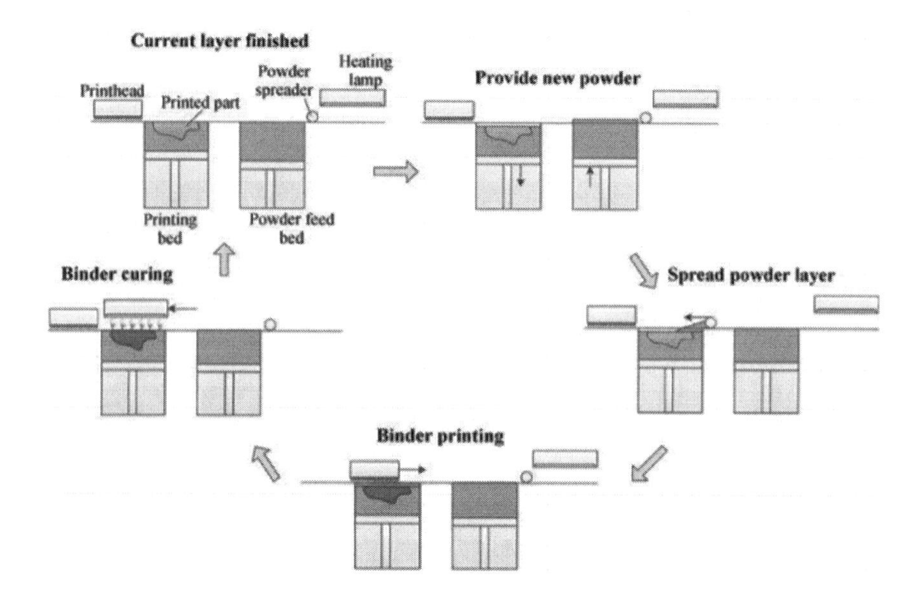

FIGURE 2.6 Binder jetting process schematic. (From Miyanaji, H. et al., *Front. Mech. Eng.*, 2018.)

processes, and direct fabrication of molds and cores. Binder jetting offers many advantages over other AM techniques that require some thermal processing, however, many binder jet polymer parts possess limited strength and often require post processing. For increased properties, parts may require the infiltration of a second material to enhance mechanical strength or color (Figure 2.6).

2.3.2 Material Jetting

Material jetting originated in the 1990s when *Solidscape* (New Hampshire, USA) was founded based on the development of the Sanders' prototype. This machine utilized the deposition of multiple thermoplastic polymers on a build substrate, one material used for the final part and the other as supports for overhangs and complex features. The key aspect to this technique is that after a few layers, a milling cutter is moved over the top of the part and cuts the spare material to level the Z-axis for further deposition. This process proved to be effective for manufacturing molds for jewelry and investment castings, laying the foundation for companies to develop this technology for thermoset polymers.

For thermosets, the jetting process involves the direct deposition of viscous-liquid monomer material, either through continuous or droplet-based techniques, outlining the current layer as specified in the slice file. After the layer is completed, a UV light is shown over the whole build area, which cures the previous layer (see schematic in Figure 2.7a). This deposition, although allowing for highly accurate builds, is limited in the number and type of materials available for the builds. Liquid viscosity plays a critical role and is a main reason why HP has become such a large player

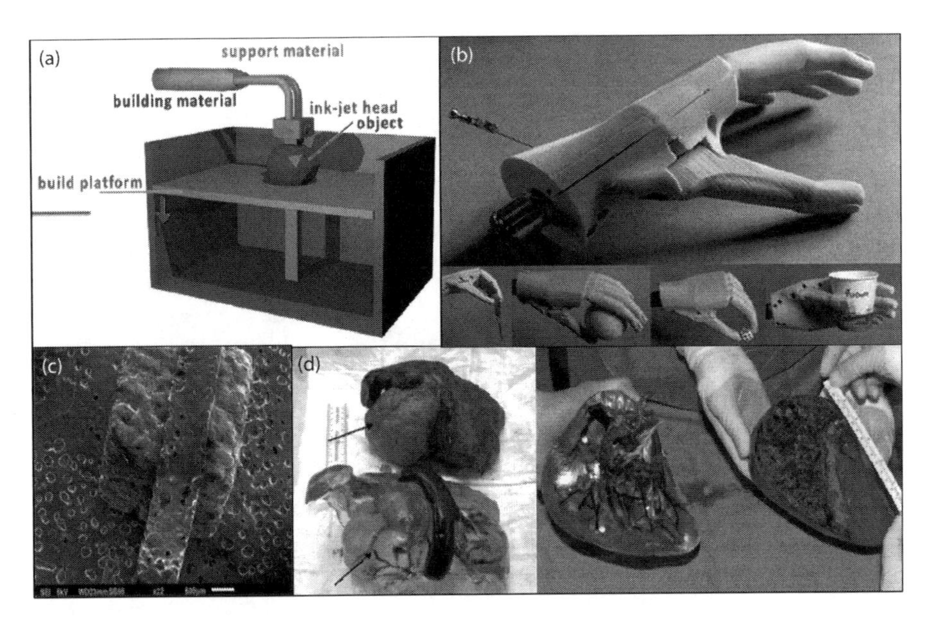

FIGURE 2.7 Material jetting process schematic and capabilities. (a) Process schematic (From Zadpoor, A.A. and Malda, J., *Ann. Biomed. Eng.*, 45, 1–11, 2017.) (b) The medical industry utilizes full color material jet printing to produce low cost prosthetics. (From Zadpoor, A.A. and Malda, J., *Ann. Biomed. Eng.*, 45, 1–11, 2017.) (c) Example of high resolution bone scaffold designed and manufactured using materials jetting. (From Velasco, M.A. et al. *J. Comput. Des. Eng.*, 3, 385–397, 2016.) (d) Example of fine surface detail and material composition throughout of liver transplant model. (From Zein, N.N. et al., *Liver Transpl.*, 19, 1304–1310, 2013.)

in machine manufacturing for this process due to their rich history in droplet-based laser-printing. Because it is a deposition-based process, multi-material and multi-color parts are easily implemented by change of deposition material. The strength and durability of material-jetted parts is also affected based on the material, which limits their functional applications. Since it is one of the only AM technologies that can combine different print materials within the same 3D printed model, it allows for the capability of constructing functional assemblies, which reduces the need for multiple builds. Some examples of these complex multi-material/multi-color builds are given in Figure 2.7b–d. These aspects give parts an added visual dimension and are a key reason why this process is great for educational purposes to understand physical phenomena or anatomy. The two leading manufacturers are *Stratasys* (Minnesota, USA) and *3D Systems. Keyence* (Japan) and *XJET* (Israel) are among several manufactures currently entering the market and conducting material jetting-based research.

Variations of material jetting have emerged out of the desire to combine the best aspects of multiple processes and are constantly under development to meet the needs of advanced industrial applications. One example of this concept is PolyJet technology, created to combine high-resolution capability with multi-material and/or color possibilities in a single part. This process was originally patented from

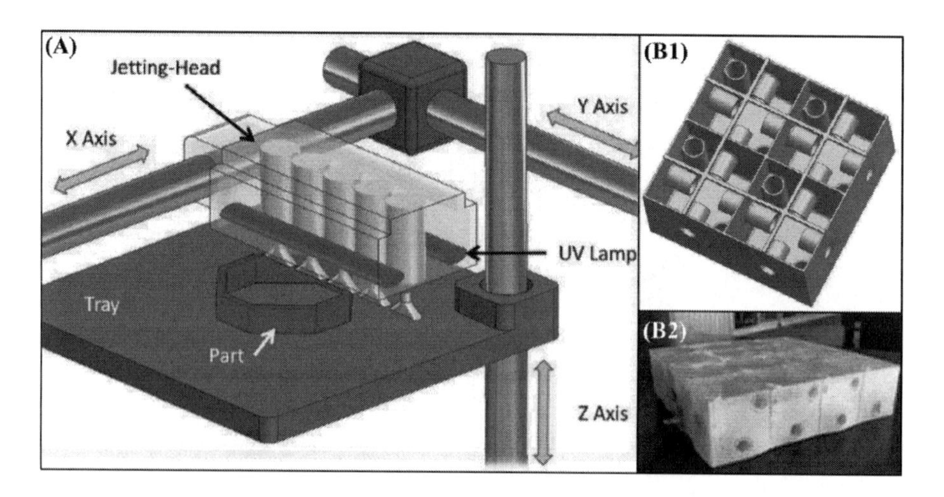

FIGURE 2.8 PolyJet processing schematic and example parts. (A) Processing schematic. (From Gay, P. et al., *Procedia Eng.*, 132, 70–77, 2015.) (B1) CAD-designed acoustic metamaterial part for PolyJet processing due to complex internal features and desire for high-resolution. (From Vdovin, R. et al., *Procedia Eng.*, 176, 595–599, 2017.) (B2) Printed part as removed from substrate. (From Vdovin, R. et al., *Procedia Eng.*, 176, 595–599, 2017.)

research out of Massachusetts Institute of Technology and eventually commercialized by *Stratasys* [29]. As shown in Figure 2.8a, this process employs a liquid-resin jetting head which uses multiple nozzles to deposit different polymer materials onto a substrate for a single layer. These materials can be different resins entirely or a combination of support material and the actual build material. After each layer has been deposited, a UV light comes across the surface to cure the deposited liquid resin. This technique is known for its ability to combine high-resolution features (Figures 2.8B1 and B2), while minimizing waste material or requiring large amounts of resin such as in SLA technology. This process is optimized for small-scale (<~5 in square) components, as SLA would be able to achieve larger components in much less time due to the mirroring functionality that enables fast production.

A technology still evolving, MultiJet, is an AM technique that originated from HP (Palo Alto, CA) that combines material jetting with binder jetting to create parts that not only are multi-color, but higher resolution than is achieved using an SLS, or other processes. Like the binder jetting process, a layer of material is drawn across a powder bed and a binder is deposited in a designed manner to create a layer. The main aspect to this process is the combination of liquid binder and "detailing" fluid which is deposited around edges of tight-tolerance in the current layer. The "detailing" fluid ensures that particles near the edge are not bound to the current layer, creating a smoother final surface finish and tighter overall tolerance for parts produced. While this technology is still evolving, it has seen significant interest from the public for the development of multi-color and visually pleasing components.

2.4 LAMINATION-BASED AM PROCESS

LOM is an AM technique that uses laminated sheets as a precursor material for layer-by-layer manufacturing. The process was originally developed by *Helisys* (now *Cubic Technologies*, San Diego, CA), who used sticky paper and a laser cutter to remove excess material around a selectively outlined layer. This technique lends itself well for fabrication of different models and parts without the need for an expensive, high-energy setup or toxic precursor materials. It is typically found in architecture, medicine, and mold-making industries, among others. As shown in Figure 2.9a, a sheet material consisting of both feedstock and an adhesive is rolled across the top surface of a build plate (using a heated roller), and then bonded to the previous layer by heating of the adhesive. A laser then traces out the current layer, removes the remaining (outer) material, rolls the remaining sheet over, and then repeats the process. This specific variation is sometimes referred to as "bond then roll" processing, as the sheets are bonded before being traced out. As can be envisioned, internal features are challenging to accomplish using this method, so a variation sometimes referred to as "roll then bond" (not shown here) has been created which enables manufacturers to trace out each layer before being bonded to the as-built structure.

Mcor (Ireland) is one of the most well-known modern manufacturers of these machines, operating a slight variant to the above described process. Their machines use white printer paper and an inkjet head that drops printer ink around the contours of the layer, as determined from the CAD file. The sheet is then glued to the previous layer, and the edges are cut from the rest of the sheet to realize the contours of the component. The process is repeated to add more sheets to a part and form intricate, colorful parts. Because the

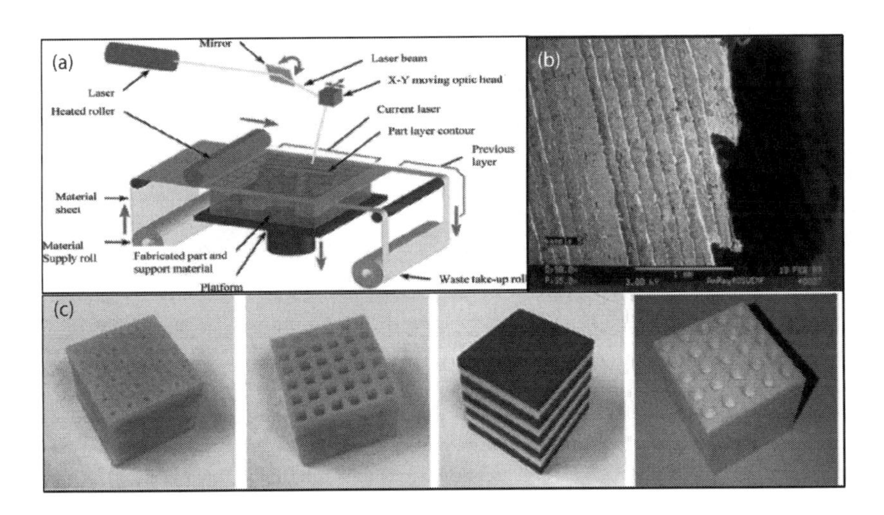

FIGURE 2.9 LOM processing. (a) Process schematic for a "bond then roll" LOM process. (From Ahn, D. et al., *J. Mater. Process. Technol.*, 212, 339–346, 2012.) (b) High-magnification image of the different layers of a LOM-processed component. (From Paul, B.K. and Voorakarnam, V., *J. Manuf. Process.*, 2001.) (c) Example of designed-porosity in a block produced via LOM technique utilizing an FGM-approach. (From Zhang, Y. and Wang, J. et al., *Procedia Manuf.*, 10, 866–875, 2017.)

contours are cut out of the paper, support structures are difficult to incorporate for large overhangs. As shown in the high-magnification image of Figure 2.9b, the layers in this process are typically well-defined and not as strongly bonded as is the case in other methods. In addition, the surface roughness is typically higher, which means using this process for functional components is challenging [32]. A main area of application, however, is in the production of color models for architecture or other visual purposes (see Figure 2.9c) where strength and integrity are less desired.

Advantages/Disadvantages Table

Process	Advantages	Disadvantages
SLA	Very fine resolution components. Established technology with wide range of machine manufacturers. Fast build times and large-part manufacturing capability.	Limited amount of commercially available resins for printing. Monomer bath requires special handling to ensure user safety. Metamaterial and polymer-ceramic composite processing is challenging.
FDM	Widely available and cost-effective machines. Able to incorporate multiple feedstocks and/or support material. Non-toxic feedstock materials, ideal for "maker-spaces" in schools.	Anisotropy introduced by large layer thickness and layer-to-layer adherence. Slow layer-by-layer build process relative to light-based processes. Resolution is not as good as other AM processes. Warping of large parts due to thermal cycling and large layer thicknesses.
SLS	Ability to process wide range of thermoplastic materials. Production of high performance polymer parts. Production of complex geometries with dimensional accuracy.	Can have rough, grainy surface finishes that require post processing. Variable powder particle size variation and distribution. No closed hollow features. Needs large amount of powder to initiate build.
Material Jetting	Creates highly accurate builds Limited functional applications depending on the material choice Hybrid techniques combine best aspects of different fundamental AM technologies: high resolution, multimaterial/color, build speed.	Limited in the number and type of materials available. Ability to combine different print materials within the same 3D printed model, reducing the need for multiple builds.
Binder Jetting	Ability to create complex, multi-material and multi-color parts.	Inherent variability in size and shape due to powder size distribution. Final parts possess limited strength and often require post processing.
LOM	Low residual stresses and shrinkage relative to other AM methods. Large parts can be fabricated rapidly. Variety of materials as well as FGM-capable (paper, sheets embedded with polymer/ceramic/metal, among others). Non-toxic materials used in processing.	Internal features challenging unless employing a "form then bond" process. Weak interfaces between layers due to adhesive bonding. Support structures typically non-existent.

2.5 FUTURE DIRECTIONS

An area that has generated significant interest in academia and industry is the concept of multi-material additive manufacturing [35]. The ability to use multiple materials in a single component enables a combination of properties and functionality that are not possible using some of the first generation AM techniques or other polymer-processing methods. Some processes such as FDM, binder jetting and material jetting were developed specifically for this functionality, whereas other processes such as SLA and SLS are challenging to achieve multi-material components. The challenges for the widespread use of multi-material AM are significant, and it is anticipated to be a large research topic over the next decade.

AM is prevalently utilized for prototypes and models, but the option to print multiple colors and compositions has led to the production of functional materials and devices. These advancements have led to the integration of 3D shapes, mechanical joints, and electronics in the building of more functional products. Future developments in three-dimensional printing (3DP) of polymers include smart polymers, nanocomposites, and the fabrication of sensors, actuators, and robots [36]. Additionally, 3D polymer parts for high throughput screening, pharmaceutical formulations, medical devices, and implants are being developed [37]. Another area of polymer-based AM that is rapidly developing is models for surgical planning. Handling big data from computed tomography or magnetic resonance imaging files to identify the desired area and then printing it in color can help physicians to plan any complex surgery. Challenges related to big-data handling and selection of multiple colors are the key to success for this application and will remain an active research area in the coming days. Although non-critical parts are regularly printed using different polymer-based AM technologies, improving mechanical properties and reducing anisotropy in performance will be another areas of research and development.

2.6 SUMMARY

Polymer-based additive manufacturing has grown immensely in recent years, and it has become imperative to understand the background and fundamentals of these techniques in order to understand the current context of additive techniques for all materials. To this end, this chapter highlights the fundamental mechanics of polymer 3D-printing processes, i.e., SLA, FDM, binder jetting, material jetting, hybrid processes, and LOM. The advantages and challenges are described in detail and some applications in which each of those techniques have been implemented. Future perspectives on the use of this technology are presented at the end of this chapter as well as suggested further reading.

SUGGESTED FURTHER READING

Ligon, S., R. Liska, J. Stamf, M. Gurr, R. Mulhaupt, Polymers for 3D printing and customized additive manufacturing, *Chem. Rev.* (2017). doi:10.1021/acs.chemrev.7b00074.

Bandyopadhyay, A., B. Heer, Additive manufacturing of multi-material structures, *Mater. Sci. Eng. R Rep.* 129 (2018) 1–16. doi:10.1016/j.mser.2018.04.001.

Gibson, I., D. Rosen, B. Stucker, *Additive Manufacturing Technologies*, Springer, Switzerland, 2015. doi:10.1007/978-1-4939-2113-3.

Bose, S., D. Ke, H. Sahasrabudhe, A. Bandyopadhyay, Additive manufacturing of biomaterials, *Prog. Mater. Sci.* (2017). doi:10.1016/j.pmatsci.2017.08.003.

QUESTIONS

1. Name three first generation polymer-based additive manufacturing techniques and discuss their advantages and challenges.
2. Briefly describe the SLA process. When would you prefer to use SLA?
3. Briefly describe the FDM process. Why is FDM the most popular polymer-based AM process?
4. Compare and contrast SLS and binder jetting-based AM processes.
5. What are the inherent advantages for Mcor or LOM-type lamination-based processes?
6. Second generation polymer AM processes such as "PolyJet" or "MultiJet" are becoming very popular. What are the inherent advantages of these processes over some of the first generation polymer-based AM processes such as SLA or SLS or binder jetting?
7. International Space Station (ISS) has an FDM machine. Why did NASA select FDM technology for ISS?

REFERENCES

1. T. Wohlers, T. Gornet, History of additive manufacturing, Wohlers Rep. (2016).
2. ASTM ISO/ASTM52900-15 Standard, Terminology for additive manufacturing – general principles – terminology (2015). doi:10.1520/ISOASTM52900-15.
3. O. Rios, Evaluation of advanced polymers for additive manufacturing (2017). https://web.ornl.gov/sci/manufacturing/docs/reports/web_PPG_MDF-TC-2014-048_Final Report.pdf.
4. J.P. Pascault, R.J.J. Williams, Thermosetting polymers, in: *Handbook Polymer Synthesis Characterization and Processing* (2013). doi:10.1002/9781118480793.ch28.
5. C.W. Hull, Apparatus for production of three-dimensional objects by stereolithography, US Patent 4,575,330. (1986) 1–16. doi:10.1145/634067.634234.
6. J.-W. Choi, H.-C. Kim, R. Wicker, Multi-material stereolithography, *J. Mater. Process. Technol.* 211 (2011) 318–328. doi:10.1016/j.jmatprotec.2010.10.003.
7. J.L. Walker, M. Santoro, Processing and production of bioresorbable polymer scaffolds for tissue engineering, in: *Bioresorbable Polymers for Biomedical Applications From Fundamentals to Translational Medicine* (2016), pp. 181–203. doi:10.1016/B978-0-08-100262-9.00009-4.
8. A. Žukauskas, M. Malinauskas, E. Brasselet, S. Juodkazis, Chapter 12.1 – 3D micro-optics via ultrafast laser writing: Miniaturization, integration, and multifunctionalities A2 – Baldacchini, Tommaso BT – Three-dimensional microfabrication using two-photon polymerization, in: *Micro Nano Technologies* (2016), pp. 268–292. doi:10.1016/B978-0-323-35321-2.00014-5.
9. R. Sodian, S. Weber, M. Markert, D. Rassoulian, I. Kaczmarek, T.C. Lueth, B. Reichart, S. Daebritz, Stereolithographic models for surgical planning in congenital heart surgery, *Ann. Thorac. Surg.* 83 (2007) 1854–1857. doi:10.1016/j.athoracsur.2006.12.004.
10. F.P.W. Melchels, J. Feijen, D.W. Grijpma, A review on stereolithography and its applications in biomedical engineering, *Biomaterials* 31 (2010) 6121–6130. doi:10.1016/j.biomaterials.2010.04.050.
11. M.P. Lee, G.J.T. Cooper, T. Hinkley, G.M. Gibson, M.J. Padgett, L. Cronin, Development of a 3D printer using scanning projection stereolithography, *Sci. Rep.* 5 (2015). doi:10.1038/srep09875.

12. Y. Sano, R. Matsuzaki, M. Ueda, A. Todoroki, Y. Hirano, 3D printing of discontinuous and continuous fibre composites using stereolithography, *Addit. Manuf.* (2018). doi:10.1016/j.addma.2018.10.033.
13. S.S. Crump, Apparatus and method for creating three-dimensional objects, US Patent 5,121,329. (1992).
14. O.A. Mohamed, S.H. Masood, J.L. Bhowmik, Optimization of fused deposition modeling process parameters: A review of current research and future prospects, *Adv. Manuf.* (2015). doi:10.1007/s40436-014-0097-7.
15. O.A. Mohamed, S.H. Masood, J.L. Bhowmik, A parametric investigation of the friction performance of PC-ABS parts processed by FDM additive manufacturing process, *Polym. Adv. Technol.* (2017). doi:10.1002/pat.4080.
16. M.F. Afrose, S.H. Masood, P. Iovenitti, M. Nikzad, I. Sbarski, Effects of part build orientations on fatigue behaviour of FDM-processed PLA material, *Prog. Addit. Manuf.* (2016). doi:10.1007/s40964-015-0002-3.
17. S. Tibbits, 4D printing: Multi-material shape change, *Archit. Des.* (2014). doi:10.1002/ad.1710.
18. S.A. Tronvoll, C.W. Elverum, T. Welo, Dimensional accuracy of threads manufactured by fused deposition modeling, *Procedia Manuf.* 26 (2018) 763–773.
19. T. Kuipers, W. Elkhuizen, J. Verlinden, E. Doubrovski, Hatching for 3D prints: Line-based halftoning for dual extrusion fused deposition modeling, *Comput. Graph.* (2018). doi:10.1016/j.cag.2018.04.006.
20. N.G. Solanki, M. Tahsin, A.V. Shah, A.T.M. Serajuddin, Formulation of 3D printed tablet for rapid drug release by fused deposition modeling: Screening polymers for drug release, drug-polymer miscibility and printability, *J. Pharm. Sci.* (2018). doi:10.1016/j.xphs.2017.10.021.
21. J. Ruan, T. Sparks, Z. Fan, J. Stroble, A. Panackal, F. Liou, A review of layer based manufacturing processes for metals (2014).
22. L. Jin, K. Zhang, T. Xu, T. Zeng, S. Cheng, The fabrication and mechanical properties of SiC/SiC composites prepared by SLS combined with PIP, *Ceram. Int.* 44 (2018) 20992–20999. doi:10.1016/j.ceramint.2018.08.134.
23. J.P. Kruth, X. Wang, T. Laoui, L. Froyen, Lasers and materials in selective laser sintering, *Assem. Autom.* (2003). doi:10.1108/01445150310698652.
24. F. Fina, A. Goyanes, S. Gaisford, A.W. Basit, Selective laser sintering (SLS) 3D printing of medicines, *Int. J. Pharm.* (2017). doi:10.1016/j.ijpharm.2017.06.082.
25. H. Miyanaji, M. Orth, J.M. Akbar, L. Yang, Process development for green part printing using binder jetting additive manufacturing, *Front. Mech. Eng.* (2018). doi:10.1007/s11465-018-0508-8.
26. A.A. Zadpoor, J. Malda, Additive manufacturing of biomaterials, tissues, and organs, *Ann. Biomed. Eng.* 45 (2017) 1–11. doi:10.1007/s10439-016-1719-y.
27. M.A. Velasco, Y. Lancheros, D.A. Garzón-Alvarado, Geometric and mechanical properties evaluation of scaffolds for bone tissue applications designing by a reaction-diffusion models and manufactured with a material jetting system, *J. Comput. Des. Eng.* 3 (2016) 385–397. doi:10.1016/j.jcde.2016.06.006.
28. N.N. Zein, I.A. Hanouneh, P.D. Bishop, M. Samaan, B. Eghtesad, C. Quintini, C. Miller, L. Yerian, R. Klatte, Three-dimensional print of a liver for preoperative planning in living donor liver transplantation, *Liver Transpl.* 19 (2013) 1304–1310. doi:10.1002/lt.23729.
29. R. Singh, Process capability study of polyjet printing for plastic components, *J. Mech. Sci. Technol.* 25 (2011) 1011–1015. doi:10.1007/s12206-011-0203-8.
30. P. Gay, D. Blanco, F. Pelayo, A. Noriega, P. Fernández, Analysis of factors influencing the mechanical properties of flat PolyJet manufactured parts, *Procedia Eng.* 132 (2015) 70–77. doi:10.1016/j.proeng.2015.12.481.

31. R. Vdovin, T. Tomilina, V. Smelov, M. Laktionova, Implementation of the additive PolyJet technology to the development and fabricating the samples of the acoustic metamaterials, *Procedia Eng.* 176 (2017) 595–599. doi:10.1016/j.proeng.2017.02.302.
32. B.K. Paul, V. Voorakarnam, Effect of layer thickness and orientation angle on surface roughness in laminated object manufacturing, *J. Manuf. Process.* (2001). doi:10.1016/S1526-6125(01)70124-7.
33. D. Ahn, J.-H. Kweon, J. Choi, S. Lee, Quantification of surface roughness of parts processed by laminated object manufacturing, *J. Mater. Process. Technol.* 212 (2012) 339–346. doi:10.1016/j.jmatprotec.2011.08.013.
34. Y. Zhang, J. Wang, Fabrication of functionally graded porous polymer structures using thermal bonding lamination techniques, *Procedia Manuf.* 10 (2017) 866–875. doi:10.1016/j.promfg.2017.07.073.
35. A. Bandyopadhyay, B. Heer, Additive manufacturing of multi-material structures, *Mater. Sci. Eng. R Rep.* 129 (2018) 1–16. doi:10.1016/j.mser.2018.04.001.
36. M. Nadgorny, A. Ameli, Functional polymers and nanocomposites for 3D printing of smart structures and devices, *ACS Appl. Mater. Interfaces* (2018). doi:10.1021/acsami.8b01786.
37. N. Scoutaris, S. Ross, D. Douroumis, Current trends on medical and pharmaceutical applications of inkjet printing technology, *Pharm. Res.* (2016). doi:10.1007/s11095-016-1931-3.

3 Additive Manufacturing Technologies for Polymers and Composites

Ranji Vaidyanathan

CONTENTS

3.1 INTRODUCTION

Additive manufacturing or AM processes, initially known as "solid freeform fabrication," "rapid prototyping," and currently described as "direct digital manufacturing," or "rapid manufacturing," "art to part," "additive layer manufacturing," or "layer manufacturing," were originally developed in the late 1980s to early 1990s (Gibson et al. 2010). This chapter will discuss some of the historical perspectives of the development of additive manufacturing technologies as related to polymer-based materials and how that progress has controlled AM process development for polymers and composites. Though the industry called these in generic terms as "rapid prototyping," an ASTM committee defined them more broadly as "additive manufacturing," under ASTM F2792 which is more descriptive of the current state of the art (ASTM-F2792). ASTM F2792 has categorized the various AM processes under seven categories (ASTM-F2792 2012; Paesano 2014). Among these, the categories that specifically relate to polymers and composites are:

- Binder jetting, where a liquid binding agent is selectively deposited to bind powder materials,
- Material extrusion, the process in which material is selectively dispensed through a nozzle or orifice,
- Material jetting, where droplets of build material are selectively deposited,
- Powder bed fusion, where particles of a polymer could be bonded together either thermally, and,
- Vat photopolymerization, where a liquid photopolymer is selectively cured by light-activated polymerization.

Table 3.1 describes the various materials and equipment manufacturers who use AM processes for polymers and composites. The earliest materials and equipment almost entirely catered to polymers and for form and fit type of applications. However, as the capabilities of the equipment as well as software have improved, various manufacturers are fabricating functional prototypes that can be directly used in actual applications (Direct-Digital-Manufacturing 2013; General-Electric 2014).

Among all the AM materials, polymers are perhaps the most advanced materials for AM techniques. A good description of the early beginnings of the technology and the current state of the art in materials is given by Gibson and others (2010), Guo and Leu (2013), and more recently, Paesano (2014). This chapter will however focus on specialized polymers, especially those reinforced with self-reinforcing fibers and chopped fibers as well as continuous fibers. The chapter will also discuss specialized polymers and blends that are required to be added to ceramics to make them suitable to be fabricated into 3D parts. Some additional discussion on polymers and composites that are used for biomedical applications different from traditional materials like polylactic acid-polyglycolic acid (PLA-PLGA) or hydroxyapatite is also presented here.

TABLE 3.1
Materials and Manufacturers of AM Technologies Using ASTM F2792 Classification for Polymers (Paesano 2014) and Composites

ASTM F2792 Classification	Materials Used for the AM Technology of Polymers and Composites	Equipment Manufacturer
Binder jetting	Polymers, powders, elastomers	3D Systems, ExOne, Z-Corp (Z-Corporation 2014)
Extrusion	Polymers, short fiber-reinforced polymers, ceramics, continuous fiber-reinforced polymers	Stratasys, MakerBot, Fab at Home, MarkForged, (MarkForged 2014) ABB, modified extrusion-based equipments
Material jetting	Polymers, waxes	3D Systems, Solidscape, Objet
Powder bed fusion	Polymers	EOS, 3D Systems
Vat photopolymerization	Photopolymers	3D Systems, EnvisionTEC

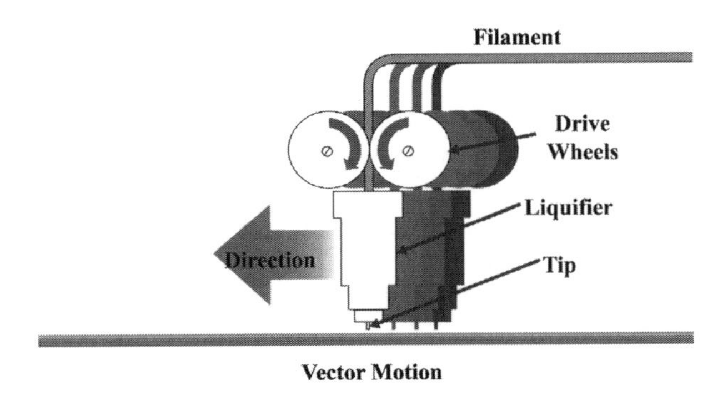

FIGURE 3.1 Schematic of the fused deposition modeling process.

Some of the earliest work for AM of polymer composites started in the mid-1990s, but is still continuing with new materials and processes, especially for higher strength thermoplastic polymers and composites. A majority of the AM processes developed for polymers and composites are extrusion-based processes, adapted to the original Stratasys equipment developed by Scott Crump (1992, 1994), shown in the schematic in Figure 3.1.

Even though the original patent by Crump and Stratasys referred to the use of various materials in the fused deposition modeling (FDM) process, such as waxes, thermoplastic resins, and metals, the process is limited to prototypes made using acrylonitrile butadiene styrene (ABS), polycarbonate, and Ultem (based on thermoplastic polyetherimide resins) and a maximum operating temperature of ~260°C. This limitation is due to two reasons: (i) temperature limit of the extruder in the FDM equipment limits the polymers that can be used for prototyping, and (ii) the support material for fabricating the support structures is also temperature limited. Even with the Ultem material, the standard soluble support material cannot be used and only a special thermoplastic support material has to be used. This support material is not water-soluble and has to be removed after the fabrication process and could be quite challenging to remove in the case of thin wall sections. This further requires that the wall thickness of the parts is built to be above a certain value so that the support structure can be removed without damaging the prototype part.

Stereolithography or SLA was the very first AM technology developed by Jacobs and 3D Systems, where a liquid photosensitive resin is converted into a solid by exposing it selectively to ultraviolet light or an ultraviolet laser (Jacobs 1992, 1995). Figure 3.2 shows a schematic of the SLA process. Variations exist in this process, where suspensions of ceramic or metal particles in a photo-curable monomer are used in the SLA process to produce metal or ceramic parts. A typical process for ceramic part manufacturing has been described by Griffith and Halloran (1996).

The selective laser sintering (SLS) technique originally developed at the University of Texas Austin by Dr. Joe Beaman and his graduate student at the time Dr. Carl Deckard can make parts out of metal and plastic powders using a high power laser (UT-Austin 2014). Parts can be created from a range of powder materials, including

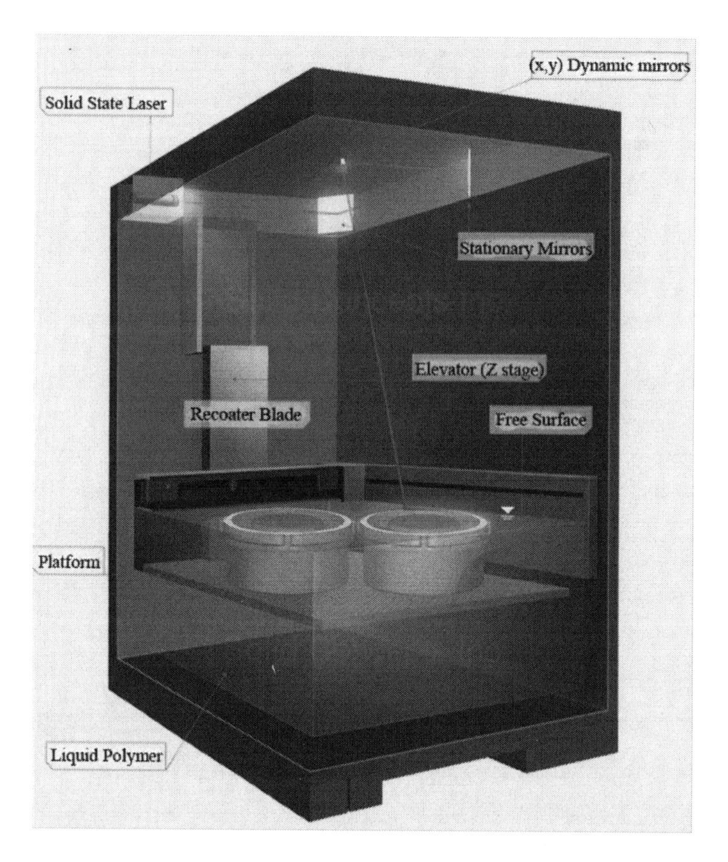

FIGURE 3.2 Schematic of the stereolithography AM process. (Courtesy of www.solidconcepts.com.)

metals, Nylon-11 and Nylon-12 polyamides, or nylons with fillers such as glass beads or carbon fibers (to enhance physical properties). SLS material properties can be comparable to those found with traditional manufacturing methods. A schematic of the SLS process is shown in Figure 3.3.

Yet another technology using polymers in the process is the Three-Dimensional Printing technology. The Three-Dimensional Printing technology (3DP™) was originally developed at the Massachusetts Institute of Technology in 1993 and is used in the Z Corporation's prototyping process, one among three such AM technologies to be developed from the original Michael Cima patent (Cima et al. 1993). Similar to the process used in other AM technologies, this technology also creates 3D physical prototypes directly from computer-aided design (CAD) models. A liquid binder is used to bind layers of deposited powders to produce the final prototype.

A schematic of the Z-Corp 3DP process is shown in Figure 3.4. In the 3DP process, the printers use standard inkjet printing technology. In this case, the parts are created layer-by-layer by depositing the liquid binder onto thin layers of powder. There is a feed piston and platform that rises incrementally for each layer, while a roller mechanism spreads the powder fed from the feed piston onto the build platform.

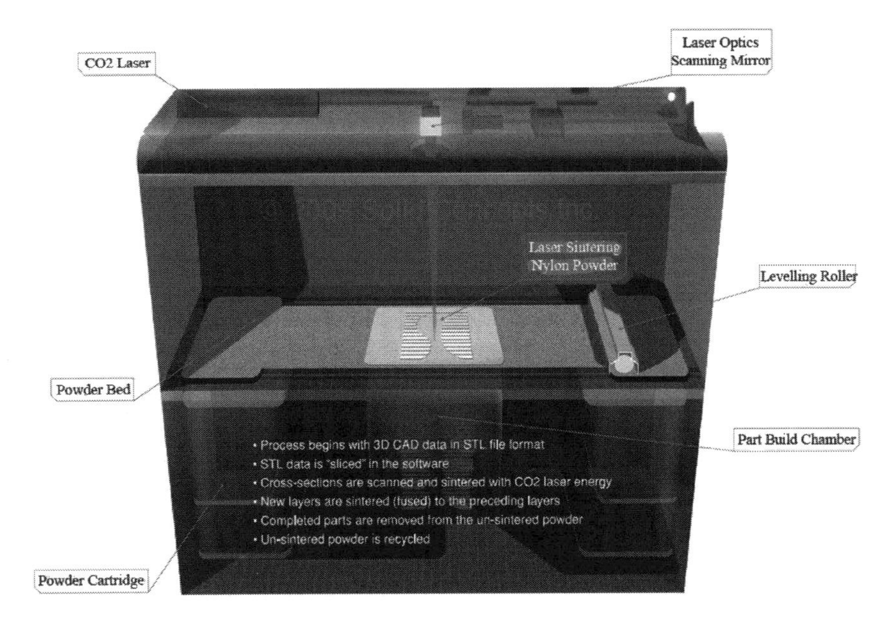

FIGURE 3.3 Schematic of the selective laser sintering AM process. (Courtesy of www.solidconcepts.com.)

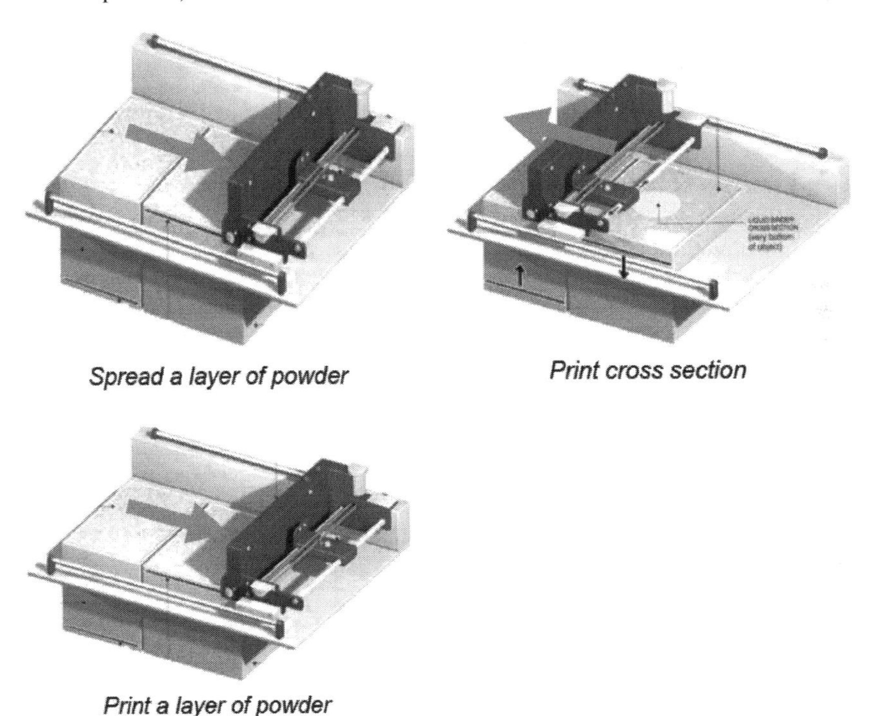

FIGURE 3.4 Process schematic for the 3DP process used in a typical binder jetting process. (Courtesy of www.3dsystems.com.)

A major advantage of this process is that it can utilize standard inkjet print heads to dispense the binder fluid onto the powder bed. It is a relatively fast process compared to other AM methods because of the multiple inkjet heads used. The inkjet print heads print in discrete locations on the powder bed, binding the powder particles together. After each layer is printed, the piston lowers by a set height and a new layer of powder is spread on top of the previous layer. After printing, the new layer is bonded to the previous layer, thus creating the final part.

3.2 ADDITIVE MANUFACTURING OF HIGH-STRENGTH THERMOPLASTICS AND FIBER-REINFORCED THERMOPLASTICS

Even though AM processes have been well established for polymers and polymer blends, similar progress in the case of high-strength engineering polymers and fiber-reinforced thermoplastic parts (both short fiber and long fiber reinforced) have been difficult to achieve, primarily due to the following issues:

- The capability to extrude a material is based on its column strength at the extrusion temperature, i.e., the amount of material capable of being extruded is dependent on the force exerted at the nozzle tip and is drastically reduced at higher extrusion temperatures. The fused deposition modeler or similar equipment uses a filament type of feeding material, limited to an extrusion temperature of 260°C and a column force of 0.35–0.4 MPa, limiting the choice of engineering polymers that can be extruded (Stuffle et al. 2000)
- The addition of fibers and their preferential alignment during deposition introduces anisotropy in properties in the part (Calvert et al. 1997)
- The choice of fibers and matching polymers is limited
- The properties of the fibers are anisotropic (thermal expansion coefficient, mechanical properties, etc.), whereas the polymers are isotropic. Since the blend of the fibers and the polymer will need to be heated prior to deposition, they will tend to expand and contract at different rates, potentially leading to cracks during the cooling step
- This problem is exacerbated in the case of a ceramic composite, which needs to go through a binder debinding and sintering step prior to consolidation. The cracking is typically observed during the cooling step after sintering due to the large difference in thermal expansion coefficient between the carbon or ceramic fibers and the matrix material. Ceramic and carbon fibers typically have a negative expansion coefficient in the thickness direction, while the matrix components can have high, positive expansion coefficients
- Support structure materials do not always match with the polymer binder or the polymer that is holding the fibers together, leading to difficulties in manufacturing parts with undercuts and overhangs.

To overcome the problems with the choice of thermoplastic polymers for AM technologies, Stuffle et al. developed a high-pressure extrusion head that was attached

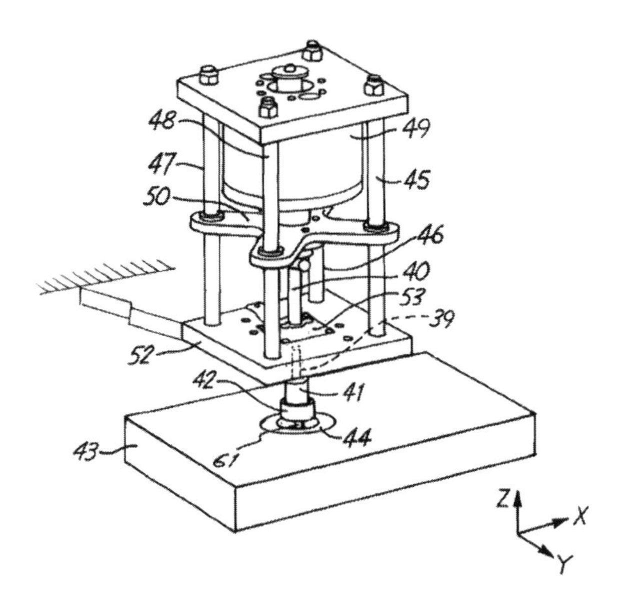

FIGURE 3.5 Schematic of a high-pressure extrusion head for AM of high-strength engineering polymers, ceramics, and metals. (From Stuffle, K.L. et al., Method and apparatus for in-situ formation of three-dimensional solid objects by extrusion of polymeric materials, US Patent No. 6,067,480, May 23, 2000.)

to an X–Y table and controlled by CAD software for AM. A schematic of the extrusion head is shown in Figure 3.5 (Stuffle et al.). Figures 3.6 and 3.7 show an FDM modeler retrofitted with the high-pressure extrusion head, while Figure 3.8 is a close-up image of an operating high-pressure extrusion head (Vaidyanathan et al. 2000). Table 3.2 shows a list of the different engineering polymers that have been used for AM with this high-pressure extrusion head (Stuffle et al. 2000). Some of the materials were true thermoplastics while others were melt processed thermoset materials.

The apparatus and process of high-pressure extrusion involves:

- Feed rod consolidation—first step
- Extrusion free forming—second step

Consolidation is the pressing of feed rods that are subsequently used in the extrusion step. The materials from Table 3.2 are typically supplied in pellet form. These pellets are then pressed in a single acting, heated cylindrical die and piston assembly at temperatures near the material's melting point under high pressure to produce a cylindrical feed rod without voids or flaws. The feed rod pressing conditions for each material are shown in Table 3.3 (Stuffle et al. 2000). The optimized temperatures and pressures for fabricating feed rods and the optimized deposition parameters are also included in Table 3.3. The rod pressing cycle is based on 10 minute hold at temperature and pressure. The deposition parameters are defined with an approximately 0.58 mm (0.023″) diameter extrusion nozzle.

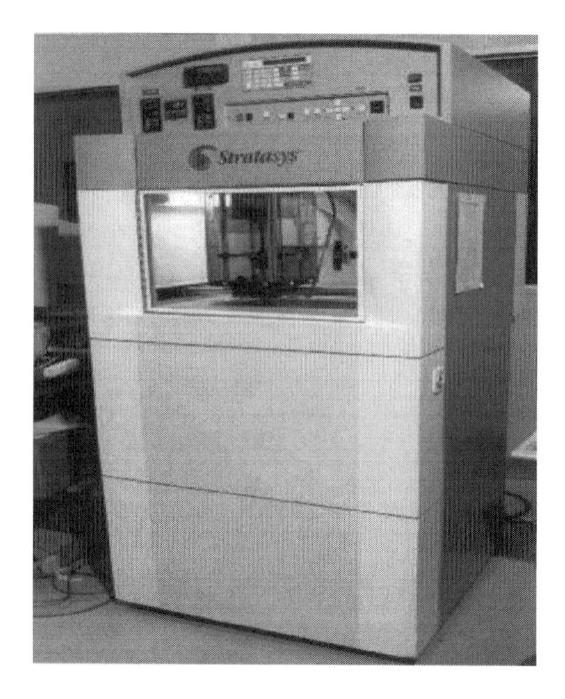

FIGURE 3.6 Retrofitted Stratasys FDM Modeler. (From Vaidyanathan, R. et al., *JOM*, 52, 34–37, 2000.)

FIGURE 3.7 Operation of FDM Modeler with high-pressure extrusion head. (From Vaidyanathan, R. et al., *JOM*, 52, 34–37, 2000.)

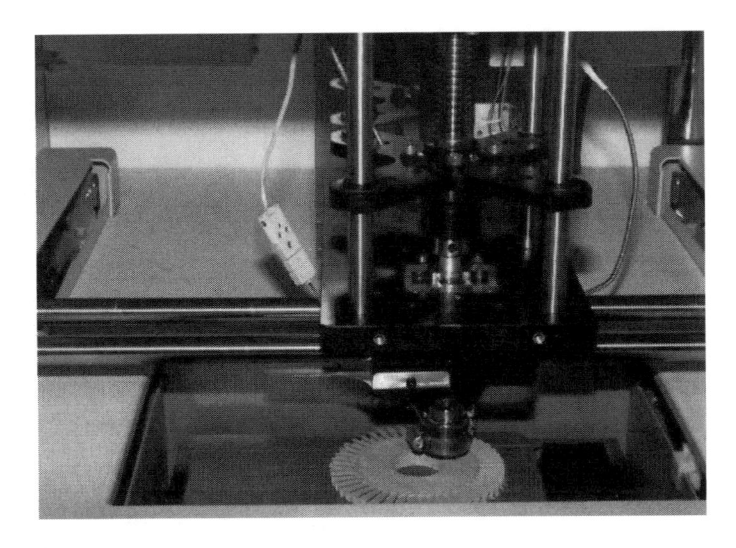

FIGURE 3.8 Close up view of high-pressure extrusion head in operation inside the FDM Modeler. (From Vaidyanathan, R. et al., *JOM*, 52, 34–37, 2000.)

TABLE 3.2
High-Strength Polymers (Reinforced and Unreinforced) Screened for AM

Polymers Screened			
Trade Name	**Current Manufacturer**	**Polymer**	**Reinforcement**
PEEK 150	Victrex USA Inc.	Polyaryletherketone	None
PEEK 450G	Victrex USA Inc.	Polyaryletherketone	None
PEEK 450CA30	Victrex USA Inc.	Polyaryletherketone	30% carbon fiber
Pellethane 2363	Dow Chemical	Polyurethane	None
Torlon	Sabic Innovative Plastics	Amide-imide	None
Lexan 141	Sabic Innovative Plastics	Polycarbonate	None
Lexan 3413	Sabic Innovative Plastics	Polycarbonate	20% glass fiber
Elvacite 2009	Lucite International Inc.	Poly (methyl methacrylate)	None

Source: Stuffle, K.L. et al., Method and apparatus for in-situ formation of three-dimensional solid objects by extrusion of polymeric materials, US Patent No. 6,067,480, May 23, 2000.

The high-pressure extrusion process works well with materials listed in Table 3.2 as well as acrylic, ABS, silicon nitride, alumina, and zirconia. It is also possible for the high-pressure extrusion head to be stationary while the base is moving or vice versa. Among all the materials from Table 3.2, PEEK 450G, PEEK450CA30, Lexan (with and without the fiber reinforcement) and Elvacite were good modeling materials and provided consistent results (Stuffle et al. 2000).

TABLE 3.3
Optimized Rod Pressing and Deposition Parameters for Extrusion-Based AM for Some Typical Engineering Thermoplastics

		Optimized Rod Pressing and Deposition Parameters				
Trade Name	**Polymer**	**T_R (°C)**	**P_R (MPa)**	**T_E (°C)**	**T_P (°C)**	**V_E (mm/s)**
PEEK 450G	Polyaryletherketone	345	4.1	400	320	0.4
PEEK 450CA30	Polyaryletherketone	345	4.1	390	320	0.4
Pellethane 2363	Polyurethane	190	4.1	210	140	0.4
Lexan 141	Polycarbonate	185	4.1	240	168	0.4
Lexan 3413	Polycarbonate	185	4.1	260	185	0.4
Elvacite 2009	Poly (methyl methacrylate)	120	4.1	185	146	0.4

Source: Stuffle, K.L. et al., Method and apparatus for in-situ formation of three-dimensional solid objects by extrusion of polymeric materials, US Patent No. 6,067,480, May 23, 2000.

T_R = Rod pressing temperature; P_R = Rod pressing pressure; T_E = Extrusion temperature; T_P = Deposition tip temperature; and V_E = Tip velocity.

Stuffle et al. also tested the AM materials [polycarbonate, poly (methyl methacrylate) and PEEK, and polycarbonate as well as poly(ether ketone) (PEEK) with fiber reinforcement] for their tensile, compressive, and fracture toughness properties (Stuffle et al. 2000). Sample densities were also measured using Archimedes principle. The test specimen geometry was of the typical "dog bone" shape. Two types of specimen orientations were tested. Type "V" samples were tested along the writing direction, while type "H" samples were tested across the writing direction. V and H refer to vertical and horizontal, which denotes the direction of material deposition with respect to the mechanical testing equipment. The equipment used was a model 1011 Instron model with a 4448 Newton load cell with vertical specimen loading and wedge-action type grips. The crosshead speed for all specimens was 5 mm/min. Tensile moduli, strength, 0.2% yield strength, and elongation and strain to fracture were calculated and reported.

Stuffle et al. (2000) reported that the measured tensile and compressive strength as well as fracture toughness values were lower than the manufacturer's reported properties. They observed that the densities of the fabricated specimens were only approximately 85% of the values reported in the literature. However, the tensile strength for PEEK with 30% carbon fiber (reported as early as in 1997) showed approximately 200% higher tensile strength (71.6 MPa versus 250 MPa) than the strength of the best thermoplastic (Ultem 9085) (Stratasys 2014) with approximately 300% improved tensile modulus (2200 MPa versus 8240 MPa). At the time these properties were reported, the high-pressure extrusion technique was still not a mature technology. With current improvements in materials and AM methods, it can be expected that the properties demonstrated by engineering thermoplastics could possibly show an improvement over the materials supplied by AM equipment manufacturers.

3.3 ADDITIVE MANUFACTURING OF HIGH-STRENGTH THERMOSETS AND THERMOPLASTICS AND CHOPPED FIBER-REINFORCED COMPOSITES

For AM of thermoset type of resin systems, there are two possibilities (Calvert 1998). The total shrinkage during curing could either be a combination of minimal shrinkage during deposition followed by large uniform post-cure shrinkage or shrinkage only during deposition that is as complete as possible before the next layer is deposited (Calvert 1998). Another method to achieve curing of the individual layers would be to internally mineralize the structures by alternating layers of gel containing dissolved salts that will cross-diffuse and precipitate. In an early demonstration of the AM process, Calvert and Liu showed that cross-linked polyacrylamide and polyacrylic acid gels could be freeformed by writing solutions of the monomer, crosslinking agent and the catalyst onto a hot plate, with the heat inducing the polymerization reaction (Calvert and Zengshe 1998). In the case of polyacrylamide, their recipe was based on 18% aqueous solution of acrylamide, methylene bisacrylamide as a cross-linker at 2%–5% of the monomer, 0.03% potassium persulfate, and 1% tetramethylenediamine as catalyst and activator. 12 wt% fumed silica was also added to make the mixture thixotropic and control its flow properties. The mixture was freeformed onto a hot plate that was kept at 60°C, with the curing occurring within 3 minutes after deposition (Calvert and Zengshe 1998). Similar shapes were also formed with polyacrylic acid. The major finding in their work was that multilayer stacks of cross-linked hydrogels would swell differently from anticipated behavior if the materials were to be taken separately. This was explained in terms of high per-chain stiffness for one of the components and a negative Poisson's ratio for the other component in a dilute base.

Calvert et al. (1997) also showed how freeforming could be used in chopped fiber-reinforced thermoplastic and thermoset composites to obtain improved properties compared to unreinforced composite materials. Even though chopped fiber composites are not as stiff as continuous fiber-reinforced composites, they are amenable to scaling up through processes such as injection molding.

Depending upon the particular processing conditions employed, it is well known that the minor phase of a polymer blend has the tendency to become deformed when subjected to shear conditions, especially when applying AM techniques to fiber-reinforced polymers. The amount of deformation experienced by these droplets is a strong function of the shear stress rate imparted to the blend, the viscosity of the individual polymers constituting the blend, and the diameter of the minor phase material. Elmendorp (1986) has proposed a relationship detailing the elongation of a minor phase droplet in a polymer blend when subjected to shear stress.

These factors are strongly dependent upon blend extrusion conditions. The polymer component viscosity and interfacial tensions are influenced by extrusion temperature while shear stress imparted to the blend increases with extrusion pressure and decreased extruder orifice diameter. Initial (quiescent) droplet diameter of the minor blend component is dependent upon relative concentration of the minor phase in the blend. The morphology of these droplets is therefore strongly influenced by the amount of shear imparted to the polymer (Fayt et al. 1987; Chin and Han 1980;

Han 1981; Kobayashi et al. 1988; Moore and Kim 1992; Wu 1987). When a blend is initially stressed, the spherical droplets become elongated into an ellipsoidal geometry (Elmendorp 1986). Increased stress causes the ellipsoids to become oriented with their major axes parallel to the polymer extrusion direction. Ultimately these ellipsoids become elongated into long continuous fibrils which are oriented parallel to the flow direction. Vanoene has discussed the transition between spherical minor phase to ellipsoids and its subsequent fibrillation while extruding a polymer blend through a nozzle (Vanoene 1972). A schematic of the fibrillation that is usually observed in a nozzle is shown in Figure 3.9. Thus, in this case, it is believed that the rheology of the polymer blend will have the property so that its minor phase will undergo fibrillation when subjected to high shear extrusion through the AM equipment (Vaidyanathan et al. 2000).

The effect of fiber content and fiber orientation in the freeformed mixture controls the mechanical properties of the AM fabricated composites, as seen in the case of thermoset epoxy composites that were freeformed using the extrusion freeform fabrication technique (Calvert, P., Extrusion Freeform Fabrication of thermoplastic and thermoplastic composites, Personal Communication, January 2001). This is shown in Figure 3.10 for Epon 828 and Araldite MY720 tetrafunctional epoxy resins. The effect of aspect ratio of the fiber size is shown in Figure 3.11, based on

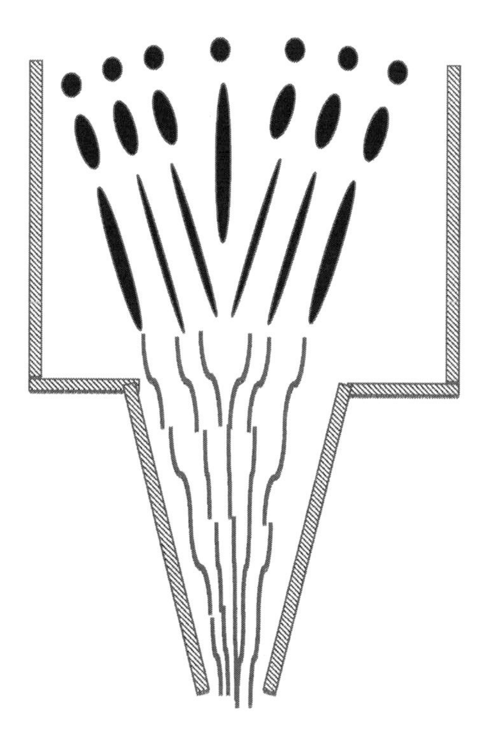

FIGURE 3.9 Schematic describing the fibrillation of a polymer blend through an extrusion orifice. (Adapted from Tsebrenko, M.V. et al., *Polymer*, 17, 831–834, 1976.)

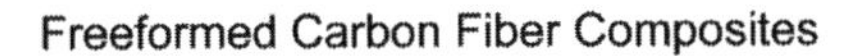

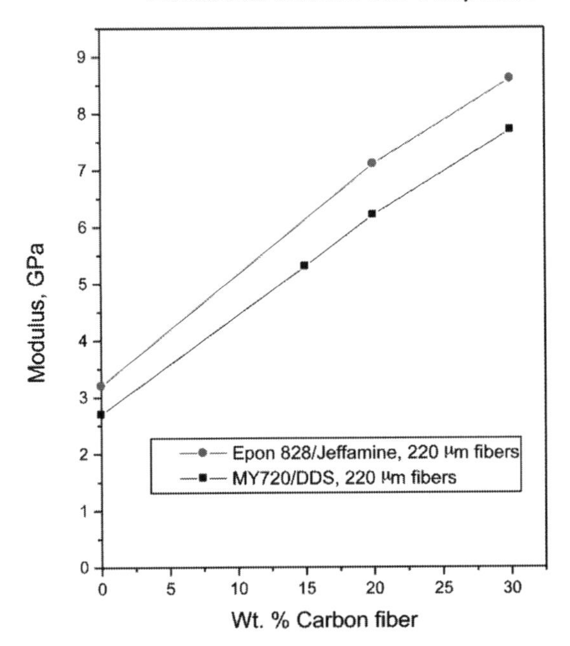

FIGURE 3.10 Correlation of fiber content on the modulus of 220 μm carbon fibers added to Epon 828 and MY720 resins.

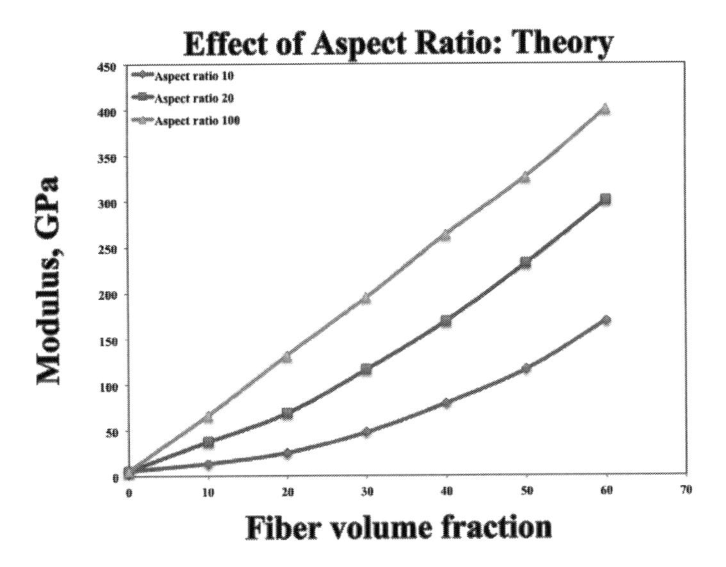

FIGURE 3.11 Predicted modulus versus aspect ratio and fiber volume fraction for Epon 828/glass fiber composites.

Halpin-Tsai equations, although the AM process will become complex and difficult beyond a certain fiber aspect ratio and fiber content.

The effects of extrusion are prominent specifically through the addition of second phase particles that have different properties compared to the polymer (Newtonian versus viscoelastic). This has been observed both in the case of fiber-reinforced thermoplastic as well as thermoset composites. Calvert et al. (1997) demonstrated the alignment effects due to extrusion of short fiber composites both for thermoplastics as well as thermoset composites (Peng et al. 1999). Table 3.4 is a listing of typical mechanical properties reported by Peng et al. showing the difference in properties when the material is deposited in directions perpendicular and parallel to the length of the bar in a thermoplastic composite material composition. There is a clear anisotropy in the properties in the two directions, which has to be taken into account while designing components to be fabricated using AM techniques. Figure 3.12 shows the fracture surface of a composite, showing alignment of the fibers as a result of extrusion in a thermoset composite material during the AM process (Calvert et al. 1997).

The effect of fiber orientation on the elastic modulus in epoxy/carbon fiber composites is shown in Figure 3.13. These samples were fabricated by writing the composites at varying angles to the axis of the test bars. It was seen that the modulus of the composite when the fiber axis is parallel to the deposition direction is double that of the composite where the material is deposited perpendicular to the testing direction.

Similar property differences have also been observed by a number of other research groups in the case of chopped fiber-reinforced thermoplastic composite materials (Kumar and Kruth 2010; Nikzad et al. 2011; Zhang et al. 2014; Goodridge et al. 2011; Hao et al. 2006). Hao et al. (2006) suggest that the presence of fibers do not provide a smooth powder bed and lead to issues in obtaining high density and strength. In their case, the best properties were obtained by coating one type

TABLE 3.4
Mechanical Properties of Thermoplastic Composite Tensile Bars

Material	Orientation	Modulus (GPa)	Tensile Strength (MPa)	Elongation (%)
PEEK	Parallel	1.7	59	3.3
	Perpendicular	1.8	88	5.3
PEEK + 30 wt% carbon fiber	Parallel	9.4	257	3.0
Polycarbonate	Parallel	1.1	64	8.7
	Perpendicular	3.6	124	3.6
Polycarbonate + 30 wt% glass fiber	Parallel	3.0	106	3.8
	Perpendicular	1.0	46	5.6
PMMA	Parallel	1.3	23	1.4
	Perpendicular	1.5	61	5.8

Source: Peng, J. et al., *Compos. Part A Appl. Sci. Manuf.*, 30, 133–138, 1999.

[a] Instron model 1011; strain rate: 5 mm s^{-1}; load cell: 4448 N; vertical specimen loading; and wedge-action type grip.

[b] PEEK = poly(ether ketone); PMMA = poly(methyl methacrylate).

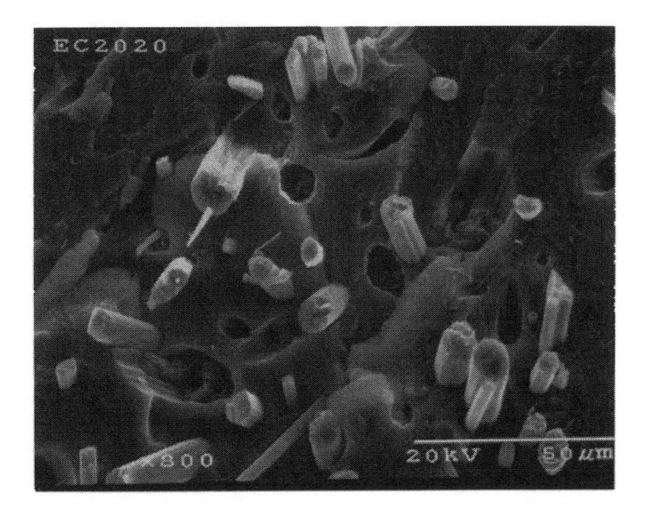

FIGURE 3.12 Fracture surface of a thermoset composite, showing alignment of fibers during extrusion in the AM process. (From Calvert, P. et al., *High Perform. Polym.*, 9, 449–456, 1997.)

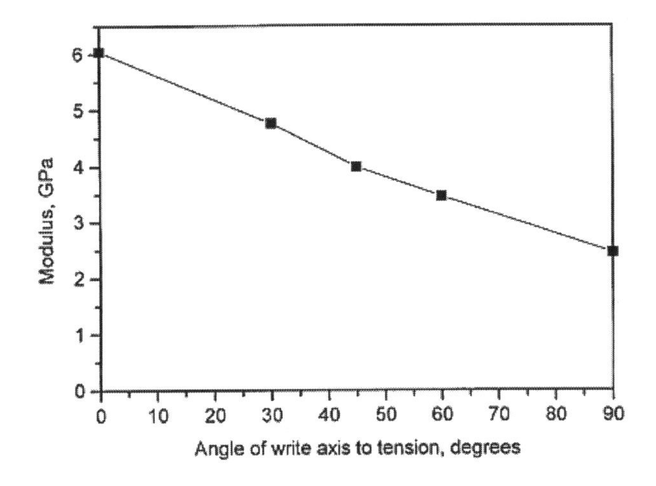

FIGURE 3.13 Measured modulus from three-point bend tests for 18 vol% glass fibers, aspect ratio 7, formed into bars with the write axis inclined to the long axis of the bar.

of powder with another type so that the composite powder could be fabricated by a standard AM technique such as selective laser sintering. Zhong et al. (2001) added chopped glass fibers to ABS to create filaments that were used as a feedstock in fused deposition modeling. They observed that the compatibility of glass fibers to the thermoplastic ABS matrix is enhanced through the addition of a compatibilizer like hydrogenated Buna-N that has butadiene and acrylonitrile groups, structurally similar to ABS (Zhong et al. 2001). It was observed that the mechanical properties, toughness, and the appearance of the filament were improved compared to blends with no glass fibers or compatibilizers.

3.4 AM PROCESSES APPLICABLE TO NANOCOMPOSITES

AM processes are especially suitable for nanocomposites that need very high forces to be extruded into complex shapes, especially since these forces may limit the amount of nanofiller that can be added to the composite beyond a certain level. Additionally, it is well known that the amount of nanofiller that can be added may be limited due to agglomeration and low surface energy of the particles (Njuguna et al. 2008).

Shofner et al. (2003) investigated the effect of an AM process (extrusion freeform fabrication) in ABS filled with single walled nanotubes (SWNT) and vapor grown carbon fibers (VGCF). The addition of 5% wt.% SWNT improved tensile modulus by 93% and tensile strength by 31%, respectively. Similarly, 5% wt.% VGCF improved tensile modulus by 44% and tensile strength by approximately 27%, respectively. This is shown in Figures 3.14 (tensile strength) and 3.15 (tensile modulus), respectively.

The effect of extrusion in the fiber alignment in a VGCF/ABS composite is shown in Figure 3.16. It can be seen that the AM process would still achieve preferred fiber orientation even when the fibers are nano-sized. Compared to an ABS blend with nanofillers that shows uniform dispersion (Figure 3.17), there is clear evidence of fiber alignment after the AM process. The nozzle size that is used for the AM process, however, would need to be approximately 50–100 times that of the diameter of the fillers, as suggested by Calvert (Calvert, P., Extrusion Freeform Fabrication of thermoplastic and thermoplastic composites, Personal Communication, 2001). Having a nozzle size that is smaller could potentially lead to clogging and rapid increase in the pressure required to extrude the polymer blend. This is one of the reasons why the standard FDM process is not capable of extruding polymer blends with high solids loading (>50% by volume) and high fiber loading (>30% by weight).

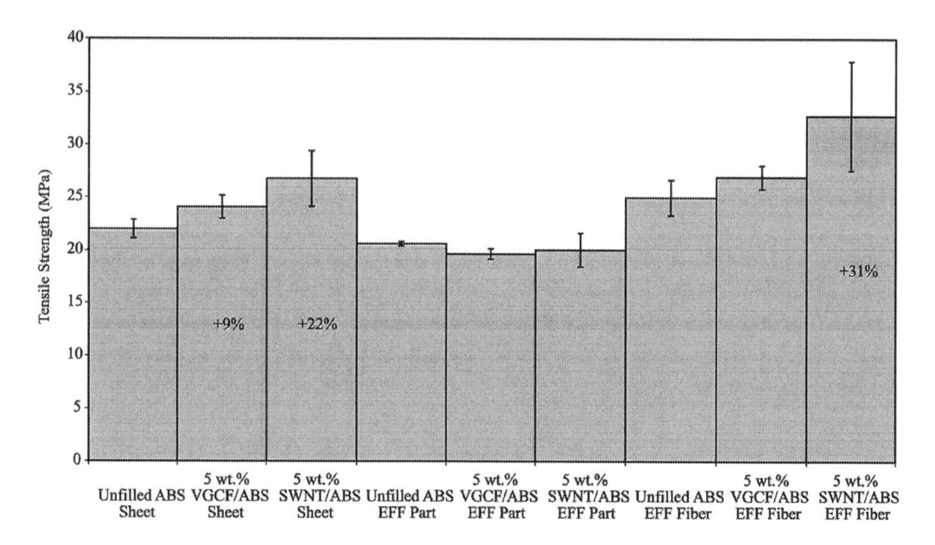

FIGURE 3.14 Tensile strength of filled ABS materials before and after AM processing. Percent changes are given for statistically significant differences.

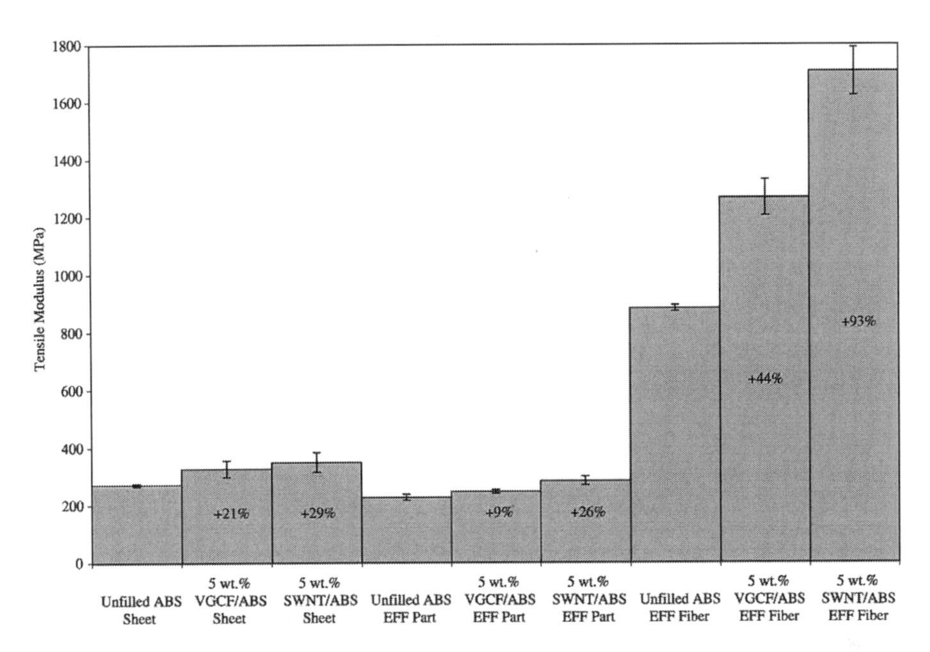

FIGURE 3.15 Tensile modulus of filled ABS materials before and after AM processing. Percent changes are given for statistically significant differences.

FIGURE 3.16 SEM image of a 5 wt% VGCF/ABS composite subjected to selective dissolution after AM processing. (From Shofner, M.L. et al., *Compos. Part A Appl. Sci. Manuf.*, 34, 1207–1217, 2003.)

Farmer et al. have recently proposed a method for combining the synthesis of aligned carbon nanotube "forests" on a substrate, curing a thermoset polymer using a UV curable resin, and building it up layer by layer (Farmer et al. 2010, 2012). Figure 3.18 is a schematic of the proposed AM process. A thin layer of UV curable thermosetting polymer is first spread on the part bed. An array of carbon nanotubes (CNTs) are

FIGURE 3.17 SEM image of a 5 wt% SWNT/ABS composite after blending in a high shear mixer, showing good dispersion and distribution. (From Shofner, M.L. et al., *Compos. Part A Appl. Sci. Manuf.*, 34, 1207–1217, 2003.)

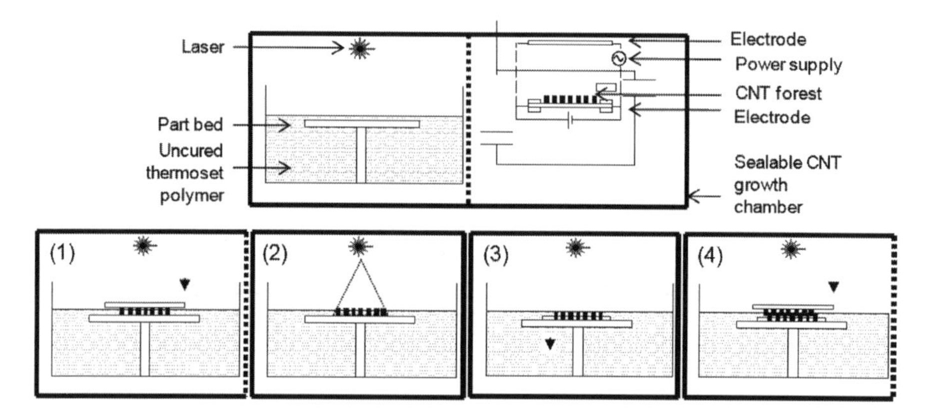

FIGURE 3.18 Schematic of the required apparatus for combined CNT growth and additive manufacturing of nanocomposites. (From Farmer, B.L. et al., Strategies to combine nanocomposite and additive layer manufacturing techniques to build materials and structures simultaneously, Paper #993 Paper read at *ECCM15—15TH European Conference on Composite Materials*, 24–28 June, Venice, Italy, 2012.)

grown in another part of the chamber and transferred onto the part bed. The CNTs are held in place either by interfacial forces or by partially curing the CNT/resin combination. A laser beam is then used to raster the resin surface and cures the resin. The part bed is then dropped such that another thin layer of resin can be introduced on the surface. A second layer of CNTs is introduced on the surface and the process is continued. However, even though the growth of CNT "forests" is a well-developed process, the combination of CNT and the AM process is still in development. The technology may become successful after it takes into account the difficulties related to the removal of

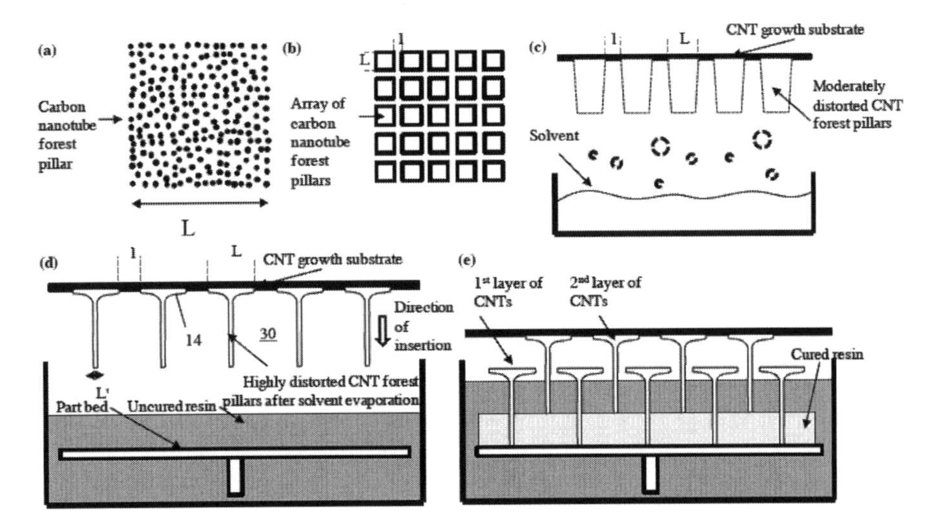

FIGURE 3.19 Process schematic for AM fabrication of polymer composites reinforced with carbon nanotube bundles.

the resin from areas where it is not needed in the individual layers as well as how support structures could be introduced into the build process.

Farmer et al. also investigated the use of partial wetting of the CNT "forest" to evaluate interleaving of CNT layers by using a patterned array to ensure through thickness continuity. This was necessary to hold the CNT bundles in place and have the resin wet and cure or bond to the bundles in place. This was done by dipping the CNT bundles in the resin matrix. In the case of thermoset resins, the partial wetting was possible till the curing temperature at which the resin viscosity dropped considerably, leading to wetted areas breaking up into pillars. Partial wetting was possible in the case of thermoplastic resins, but it was a challenge to confirm that partial wetting was obtained (Farmer et al. 2012).

Farmer et al. also proposed a modification to the procedure described in Figure 3.18 to increase the volume loading of the reinforcements (Farmer et al. 2012). A schematic strategy for this approach is shown in Figure 3.19. Additional modifications are possible where fiber orientation control would be possible both interlayer and intralayer.

3.5 AM PROCESSES FOR CONTINUOUS FIBER-REINFORCED COMPOSITES

The AM processes that have been developed so far for continuous fiber-reinforced composites are variations of the automated tow placement process originally developed for laying down different composite prepreg layups. One of the earliest processes developed and demonstrated by Don Klosterman and others at University of Dayton Research Institute (UDRI) was based on an improvement of the laminated object manufacturing (LOM) technique for designing and manufacturing ceramic matrix composites (Meilunas et al. 2002; Klosterman et al. 1998, 1999; Meilunas 2001).

This was funded from among the second set of solid freeform fabrication projects funded by Defense Advanced Research Projects Agency (DARPA) and the Air Force and carried out by a team from Northrop Grumman, Helisys, and University of Dayton Research Institute (Meilunas 2001). This team was put together by a set of individuals who had worked on previous DARPA funded projects related to low cost ceramic composites and solid freeform fabrication of ceramics from the early to mid-1990s. The AM process that was developed was a follow up project from the low cost ceramic matrix composites (LC3) program (Gonczy and Sikonia 2005) funded by DARPA from 1991–1997.

The basic issue with AM of continuous fiber-reinforced composites is that none of the AM processes are really capable of accommodating continuous fiber preforms or prepregs or woven mats, including the use of a laser to accurately machine the ends of the fibers after each layer is laid down over the previous layer. It is difficult or impossible for AM processes to take into account the geometrical issues such as fiber orientation and fiber continuity. For the first time, this group demonstrated that it was possible to modify the LOM process to include fiber-reinforced structures, especially thin, curved shell components. It should be noted that this process was developed when the AM process development was still in its infancy and the software and hardware capabilities were still being resolved. The process needed to take the following issues into account (Meilunas 2001):

- The curved LOM process (previously developed at UDRI under DARPA funding) (Klosterman et al. 1999) had several limitations to generate complex curved structures
- The curved LOM system as set up originally had several hardware and software inadequacies that impeded further hardware/software upgrades. Additionally, the build envelope of the curved LOM machine after modifications was smaller compared to required part sizes. The curvatures that could be introduced into the component could not be built without introducing wrinkles during the lay-up process
- This limited the commercial viability of the LOM process to be adapted for continuous fiber-reinforced composites and components.

In order to account for the possible complex geometries and sizes that could be encountered in a component, it was found necessary to modify the build sequence in the curved LOM so as to avoid the potential to introduce wrinkles in the part during the lay-up process. The original curved LOM equipment had a scanning galvanometer-based mirror system that had to be changed to a galvo scanner, which provided better laser positioning accuracy and more uniform corner cutting and better edge definition in finished parts (Meilunas 2001). Due to several of these problems, the UDRI-Northrop Grumman team decided to procure a new curved LOM system that was capable of handling all the technical issues raised during the initial stages of the project.

The modification to the original plan combined the commercial 2D LOM build process with final composites forming step resulting in curved composite components, as shown in Figure 3.20. The final step could be achieved by using either a matched mold or by using a diaphragm to compact the laminate as the final step.

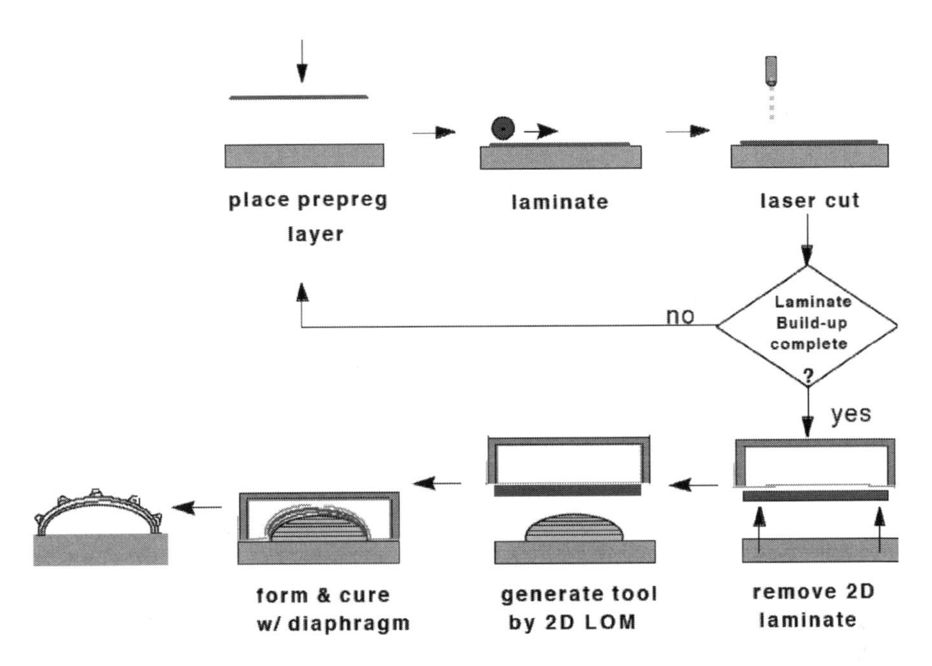

FIGURE 3.20 Schematic of curved LOM/composite forming process developed by UDRI/ Northrop Grumman. (From Meilunas, R., Laminated Object Manufacturing-Based Design Ceramic Matrix Composites, Final Report No. AFRL-ML-WP-TR-2001-4074 to DARPA/ Wright Patterson AFB, 2001.)

The benefit of this process is that it can decrease manufacturing costs and obtain consistent properties by eliminating any unnecessary or time-consuming hand lay-up procedures (Meilunas 2001).

In the process schematic shown above, a series of ply stacks are generated using commercially available CAD/computer-aided manufacturing (CAM) software or other software such as FiberSIM. The ply stack data are then utilized by the LOM system to fabricate individual flat 2D laminate preforms incorporating any lay-up sequence. Utilizing this ply stack data, the LOM system can now fabricate individual (2D) laminate preforms incorporating any cuts or darts in individual ply layers that can be used to drape the laminate over a curved tool. The final step is to place a conformable vacuum diaphragm on top of the tooling and the prepreg layers and to cure it in place. This curing can be done either on the LOM machine itself or in a separate oven. Several other modifications were also made.

The major advantages of this technique are (Meilunas 2001):

- The laser positioning equipment installed in the modified LOM machine allowed increased accuracy and speed
- The LOM machine allowed the integration of fiber-reinforced composites analysis software to be integrated to the lay-up selected and built
- Since it is possible to modify the geometric shape of the individual plies prior to the final consolidation step, this allows extremely complex geometries to be fabricated using AM processes.

A schematic of the composites forming cell and the actual forming cell built under this project is shown in Figures 3.21, while Figure 3.22 shows the actual set up. It consists of a vacuum box of approximate internal dimensions of 775 × 550 × 200 mm. There are two 1.9 mm thick silicone rubber diaphragms that are bonded to individual 76 mm wide aluminum frames and a heat lamp array of six 375-watt infra red (IR) lamps that are mounted on adjustable sockets. The bottom silicone rubber is opaque

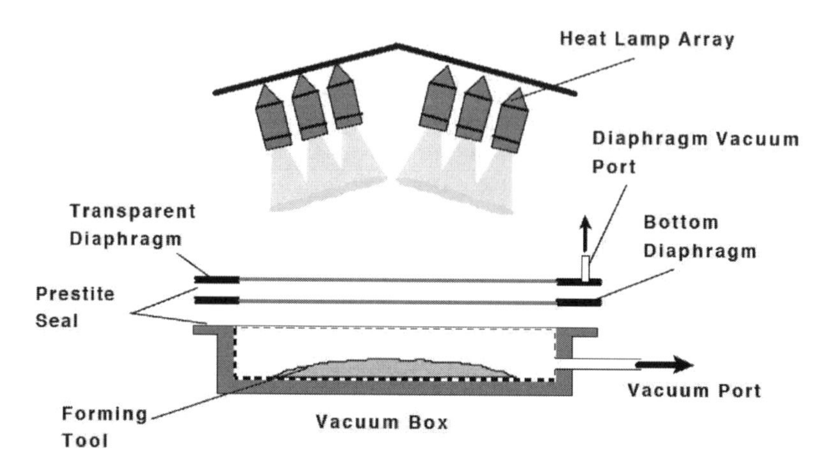

FIGURE 3.21 Schematic of composites forming cell for LOM. (From Meilunas, R., Laminated Object Manufacturing-Based Design Ceramic Matrix Composites, Final Report No. AFRL-ML-WP-TR-2001-4074 to DARPA/Wright Patterson AFB, 2001.)

FIGURE 3.22 LOM composites forming system. (From Meilunas, R., Laminated Object Manufacturing-Based Design Ceramic Matrix Composites, Final Report No. AFRL-ML-WP-TR-2001-4074 to DARPA/Wright Patterson AFB, 2001.)

FIGURE 3.23 Modified LOM2030H system with laser scanner subsystem. (From Meilunas, R., Laminated Object Manufacturing-Based Design Ceramic Matrix Composites, Final Report No. AFRL-ML-WP-TR-2001-4074 to DARPA/Wright Patterson AFB, 2001.)

while the top silicone rubber is transparent so as to provide efficient heat transfer to the curing process. There is also a vacuum port on the aluminum frame attached to the top diaphragm for pulling a vacuum during the curing process (Meilunas 2001).

A modified LOM equipment used to fabricate the demonstration component is shown in Figure 3.23, while the demonstration component made using the LOM/Composite Forming technology is shown in Figure 3.24.

Yet another method developed in the early 2000s was a technique based on the automated tow placement (ATP) for polymer composites (Don et al. 1997) and adapted for ceramic matrix composites such as C/SiC and C/ZrC (Vaidyanathan et al. 2005). The ATP process is an enabling technology, developed originally for thermoset composites and more recently for in-situ non-autoclave consolidation of large-scale thermoplastic composite materials for high speed civil transport applications. The knowledge base of the ATP process for thermoplastic prepregs can be utilized to lay down ceramic tows in the desired configuration thus allowing the use of a proven technology for low cost, rapid fabrication of large complex ceramic parts.

The automated tow placement system employs two hot-gas nitrogen torches to heat the material and two rollers to provide the pressures required for consolidation. The purpose of the first torch and roller is to preheat the composite surface and incoming tow together. The material is thus "tacked" to the surface with this roller. This tacking procedure is useful in that the fed material is carefully bonded to the surface and not pulled with the main consolidation roller. This tacking approach also aids in improving the efficiency of the cut and re-feed mechanism. The second torch (main heater) provides supplemental through thickness heating to facilitate consolidation and bonding of the tow and substrate under the consolidation roller.

FIGURE 3.24 200 mm × 200 mm Nextel 312/Blackglas resin composite fabricated using LOM/composite forming AM process. (From Meilunas, R., Laminated Object Manufacturing-Based Design Ceramic Matrix Composites, Final Report No. AFRL-ML-WP-TR-2001-4074 to DARPA/Wright Patterson AFB, 2001.)

These rollers provide the necessary forces to achieve complete intimate contact across the tow interface and as a boundary pressure for preventing any internal void development. The forces applied to both rollers are controlled independently using a series of pneumatic actuators. The composite tows can be placed in a regular repeating pattern or with brick-face symmetry. The brick-face geometry has the advantage that more homogeneity is achieved throughout the composite structure. An image of the ATP equipment used is shown in Figure 3.25, while a schematic of the process is shown in Figure 3.26.

The ATP process for thermoplastics lays down prepreg tows, typically 0.125–0.2 mm in thickness, with the tow width depending on the hardware. The ATP head can lay down 6.25-mm wide tows, while industrial machines, such as Cincinnati Milacron's Gantry System can lay down tows as wide as 150 mm. The modified ATP or ceramic composite ATP (CCATP) was primarily developed for a class of materials termed fibrous monoliths (Kovar et al. 1997) reinforced with carbon fiber tows. However, issues such as difference in thermal expansion coefficient between the matrix and the fibers are still outstanding, and the technique is not yet fully developed. For example, the large mismatch between the coefficient of thermal expansion of ultra high-modulus (UHM) carbon fiber (-0.5×10^{-6} ppm/K) and the SiC matrix (3.6×10^{-6} ppm/K) will result in residual stresses and cracks within the post-consolidated matrix. Incorporating a suitable interfacial material is crucial in reducing or eliminating matrix cracking in composites caused by the coefficient of thermal expansion mismatch between matrix and fiber. Composite strength and toughness are improved when the interface material deflects matrix cracks. Accordingly, a boron nitride interface was applied between the fibers and the matrix prior to its introduction into the matrix.

FIGURE 3.25 A typical ATP set up used. (From Yarlagadda, S., 2014.)

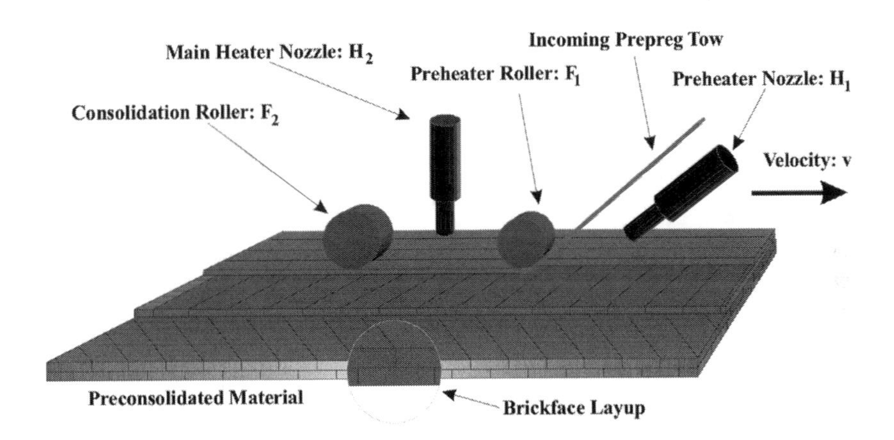

FIGURE 3.26 Schematic of the ATP process. (From Yarlagadda, S., 2014.)

Typical process parameters for tow placement of thermoplastics are listed in Table 3.5. These parameters are optimal for an APC/PEEK system, which has a glass transition temperature of 156°C and a melting temperature (PEEK is semi-crystalline) of 332°C. In contrast, the thermoplastic binder used in a ceramic matrix system has a processing temperature in the range of 100°C to 140°C.

The critical process parameters are torch temperatures, torch heights, head velocity, and consolidation force. Initial estimates on suitable operating ranges for these parameters were based on thermal models developed for the ATP process. Based on

TABLE 3.5
ATP Process Parameters for Carbon/PEEK Prepreg Tows

Process Parameter	Value
Initial thickness of tow	0.1778 mm
Width of tow	6.35 mm
Radius of roller (1)	15.8 mm
Radius of roller (2)	19.0 mm
Roller (2) location	80 mm from tacking Roller
Initial composite temperature	100°C
Initial roller temperature	25°C
Ambient air temperature	25°C
Gas flow rate in torches	50 liters/min
Location of preheater torch	75 mm from nip point location
Location of main torch	35 mm from nip point
Torch temperatures	850°C
Torch heights	Variable
Head velocity	Variable (up to 100 mm/s)
Consolidation force	300 N
Panel size	150 mm × 150 mm
Fiber orientation	Full range (−90 to +90)

these models, process maps relating material temperature, head velocity, and torch temperatures are generated.

Several process modifications were performed to obtain better quality material and are listed below:

- The preheater torch is set to operate at 500°C
- Roller 1 is disabled
- The main heater torch is positioned to actively cool Roller 2
- Gas flow rate for preheater torch is at 50 liters/min. and for main heater torch (now actively cooling) at 25 liters/min.

The modified setup is shown in Figure 3.27. Roller 1 has been disabled completely, by positioning it above Roller 2. During tow placement of thermoplastic tows, Roller 1 was used to tack hot prepreg to the laminate and prevent it from being pulled by the consolidation roller (Roller 2). In the present case, the ceramic tows have a much lower processing temperature, so that one heater torch and one roller are sufficient to achieve good tack and consolidation. The main heater torch has been positioned above Roller 2 to actively cool it and maintain its temperature below the glass transition temperature of the thermoplastic binder. This prevents the matrix of the tows from adhering to the roller and creating bare fiber spots and inconsistent quality.

A turntable/winding mandrel system (Figure 3.28) with an accuracy of ±1° was used. The system can be used in a horizontal mode (for flat components) or vertical mode (for axisymmetric components such as thrusters and cylinders). This device

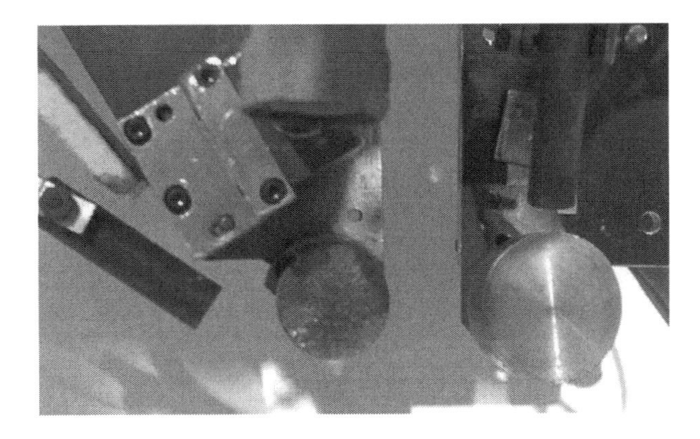

FIGURE 3.27 Modified ATP head configuration for ceramic tow placement, with Roller 1 disabled (raised above Roller 2). (From Vaidyanathan, K.R. et al., Continuous fiber reinforced composites and methods, apparatuses, and compositions for making the same, US Patent No. 6,899,777 B2, May 31, 2005.)

FIGURE 3.28 ATP fabrication of a composite panel and a blisk component using a turntable/winding mandrel consolidating ceramic tows for the laminate. (From Vaidyanathan, K.R. et al., Continuous fiber reinforced composites and methods, apparatuses, and compositions for making the same, US Patent No. 6,899,777 B2, May 31, 2005.)

can interface directly with robot programming software to perform the CAD/CAM implementation on the ATP head. This modification can also improve the life of the robot. The RobotStudio software aids in simulating the tow placement process and provides the tool path information to the ABB robot. An image of the simulation from consecutive layers is shown in Figure 3.29.

The process parameters for ceramic composite ATP are listed in Table 3.6. Some typical components fabricated are shown in Figure 3.30. The entire robot movement sequence is setup by computer programs developed for the ATP thermoplastic tow placement experiments. The torch parameters (temperatures, heights, and gas

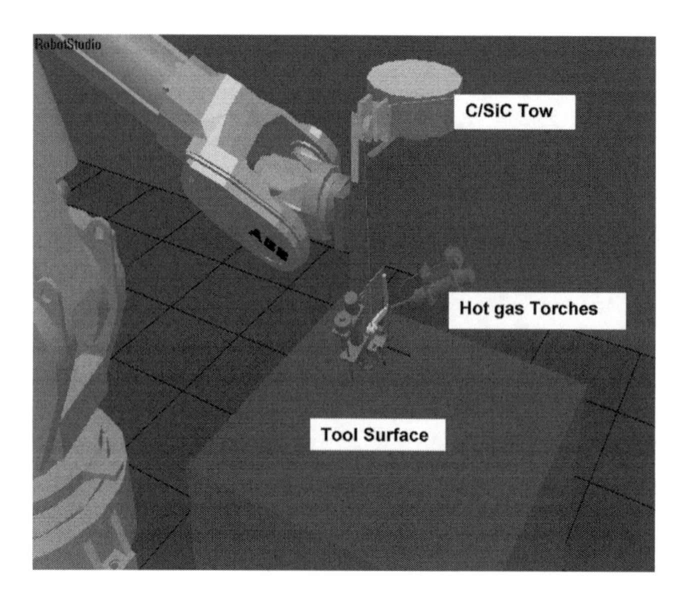

FIGURE 3.29 A simulation of the tool path from RobotStudio software. (From Yarlagadda, S., 2014.)

TABLE 3.6
Process Parameters for Ceramic Tow Placement of Ceramic Matrix Composites Laminates

Process Parameter	Value
Initial thickness of tow	1.0 mm
Width of tow	2.0 mm
Roller (1)	Disabled
Radius of roller (2)	19.0 mm
Roller (2) location	80 mm from Roller 1
Gas flow rate in preheater torch	50 liters/min
Gas flow rate in main torch	25 liters/min
Location of preheater torch	75 mm from nip point location
Location of main torch	Above Roller 2
Preheater torch temperature	500°C
Main torch temperature	25°C
Preheater torch height	Maximum (12 mm)
Head velocity	50 mm/s
Consolidation force	Minimum (190 N)

flow rates), consolidation force, and head velocity can be controlled on the fly as inputs to the computer program. Final panel dimensions and lay-up sequences are also inputs to the program. Once these inputs are given to the program, it can operate the robot in automatic mode and lays down the tows as specified. Green ceramic

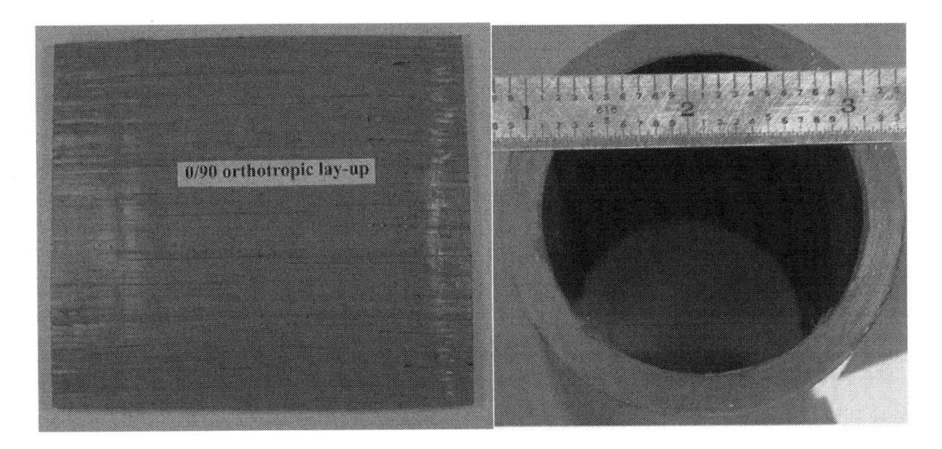

FIGURE 3.30 Some typical components fabricated using the CCATP process. (From Vaidyanathan, K.R. et al., Continuous fiber reinforced composites and methods, apparatuses, and compositions for making the same, US Patent No. 6,899,777 B2, May 31, 2005.)

matrix laminates of any size (within limits of the robotic work cell), fiber orientation, and material system can be fabricated by this technique. The work in the project was not focused on optimizing process parameters, but on demonstrating the feasibility of rapid, low cost fabrication of fiber-reinforced ceramic composites.

As can be seen from the images of the ATP fabricated panels (Figure 3.30), the tow-placement mechanism experienced some difficulties regarding the nature of the juxtaposed tow placement. Some tows were not placed in a straight, in line fashion, resulting in some gaps between adjacent tows. There were four major issues to be addressed if the CCATP was to be adopted by the AM enthusiasts. These are:

- Tape width
- Tape overlap requirements in the planar and thickness directions
- Gap between adjacent radial fibers in a polar weave configuration
- Rotation speed.

Jang et al. also have received a patent in 1999 for AM processing of fiber-reinforced composites (Jang et al. 1999). However, this is more a description of a process rather than the actual examples of continuous fiber-reinforced composites built by AM processes. Similarly, Ryan Dehoff and Lonnie Love at Oak Ridge National Laboratory run a manufacturing demonstration facility that appears to be more on the lines of the extrusion freeform fabrication of high-strength polymers filled with chopped carbon fibers (Dehoff 2014). MarkForged is one of the many 3D printing machines that can handle carbon fiber-reinforced composites, although this also seems to be a polymer filled with chopped carbon fibers (MarkForged 2014). What is exciting is that the technology developed at Oak Ridge National Lab is now being used to produce the world's first 3D printed automobile that was test run at the International Manufacturing Technology Show (IMTS) show in Chicago in September 2014. Reportedly, the entire car body was built in 44 hours on-site using a huge 3D printing machine. If AM technologies have to truly take off, such acceptance needs to come from the public so that the machine manufacturers as well as users can adopt and adapt to new materials based on fiber-reinforced composites. The challenge of continuous fiber-reinforced composites is still remaining and needs to be resolved for it to be adapted to AM.

3.6 ROLE AND SELECTION OF APPROPRIATE BINDERS FOR AM PROCESSES

In an AM freeforming system for ceramics, the feedstock, consisting of ceramic powders plus wax binder, melts during deposition, especially if it is based on extrusion. After extrusion, the binder has to be removed very carefully and the part is then sintered. The feedstock typically comprises a multi-component blend of waxes and polymers with greater than 50 vol% ceramic powder (typical volume percentages). The feedstock's melting and flow properties and the binder behavior during removal are crucial to the process and need to be characterized for both process development and quality control (QC) purposes. The rheology of particle-filled polymers is very sensitive to particle content, particle surface chemistry, and the binder chemistry at these high volume fractions. Therefore, it is critical that the appropriate binder be chosen based on the type of AM processing methodology chosen.

For example, the quality of the green ceramic feedstock has a strong influence upon the robustness of the process and its ability to reproducibly fabricate high strength, dimensionally accurate ceramic components using the extrusion based processes such as fused deposition of ceramics (FDC) (Agarwala et al. 1996) and the extrusion freeforming (EFF) (Vaidyanathan et al. 2000). A high degree of homogeneity is desirable in order to minimize density gradients between the binder and ceramic powders. If density gradients are present in the feedstock, it could lead to non-uniform firing shrinkage and formation of defects within the freeformed ceramic bodies. The feedstock should also possess a reproducible rheology so that it can be accurately freeformed into the desired green ceramic component. Further requirements for the rheology of EFF feedstock are a low melt viscosity (extrudable at low pressures) as well as the ability to undergo rapid solidification upon deposition (enabling more rapid part build rates). The binder should be easily removable from the freeformed green bodies under controlled conditions and leave minimal pyrolysis residue. The ease of binder removal is determined by the binder burnout schedule, which in turn is defined by the part thickness. Finally, the resulting bodies should be readily sinterable into dense ceramic components.

The following is an example of how the rheological properties will affect the AM process and how the polymer binder composition will affect the final component features and properties. Adding a ceramic powder such as Si_3N_4 to the binder increases its viscosity quite considerably, even at very low shear rates (1–10 sec^{-1}). There also appears to be a critical solid loading content, beyond which the viscosity of the filled systems would increase dramatically. Figure 3.31 is a representation of the viscosity of the filled systems as a function of solids loading content at a constant shear rate of approximately 1 sec^{-1}. The critical viscosity limit for the binder used was obtained at approximately 15% by weight, in the case of Si_3N_4.

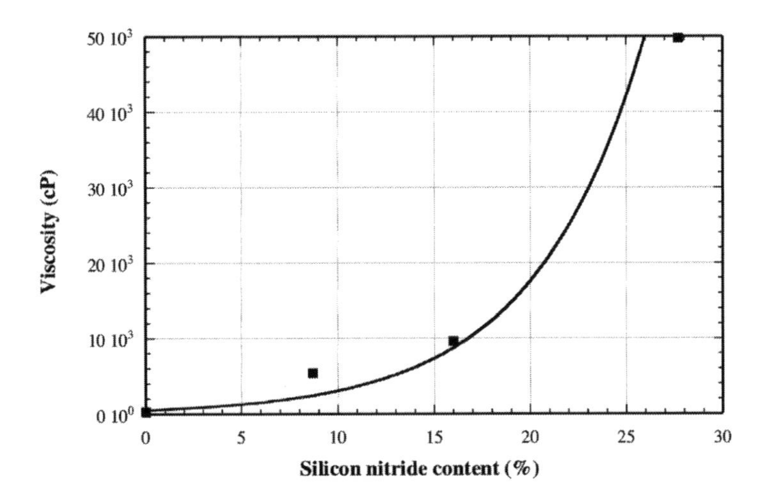

FIGURE 3.31 Variation of viscosity vs. silicon nitride content in a typical polyethylene co-ethacrylate (EEA) based binder system.

Dynamic rheological measurements are another powerful method for predicting the extrusion behavior of a ceramic formulation that is used in an extrusion-based AM process. Dynamic rheological characterizations presented below were made using a Rheometrics model controlled-stress rheometer. Dynamic measurements characterize the flow of materials at near equilibrium states by applying a small strain. There are two types of dynamic measurements. In the first type, external sinusoidal conditions of known frequency and amplitude are imposed on the fluid to induce an oscillatory flow. The frequency, amplitude, and phase of the response of the fluid are measured. The second type is relaxational flow where under external conditions such as force or strain, the fluid undergoes a rapid change from one steady state to a second steady state. The response of the fluid as it approaches a new equilibrium state is then measured.

Dynamic viscosity response as a function of frequency is essentially similar to the steady shear viscosity vs. shear rate response for unfilled polymers (Bigg 1982). The basic equations are given below.

$$\gamma = \gamma_o \Sigma \iota \nu \omega \tau \tag{3.1}$$

$$\gamma' = \gamma_o \omega \mathrm{X} o \sigma \omega \tau \tag{3.2}$$

$$\tau = \tau_o \Sigma \iota \nu (\omega \tau + \delta) \tag{3.3}$$

$$\Gamma' = (\tau_o / \gamma_o) \chi o \sigma \delta \tag{3.4}$$

$$\Gamma'' = (\tau_o / \gamma_o) \sigma \iota \nu \delta \tag{3.5}$$

$$\eta^* = \left[(\Gamma'/\omega)^2 + (\Gamma''/\omega)^2 \right]^{0.5} \tag{3.6}$$

Dynamic measurements are frequently applied at low shear rates approaching zero to characterize polymer structures (Bigg 1983). For uncross-linked polymers, at very low frequencies, η^* approaches to η (steady flow viscosity), but for cross-linked polymers, increasing frequency causes η^* to fall monotonously. The rheological response of a suspension depends on the degree and strength of particle-particle interactions and particle-matrix interactions. Most suspensions could be classified into three types, based on the concentration. The first type is a dilute concentration, where the individual filler particles do not interact with each other. The response of such a suspension is primarily that of the matrix. At a certain concentration, the individual filler particles begin to interact with neighboring particles. Such interactions are limited to particles in a local neighborhood. This is the second type of suspension. The concentration at which these particle-particle interactions begin depends on the geometry and surface activity of the filler particles. The third type of suspension is where the filler particles form a complete network within the matrix. In such a suspension, the movement of each particle affects the position of all the others, the effect diminishing with distance away from the particle in question. This also depends on the nature of the particles. In an extrusion-based AM process, these forces and interactions are critical and need to be taken into account.

Dynamic measurements should be made at a temperature close to the extrusion temperature. In the case of silicon nitride formulations used in the EFF process (Vaidyanathan et al. 2000), these were made at 150°C, since the extrusion of the standard silicon nitride formulation on the high-pressure extruder head is also performed at 150°C. Figure 3.32 shows the measured viscosity as a function of frequency. It can be seen that the viscosity decreases with increasing frequency. This suggests that the formulation has a very shear dependent behavior over the range of 0.1 to 100 rad/sec. Additionally, the formulations are also non-Newtonian and shear thinning. The non-Newtonian and shear thinning nature is dependent on the particle size of the fillers (Kamal and Mutel 1985).

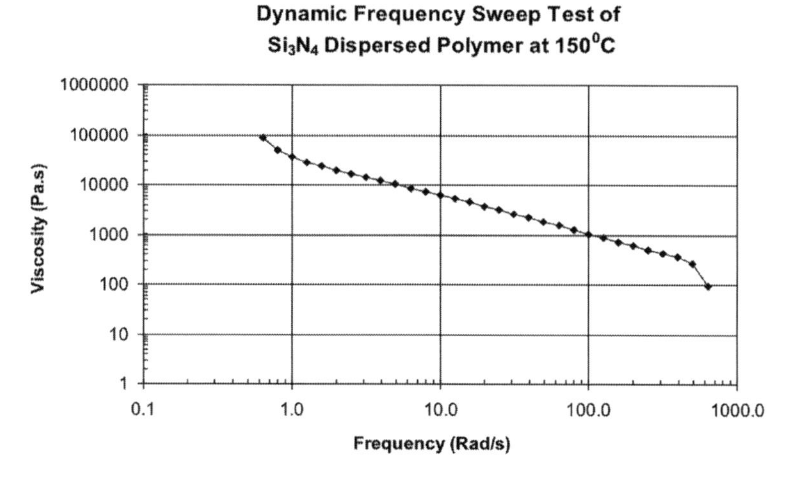

FIGURE 3.32 Viscosity vs. frequency response of standard high pressure extruder-head ceramic feedstock composition.

The free-formable slurry evaluated had 55 vol.% ceramic powder (Starck M-11 silicon nitride). Measurements shown below were done in dynamic oscillatory shear. At 150°C and 2% strain, the viscosity was 36 kPa.s (360 kPoise) at 1 rad/sec and dropped by a factor of 6 per decade of frequency increase (Figures 3.33 and 3.34). This compares with 50 Pa.s at 25 wt% silicon nitride loading and 1 rad/sec frequency. The viscosity also drops roughly in proportion to the reciprocal of the strain. Thus, this material is non-ideal and there are strong particle-particle interactions.

Simple polymer-fluid models predict a viscosity that is roughly constant up to a shear rate equal to the reciprocal of the characteristic relaxation time, after which the viscosity drops. The drop is seen, but since there is no plateau at low shear rates, the relaxation time is apparently more than 1 second. The implication is that particle-particle interactions dominate the viscous flow, but reform slowly once the melt is sheared.

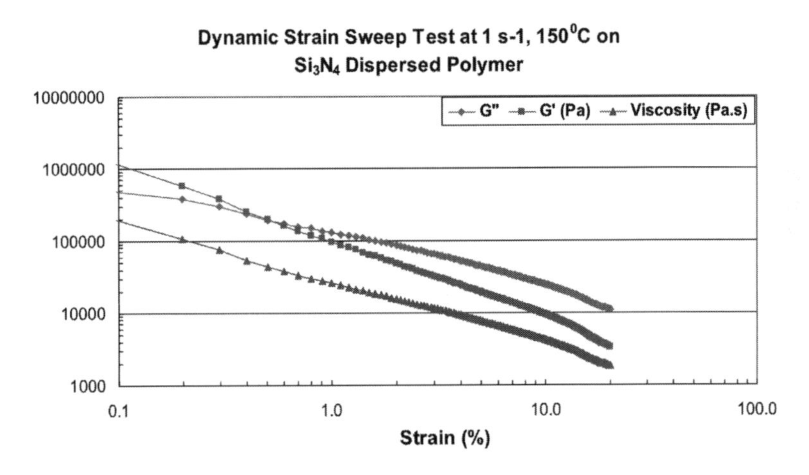

FIGURE 3.33 Storage modulus (G′), loss modulus (G″), and viscosity as a function of strain at a constant frequency of 1 rad/sec and 150°C.

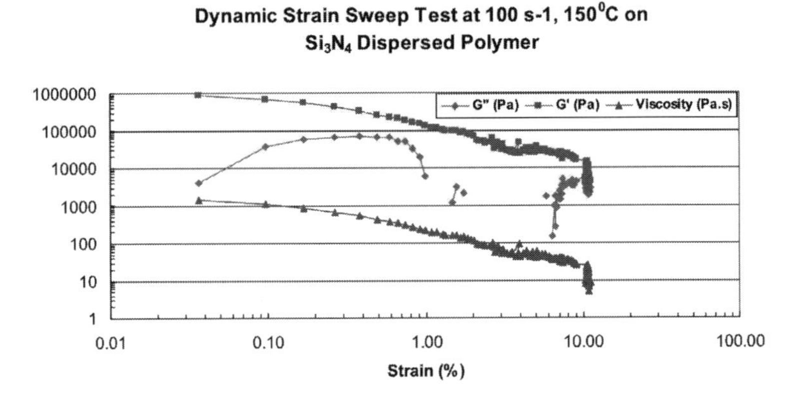

FIGURE 3.34 Storage modulus (G′), loss modulus (G″), and viscosity as a function of strain at a constant frequency of 100 rad/sec and 150°C.

It is sensible to compare this data with the expected range of operation of this material during freeforming. Taking a capillary of 0.2 mm diameter delivering slurry at 1 cc/minute, a strain rate of about 10^4/sec could be expected. The viscosity data at 150°C extrapolates to about 30 Pa.s at this shear rate. Given a nozzle length of about 3-mm, the drive pressure needed would be about 3.9 MPa, or 600 psi. However, should the flow stop, the viscosity and pressure would rise by about three orders of magnitude, as there are no shear forces on the material flowing through the capillary. While this may not be the source of any instability that might be seen, it is a plausible cause. Another possible reason for the instabilities could be the extensional thickening of the formulation near the capillary opening. The implication is that dispersing agents will need to be added to a ceramic or metal powder formulation that will lead to much lower viscosities at similar particle loadings.

At low frequencies, the response of the polymer dominates the behavior of the system. This accounts for the lower shear modulus at low frequencies. The time scale of the experiment allows for movement of the particles in the molten polymer. From the G′ vs. ω and G″ vs. ω plots (Figure 3.35), it can be seen that the loss modulus response as a function of frequency is greater than the storage modulus response. This indicates that the polymer entanglements are not dominating the elastic behavior of the system. If polymer entanglements were to dominate the elastic behavior of the system, the storage modulus response would be greater than the loss modulus response (Dealy 1990). The overall amount of polymer in the binder is lowered or diluted by the addition of a typical plasticizer (butyl oleate) and the wax (AL3). This in turn possibly causes the effect of entanglements that are directly attributable to the presence of polymers, to be lowered. At higher frequencies, the particles have less time to move and their motion is further hindered by the close proximity of the neighboring particles. The material behaves more like a solid and G′ increases with ω. However, we can see a strong dependence of viscosity and G″ on frequency (Dealy 1982; Ferry 1982; Middleman 1968).

Capillary rheology measurements made using an Instron Model 3211 capillary rheometer are shown here. Viscosity measurements were performed at temperatures

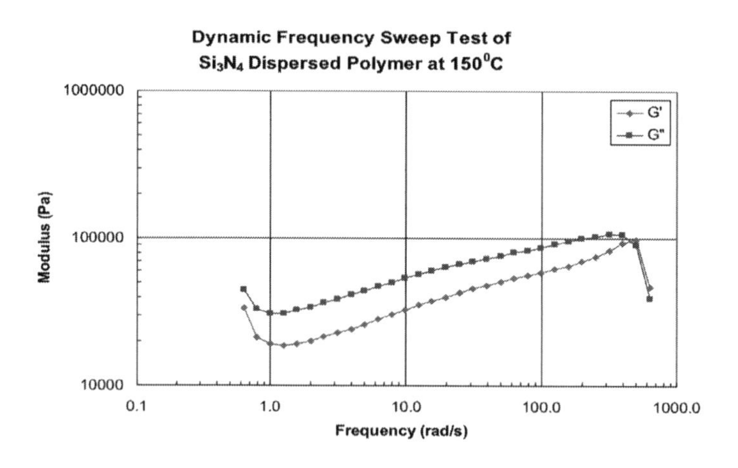

FIGURE 3.35 G′ and G″ vs. ω response of standard silicon nitride feed stock material at 150°C.

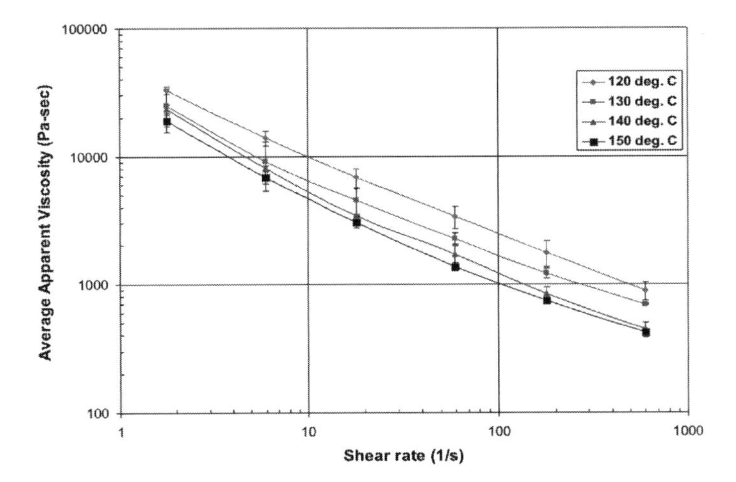

FIGURE 3.36 Shear rate-viscosity relationship for an EEA based binder system with 55 vol% Si_3N_4 between 120°C and 150°C.

between 120°C and 150°C. Figure 3.36 shows the viscosity-shear rate response of an poly(ethylene–ethacrylate) (EEA)-based binder formulation with 55 vol% Si_3N_4 between 120°C and 150°C. Increasing the temperature from 120°C to 150°C is seen to decrease the viscosity and shear stress by approximately one order of magnitude. These results suggest that the EFF formulations are highly non-structured and non-Newtonian. Further, the capillary rheometer results generally support the dynamic rheology results.

The viscosity of the ceramic feedstock material used in the EFF process is higher than that of the ceramic feedstock material by approximately a half order of magnitude in the shear rate ranges investigated. It should be pointed out that 70°C happens to be the FDM temperature of the pure binder system used for fused deposition of ceramic formulations, due to the reason that the feedstock is in the form of a filament rather than a ceramic powder/polymer binder blend. Therefore, the viscosity of the pure FDC binder system provides a sort of viscosity limit for the FDM process for successful FDC. Since the EFF Si_3N_4 formulation is capable of being freeformed successfully, this suggests that the EFF process can handle the increased viscosity of the ceramic binder systems effectively. This also suggests that increased solids loading in the binder system is possible with the EFF system. A typical feedstock material developed for the EFF process is shown in Table 3.7.

TABLE 3.7
Typical Green Ceramic Feedstock Composition for EFF

Component	Concentration (Volume %)
Silicon Nitride or similar ceramic powder	≈55
Saturated Elastomer	≈25
Fatty Acid Ester Plasticizer	≈10
Paraffin Wax	≈5
Acryloid Additive	≈5

The next step to be taken for a ceramic part to be built with sufficient mechanical properties is to derive an adequate binder burnout cycle to remove the polymer binder. Binder burnout is one of the crucial steps in ceramic processing (Evans and Edirisinghe 1991; Calvert and Cima 1990). The binder is an essential component in ceramic processing, particularly in extrusion freeform fabrication imparting strength to the green part (Calvert and Cima 1990). A better understanding and optimization of burnout could allow the processing of larger ceramic parts by AM processing and prevent defects from being introduced during the decomposition of the binder.

A systematic study of the binder removal process has started only in the last couple of decades (Evans and Edirisinghe 1991). Thus, there is no general basis behind the several non-linear binder removal temperature-time schedules quoted in the literature, except that the heating rate up to the softening point of the powder-binder formulation is rapid in comparison with that used during the actual pyrolytic degradation of the binder system (Evans and Edirisinghe 1991). It is necessary to modify temperature-time pyrolysis schedules to suit the binder system used, and the powder, which could in some instances assist the thermal decomposition.

A typical process that can be used to optimize the binder burnout cycle, especially those containing a ceramic powder and polymer blend as the binder is described in the following section. Feed-rods sectioned into pieces with thickness varying from 4 mm to 20 mm can be used for the optimization of the binder burnout cycle.

The samples are packed in ceramic or graphitic powder bed, placed in high-purity alumina crucibles, and heated to 600°C at controlled ramp rates. In most cases, the binder is completely burnt out prior to this temperature. Ramp rates varying from 2, 1, 0.5, 0.2, 0.1, 0.05, 0.03, 0.02, and 0.01°C/min are chosen. In the experiments conducted, five samples of the same thickness were placed in the crucible and burnt out at each different ramp rate. Samples were taken out of the furnace at regular intervals. The total temperature range from room temperature to 600°C for complete binder burnout was divided into equal temperature zones. All samples were weighed initially before starting the experiment. One sample was drawn during each temperature zone, cooled, and weighed. The weight of the individual samples was normalized to account for the initial weight of the samples. The percentage weight loss was calculated for each temperature zone. The rate of percentage weight loss was calculated by considering the time interval between two temperature set points.

The normalized difference table is shown in Table 3.8 for a sample thickness equal to 5 mm with a ramp rate of 2°C/min.

The major issues to be concerned with during binder burnout are the prevention of cracks, warpage, delamination, and oxidation. This occurs due to insufficient time for the decomposed products to diffuse out to the component surface and subsequently vaporize. Porosity development during this stage can help prevent cracking and warpage. In the EFF process, 3% microcrystalline wax was added as an ingredient. This 3% microcrystalline wax added to the binder may prevent the cracking due to the formation of porous channels. Vaporization at

TABLE 3.8
The Normalized Difference Table for a Sample Thickness Equal to 5 mm for a Ramp Rate of 2°C/min

Temperature (°C)	%W	d(%W)/dt
35	0	0
85	0	0
135	0	0
185	0.1266	0.005064
235	0.4136	0.001148
285	1.3016	0.03552
335	2.181	0.03517
415	5.2867	0.124288
465	10.6625	0.21503
525	20.5645	0.33
575	20.687	0.0049

early stages may leave porous channels and help high-temperature degradation products to escape through these channels.

A typical thermo-gravimetric analysis (TGA) plot showed two prominent peaks at 160°C–250°C and 350°C–450°C, as shown in Table 3.9. These data were obtained by heating the sample in a small platinum pan in flowing nitrogen of 100°C/min at a ramp rate of 2°C/min.

The total percentage weight loss in Table 3.9 corresponds to approximately the total initial polymer content in the binder system, suggesting that binder burnout will be complete at 450°C. However, when the approximately same weight of the sample (considering same cross sectional area) was heated at 10°C/min, the peaks shifted 30°C higher (Figure 3.36). The total area under each peak seemed to be constant. So if the ramp rates are decreased by 5 times (0.4°C/min), the peaks will shift by 30°C to the left, but this could be compensated with an increase in the thickness of the sample. It was also seen that the individual components are completely burned out by 600°C (Figure 3.37).

Binder evolution events are observed to shift to higher temperatures with increasing ramp rates. Evolution events also should shift to higher temperature with increasing part size (Evans and Edirisinghe 1991). This is mainly because sufficient time is not given to events to reach thermal equilibrium. Due to this thermal excursion, bloating and cracks are seen in the samples that have undergone

TABLE 3.9
Prominent Peak Information from TGA Plot Shown in Figure 3.37

Peak No.	Starting Temp. (°C)	End Temp. (°C)	% Wt. Loss Associated
#1	160	250	4.2
#2	350	450	13

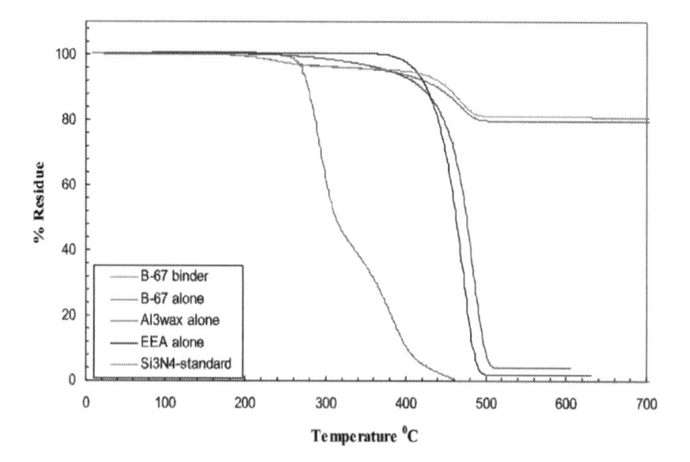

FIGURE 3.37 Thermal decomposition of the binder components and ceramic formulations for AM processes.

higher heating rates, or larger samples where sufficient time is not given for decomposed products to escape to surface (Figure 3.37).

With many factors affecting the binder removal process, the atmosphere surrounding the samples in the furnace is a very important parameter to be considered. The thermogravimetric traces obtained are shown in Figure 3.38. These allow the comparison of thermal decomposition kinetics of the binder system in the different atmospheres used. In the presence of flowing nitrogen, thermal degradation occurs with the major weight loss occurring between 400°C and 500°C. One third of the binder is lost before 300°C.

In the presence of flowing air and flowing oxygen (Figure 3.39), oxidative degradation was observed. The major weight loss occurs at a much lower temperature range.

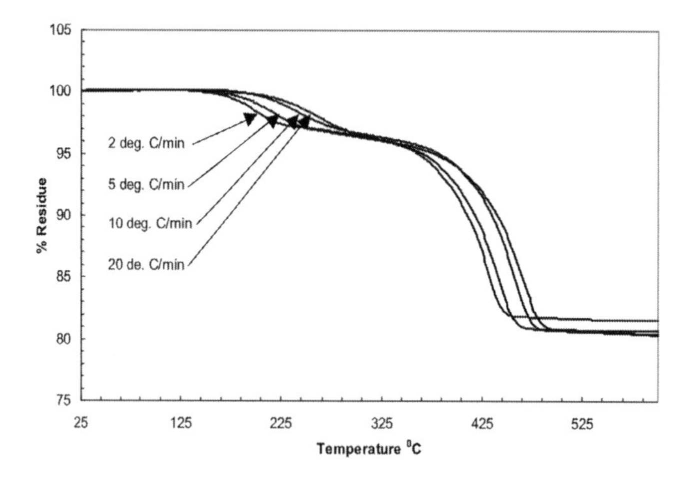

FIGURE 3.38 Effect of ramp rates on the binder burnout of silicon nitride formulations used for extrusion-based AM processes.

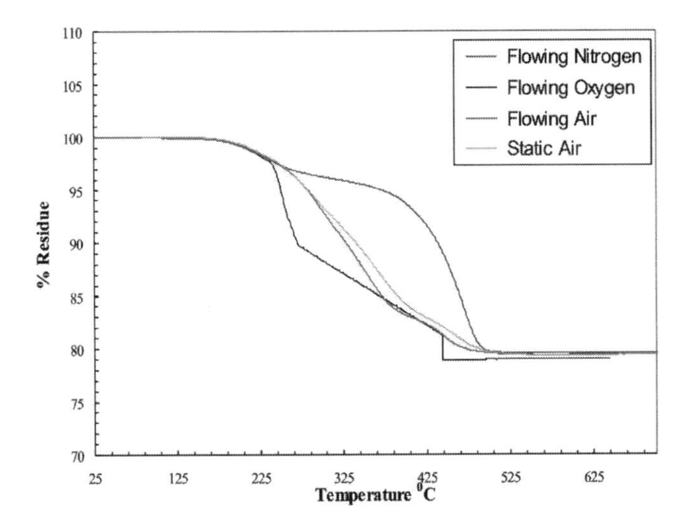

FIGURE 3.39 Effect of binder burnout atmosphere on binder burnout behavior of the silicon nitride formulations used for the AM process.

It is also clear that compared with static air, the use of flowing air accelerated the binder degradation processes appreciably, due to the efficient removal of decomposition products. In flowing oxygen, at about 240°C, rapid loss of binder is accompanied by a sudden increase in temperature, suggesting that combustion could have occurred. Combustion could lead to the disintegration of the specimen.

In the presence of flowing air, it may be very hard to control a rapid rate of weight loss even with the slowest ramp rates. The next choice is static air, but oxidation can be a problem. Therefore, flowing nitrogen can be chosen as a safe binder burnout atmosphere with reasonable binder burnout schedules.

It was already explained previously how to generate a normalized rate of weight loss table from data collected on actual samples. The process can be repeated with different heating rates on 5 mm thick-section ceramic samples in a nitrogen atmosphere. A binder burnout cycle can be generated by assuming d(%W)/dt being equal to say 0.003, which may correspond to 0.01 mg/min weight loss, which is the boundary value between a crack free and cracked sample. This means that if the rate of weight loss is more than this value, the samples would crack. If this is an acceptable rate of weight loss, which permits to burnout in a reasonable amount of time, then a heating rate cycle can be developed as given in Table 3.10.

The superimposition of this heating rate diagram on the TGA for the silicon nitride binder is shown in Figure 3.40. It is clear from this that the higher the rate of weight loss, the slower is the optimum ramp rate.

The total burnout cycle time developed in this case was for 5.19 days. This schedule was tested on actual samples. In this case, it was observed that samples with a thickness less than 5 mm did not crack and 10 mm or more cracked. This implied that the boundary line for crack prone and crack free zone is in between 5 mm and 10 mm section thickness for this binder burnout schedule. The above burnout cycle seemed to work successfully for sample thickness less than or equal to 5 mm.

TABLE 3.10
A Binder Burnout Schedule Developed for a 5 mm Ceramic Sample

Temperature (°C)	Ramp Rate (°C/min)	Duration Time (min)
35–135	2	50
135–185	2	50
185–325	0.1	1500
325–375	0.05	1000
375–475	0.03	3332
475–525	0.05	1000
525–575	0.1	500
575–625	1	50

For defining binder burnout variation with respect to the thickness, we could assume a linear dependency or a parabolic rate of weight loss with the section thickness. According to a linear weight loss model, the section thickness is directly proportional to burnout time. By decreasing the time interval at each segment or increasing the ramp rate to four times to its original value in the burnout cycle given in Table 3.10, crack free samples with section thickness 1.25 mm or less could be burnt out. Similarly, if the section thickness depends on a square law with burnout time, decreasing the cycle time to 1/4th its original value should allow to burnout and produce crack-free samples with section thickness 2.5 mm or less.

A typical graph between log (thickness) vs. log (time) for the linear and square law, which has slopes 1 and 2, respectively, are shown in Figure 3.41. The graph can be divided into crack prone and crack free zones.

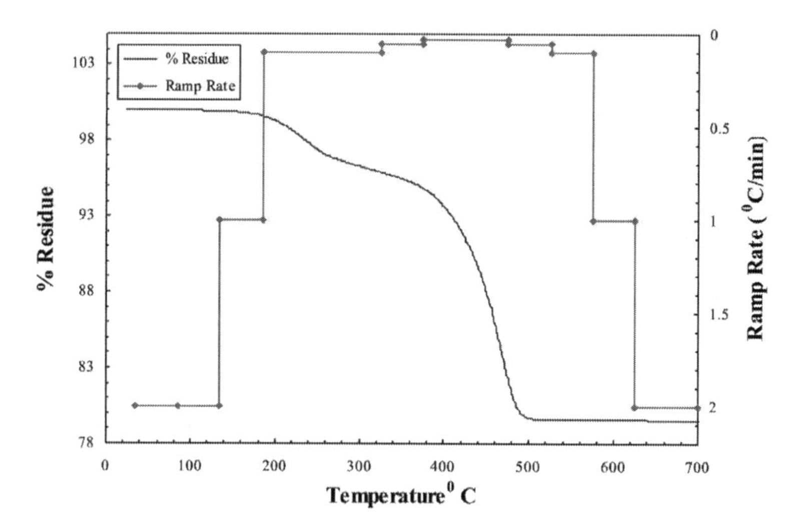

FIGURE 3.40 Superposition of the heating cycle on the TGA curve for a typical AM binder system.

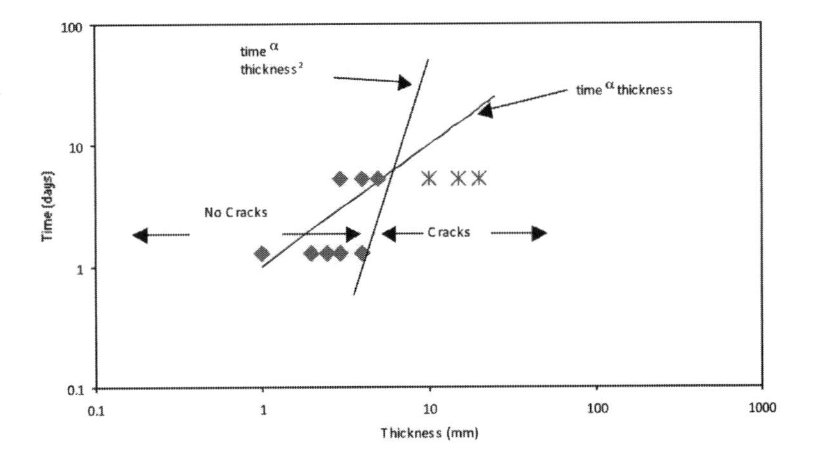

FIGURE 3.41 Crack-free and crack-prone samples for the linear and square-law models for a typical formulation used for AM process of ceramics.

A comparison of the linear and parabolic rate loss models is shown in Figure 3.41, which reveals that section thickness dependency follows a square law. This allows to burnout thick samples in less time.

3.7 SPECIAL CASES—IN SITU FIBER REINFORCEMENT DURING ADDITIVE MANUFACTURING

In some cases, during the AM process, second phase particles may coalesce due to the action of heat and pressure and combine to create in-situ fibers, leading to strengthening during the AM process itself. For example, Lombardi et al. conducted extrusion freeforming experiments in polymer blends having controlled microstructures (Lombardi et al. 1998). These blends are composed of at least two immiscible polymer components where its major phase is present in at least two-fold excess compared to the minor phase. The major phase is composed of poly-2-ethyl-2-oxazoline (PEOx), a water soluble thermoplastic reinforced with fine talc filler while the minor phase is composed of a high glass transition temperature styrenic copolymer (Tg ca. 140°C). In this case, the styrenic copolymer was added to the (PEOx) to increase the heat distortion temperature of the polymer blend as well as reduce its sensitivity to ambient humidity (Lombardi et al. 1998).

As can be seen from the SEM micrograph in Figure 3.41 that shows the feedstock microstructure, the styrenic minor phase is present as spherical droplets uniformly dispersed throughout the PEOx major phase. This type of microstructure is typically encountered in blends composed of two immiscible polymer phases where the minor phase adopts a spherical morphology to minimize its surface area and energy (Utracki 1990; Sperling 1997; Tsebrenko et al. 1976; Vanoene 1978, 1972). Coalescence is suppressed within the blend by the presence of a small amount of a third compatibilizing polymer that is miscible in both the styrenic and PEOx polymers by decreasing the compositional gradient and interfacial energy between the minor and major polymer phases (Fayt et al. 1987). Sufficient plasticizer has also been added to modify the rheology of the PEOx phase of these blends such that it can be accurately freeformed using the extrusion-based AM techniques. Heat-treating the feedstock results in a microstructure with perfectly circular second phase structures, which clearly shows the effect of extrusion on the preferred alignment as shown in Figures 3.42 and 3.43.

The freeformed blend can also function as a water-soluble support structure and was demonstrated for the fabrication of intricate ABS polymer prototype components using extrusion-based AM techniques, as shown in Figure 3.44. This was one of the very first water-soluble support structure materials successfully demonstrated for the FDM process. In this case, the blend was formulated as a filament 1.778 mm in diameter and extruded through a second nozzle while the ABS filament was extruded in an FDM 1600 Modeler (Lombardi et al. 1998; Artz et al. 2000). Figure 3.45 shows the ability to wash out the support material from the AM part. This material could be blended either as a filament or a feed rod, making it suitable for either the FDM or the EFF AM process.

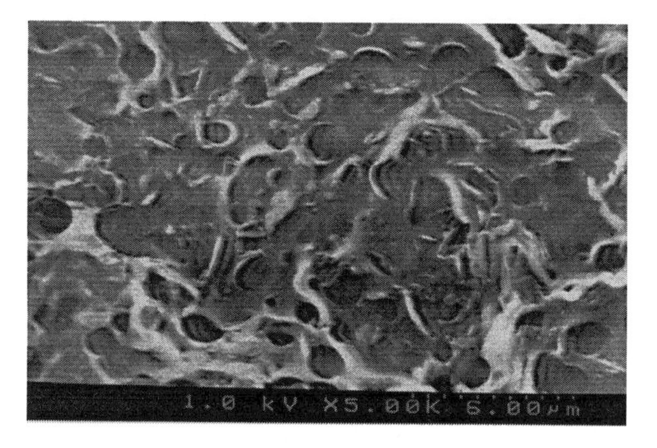

FIGURE 3.42 Microstructure of PEOx/styrenic polymer blend feedstock for extrusion-based AM process.

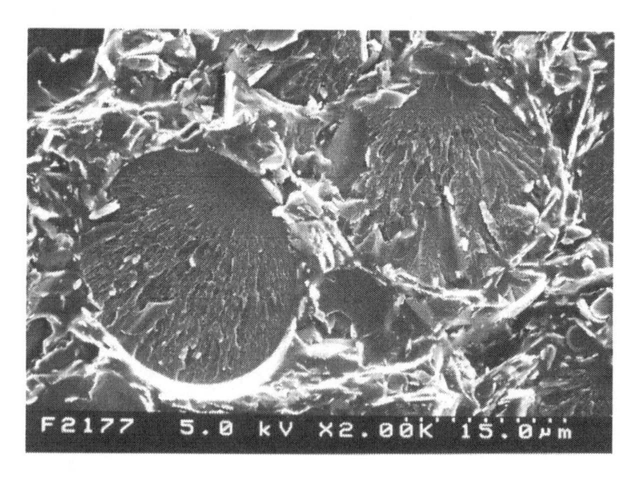

FIGURE 3.43 Microstructure of PEOx/styrenic polymer blend EFF feedstock after heat treatment at 140°C.

FIGURE 3.44 Fabrication of intricate shapes in the FDM process using a water-soluble material as the support material.

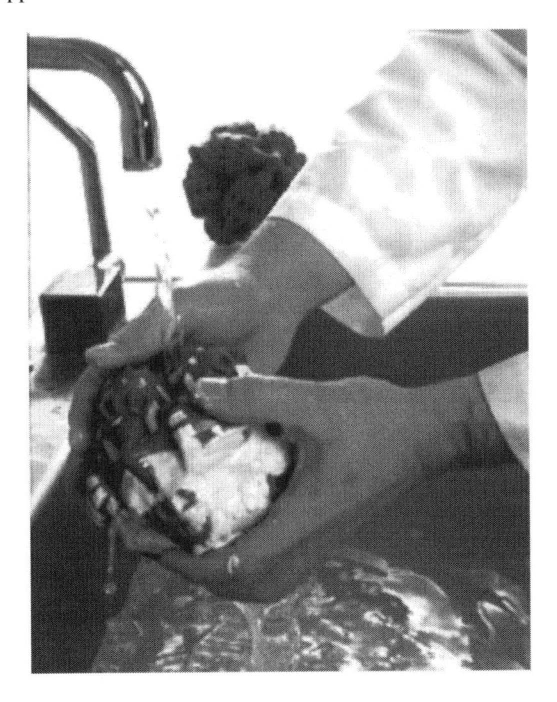

FIGURE 3.45 Ability to wash out the support structure from a complex-shaped FDM component using a PEOx/styrenic polymer blend feedstock.

3.8 CURRENT CHALLENGES AND FUTURE TRENDS

It is an accepted fact that AM processes are well developed for pure polymers or polymer blends and any improvements might be incremental. However, we still have not seen enough enhancements in polymers used in the tissue engineering area. For AM processes to be truly ground breaking, especially for composite materials, it is necessary that new materials, fibers, and interfacial coatings will need to be developed. Since most of the polymer and composite technologies involves the heating and cooling of the ingredients, it is necessary that the thermal expansion and shrinkages need to match or at the very least, the difference in properties need to be mitigated with the help of appropriate interfacial coatings. At some point, the automated tow placement technique that is being used to produce components that can be used in real applications will need to be merged with an AM process or the ATP process will have to be modified to become a truly AM process.

The other path-breaking development will be in the chopped fiber composites area, although we have started to see the developments in 3D printed cars and companies like MarkForged. Perhaps these will overlap into the tissue engineering area, leading to new materials, processes, and applications.

The most important requirement that will be needed uniformly by every AM technique would be to obtain the reliability of many parts while only building one or two parts. It is believed that this would be the greatest challenge for any AM technology and parts made by those techniques.

ADDITIONAL READING MATERIALS

Campbell, Thomas A, and Olga S Ivanova. 2013. 3D printing of multifunctional nanocomposites. *Nano Today* 8 (2):119–120.

Gao, Wei, Yunbo Zhang, Devarajan Ramanujan, Karthik Ramani, Yong Chen, Christopher B Williams, Charlie CL Wang, Yung C Shin, Song Zhang, and Pablo D Zavattieri. 2015. The status, challenges, and future of additive manufacturing in engineering. *Computer-Aided Design* 69:65–89.

Hofstätter, Thomas, David B Pedersen, Guido Tosello, and Hans N Hansen. 2017. Applications of fiber-reinforced polymers in additive manufacturing. *Procedia CIRP* 66:312–316.

Lee, Jian-Yuan, Jia An, and Chee Kai Chua. 2017. Fundamentals and applications of 3D printing for novel materials. *Applied Materials Today* 7:120–133.

de Leon, Al C, Qiyi Chen, Napolabel B Palaganas, Jerome O Palaganas, Jill Manapat, and Rigoberto C Advincula. 2016. High performance polymer nanocomposites for additive manufacturing applications. *Reactive and Functional Polymers* 103:141–155.

Ligon, Samuel Clark, Robert Liska, Jürgen Stampfl, Matthias Gurr, and Rolf Mülhaupt. 2017. Polymers for 3D printing and customized additive manufacturing. *Chemical Reviews* 117 (15):10212–10290.

Ning, Fuda, Weilong Cong, Jingjing Qiu, Junhua Wei, and Shiren Wang. 2015. Additive manufacturing of carbon fiber reinforced thermoplastic composites using fused deposition modeling. *Composites Part B: Engineering* 80:369–378.

Parandoush, Pedram, and Dong Lin. 2017. A review on additive manufacturing of polymer-fiber composites. *Composite Structures* 182:36–53.

Quan, Zhenzhen, Zachary Larimore, Amanda Wu, Jianyong Yu, Xiaohong Qin, Mark Mirotznik, Jonghwan Suhr, Joon-Hyung Byun, Youngseok Oh, and Tsu-Wei Chou. 2016. Microstructural design and additive manufacturing and characterization of 3D orthogonal short carbon fiber/acrylonitrile-butadiene-styrene preform and composite. *Composites Science and Technology* 126:139–148.

Singh, Sunpreet, Seeram Ramakrishna, and Rupinder Singh. 2017. Material issues in additive manufacturing: A review. *Journal of Manufacturing Processes* 25:185–200.

Tekinalp, Halil L, Vlastimil Kunc, Gregorio M Velez-Garcia, Chad E Duty, Lonnie J Love, Amit K Naskar, Craig A Blue, and Soydan Ozcan. 2014. Highly oriented carbon fiber–polymer composites via additive manufacturing. *Composites Science and Technology* 105:144–150.

Wang, Xin, Man Jiang, Zuowan Zhou, Jihua Gou, and David Hui. 2017. 3D printing of polymer matrix composites: A review and prospective. *Composites Part B: Engineering* 110:442–458.

SUGGESTED CHAPTER END EXERCISES

Component design and manufacturing, which could be posted at the end of the class rather than after each chapter.

1. In this book, we discussed different methods of additive manufacturing. Suppose you are working in the automotive industry and you are in-charge of making the first 50–100 prototypes of a double outlet fan ducting shown below. There are many ways to make this part by additive manufacturing. Most common way would be to make the mold for injection molding this part (which could be made from a high-strength plastic capable of withstanding high temperatures).

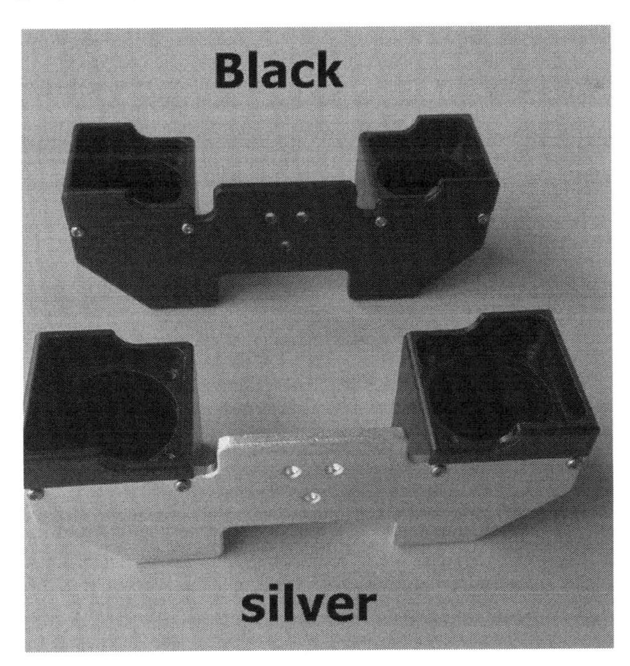

There would be 3–4 steps required to achieve this part most economically. In your opinion, (1) What would be the required steps to make the mold to injection mold this part? It would be recommended to make the part in SolidWorks or some such approach. (2) What are the tolerances and allowances that you need to add to the mold? (3) If the final part needs to withstand 100°C, what are two or three materials that can be used for this part?

2. From the web/YouTube, find examples/images of parts that can be made from composite materials. (1) In your opinion, what advantages would be provided by AM compared to traditional manufacturing methods for fabricating these components? (2) What would be the disadvantages for AM compared to traditional manufacturing methods?
3. Based upon an Internet search, provide three instances where AM and polymer composites are being used for actual applications.
4. Make a list of all the commercially available units for fabricating polymer and composite articles. What are their advantages and disadvantages compared to commercial units fabricating metal AM parts?
5. Can there be parts made with different fiber loading in different parts of a component? If no such units exist today, how can you create such a unit? Would it be a commercially viable product?
6. Are there any commercial units that can combine deposition and subtraction processes for polymer and composite materials? What would be the advantages and disadvantages of such a unit?

REFERENCES

Agarwala, Mukesh K, Amit Bandyopadhyay, R van Weeren, Ahmad Safari, Stephen C Danforth, Noshir A Langrana, Vikram R Jamalabad, and Philip J Whalen. 1996. FDC, rapid fabrication of structural components. *American Ceramic Society Bulletin* 75 (11):60–66.

Artz, Gregory John, John Lang Lombardi, and Dragan Popovich. 2000. Water soluble rapid prototyping support and mold material. US Patent No. 6,070,107. May 30, 2000.

ASTM-F2792. 2012. F2792. 2012. Standard Terminology for Additive Manufacturing Technologies. In ASTM F2792-10e1.

Bigg, DM. 1982. Rheological analysis of highly loaded polymeric composites filled with non-agglomerating spherical filler particles. *Polymer Engineering & Science* 22 (8):512–518.

Bigg, DM. 1983. Rheological behavior of highly filled polymer melts. *Polymer Engineering & Science* 23 (4):206–210.

Calvert, Paul. 1998. Freeforming of polymers. *Current Opinion in Solid State and Materials Science* 3 (6):585–588.

Calvert, Paul, and Liu Zengshe. 1998. Freeform fabrication of hydrogels. *Acta Materialia* 46 (7):2565–2571.

Calvert, Paul, and Michael Cima. 1990. Theoretical models for binder burnout. *Journal of the American Ceramic Society* 73 (3):575–579.

Calvert, Paul, Tung Liang Lin, and Hogan Martin. 1997. Extrusion freeform fabrication of chopped-fibre reinforced composites. *High Performance Polymers* 9 (4):449–456.

Chin, Hong Bai, and Chang Dae Han. 1980. Studies on droplet deformation and breakup. II. Breakup of a droplet in nonuniform shear flow. *Journal of Rheology (1978-present)* 24 (1):1–37.

Cima, Michael J, John S Haggerty, Emanuel M Sachs, and Paul A Williams. 1993. Three-dimensional printing techniques. Canadian Patent No. CA2031562C, November 22, 1994.

Crump, S Scott. 1992. Apparatus and method for creating three-dimensional objects, US Patent No. 5,121,329, June 9, 1992.

Crump, S Scott. 1994. Modeling apparatus for three-dimensional objects, US Patent No. 5,340,433, August 23, 1994. Google Patents.

Dealy, John M. 1982. *Rheometers for Modern Plastics*. New York: Van Nostrand Reinhold.

Dealy, John M. 1990. *K. F. Wissbrun. Melt Rheology and Its Role in Plastics Processing*. New York: Van Nostrand Reinhold.

Dehoff, Ryan. 2014. Made in Chicago: World's first 3D printed car. Available at http://wgntv.com/2014/09/13/made-in-chicago-worlds-first-3d-printed-electric-car/. Accessed on September 15 2014.

Direct-Digital-Manufacturing. 3D Printing Series #4 – Paramount Industries: A 3D Systems Company, an Additive Manufacturing Success Story - available http://www.dvirc.org/3d-printing-series-4-paramount-industries-a-3d-systems-company-an-additive-manufacturing-success-story-2/ - sthash. IqiJjOTV.dpuf, August 13, 2013. 2013 [cited September 2013].

Don, Roderic C, John W Gillespie Jr, and Steven H McKnight. 1997. Bonding techniques for high performance thermoplastic compositions. US Patent No. 5,643,390. July 1.

Elmendorp, Jacob Jacobus. 1986. A study on polymer blending microrheology. *Polymer Engineering & Science* 26 (6):418–426.

Evans, Julian RG, and Mohan J Edirisinghe. 1991. Interfacial factors affecting the incidence of defects in ceramic mouldings. *Journal of Materials Science* 26 (8):2081–2088.

Farmer, Benjamin L, Mark A Beard, Oana Ghita, Robert Allen, and Ken E Evans. 2010. Assembly strategies for fully aligned and dispersed morphology controlled carbon nanotube reinforced composites grown in net-shape. Paper read at *2010 MRS Fall Meeting – Symposium Z – Hierarchical Materials and Composites—Combining Length Scales from Nano to Macro*. Eds. J.H. Moon, G.M. Odegard, M.S.P. Shaffer, B.L. Wardle. Vol. 1304, at Boston, MA.

Farmer, Benjamin L, Robert J A Allen, Oana Ghita, Mark A Beard, and Ken E Evans. 2012. Strategies to combine nanocomposite and additive layer manufacturing techniques to build materials and structures simultaneously, Paper #993 Paper read at *ECCM15—15TH European Conference on Composite Materials*, 24–28 June, at Venice, Italy.

Fayt, Roger M, Robert Jerome, and Philippe J Teyssie. 1987. Characterization and control of interfaces in emulsified incompatible polymer blends. *Polymer Engineering & Science* 27 (5):328–334.

Ferry, John D. 1982. *Viscoelastic Properties of Polymers*. New York: John Wiley & Sons Inc.

General-Electric. GE Reports, 2014, This electron gun builds jet engines, available at http://www.gereports.com/post/94658699280/this-electron-gun-builds-jet-engines, accessed on August 17, 2014, 8-17-2014. Available from http://www.gereports.com/post/94658699280/this-electron-gun-builds-jet-engines.

Gibson, Ian, David W Rosen, and Brent Stucker. 2010. *Additive Manufacturing Technologies*. New York: Springer.

Gonczy, Stephen T, and John G Sikonia. 2005. Nextel™ 312/Silicon Oxycarbide Ceramic Composites. In *Handbook of Ceramic Composites*. New York: Springer.

Goodridge, Ruth Davina, Meisha L Shofner, Richard JM Hague, Michael McClelland, MR Schlea, RB Johnson, and Christopher J Tuck. 2011. Processing of a Polyamide-12/carbon nanofibre composite by laser sintering. *Polymer Testing* 30 (1):94–100.

Griffith, Michelle L, and John W Halloran. 1996. Freeform fabrication of ceramics via stereolithography. *Journal of the American Ceramic Society* 79 (10):2601–2608.

Guo, Nannan, and Ming C Leu. 2013. Additive manufacturing: Technology, applications and research needs. *Frontiers of Mechanical Engineering* 8 (3):215–243.

Han, Chang. 1981. *Multiphase Flow in Polymer Processing*. New York: Academic Press.

Hao, L, MM Savalani, Y Zhang, KE Tanner, and Russell A Harris. 2006. Effects of material morphology and processing conditions on the characteristics of hydroxyapatite and high-density polyethylene biocomposites by selective laser sintering. *Proceedings of the Institution of Mechanical Engineers, Part L: Journal of Materials Design and Applications* 220 (3):125–137.

Jacobs, Paul F. 1995. *Stereolithography and Other RP&M Technologies: From Rapid Prototyping to Rapid Tooling*. New York: Society of Manufacturing Engineers.

Jacobs, Paul Francis. 1992. *Rapid Prototyping & Manufacturing: Fundamentals of Stereolithography*. Society of Manufacturing Engineers.

Jang, Bor Z, Erjian Ma, and C Jeff Wang. 1999. Apparatus and process for producing fiber reinforced composite objects. US Patent No. 5,936,861, August 10, 1999.

Kamal, MR, and A Mutel. 1985. Rheological properties of suspensions in Newtonian and non-Newtonian fluids. *Journal of Polymer Engineering* 5 (4):293–382.

Klosterman, Donald A, Richard P Chartoff, Nora R Osborne, George A Graves, Allan Lightman, Gyoowan Han, Akos Bezeredi, and Stan Rodrigues. 1999. Development of a curved layer LOM process for monolithic ceramics and ceramic matrix composites. *Rapid Prototyping Journal* 5 (2):61–71.

Klosterman, Donald, Richard Chartoff, George Graves, Nora Osborne, and Brian Priore. 1998. Interfacial characteristics of composites fabricated by laminated object manufacturing. *Composites Part A: Applied Science and Manufacturing* 29 (9):1165–1174.

Kobayashi, Shiro, Mureo Kaku, and T Saegusa. 1988. Miscibility of poly (2-oxazolines) with commodity polymers. *Macromolecules* 21 (2):334–338.

Kovar, Desiderio, Bruce H King, Rodney W Trice, and John W Halloran. 1997. Fibrous monolithic ceramics. *Journal of the American Ceramic Society* 80 (10):2471–2487.

Kumar, Sanjay, and J-P Kruth. 2010. Composites by rapid prototyping technology. *Materials & Design* 31 (2):850–856.

Lombardi, John L, Greg J Artz, Dragan Popovich, Ranji Vaidyanathan, and S Boggavarapu. 1998. Issues Associated with the development of a Water Soluble Support Material for use in Extrusion Freeforming & Fused Deposition Modelling. Paper read at *9th Annual Solid Freeform Fabrication Symposium Proceedings*, at Austin, TX.

MarkForged. 2014. available at http://www.extremetech.com/extreme/175518-worlds-first-carbon-fiber-3d-printer-demonstrated-could-change-the-face-of-additive-manufacturing-forever, accessed on June 28, 2014.

Meilunas, Raymond. 2001. Laminated Object Manufacturing-Based Design Ceramic Matrix Composites. Final Report No. AFRL-ML-WP-TR-2001-4074 to DARPA/Wright Patterson AFB.

Meilunas, Raymond J, Gregory P Dillon, and Jerrell A Nardiello. 2002. System for constructing a laminate, US Patent No. 6,484,776 B1, November 26.

Middleman, Stanley. 1968. *The Flow Behavior of High Density Polymers*. New York: John Wiley & Sons.

Moore, JA, and Ji Heung Kim. 1992. Blends of poly (enamino nitrile). *Macromolecules* 25 (5):1427–1432.

Nikzad, Mostafa, Syed Masood, and Igor Sbarski. 2011. Thermo-mechanical properties of a highly filled polymeric composites for fused deposition modeling. *Materials & Design* 32 (6):3448–3456.

Njuguna, James, Krzysztof Pielichowski, and Sahil Desai. 2008. Nanofiller-reinforced polymer nanocomposites. *Polymers for Advanced Technologies* 19 (8):947–959.

Paesano, Antonio. 2014. Polymers for additive manufacturing: Present and future. *SAMPE Journal* 50 (5):34–43.

Peng, Jiong, Tung Liang Lin, and Paul Calvert. 1999. Orientation effects in freeformed short-fiber composites. *Composites Part A: Applied Science and Manufacturing* 30 (2):133–138.

Shofner, Meisha, Francisco Javier Rodríguez-Macías, Ranji Vaidyanathan, and Enrique V Barrera. 2003. Single wall nanotube and vapor grown carbon fiber reinforced polymers processed by extrusion freeform fabrication. *Composites Part A: Applied Science and Manufacturing* 34 (12):1207–1217.

Sperling, Leslie Howard. 1997. *Polymeric Multicomponent Materials: An Introduction.* New York: John Wiley & Sons.

Stratasys. 2014. Properties for Ultem 9085. Available at http://www.stratasys.com/~/media/Main/Secure/Material Specs MS/Fortus-Material-Specs/Fortus-MS-ULTEM9085-01-13-web.ashx, accessed on August 17, 2014.

Stuffle, Kevin L, Peter J Creegan, John L Lombardi, Paul D Calvert, John A O'Kelly, Robert A Hoffman, and Gabriel C Chambers. 2000. Method and apparatus for in-situ formation of three-dimensional solid objects by extrusion of polymeric materials. US Patent No. 6,067,480. May 23, 2000.

Tsebrenko, MV, AV Yudin, TI Ablazova, and GV Vinogradov. 1976. Mechanism of fibrillation in the flow of molten polymer mixtures. *Polymer* 17 (9):831–834.

UT-Austin. 2014. Selective Laser Sintering, Birth of an Industry 2012 [cited August 24 2014]. Available from http://www.me.utexas.edu/news/2012/0712_sls_history.php.

Utracki, Leszek A. 1990. *Polymer Alloys and Blends: Thermodynamics and Rheology.* New York: New York: Hanser Publisher.

Vaidyanathan, K Ranji, Joseph Walish, Mark Fox, John W Gillespie Jr, Shridhar Yarlagadda, Michael R Effinger, Anthony C Mulligan, and Mark J Rigali. 2005. Continuous fiber reinforced composites and methods, apparatuses, and compositions for making the same, US Patent No. 6,899,777 B2, May 31, 2005.

Vaidyanathan, R, Joseph Walish, John L Lombardi, Sridhar Kasichainula, Paul Calvert, and Kenneth C Cooper. 2000. The extrusion freeforming of functional ceramic prototypes. *JOM* 52 (12):34–37.

Vanoene, H. 1972. Modes of dispersion of viscoelastic fluids in flow. *Journal of Colloid and Interface Science* 40 (3):448–467.

Vanoene, H. ed. 1978. *Polymer Blends.* Edited by D. R. P. A. S. Newman. 2 vols. Vol. I. New York: Academic Press, pp. 295–349.

Wu, Souheng. 1987. Formation of dispersed phase in incompatible polymer blends: Interfacial and rheological effects. *Polymer Engineering & Science* 27 (5):335–343.

Yarlagadda, Shridhar. 2014. Automated tow placement of composites.

Z-Corporation. Z Corporation 3D Printing Technology [pdf]. Z-Corporation 2014 [cited August 15, 2014. Available from http://www.zcorp.com/documents/108_3D Printing White Paper FINAL.pdf.

Zhang, Yi, Jonathan Stringer, Richard Grainger, Patrick J Smith, and Alma Hodzic. 2014. Improvements in carbon fibre reinforced composites by inkjet printing of thermoplastic polymer patterns. *Physica Status Solidi (RRL)-Rapid Research Letters* 8 (1):56–60.

Zhong, Weihong, Fan Li, Zuoguang Zhang, Lulu Song, and Zhimin Li. 2001. Short fiber reinforced composites for fused deposition modeling. *Materials Science and Engineering: A* 301 (2):125–130.

4 Additive Manufacturing of Metals Using Powder Bed-Based Technologies

M. J. Mirzaali, F. S. L. Bobbert, Y. Li, and A. A. Zadpoor

CONTENTS

4.1 INTRODUCTION

Metal additive manufacturing (AM) techniques are solid free-form fabrication processes that are used for fabrication of three-dimensional objects from metals and their alloys. Recent developments in metal AM techniques have enabled fabrication of parts with reliable and reproducible properties. The layer-by-layer nature of AM processes provides a platform to produce arbitrarily complex geometries with tailor-made mechanical properties that cannot be realized using conventional manufacturing techniques [1,2]. AM processes have been therefore enjoying a great deal of academic attention as well as industrial interest.

AM processes have been used in various industries particularly high added value industries such as biomedical [3,4] and aerospace [5–7]. As opposed to the initial days of AM where it was primarily used for fabrication of prototypes and models [8], currently available AM processes are mature enough to be used for fabrication of functional and ready-to-use products [9]. As a result, the industrial interest in AM is rapidly growing with more than $1.25 billion investment reported in 2018 according to the Wohlers Reports [10].

Among different AM processes used for fabrication of metals and their alloys, the most important ones are based on powder bed fusion (PBF) (e.g., selective laser melting (SLM) and electron beam melting (EBM)) DSL (VDSL), directed energy deposition (DED), sheet lamination, metal extrusion, and binder jetting (previously referred to as 3D printing). Of these methods, PBF and binder jetting use metal powders [11–13], metal extrusion techniques use wires or rods [14–16], and sheet lamination processes use metal sheets [17]. In DED processes, wires or powder metals can be used as the raw material [18–20]. In this chapter, we will specifically focus on the different types of powder-based AM techniques (e.g., SLM, EBM, and DED). While we cover the various types of materials and processes used in different industrial sectors, special attention will be paid to the biomedical applications of powder-based metal AM techniques. Moreover, we will cover the processes, materials, and equipment parameters that highly influence the quality of the final product and whose optimization can be time-consuming and costly. Furthermore, we discuss the most critical aspects influencing the goodness of the final products. We conclude with a discussion of the challenges, limitations, and future outlooks for metal powder-based AM.

4.2 METAL AM PROCESSES

Various types of AM processes are currently available for producing metallic AM parts. The differences between these processes mainly concern the ways in which layers are deposited on top of each other to create the parts. Methods such as SLM or direct laser metal sintering, selective laser sintering, and EBM use the PBF technologies. The layer-by-layer melting of the areas of interest within the powder bed is the basis of these techniques (Figure 4.1a). The differences between these methods include those related to the source of energy (e.g., laser or electron beam) for either sintering or melting the metal powder [21,22]. In contrast, DED techniques feed powder or wire into the melt pool (Figure 4.1a). When the nozzle (tooltip) moves back and forth from the melt pool, it solidifies and creates the metal layers. The layer-by-layer deposition continues until the entire part is fabricated [18]. DED techniques are particularly useful when near net shape parts are desired, when the part is very large and cannot easily fit into the build volume of PBF process, or when AM is used for the repair of damaged/worn out parts or for adding features to existing parts [2].

There are individual processing limitations, pros, cons, and specific capabilities for each of these AM technologies. These include material availability, design accuracy, part quality (geometrical fidelity), build volume, and scanning speeds as the main pre-processing parameters. The types of appropriate post-processing treatments are the other important factor when deciding what type of process to use for a particular application, as post-treatment processes may be required to enhance the mechanical properties and functionalities [23] of AM parts. These issues will be addressed in the following sub-chapters.

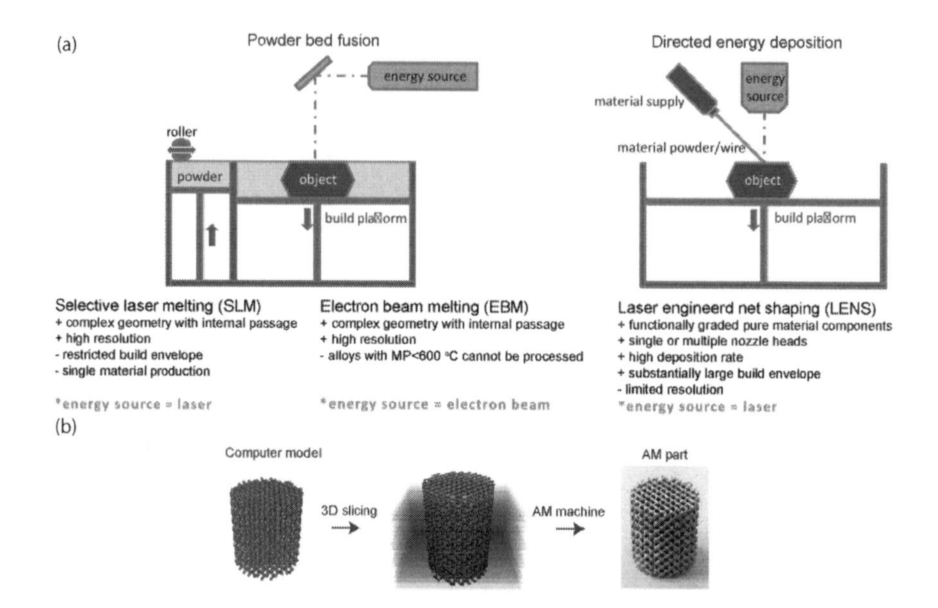

FIGURE 4.1 (a) Schematic drawings of metal powder-based AM processes with essential differences, pros, and cons. (b) The different steps from a CAD model to an AM part.

4.2.1 Powder Bed Fusion

In PBF processes, the energy is delivered to the build area through the source (e.g., laser or electron beam), resulting in the melting or sintering of the powder and creating the desired shape. This process continues by racking the metal (or metallic alloy) powders in the build area for making the next layer until the final three-dimensional object is printed [11]. One of the advantages of PBF processes is their ability to produce complex geometries with internal passages at high resolutions [21]. Some of their limitations include a restricted build envelop and the complexities associated with changing the processed material, meaning that in most cases only single material per part could be used [18]. *In-situ* alloying of powders is, however, possible in certain cases (Section 3.5).

4.2.1.1 Selective Laser Melting (SLM)

In SLM, a focused, high power-density laser is used to melt and fuse metallic powder (Figure 4.1a). The SLM process provides the ability to make organic geometries with complex internal features and passages that cannot be produced by traditional manufacturing processes such as casting or molding [11]. The powder bed may be heated up to 250°C to better manage the thermal gradients and lessen the adverse effects of thermal stresses [24]. The powder size is usually in the range of 10–60 μm [24]. SLM usually operates under an inert atmosphere [25].

To create objects using this technique, 3D computer-aided design (CAD) data with slice thicknesses in the range of 20–100 micrometers are usually required to process the successive layers. The CAD designs may be transformed into the standard tessellation language format at some point to facilitate communication with the different types of intermediate software packages and to create the input for the software package installed on the SLM machine. Having the input files ready, the corresponding process parameters and the design of the support structures can be set, and final AM parts can be produced (Figure 4.1b). In some cases, particularly when fabricating lattice structures, it is also possible to use a vector-based approach [26] where the laser scanning lines are directly related to the geometry of the part that is going to be printed without the need for an intermediate standard tessellation language file.

4.2.1.2 Electron Beam Melting

The EBM process is similar to SLM with a few major differences. Unlike SLM that uses a laser as the source of energy, an electron beam is used in EBM as the energy source (Figure 4.1a). Moreover, the electron beam consolidates the metal powders into a solid mass under a high vacuum condition [18,25]. The powder used in EBM is usually larger in size (as compared to SLM) and is in the range of 60–105 μm [27]. A larger powder size means that the minimum feature size and the surface roughness tend to be larger in EBM, as compared to SLM [28].

Another difference between SLM and EBM is their cooling rates. The build plate in EBM is maintained at a much higher temperature of 650–700°C. After the completion of the print job, the material is allowed to slowly cool down from 700°C to room temperature inside the chamber. The cooling rate in SLM is, however, in the range of 10^4–10^6 K/s [29–31]. Such differences in the cooling rate can result in different microstructures in the final product and profoundly influence the resulting mechanical properties. Furthermore, due to the high temperature of the substrate plate in EBM, it is not possible to fabricate alloys with a melting point lower than 600°C, while this is not an issue in SLM, which is why SLM can be used for a broad spectrum of metallic alloys [11].

4.2.2 Directed Energy Deposition

In DED, a concentrated energy source is under continuous interaction with a feeding layer of powder or wire (Figure 4.1a) [32]. One or more nozzles can be used for delivering the same or different types of powder material. Although different technologies can be categorized as being a DED technique [19,33], laser engineered net shaping (LENS) is the prevalent process that uses multiple nozzles for powder deposition. The cooling rate of the LENS process is between 10^3 and 10^4 K/s [34,35].

The ability to simultaneously process multiple materials [18], the ability to repair [36] or clad existing parts [19], higher deposition rates, and a more comprehensive process window for the construction of larger parts [37] are some of the advantages of DED processes. DED techniques, however, have limitations that make them less suitable for fabrication of parts with finer geometries and with hollow cooling passages [18].

4.2.3 Other Techniques

Other AM techniques include combinations of metal printing with other AM techniques (e.g., extrusion-based techniques) to create hybrid materials [38] or to supplement AM techniques with other (conventional) manufacturing processes (e.g., machining) to develop the so-called hybrid manufacturing processes [39,40]. The hybrid manufacturing can simultaneously benefit from the best features available in additive and conventional manufacturing technologies [41].

4.3 POWDER CHARACTERISTICS

Raw powder is used as the input material for AM processes, and its quality influences the mechanical properties of the final product [42–44]. Therefore, the properties of the powder are highly relevant in order to create parts with consistent, repeatable, and predictable properties. In this context, the term powder quality refers not only to the virgin powder, but also to the powder recycled from previous AM build cycles [45].

In this section, we discuss the different aspects that must be taken into account for the selection of raw powder and the metrology methods that can be used to quantitatively measure them. Some of these essential properties include the powder manufacturing techniques, the chemical composition of the powder, the powder morphology (e.g., powder size and shape), powder density, and powder flowability.

4.3.1 Powder Manufacturing Technique

The characteristic of the feedstocks in the AM processes can significantly influence the properties of the final product. Therefore, the manufacturing of these powder particles is of high importance, as it dictates the quality of the feedstock. A few well-established techniques, i.e., high-pressure water jet, gas-, plasma-, and rotary-atomization, are currently available for the production of the powder used in metal AM (Figure 4.2). The characteristics (e.g., size and morphology) and compositions of the powder produced by these techniques differ from each other [22]. That affects the density of the final AM parts and ultimately impacts their mechanical properties.

The water atomization methods [46] are suitable for the powder manufacturing of non-reactive materials (e.g., steel). This method is a low-cost process in which liquid metal is atomized by high-pressure water jets. The final powders have an irregular shape with sizes varying between a few μm and 500 μm [47] (Figure 4.2). The powder manufactured by this method also exhibits rougher surfaces [47].

The gas atomization technique [48] can be used for producing powder from both reactive (e.g., titanium alloys) and non-reactive metals. The particles produced by the gas atomization techniques have more homogeneous spherical shapes (Figure 4.2) with dimpled surfaces and lower oxygen content as compared to the ones produced by water atomization techniques [27,47]. The particles with more oxygen uptake can create an oxide layer on the melt pool, which has undesired effects on the flowability, chemical compositions, and final mechanical properties

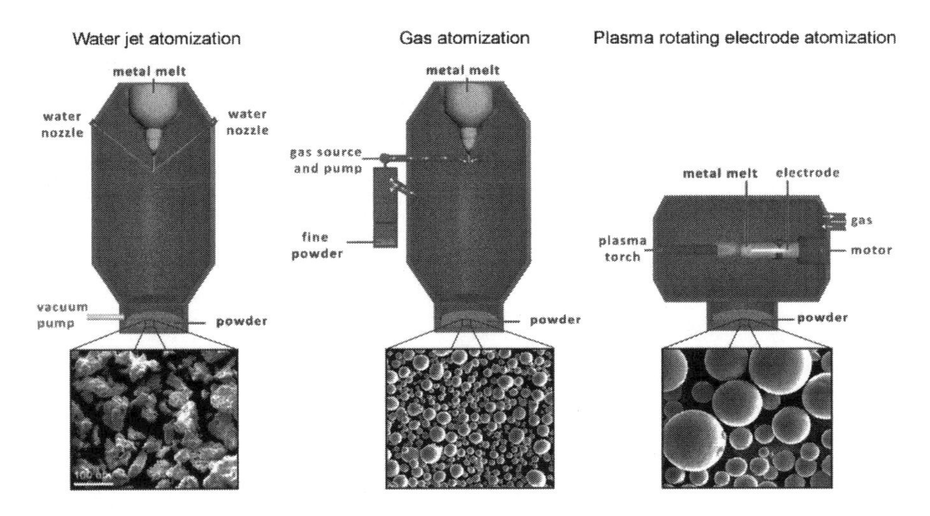

FIGURE 4.2 Powder atomization processes. Powder particles obtained from water-jet (With kind permission from Springer Science+Business Media: *Int. J. Adv. Manuf. Technol.*, Direct additive laser manufacturing using gas-and water-atomised H_{13} tool steel powders, 25, 2005, 471–479, 2008, A. J. Pinkerton, A.J. and Li, L.), gas atomization, and plasma rotating electrode atomization. (The gas and plasma atomization figures are reprinted with kind permission from Springer Science+Business Media: *JOM*, Review of the methods for production of spherical Ti and Ti alloy powder, 69, 2017, 1853–1860, 2017, Sun, P. et al.)

of AM materials [49]. In order to avoid the formation of an oxide layer on the melt pool, an inert gas (e.g., argon or nitrogen) can be used in the gas atomization process. The selection of the inert gas can influence the microstructure and phase composition of the powder particles [50,51]. These are the main advantages of gas atomizations over water atomization processes. In contrast, the porosity in the form of entrapped bubble gases can be created in the feedstock manufactured by gas atomization [52].

The plasma rotating electrode process atomization method [53,54] can be performed using plasma torches and an inert gas such as argon. The finer particle powders (40 μm) with homogeneous spherical shapes and smooth surfaces can result from this process [27,55,56] (Figure 4.2).

There are also other processes such as electrolytic, hydride-dihydride, and metallothermic processes (e.g., the TIRO process [57]) for the production of powder particles of Ti-based alloys that are currently under development [58].

4.3.2 Powder Shape and Size

X-ray, scanning electron microscopy (SEM), and micro-computed tomography can be used to obtain information on the shape and surface morphology of the powder particles in 3D [45]. Using some of those techniques (e.g., micro-computed tomography), stacks of images with a voxel size of a few micrometers can be collected, and via image analysis, the important morphological parameters can be calculated. The information on the particle size distribution can be calculated using the laser diffraction method. The output of such an analysis is the particle size distribution, which gives the volume fraction occupied by the particles of different sizes [45].

In order to achieve better mechanical properties of AM products, the best shape for the powder particles is spherical. Star-like and convex shapes are the other possibilities, where "star-shaped" particles show weaker performance as compared to the other shapes [45].

The shape and size of the powder can be changed after repeated use in AM build cycles, as the particles may fuse. Sieving can be used to adjust the size distribution of the powder particles [43,45]. Some studies recommend that the reused particles should pass through an 80 μm sieve [45]. Moreover, it is important to understand if the powder undergoes any chemical or physical changes and for how many times a specific powder stock can be reused. The particles that are close to the molten areas during the AM process are more likely to be exposed to high levels of energy, meaning that they may partially melt or sinter together and, thus, not pass through the sieve.

4.3.3 Density

Calculating the particle density is important, as the density of metal powder provides information about the porosity of the built part. The size of the metal powder particles is in the micrometers range, meaning that single measurements of their mass and volume are time-consuming and challenging. Helium pycnometry is therefore often used to measure the density of the metal powder [45]. In this method, a sample of the powder with a known weight is placed in a known-volume-pycnometry container. Helium, under atmospheric pressure and controlled temperature, flows into the pycnometry container and replaces the air. Based on the ideal gas law, the volume of the helium filling the container can be calculated. By subtracting this volume from the volume of the container, the density of the particles can be quantified. The powder particles themselves may have internal porosities. If this porosity is on the surface, it can be included in the overall density of the solid material. Otherwise, there will be a discrepancy between the overall density of AM objects and that of the powder particles.

4.3.4 Flowability

The manufacturing techniques used for the production of feedstocks can influence the flowability of the powder particles during AM processes. Finer powder particles and a more uniform size distribution usually lead to a more homogeneous melting and better flowability [59]. However, high-quality powder particles with good surface finish can be quite expensive due to the high costs of atomization techniques involved in their manufacturing [27]. The flowability can be measured using the Hall flow meter [45,60].

4.3.5 *In-Situ* Alloying and Composites

In-situ alloying refers to the processes where different feedstocks with different compositions are simultaneously fed into the melt pool. This can provide new opportunities in material design [61]. Such a mixture allows for building parts with tailored compositions/properties that are otherwise difficult to achieve. An example is *in-situ* composite LENS-processed Ti-6Al-4V-TiB, which are composed of pre-alloyed Ti6Al4V and boron powders [62]. Mixing these two components in a composite material can create a material with high strength and stiffness due to the existence of the borides together with high toughness and damage tolerance due to the existence of the Ti-alloy matrix. Other examples are SLM-processed *in-situ* Ti–26Nb alloy for biomedical applications [63] and anchorless SLM AlSi12 *in-situ* alloy, which can be used to reduce the residual stresses developed during the AM process [64].

4.4 EFFECTS OF PROCESSING PARAMETERS AND QUALITY MEASUREMENTS

The processing parameters of AM define the success of the printing process. There are different processing parameters involved depending on the type of the AM process. Some of these processing parameters are beam/laser energy power, beam/laser diameter, beam/laser line spacing, beam/laser scanning speed, scanning strategy, and also a build-plate pre-heat temperature [11]. These processing parameters should be controlled and adjusted in order to ensure the quality, consistency, and reliability of AM materials.

In order to accurately select the processing parameters, different process maps [65–68] have been developed in several research groups worldwide. These design maps help in the proper identification of AM processing parameters including laser/electron beam power and powder feed rate. Finding these design maps are often costly and time-consuming and may be both design- and machine-specific.

The print job is considered to have been successful, if the following three objectives are satisfied. First, the built AM part is defect-free and has the same composition as the parent material. Second, the geometry of the printed part matches the CAD design (geometric fidelity). Finally, the final AM parts should satisfy certain requirements regarding microstructure, surface roughness, and mechanical properties. When appropriately used, a powder-based metal AM process could result in mechanical properties that are similar to or are even higher than the ones achieved by conventional manufacturing techniques [2]. These aspects are discussed in the following sub-chapters.

4.4.1 DENSITY

The first aim of a successful AM process is to fabricate highly dense parts, typically more than 99.5%. However, manufacturing defects can occur during the AM processes that result in a part with a lower relative density. One type of these defects is caused by gas entrapment due to rapid cooling rates. This can lead to the formation of microporosity (10–50 μm with less than 1% volume) [69] in AM materials. These microporosities can promote the propagation of cracks and reduce the mechanical properties of AM parts. Among different processing parameters, the energy density (too low or too high level of energy) has significant effects on the formation of these microporosities and may consequently decrease the density of AM materials [70].

4.4.2 Geometric Fidelity

The dimensional accuracy of AM parts also known as geometric fidelity depends on the AM processing parameters and feedstocks properties. The geometric fidelity can be quantified statistically by comparing the deviation of the actual geometry of the printed structures from the CAD design [71,72].

The geometric accuracy of parts fabricated with different AM processes can differ from each other. For example, PBF processes (e.g., SLM, EBM) produce parts with a higher degree of dimensional accuracy as compared to DED processes [73]. That is partially due to the fact that DED processes are capable of producing substantially larger specimens.

Geometric fidelity is particularly important when porous AM materials (e.g., lattice structures) are produced. Two main types of manufacturing irregularities can be observed during the AM processes of these porous materials. First, the cross-section of the struts exhibits certain levels of irregularity. Second, accidental fusions of powder particles may create imperfections and cause the porous material to show a smaller porosity at the macro-scale than intended [74]. These types of irregularities and imperfections can significantly influence the quasi-static mechanical and fatigue behavior of AM cellular structures [74].

4.4.3 Surface Roughness

Surface roughness is one of the most important features affecting the quality of AM parts. The surface morphology can be analyzed using a surface profilometer or an SEM to measure the surface roughness [75]. There are a few empirical equations that measure the height of peaks and valleys on the rough surface at different locations within the specimen [75]. An average surface roughness of <1 mm is sometimes considered the minimum requirement for AM components [76]. Post-treatment processes such as surface machining, shot peening, chemical polishing, grinding, and hot isostatic pressing (HIP) can be used to improve the quality of the surfaces [77,78].

Different AM technologies can achieve distinct surface characteristics. For example, the surfaces of SLM parts are relatively smooth as compared to the ones fabricated by EBM or DED [28]. This is due to the thinner layers (30 μm in SLM as compared to 70 μm in EBM), slower scan speeds, and finer powder sizes [11,28] used in SLM. For SLM parts, a small particle size of <20 μm may result in a more smooth surface finish [59]. A thinner layer, however, results in a lower rate of production.

The manufacturing process of the feedstocks can also influence the surface roughness. As an example, the powder particles manufactured by high-pressure water jet tend to create AM parts with higher surface roughnesses as the particles obtained from this process exhibit coarser surfaces as compared to the particles obtained from atomization techniques [47]. The quality of re-used powder can also influence the surface quality of the final product [79].

The other causes of rough surfaces include the formation of defects (e.g., unmolten powder particles) and balling phenomena during AM processes [80–82]. This happens when the delivered energy is not high enough to fully melt the particles. The unmolten particles can then stick to the surface of the AM part and roughen its surface.

The surface quality and the well-definedness of the design features are also linked to the build rate. The feature quality/resolution is inversely correlated to the build rate, meaning that it may be necessary to apply post-AM surface treatment to the parts fabricated with a high deposition rate [21]. The surface condition also affects the mechanical properties of the AM parts, particularly their fatigue resistance. In order to achieve the desired surface finish, AM parts may need to be subjected to surface treatments such as machining.

4.4.4 Microstructures

The microstructure of AM products can be characterized by optical microscopy, SEM equipped with an electron-backscatter-diffraction system, and scanning transmission electron microscopy [83,84]. In order to detect grain boundaries, AM materials can be immersed in the proper etchant, which is dependent on the type of the material.

The microstructure of AM parts is generally finer than as-cast or as-wrought parts [85]. The exact specifications of the microstructure can be strongly influenced by the type of the applied AM process. That is due to the fact that the cooling rates and thermal cycles involved in various AM processes could be very different [86,87]. The cooling rate can be controlled by the power and velocity of the energy source as well as the presence (and temperature) of substrate pre-heating [88]. For example, $\alpha + \beta$ lamellar microstructures may be achieved in the case of laser-based AM of Ti-6Al-4V (Section 6.1) [67,89,90].

Furthermore, different solidification mechanisms influence the final microstructures of AM parts. Equiaxed grains that are heterogeneously nucleated on the particles that are partially melted and/or epitaxial growth of parent grains are examples of such microstructural features [19]. The rapid solidification can also lead to the formation of a metastable phase that finally results in higher elastic properties [69].

Due to the layer-by-layer nature of AM processes, the parts are subjected to repeated thermal cycles and undergo different liquid-solid transformation phases. This can result in the formation of different microstructures within each deposition layer and lead to the so-called microstructural banding [91,92]. In addition, the thermal history during AM processes has a great influence on the homogeneity of the microstructures of AM materials.

Moreover, the directional heat flow in AM processes can lead to the formation of microstructures with preferred directionality (anisotropy) in grain growth (e.g., columnar microstructures) due to the temperature gradients experienced by the components [86,93–95].

The scanning speed and laser power control the morphology and shape of the microstructure. A combination of high scanning speed and low laser power can create finer microstructures, whereas low scanning speeds and high laser powers may lead to coarser microstructures with more columnar shapes [96,97]. Post-processing techniques (e.g., heat treatment and HIP) could also influence the microstructures of AM materials [98].

In the case of AM porous materials, the quality of the resulting AM material may be dominated by the effects of the orientation of the struts relative to the build direction, with horizontal struts being the most prone to imperfections [99]. These imperfections mostly concern the porosity of AM porous materials.

4.4.5 Chemical Composition

The chemical composition of the metallic alloys may be dependent on environmental (e.g., humidity) and atmospheric (e.g., the oxygen level) conditions. Oxidation during the AM process may lead to the formation of an oxide layer or inclusions in the AM parts, thereby reducing the mechanical properties [61]. In addition to these environmental conditions, the AM processing parameters such as higher energy levels and temperatures can speed up the chemical degradation and oxidation of AM parts [100]. Repeated recycling of powders is one of the causes of changes in the chemical composition of the powder material and may influence the mechanical properties of AM parts [45].

4.5 DESIGN FOR METAL AM

Design for AM approaches facilitate the realization of "designer" materials and "custom-made designs" and are assisted by the "batch-size-indifference" and "complexity-for-free" features of AM processes [2,101]. The designer materials are also known as metamaterials and are a class of advanced functional materials whose macroscopic properties and functionalities derive from their design at the micro-scale [102]. The rational design of the microstructure of the designer materials can result in novel mechanical [103–105], thermal [106], acoustic [107], or biomedical [108] properties (Figure 4.3).

Custom-made parts such as patient-specific AM biomaterials and implants are also of importance particularly when no single design matches all the requirements and applications. Custom-made orthopedic implants are examples of the parts were due to the difference in the size, shape, and geometry of individuals' bones, specific designs need to be considered [109–111] (Figure 4.3f).

The CAD model of a part (e.g., an implant) is the primary input for AM processes. These models can be the result of a generic design process or may be based on the specific images obtained using an imaging modality such as computed tomography or magnetic resonance imaging [23]. An important aspect of design for AM processes relates to the procedures used for the development of such CAD models to make, for example, patient-specific implants.

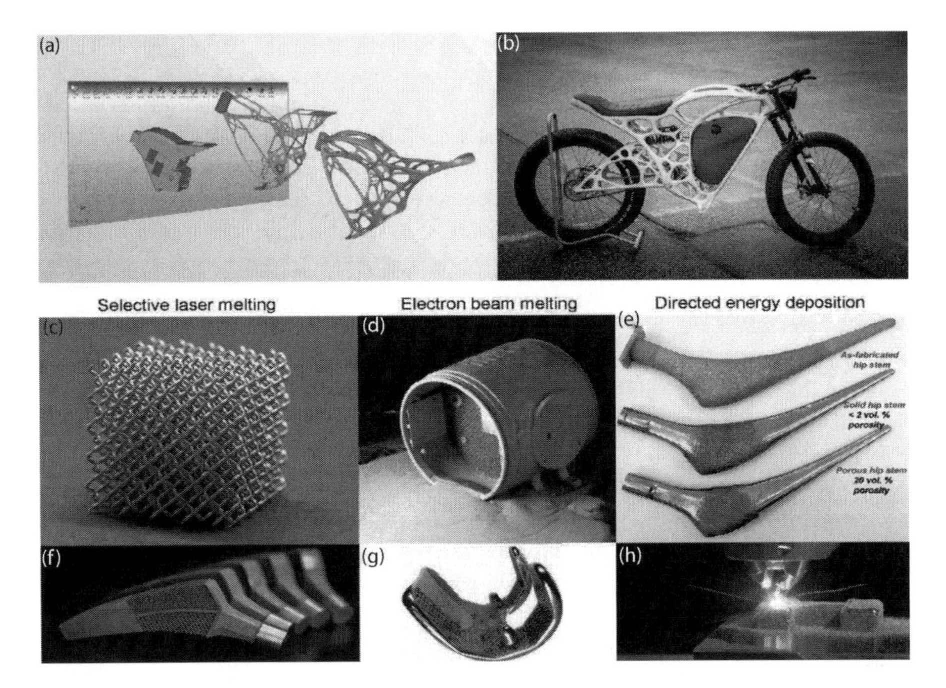

FIGURE 4.3 Topology optimization for AM, (a) Simulation driven design process with Altair's OptiStruct ©Altair, (b) Airbus APWorks' Light Rider–The world's first 3D printed motorcycle ©APWORKS, (c) A pentamode lattice structure (Ti-6Al-4V) (Reprinted with permission from Hedayati, R. et al., *Appl. Phys. Lett.*, 110, 091905, 2017. Copyright 2017 by the American Institute of Physics.), (d) a hydraulic manifold (titanium) (Reprinted from *Titanium Powder Metallurgy*, Dutta, B. and Froes, F.H.S., The additive manufacturing (AM) of titanium alloys, 447–468, Copyright 2015, with permission from Elsevier.), (e) net shape functional hip stems with designed porosity fabricated using LENS™ (titanium) (Reprinted from *Mater. Sci. Eng. C*, 30, España, F.A., Design and fabrication of CoCrMo alloy based novel structures for load bearing implants using laser engineered net shaping, 50–57, Copyright 2010, with permission from Elsevier.), (f) hybrid meta-biomaterials incorporating both auxetic and conventional meta-biomaterials (Ti-6Al-4V) (From Kolken, H.M. et al., *Mater. Horiz.*, 5, 28–35, 2018.), (g) partially finished EBM femoral implant (Co-base) (Reprinted from *J. Mater. Sci. Technol.*, 28, Murr, L.E. et al., Metal fabrication by additive manufacturing using laser and electron beam melting technologies, 1–14, Copyright 2012, with permission from Elsevier.), and (h) deposition of Ti6Al4V within the LENS system (Reprinted from *Addit. Manuf.*, 6, Nassar, A.R. et al., Intra-layer closed-loop control of build plan during directed energy additive manufacturing of Ti–6Al–4V, 39–52, Copyright 2015, with permission from Elsevier.)

4.5.1 Topology Optimization

Topology optimization is a powerful approach for finding the optimum material distribution within a design space. Often, the optimized topology is too complex and costly to manufacture using conventional manufacturing processes. However, AM can deal with these complex shapes given its "complexity-for-free" characteristic. There are, of course, some limitations regarding the geometries that could be realized with AM. For example, the overhang length is often limited, and not all geometries could be printed without internal support structures. The space of realizable geometries is, nevertheless, much larger than the conventional processes.

Finite element analysis combined with optimization algorithms can be used for optimizing the internal geometry and, thus, the density distribution of AM parts. In this approach, an objective function is defined and the optimized topology is determined such that the objective function is minimized. The objective function can, for example, be the compliance of the AM part. TOSCA [112], Pareto Works [113], and PLATO [114] are some of the optimization tools used for such optimization purposes [115]. There are also free codes available for topology optimization purposes [116] that can be implemented in AM design processes. An example of the topology optimization for an AM bike frame is presented in Figure 4.3.

The topology optimization can also include the constraints dictated by the AM process [117,118]. Fully integrating the various aspects of AM including the effects of the processing parameters on the microstructures, mechanical properties, and geometric fidelity in topology optimization algorithms is the ultimate goal of such approaches to ascertain the designed parts are not only optimized topologically, but are also optimized from the performance and printability viewpoints [119,120].

4.5.2 Lattice and Other Architected Structures

AM processes can be utilized in the fabrication of cellular lattices with complex microstructural geometries. PBF processes can only build cellular structures with open-cell micro-architectures (it should be possible to remove the residual powder). The shapes of the cellular lattices can vary from homogeneous structures with regular and repeated unit cells to structures with graded and/or irregular shapes [121]. The geometry of these cellular structures can come from a CAD design, from image-based designs, from implicit surface modeling, or simply be the output of a topology optimization process [122]. In order to meet the printability requirements of such lattices, the angle of the struts in the lattices should be within specific ranges. Sacrificial support structures may also be needed to hold the part during the manufacturing process. The design and use of these support materials add more constraints to the AM processes and can profoundly influence the manufacturability, time, cost, and energy utilization. Therefore, efficient and optimized design of support structures can save material and lead time [123,124].

Lattices with various internal microstructures are of interest in many engineering fields such as aerospace and biomedical engineering given their favorable properties such as their high load-bearing capacities (Figure 4.3). The micro-architectural design of lattice structures directly controls their physical and mechanical properties [122]. For example, in the case of AM porous biomaterials, the topological design of cellular lattices can be adjusted to achieve specific values of porosity, pore size, and interconnectivity, thereby controlling the mass transport properties (e.g., permeability) of the resulting scaffolds [125–128]. The topological design at the small scale also affects tissue in-growth into the open porous structures, which in turn results in improved implant fixation and osseointegration [129,130] necessary to boost the bone regeneration performance [131] and longevity of orthopedic implants. An increased surface area is another advantage of such AM porous biomaterials, which could be exploited for bio-functionalization purposes [122,132,133] that are aimed to, for example, prevent implant-associated infections [134,135].

4.6 MATERIALS

Different metallic materials including Ti alloys, superalloys, steels, and Al alloys have been already processed with AM, while attempts to process other metallic materials with AM are ongoing [61]. Based on the different processing parameters involved in each AM process and the inherent properties of the specific metallic alloy being processed, there are certain considerations regarding the fabrication of metallic material via powder-based AM processes. One of these considerations is the atmospheric conditions as different metallic alloys interact differently with the atmosphere. An example is the formation of an oxide layer on top of the melt pool for aluminum, 18Ni300 maraging steel [136], and Inconel 718 [137]. Moreover, some metallic alloys (e.g., Mg alloys [138]) are highly reactive, meaning that even minuscule amounts of free oxygen should be taken into account when designing the AM process.

Furthermore, due to the high thermal conductivity and high reflectivity of some metallic alloys (e.g., aluminum), their processing with AM is challenging. In this section, we shortly review some of the metallic materials processed with AM and discuss some material-specific aspects such as the properties of the different metallic alloys processed with AM.

4.6.1 Titanium and Its Alloys

Titanium alloys particularly Ti-6Al-4V [87,139] are some of the most extensively studied materials for AM purposes. Ti-6Al-4V has a high specific strength (the ratio of the strength to weight) and excellent corrosion resistance, good formability, and is widely used in the aerospace and biomedical (e.g., in orthopedic implants) industries [140].

Generally, the properties of AM Ti-6Al-4V are dependent on its thermal history and AM processing parameters. The microstructure of AM Ti-6Al-4V is usually composed of primary and secondary α grains along with stabilized β phases [141], and the thermal history can influence the formation of each phase. In addition, different manufacturing processes and parameters can change the microstructure of AM Ti-6Al-4V [142] and consequently its mechanical properties. The amount of α and β phases can, therefore, vary in AM Ti-6Al-4V with higher amounts of the α phase leading to a higher elastic modulus [143,144]. The rapid cooling of Ti-6Al-4V during

AM may also create martensitic needle-shaped martensite α' phases that increase the strength of the alloy and decrease its ductility [28,145]. This is particularly the case for laser-based PBF and DED processes. Ti-6Al-4V processed with EBM is more likely to show a mixture of α and β phases due to the lower cooling rates experienced during this process, in turn leading to a reduced tensile strength, but increased ductility [18,28,146].

Post-AM heat-treatment processes could be used to modify the microstructure of SLM Ti-6Al-4V alloy [147]. Heat treatments at temperatures higher than the β transus temperature can result in a more homogeneous structure at the macro level. Full annealing at temperatures of >850°C has been recommended for SLM Ti-6Al-4V to increase its fracture strain and ductility [147]. In most cases, the mechanical properties of AM Ti-6Al-4V are higher than those of wrought titanium alloys while exhibiting lower ductility. Moreover, slight anisotropies in the yield and tensile strengths related to the build directions have been reported [141]. Adding other (biocompatible) elements (e.g., Ta, Nb, Zr, Mo) can result in lower values of the elastic modulus as compared to Ti-6Al-4V, as these elements are often β-stabilizing elements [122,148]. The examples of these Ti-based alloys with improved biocompatibility are Ti-6Al-7Nb [149] and Ti-24Nb-4Zr-8Sn [150].

In addition, SLM of (porous) pure titanium (particularly grade 2) has also drawn a great deal of attention [151–154]. Unlike titanium alloys that show a high specific strength, pure titanium has a lower specific strength making it more useful for non-load-bearing components (e.g., cranio-maxillo-facial implants) [155,156]. There are also a few advantages for pure titanium over titanium alloys (particularly Ti-6Al-4V) such as, absence of hazardous elements (e.g., Al or V) [157], excellent *in-vivo* performance, higher ductility, lower cost, and higher levels of normalized fatigue resistance [154].

4.6.2 Steels

Not all AM techniques are suitable for processing the various types of steels. Indeed, different cooling rates in AM processes can create different phase compositions (e.g., austenite or martensite) [50]. The laser-based AM has been mainly used for fabrication of austenitic stainless steels (e.g., AISI 316L [158,159] and AISI 304L [160]), 18Ni300 maraging steels [161,162], tool steels (H13), precipitation-hardenable stainless steels (AISI: 630), and soft magnetic high-silicon steel [163]. EBM, on the other hand, has been used for fabrication of an austenitic stainless steel (AISI: 316L) and tool steels (H11) [22,164].

The cooling rate experienced during AM processes can affect the microstructure of steels. Different cooling rates may lead to the presence or absence of the martensitic phase in stainless steels. For example, SLM 316L stainless steels have been found to exhibit a refined microstructure with the absence of the martensitic phase as compared to conventionally processed ones. These microstructural differences result in higher tensile strengths, but lower elongations when compared with those fabricated conventionally [165].

4.6.3 Al Alloys

There are a limited number of Al alloys available for AM, given the fact that most Al alloys are machinable while suffering from low weldability. There is therefore limited commercial advantage to AM processing of some Al alloys. Furthermore, the existence of volatile elements such as Zn in hardenable Al causes turbulence and splattering in the melt pool that can create porosity in the AM parts. Moreover, some other elements (e.g., Mg and Li) that are present in Al alloys tend to annihilate under vacuum. Finally, Al alloys show a high level of reflection when in contact with a laser wavelength, making it challenging to utilize them in laser-based techniques [166]. On the plus side, the high thermal conductivity of Al alloys materials allows for faster reduction of thermal stresses and consequently higher rates of production [22].

Among different Al alloys, the eutectic AL-Si and hardenable Al-Si-Mg alloys (e.g., ALSi10Mg and AlSi12) are the most commonly used materials for laser-based AM processes [27,167]. In addition, DED approaches have been implemented for the fabrication of Al-Cu alloys [168]. AM Al alloys display finer-grained microstructures and consequently higher tensile strengths as compared to the ones fabricated by conventional methods. Indeed, the thermal history experienced by Al alloys during AM processes can lead to the formation of finer microstructures different from those of cast or wrought counterparts [83]. Depending on the composition and solidification process, the microstructure of Al-Si alloys may consist of primary aluminum with (fully) eutectic or a mixture of primary silicon and eutectic. Typically, the microstructure of Al-Si alloys contains columnar dendrites and equiaxed dendrites [27].

4.6.4 CoCr Alloys

CoCr alloys show a good level of biocompatibility, which is why they have been widely used for biomedical applications [169–171]. AM Co-Cr has been shown to have high levels of crystallographic anisotropy, which can be the main cause of high variations in the elastic modulus reported in the literature [172–174]. SLM Co-Cr alloy has a minimal columnar grain size as compared to conventionally manufactured alloys. The rapid cooling of Co-Cr alloys during the SLM process can create finer and more irregular columnar dendritic microstructures, which result in higher yield strengths [175].

4.6.5 Superalloys

Nickel-based alloys (e.g., IN625 and IN718) are examples of superalloys appropriate for high-temperature applications (e.g., gas turbines). Both laser-based techniques [176–179] and EBM [94,180] have been used for the AM of Ni-based superalloys.

During AM processes, columnar architectures may be formed in the microstructure of IN625 and IN718 due to the formation of γ' and γ'' precipitates [85]. Similar features cannot be found in cast or wrought alloys. These unusual columnar precipitations can increase the elongation of Ni-based superalloys, although their effect can be eliminated by HIP [92].

4.6.6 Biodegradable Metals

Biodegradable metals can be used in the fabrication of orthopedic and cardiovascular implants [181]. These materials gradually break down in the human body and are absorbed or excreted [182]. Therefore, these metallic implants can disappear when the healing process is finished. Magnesium, zinc, iron, and their alloys are the most common metals used as biodegradable implants due to their appropriate biodegradation rates, biocompatibility, their potential as drug delivery vehicles, and bioactivity [183,184]. Examples of the applications of Mg-based alloys are biodegradable cardiovascular stents [185] and orthopedic implants [186].

AM of biodegradable metals particularly design and manufacturing of topologically ordered porous structures has just started [187,188]. In particular, AM of Mg is extremely challenging due to its extremely high inflammability, which is why the first ever report of Mg-based porous structures realized by direct metal printing has just appeared in the literature [188]. AM of other biodegradable metals is not as challenging. However, until recently, not much work was performed on those materials either.

The degradation rate and profile are vital parameters that need to be controlled in the design of biodegradable metals. Pure Mg has a rapid degradation rate and can release high volumes of hydrogen when being corroded inside the body. The rate of degradation can be adjusted by either alloying (e.g., adding elements such as Y, Sr, Zn, Zr, and Ca) or by using amorphous structures such as metallic glasses (e.g., MgZnCa [189,190]). Such a combination not only can decrease the biodegradation rate, but may also increase the mechanical properties of the material [191].

Iron is another biodegradable metal that shows lower biodegradation rates than Mg alloys. Porous iron-based implants produced by direct metal printing have shown biocompatibility, appropriate degradation rate, and an acceptable range of mechanical properties after several weeks of biodegradation [187].

Zn is another biodegradable metal with a low rate of biodegradation. Mainly SLM techniques have been used for the fabrication of porous Zn-based materials [192,193]. Zn metals have a low melting point. Therefore, their successful processing is only possible within a narrow window of processing parameters. Sub-optimal processing parameters result in highly porous materials [192].

4.6.7 Shape Memory Alloys

Shape memory alloys are materials that are capable of recovering their initial shape once triggered by an external stimulus (e.g., high temperatures). This unique property together with good biocompatibility, corrosion resistance, and high ductility [194] make them promising candidates for fabrication of medical devices such as surgical tools, stents, orthodontic wires, and staples for bone fractures [122]. Nickel-titanium (NiTi) alloys (e.g., Nitinol with 50% Ni and 50% Ti) are the most commonly used types of shape memory alloys. The martensite/austenite phase transformation is responsible for the shape memory property of shape memory alloys. While the martensite phase is stable under low-temperature conditions, the austenitic is stable at high temperatures [195]. Solid and porous NiTi materials have been produced by laser-based AM techniques [196,197]. Functionalized AM porous nitinols have shown high levels of tissue regeneration performance. Therefore, these materials can be used as a basis for development of deployable orthopedic implants [198].

4.6.8 Tantalum

Porous tantalum is one of the metallic materials used for the fabrication of medical devices and has been made by SLM [95,169,199], LENS [200], and EBM [201]. Porous tantalum exhibits excellent biocompatibility, and its mechanical properties are similar to those of the human bone, resulting in a highly osteoconductive behavior. Tantalum can also be used as a coating on the surface of other porous AM biomaterials such as Ti-based alloys [202,203]. The osteoblast cells that are responsible cells for bone growth show higher levels of attachment, differentiation, and proliferation on the surface of Ta as compared to Ti-6Al-4V porous AM biomaterials [204], making them a potential replacement for Ti-based scaffolds. However, Ta is an extremely expensive material, which is why Ti-Ta alloys have received much attention recently [199,205].

4.7 POST-AM TECHNIQUES

Post-AM techniques can be implemented to improve the microstructure and mechanical properties of AM materials while achieving new functionalities and reducing the residual stresses. These techniques can reduce or eliminate the process-induced defects such as microporosity and lack of fusion, although they might also affect the cost-effectiveness of the AM process [206]. The materials processed with SLM and DED require stress relief treatments the most, while those processed with EBM may not require that.

4.7.1 Residual Stress

Residual stress is the stress present in a body when no external forces are applied and the other mechanical and thermal loads are in equilibrium (Figure 4.4a). Residual stresses are the results of local and global thermal gradients and the high cooling/heating rates experienced during and after the printing process [207]. Therefore, the magnitude of the residual stresses is dependent on the thermal history experienced during the AM processes. The presence of residual stresses has a significant effect on the mechanical performance (e.g., tensile strength and fatigue resistance) and lack of geometric accuracy in AM parts [208].

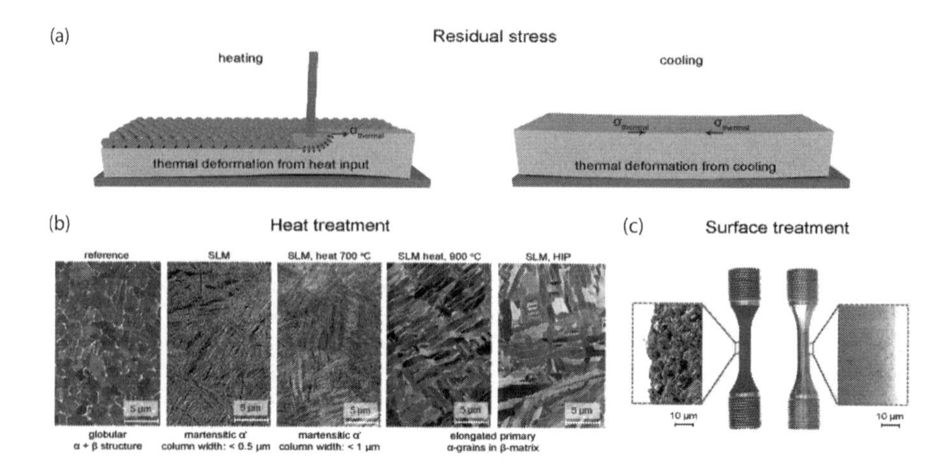

FIGURE 4.4 The residual stresses and effects of post-AM treatments. (a) the way in which residual stresses are created. (b) SEM micrographs of etched SLM Ti-6Al-4V surfaces. From left to right: Globular "reference," lamellar SLM, structures of Ti-6Al-4V alloy and the effects of thermomechanical treatment on the microstructure of SLM specimens. (c) Test specimen surface treatment. As-built (left) and machined (right). (Reprinted from *J. Mater. Process. Technol.*, 220, Kasperovich, G. and J. Hausmann, J., Improvement of fatigue resistance and ductility of TiAl6V4 processed by selective laser melting, 202–214, Copyright 2015, with permission from Elsevier (Figure b and c).)

Among all processing parameters, the laser power, powder feed rate, scanning strategy, and scanning speed are the most important parameters that can affect the level and type of residual stresses in the AM materials [207,209]. In addition to these processing parameters, the properties of the base material such as the thermal conductivity, thermal expansion coefficient, Young's modulus, and yield stresses together with the geometry of the part can influence the level of residual stresses [210,211].

The residual stresses can be reduced by optimizing the processing parameters, by using a proper scanning strategy, and by preheating the substrate or the last deposited layer [66,209,211]. Proper placements of supports within the design can be also helpful in minimizing the distortions of AM parts. This can be guided by computational models that provide us with a better understanding of the thermal stresses developed in the AM part [27]. Due to the presence of substrate preheating in EBM, AM materials fabricated by this method show lower magnitudes of residual stresses as compared to SLM and DED materials [18].

4.7.2 Heat Treatments

Heat treatment processes can be used to improve the microstructures of AM parts. These thermal treatments affect the grain size and precipitates [212,213] (Figure 4.4b).

After heat treatments, the residual stresses in the samples are relieved due to the atomic diffusion. To remove the residual stresses, SLM and DED parts are usually annealed [212]. There are standard treatment processes available in the literature that are commonly used for Ti-6Al-4V and IN718 [214]. The duration of the heat treatment processes can also influence the microstructural features of AM materials and must be taken into account.

Due to the spatial distribution of residual stresses in AM parts, local heat treatments may be required, as a global heat treatment can affect the geometric accuracy and mechanical properties of AM materials. As previously discussed, the exact location of the residual stresses can be determined by computational modeling on which basis local heat treatment processes can be defined to remove the undesired stresses at specific parts of the AM part.

4.7.3 Surface Treatments

Post-AM surface treatments including mechanical (e.g., machining (Figure 4.4c) and polishing) [78] or chemical (e.g., etching) might be required to improve the quality of AM parts. In addition to being an important manufacturing requirement, surface properties could significantly influence the mechanical properties of AM parts. For example, the surface roughness is inversely correlated to the fatigue life of AM materials [215].

The required type of surface treatment may be dependent on the type of the AM process. For example, using a finer powder (less than 20 μm) particularly in SLM can create smoother surfaces that do not require post-AM treatments, whereas parts fabricated with DED or EBM usually have rougher surfaces that require post-AM treatments. The surface treatment can also be performed to increase the bioactivity and biocompatibility of (porous) AM biomaterials [140,216], for example, to improve their tissue regeneration performance [4,217] or to prevent implant-associated infections [134,135]. An example is covering the surface of the Ti-alloys by an oxide layer (3–7 nm) to increase their corrosion resistance and biocompatibility [122].

To improve the bioactivity of AM porous biomaterials, surface modifications, i.e., surface coating and surface corrosion, can be performed. One of the well-established surface coating methods is sol-gel processes in which an oxide layer at low temperature is applied to the surface of AM porous biomaterials. Such sol-gel processes are low-cost processes, which can create a coating layer with a uniform and homogenized microstructure and with bioactive properties [218,219]. The surface treatments involve the interaction of the surface of AM porous biomaterials with a (corrosive) chemical solution. Some of the examples of these corrosion-related treatments are alkali treatment [220–222], acid etching [223], and anodization [224].

4.7.4 HIP

The HIP process simultaneously applies high temperatures and high pressures to close the micro-porosities and imperfections created during AM processes [225].

This also changes the microstructural features of AM materials (Figure 4.4b). Based on the material type, different protocols can be implemented. Through the elimination of (some of the) manufacturing imperfections, HIP can enhance the fatigue and static mechanical properties of AM materials [226]. The HIP can also increase the ductility, although it might decrease the yield strength of the material due to certain levels of annealing [226,227].

4.8 MECHANICAL PROPERTIES

The specimens used for mechanical testing of AM metals (i.e., cylindrical or flat specimens) usually follow the guidelines suggested by ASTM (or other comparable) standards (e.g., ASTM, E8) [228]. Tension, compression, hardness, fracture toughness, and creep tests are the most common types of mechanical tests used for the mechanical characterization of the AM materials. The elastic modulus, yield strength, ultimate tensile strength, and elongation to failure, compressive strength, and stress intensity factor are some of the mechanical properties resulting from these experiments. The mechanical properties of AM materials are dependent on the build direction as well as their position within the build envelope [206,229]. The mechanical properties of AM parts therefore show significant levels of anisotropy, with the tensile mechanical properties perpendicular to the build plate being usually the lowest [21]. It is therefore important to mention the build direction when reporting the mechanical properties of AM materials.

The chemical composition and microstructures (crystalline orientation) are among the most important factors influencing the mechanical properties of both solid and lattice AM materials. In addition, other laser/energy beam parameters such as laser power, scanning speed, and the quality of powder can profoundly affect the mechanical properties of these materials [21,73,206,230]. With proper choice of the processing parameters, the mechanical properties of AM components can reach (and even exceed [2,22,27]) those of conventionally fabricated ones.

It should be also noted that as AM materials experience high rates of cooling/heating, they usually include microporosity and inclusions that can reduce their ductility as compared to the ones fabricated conventionally [21,206]. Moreover, the levels of oxygen, moisture, and nitrogen can influence the mechanical properties of AM materials. For example, a higher level of oxygen has been shown to increase the strength while decreasing the ductility of Ti-6Al-4V [231]. Post-treatment processes such as heat treatments (to relieve the residual stresses) and HIP (to remove the manufacturing imperfection) can significantly increase the ductility of both porous and bulk AM materials [206].

4.8.1 Solid Specimens

The mechanical properties of solid AM materials strongly depend on the orientation in which the specimens are printed [206]. Their properties also depend on the type of the AM processes as the resulting microstructures, the level of residual stress, and manufacturing imperfections may be different for each technique [232].

To put the mechanical properties of AM material in perspective, we have summarized some of the values found in the literature regarding the elongation and ultimate tensile strength of AM Ti-6Al-4V (Figure 4.5). We only used the data pertaining to the specimens printed horizontally. There is significant scatter in the data presumably due to the very different processing parameters that may have been used in different studies and are not always reported. This data also show that the mechanical properties of as-built AM Ti-6Al-4V may approach those of conventionally fabricated Ti-6Al-4V in certain cases (Figure 4.5a). There are also marked differences between the elongation and tensile strength of parts manufactured by various AM techniques including EBM, SLM, and DED. In general, EBM-processed Ti-6Al-4V shows higher ductility as compared to the material processed by SLM and DED (Figure 4.5).

We also collected data regarding the effects of post-AM treatments on the mechanical properties of AM materials (Figure 4.5b–d). The collected data clearly show that HIP of Ti-6Al-4V can increase the ductility of the material regardless of the AM process while machining increases the strength of AM materials and decreases the ductility of EBM materials [28].

4.8.2 Lattice and Architected Materials

AM lattice and architected materials have a wide range of potential and actual applications in the load-bearing components used in the biomedical and aerospace industries. The mechanical properties of porous structures are directly related to their topological design at the micro-scale [233–236]. It is often assumed that there are no modulations between the topological design and material type in determining the mechanical properties of porous AM materials [142,237,238] (Figure 4.6a and b). However, recent research has shown that this assumption, while being a good first approximation, is not entirely valid [239]. AM processing parameters such as the exposure time and energy may also affect the mechanical properties of porous AM materials, with an increased energy input resulting in higher mechanical properties (up to a certain limit) [26].

Different types of mechanical loads (e.g., tensile, compression, and torsion) may be applied to cellular structures. The elastic modulus, yield stress, yield strain, and energy absorption are the most common mechanical properties that are often reported. As for the compression tests, the typical strain-stress curves of most AM porous materials demonstrate three stages including a linear elastic region followed by a plateau stage where the stress levels stay constant while the strain increases. Finally, in the densification stage, the stress starts to increase exponentially, as the self-contact of the struts creating the pores causes the stress level to increase. The plateau stress and energy absorption, in particular, can be highly affected by the type of the unit cell used in the design of the lattice structure as well as the material types (due to highly different ductilities of different materials) [233,239].

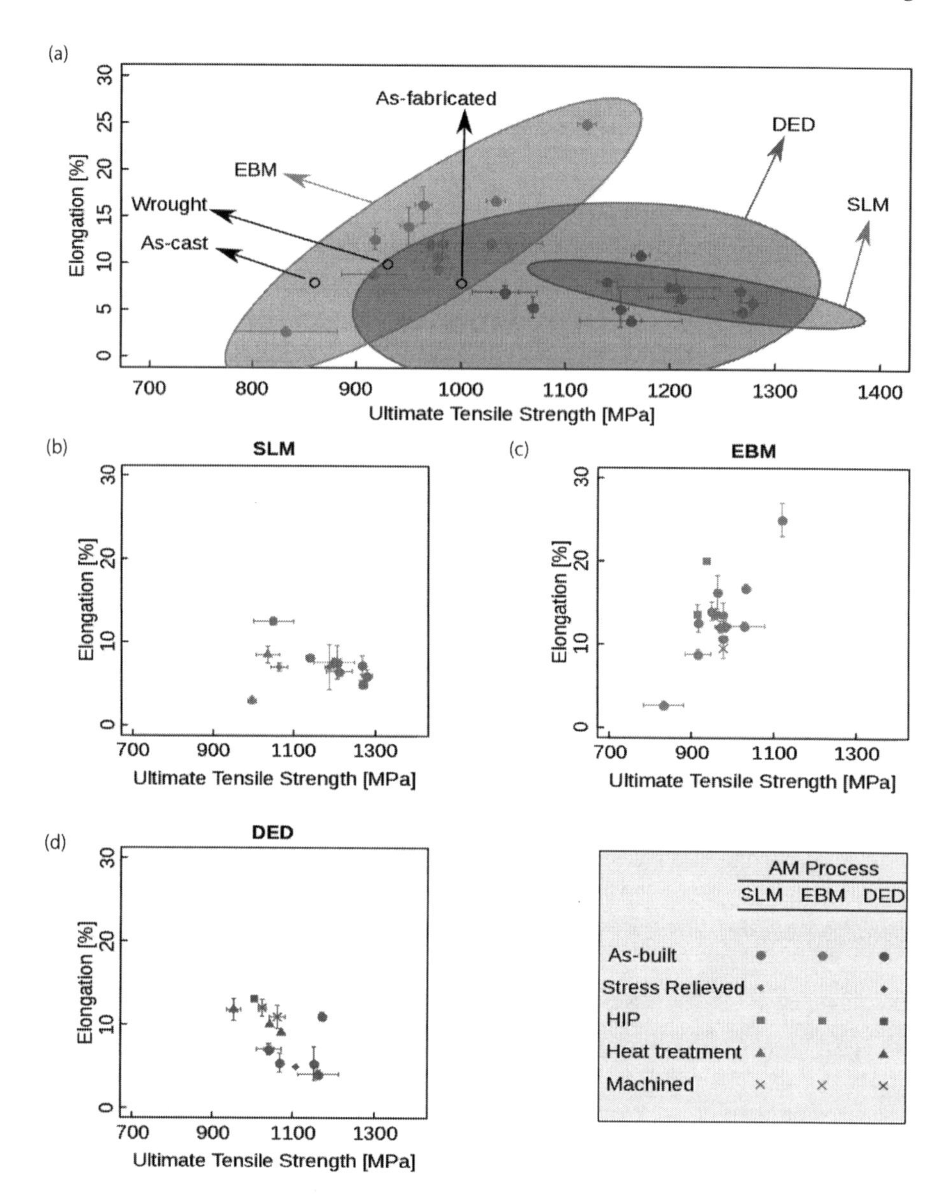

FIGURE 4.5 Elongation vs. ultimate tensile strength for AM Ti-6Al-4V. (a) as-built, (b) comparison of as-built and post-processed SLM, (c) EBM, and (d) DED. As-built data were collected from Rafi et al. (2013), Vilaro et al. (2011), Simonelli et al. (2014), Cain et al. (2015), Facchini et al. (2010), Hollander et al. (2006), and Vrancken et al. (2012) for SLM, from Hrabe and Quinn (2013), Tan et al. (2015), Ackelid and Svensson (2009), Edwards et al. (2013), Mohammadhosseini et al. (2013), Murr et al. (2009), Rafi et al. (2012), Rodriguez et al. (2015), and Wang et al. (2016) for EBM, and from Alcisto et al. (2011), Amsterdam and Kool (2009), Dinda et al. (2008), and Keicher et al. (1998) for DED. *(Continued)*

In the case of AM porous biomaterials used in orthopedic implants, the mechanical properties of the lattice structure should be close to those of the bone it is connected to [240,241] (Figure 4.6c), to prevent the stress shielding. Rational design of the topology of the unit cells could help in adjusting the mechanical properties of the resulting lattice structures and meeting the above-mentioned criterion [220,230,242]. It is important to realize that the initial values of the mechanical properties of AM porous biomaterials may change over time, as the bone tissue grows into the porous structure. Indeed, the effects of bone tissue in-growth on the mechanical properties of AM porous biomaterials have been investigated using polymeric fillers with elastic moduli close to those of the bone (i.e., 0.7 ~ 1.5 GPa) [26]. The study found that the quasi-static mechanical properties and fatigue resistance of AM porous lattice structures may experience a multi-fold increase, as the bone tissue grows and fills the pores of these structures.

In addition to the role of topological design, microstructural characteristics such as grain size, anisotropy, and the distribution of a second phase can highly affect the yield stress and strength of AM porous materials [21,22,122]. In general, smaller grain sizes and uniform distributions of the second phase can increase the yield strength of AM lattice structures [239]. The uniform distribution of the second phase can effectively prevent the dislocation movements, thereby limiting the plasticity of AM lattice structures [243].

The type of the unit cell on which a lattice structure is based can influence the failure mechanism of AM cellular (bio-)materials under monotonic mechanical testing. In general, different unit cell designs can be categorized as being either stretch-dominated (e.g., cubic) or bending-dominated (e.g., diamond). The failure mechanisms in stretch-dominated structures have been found to be primarily layer-by-layer, while bending-dominated structures are found to fail following the formation of 45° shearing bands [244].

Computational [74,245] and analytical models [236,246–248] have been developed recently to further study the effects of the topological design and the imperfections created during AM processes on the mechanical properties of porous AM materials. These studies generally find that both topological design and manufacturing imperfection can strongly affect the mechanical properties of AM materials. The effects of topological design and manufacturing imperfections on the spatial distribution of micro-strains during mechanical testing of lattice structures can be studied experimentally in 3D using high-resolution microscopic digital image

FIGURE 4.5 (Continued) (a) Data were collected from Simonelli et al. (2014) and Cain et al. (2015) for stress-relieved SLM materials, from Rekedal and Liu (2015), Mower and Long (2016), and Qiu et al. (2013) for HIP materials, from Vilaro et al. (2011) and Cain et al. (2015), for heat-treated materials, and from Rafi et al. (2013) and Vilaro et al. (2011) for machined materials. (b) The collected data for HIP EBM materials are from Ackelid and Svensson (2009), Mohammadhosseini et al. (2013), and Christensen et al. (1998) and while the data for machined materials originate from Rafi et al. (2013) (c). Post-processed DED data were adopted from Kobryn and Semiatin (2001) for stress-relieved and HIP, from Alcisto et al. (2011), Zhai et al. (2016) for heat-treated, and from Carroll et al. (2015), Qiu et al. (2013), for machined processed. Data were collected from the studies where the specimens have been fabricated horizontally. The ellipses in (a) show 95% confident interval. The data pertaining to as-cast International (UNS R56400) (2014), wrought International (UNS R56406) (2014), and as-fabricated Donachie (2000) conditions are presented for comparison.

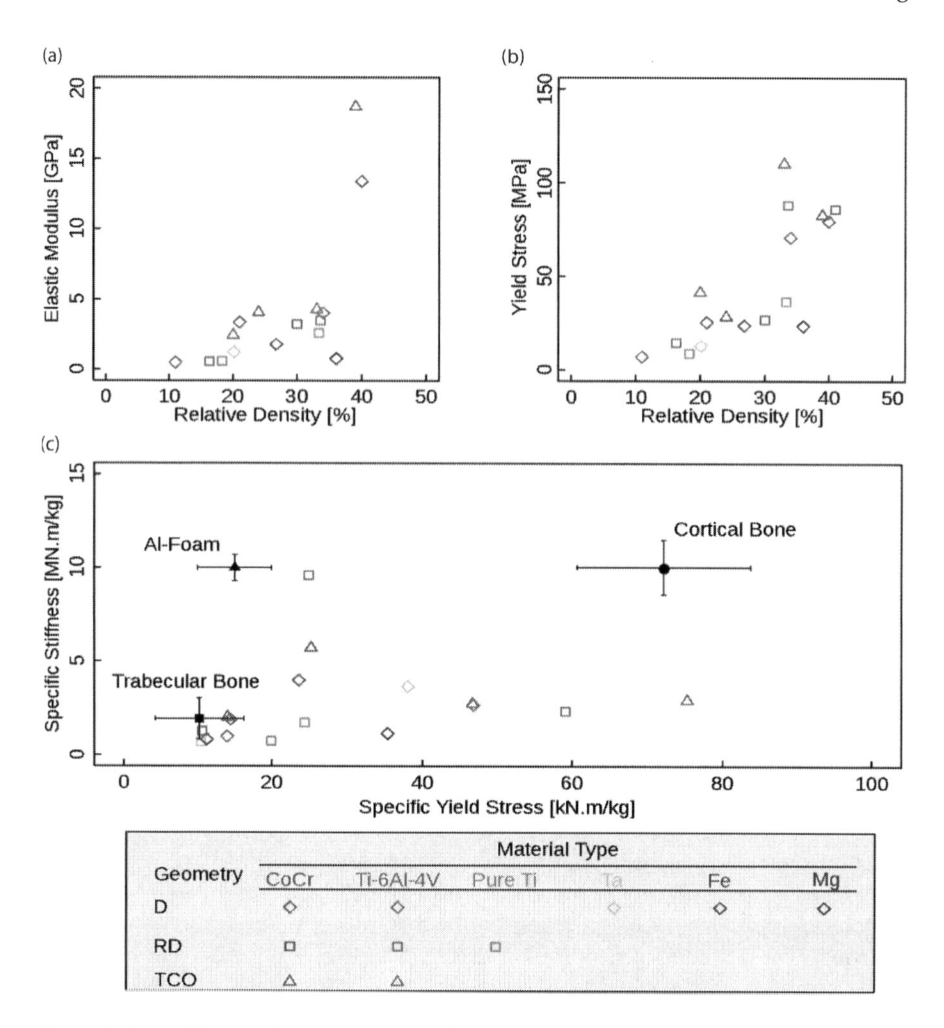

FIGURE 4.6 The elastic properties (elastic modulus (a) and compressive yield stress (b)) of AM porous materials with respect to their relative densities. The data were collected for CoCr Hedayati et al. (2018), Ti-6Al-4V Ahmadi et al. (2014), pure titanium (Ti) Wauthle et al. (2015), tantalum (Ta) Wauthle et al. (2015), iron (Fe) Li (2018), and magnesium (Mg) Li et al. (2018). Wherever possible, different unit cells types such as diamond (D), rhombic dodecahedron (RD), and truncated cuboctahedron (TCO) are included. The specific properties (i.e., the ratio of the elastic properties to the density of porous structures) are compared with that of natural materials [e.g., cortical Carter and Spengler (1978), Mirzaali et al. (2015, 2016) and trabecular Goldstein (1987), Mirzaali et al. (2017, 2018) bone] and aluminum foams Andrews et al. (1999), Miyoshi et al. (2000), Mirzaali Mazandarani et al. (2016) (c).

correlation (DIC) (Figure 4.9b) [249]. The results of such studies have indicated that the presence of manufacturing imperfections can lead to early nucleation and localization of strains and later be the basis for the specimen failure [249]. This type of 3D full-field strain measurements could shed light on the failure mechanisms occurring at the micro-scale and can be also used for the validation of micro-finite element analyses.

4.9 FATIGUE BEHAVIOR

Fatigue testing has been widely employed for the characterization of AM materials under cyclic loading. The ASTM standard E466 defines the orientation and dimensions of the fatigue crack growth and fracture tests [250]. The applied cyclic loading can be described by the minimum stress, maximum stress, stress ratio (i.e., the ratio of the maximum stress to the minimum stress), and the loading frequency. The main outputs of such tests are the fatigue strength and fatigue life cycle curves (S-N curves). The fatigue testings of AM parts have been usually conducted under load-controlled conditions with constant amplitudes and different stress ratios (mostly –1 and 0.1) [215,243,251–254].

Due to the different microstructures resulting from different AM processes, the fatigue properties of AM parts can be significantly different. For example, the formation of unstable martensitic α' can decrease the fatigue life of Ti-6Al-4V-fabricated by LENS [146].

The surface roughness and residual stresses can dominate the fatigue behavior of AM materials [255]. Furthermore, process-related defects such as unmolten particles on the surface of AM materials can significantly reduce the fatigue life [255,256]. Such irregularities can act as stress concentration points and result in early failures. Surface machining, post-AM heat treatments (e.g., HIP), and process optimization can improve the fatigue behavior of AM materials [206,257,258] by eliminating such process-related defects (Figure 4.7).

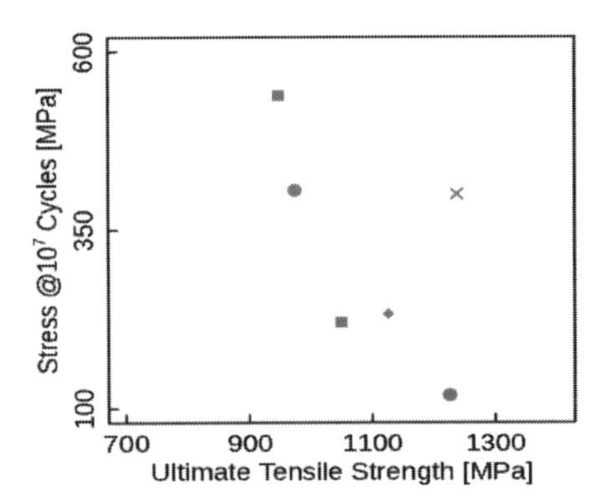

	AM Process	
	SLM	EBM
As-built	●	●
Stress Relieved	◆	
HIP	■	■
Machined	×	

FIGURE 4.7 The endurance limit at 10^7 cycles vs. ultimate tensile strength of Ti-6Al-4V processed with SLM and EBM. The corresponding studies are: SLM (as-built) Baca et al. (2015), HIP Rekedal (2015), stress-relieved Rekedal and Liu (2015), machined Mower and Long (2016), and EBM (as-built) and HIP EBM Frazier (2014). Ultimate tensile strength is calculated based on the mean data presented in Figure 4.5.

4.9.1 Solid Specimens

Process-related defects can act as stress risers or stress concentration points, thereby stimulating the initiation of cracks originating from the interior or the sub-surface of AM materials. The fatigue performance of parts fabricated by different AM processes differs from each other. For example, the fatigue resistance of SLM Ti-6Al-4V is better than that of EBM Ti-6Al-4V [28]. This can be due to the formation of different grain microstructures and different thermal cycles experienced in these processes. Moreover, the fatigue life of AM bulk materials depends on the specifications of the stresses applied in the fatigue experiments [257]. All these process-related parameters, various microstructural formations, and induced imperfections during different AM processes have resulted in a broad spectrum of fatigue life data specifically for AM Ti-6Al-4V [259]. One should therefore be careful when interpreting the results presented in Figure 4.7 as the SLM and EBM data presented here are not directly comparable, and more significant variations in the fatigue lives of these materials are expected.

4.9.2 Lattice and Architected Materials

Because of the application of AM porous biomaterials in the fabrication of bone-mimicking orthopedic implants, the fatigue behavior of these materials has been primarily studied under compression-compression high cycle and low cycle fatigue loadings [252,254,260]. During the cyclic compression loading, some of the struts of AM porous (bio-)materials can undergo bending or stretching depending on the topological design of the lattice structure. This means tensile stresses may be developed inside the porous structure. These tensile stresses are the main source of the microcracks opening and propagating in these cellular structures as a consequence of cyclic loading.

The S-N curve of AM lattice structure has been shown to be dependent on the geometry of the lattice structure (e.g., lattices with triply periodic minimal surfaces [125]), on the applied stress ratios [261], and on the material type [243]. The fatigue resistance of AM porous materials increases with the relative density [254,260]. The fatigue life of these materials is also different in the high cycle and low cycle regions [243], and is affected by the plastic behavior of AM material that influences crack tip plasticity and, thus, crack initiation and propagation mechanisms [243].

A comparison between the endurance limits of different AM porous materials made using different topological designs and material types (e.g., CoCr and Ti-6Al-4V), natural materials (e.g., cortical bone), and aluminum foams are presented in Figure 4.8. Although the fatigue lives of AM porous structures are within the range of the fatigue lives observed for natural and synthetic materials, it is clear that the topological design and material type significantly influence the fatigue life of these materials (Figure 4.8).

Another approach that has been used before is to calculate the normalized S-N curves of lattice structures. In this approach, the S-N curve of each lattice structure is normalized with respect to its yield or plateau stress [243,254,260]. Using this

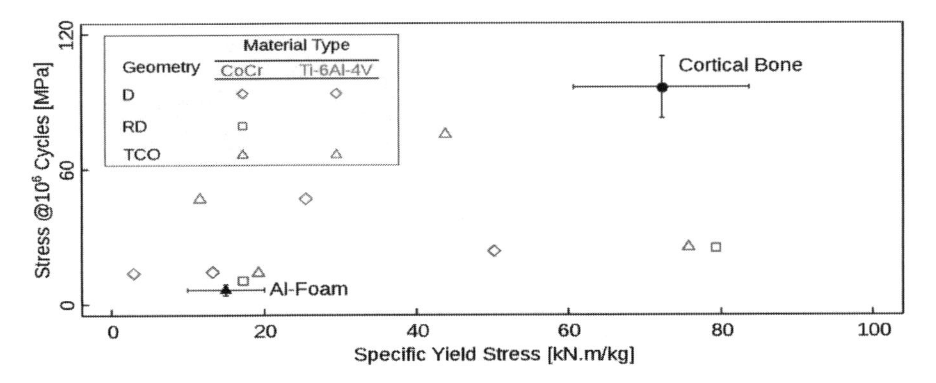

FIGURE 4.8 The endurance limit at 10^6 cycles vs. specific compressive yield stress of SLM-processed Ti-6Al-4V Yavari et al. (2015) and CoCr Ahmadi et al. (2018). The specific yield stress is the ratio of the yield stresses obtained from monotonic compressive loading of the AM porous materials to their individual densities. The stresses at 10^6 cycles are compared to those of cortical bone and aluminum foam at 10^7 cycles.

approach enables one to isolate the effects of quasi-static mechanical properties from the specific dynamic effects that influence the fatigue life of the lattice structure. Interestingly, it has been shown that the material type may play a more important role in determining the fatigue life of lattice structures than topological design [243]. Moreover, for the same type of the unit cell and material, the normalized S-N data of different lattice structures have been found to largely overlap with each other irrespective of their porosities [254,260]. This is important because separate experiments may not be required to determine the S-N curve for each porosity. Due to the expensive and time-consuming nature of fatigue tests, computational approaches have also been developed to predict the fatigue life [262] and fatigue crack propagation of AM materials [263,264] (Figure 4.9a).

To visualize the damage accumulation and gradual deformation in the porous structures under cyclic fatigue loading, optical deformation-tracking techniques such as digital image correlation can be used (Figure 4.9c). This technique has been used before to investigate the effects of stress ratio on the local deformation of SLM lattice structures made from Ti-6Al-4V based on the diamond unit cell [261]. Such approaches can be useful for understanding the mechanisms deriving the fatigue behavior of AM lattice structures and for validating the computational models describing and predicting the fatigue behavior of such materials.

As previously discussed, post-treatment processes can significantly increase the fatigue life of AM materials including those of AM lattice structures. For example, a recent comprehensive study has shown that HIP and surface treatments such as sand-blasting can be used independently and in combination with each other to increase the fatigue resistance of SLM Ti-6Al-4V limit [227]. In this case, HIP can close the micro-porosities present inside the specimens and transform the microstructure of

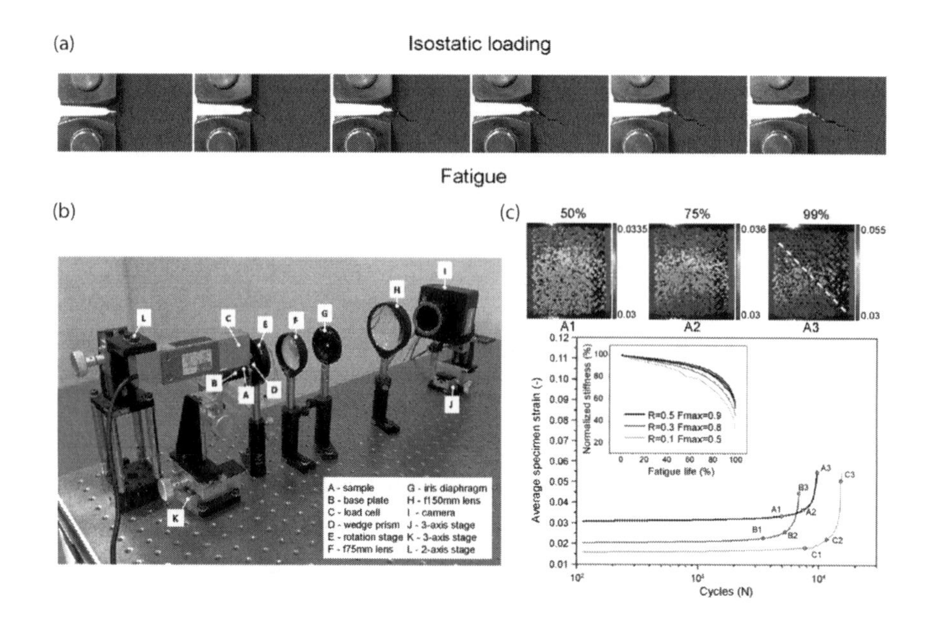

FIGURE 4.9 Isostatic loading and fatigue. (a) Crack propagation in an AM cellular structure (Reprinted from *Int. J. Fatigue*, 113, Hedayati, R. et al., Multiscale modeling of fatigue crack propagation in additively manufactured porous biomaterials, 416–427, Copyright 2018, with permission from Elsevier.), (b) the optical setup used for high-magnification (i.e., microscopic) full-field strain measurement during mechanical testing of AM lattice structures (Reprinted from *J. Mech. Behav. Biomed. Mater.*, 69, Genovese, K. et al., Microscopic full-field three-dimensional strain measurement during the mechanical testing of additively manufactured porous biomaterials, 69, 327–341, Copyright 2017, with permission from Elsevier.), and (c) strain distribution during fatigue tests at equal amplitudes and different mean stresses (0.675 σ_y (A), 0.52 σ_y (B), 0.275 σ_y (C)) (Reprinted from *J. Mech. Behav. Biomed. Mater.*, 70, de Krijger, J. et al., Effects of applied stress ratio on the fatigue behavior of additively manufactured porous biomaterials under compressive loading, 7–16, Copyright 2017, with permission from Elsevier.)

Ti-6Al-4V to a more ductile combination of α and β phases, while sandblasting can smoothen the surface of the specimens and induce more compressive stresses on the surface of the struts of the lattice structure.

4.10 CONCLUDING REMARKS AND FUTURE DIRECTIONS

Within the past decades, metal AM processes have received tremendous attention from both academic researchers and industry. Nowadays, AM techniques are mature enough to make functional materials with reliable, reproducible, and predictable properties thanks to the efforts of researchers from different disciplines including materials, mechanical, chemical, aeronautical, and biomedical engineers and scientists. In this book chapter, we addressed some of the critical aspects of powder-based AM metallic materials such as the effects of processing parameters and material types on the quality and properties of the final products.

The material type, geometrical accuracy, desired properties, final (surface) qualities, post-processing, and lead time are some of the important factors that are considered as crucial criteria for selection of the most appropriate AM process for each application. In order to achieve successful, consistent, and reproducible results, it is suggested to report the processing parameters including the build direction, scanning speed, scanning patterns, laser/beam power, and laser scanning speed, as each of these parameters can highly affect not only the quality, but also the mechanical properties of AM parts.

Despite the significant achievements in AM techniques, there are still some limitations that are summarized below. Future research should address these aspects to make powder-based metal AM more widely applicable.

One of the major limitations of powder-based metallic AM processes is the smallest feature size that can be reliably fabricated using these techniques. Since the size of the features that can be fabricated is dependent on the size of the powder particles and the flowability of the particles is dependent on their size as well, achieving finer resolutions may require complete re-thinking of AM processes. The second limitation concerns multi-material AM that is currently very challenging using metal AM processes. Another important direction is hybrid manufacturing where AM technologies can be combined with subtractive and formative manufacturing processes to improve both the cost-effectiveness and quality of the final products while being able to fabricate parts with more complex geometries. Such a hybrid processing approach can affect the lead time, surface quality, cost, tooling life, and geometric fidelity of AM parts.

The materials discussed here are some of the most widely processed metals with powder-based AM. Many more materials are being researched to be used for fabrication of AM parts. One of the main aims of future research should be a further expansion of the materials available for AM to meet all industrial and research needs and to achieve higher (mechanical) performance and advanced functionalities. This is an open and active area of research that is rightly receiving much attention and is expected to be one of the most fertile areas of research in metal AM.

Another important limitation is the need for costly and time-consuming processes of parameter optimization. Development of advanced process monitoring techniques [73,265] and computational tools [1,73] to facilitate (on-the-fly) process optimization can shorten the manufacturing time while enhancing the geometrical accuracy, quality, and performance of the final products.

Post-treatment processes can eliminate much of process-related defects such as microporosity, unmolten particles, and manufacturing irregularities. These defects can highly affect the surface quality of AM parts and consequently influence their quasi-static mechanical properties and fatigue life. The heat treatment processes can also relieve the residual stresses induced during AM processes. Development of protocols for post-AM treatments including HIP, polishing, shot peening, machining, and electrochemical treatments should be another priority in future research. Special attention should be paid to the effects of material type on the specifications of the protocols, as different materials respond differently to these treatments.

Finally, the design freedom offered by AM processes provides this exciting opportunity to create designer and custom-based materials. This opportunity should be exploited to the maximum possible extent to create advanced functional materials with unprecedented properties and novel functionalities.

SUGGESTED QUESTIONS FOR INSTRUCTORS

1. Which metal AM technologies use powder materials?
2. What are the main differences, applications, and limitations of SLM, EBM, and DED techniques?
3. What is a hybrid manufacturing process, and how can it improve upon AM processes?
4. What are the most important processing parameters affecting the quality of final AM products?
5. What are the required powder characteristics in AM processes? How can these powder characteristics be measured?
6. What are the feature and morphological differences of the powder materials produced by water-jet, gas, and plasma atomization techniques?
7. How can AM techniques be used for the fabrication of custom-made parts?
8. How can the surface roughness affect the mechanical properties of AM parts?
9. What are the essential post-treatment processes for each of the AM processes covered in this chapter, and how can these processes influence the mechanical properties of AM parts?
10. Why fatigue properties of AM processed parts are important?
11. What can be done to improve the fatigue response of AM processed parts?

SUGGESTED FURTHER READINGS

Metal AM Processes

1. A. A. Zadpoor, "Frontiers of additively manufactured metallic materials." *Materials*, vol. 11, p. 1566, 2018.
2. P. K. Gokuldoss, S. Kolla, and J. Eckert, "Additive manufacturing processes: Selective laser melting, electron beam melting and binder jetting—Selection guidelines," *Materials*, vol. 10, no. 6, p. 672, 2017.
3. W. E. Frazier, "Metal additive manufacturing: A review," *Journal of Materials Engineering and Performance*, vol. 23, no. 6, pp. 1917–1928, 2014.
4. D. Herzog, V. Seyda, E. Wycisk, and C. Emmelmann, "Additive manufacturing of metals," *Acta Materialia*, vol. 117, pp. 371–392, 2016.
5. L. E. Murr, S. M. Gaytan, D. A. Ramirez, E. Martinez, J. Hernandez, K. N. Amato, P. W. Shindo, F. R. Medina, and R. B. Wicker, "Metal fabrication by additive manufacturing using laser and electron beam melting technologies," *Journal of Materials Science & Technology*, vol. 28, no. 1, pp. 1–14, 2012.
6. C. Selcuk, "Laser metal deposition for powder metallurgy parts," *Powder Metallurgy*, vol. 54, no. 2, pp. 94–99, 2011.
7. N. Guo and M. C. Leu, "Additive manufacturing: Technology, applications and research needs," *Frontiers of Mechanical Engineering*, vol. 8, no. 3, pp. 215–243, 2013.

Effects of AM Processing Parameters

8. S. Ahmadi, R. Hedayati, R. A. K. Jain, Y. Li, S. Leeflang, and A. Zadpoor, "Effects of laser processing parameters on the mechanical properties, topology, and microstructure of additively manufactured porous metallic biomaterials: A vector-based approach," *Materials & Design*, vol. 134, pp. 234–243, 2017.
9. T. DebRoy, H. Wei, J. Zuback, T. Mukherjee, J. Elmer, J. Milewski, A. Beese, A. Wilson-Heid, A. De, and W. Zhang, "Additive manufacturing of metallic components–process, structure and properties," *Progress in Materials Science*, 2017.
10. J. Beuth, J. Fox, J. Gockel, C. Montgomery, R. Yang, H. Qiao, E. Soylemez et al., "Process mapping for qualification across multiple direct metal additive manufacturing processes," in *Proceedings of SFF Symposium*, Austin, TX, August, 2013, pp. 12–14.
11. A. Strondl, O. Lyckfeldt, H. Brodin, and U. Ackelid, "Characterization and control of powder properties for additive manufacturing," *JOM*, vol. 67, no. 3, pp. 549–554, 2015.
12. J. Karlsson, A. Snis, H. Engqvist, and J. Lausmaa, "Characterization and comparison of materials produced by Electron Beam Melting (EBM) of two different Ti–6Al–4V powder fractions," *Journal of Materials Processing Technology*, vol. 213, no. 12, pp. 2109–2118, 2013.
13. J. A. Slotwinski, E. J. Garboczi, P. E. Stutzman, C. F. Ferraris, S. S. Watson, and M. A. Peltz, "Characterization of metal powders used for additive manufacturing," *Journal of Research of the National Institute of Standards and Technology*, vol. 119, p. 460, 2014.
14. R. J. Hebert, "Metallurgical aspects of powder bed metal additive manufacturing," *Journal of Materials Science*, vol. 51, no. 3, pp. 1165–1175, 2016.
15. L. Murr, "Metallurgy of additive manufacturing: Examples from electron beam melting," *Additive Manufacturing*, vol. 5, pp. 40–53, 2015.

Design for Metal AM

16. A. Zadpoor, "Design for additive bio-manufacturing: From patient-specific medical devices to rationally designed meta-biomaterials," *International Journal of Molecular Sciences*, vol. 18, no. 8, p. 1607, 2017.
17. A. A. Zadpoor, "Mechanical meta-materials," *Materials Horizons*, vol. 3, no. 5, pp. 371–381, 2016.
18. D. Brackett, I. Ashcroft, and R. Hague, "Topology optimization for additive manufacturing," in *Proceedings of the Solid Freeform Fabrication Symposium*, Austin, TX, 2011, vol. 1, pp. 348–362.
19. Y. Saadlaoui, J.-L. Milan, J.-M. Rossi, and P. Chabrand, "Topology optimization and additive manufacturing: Comparison of conception methods using industrial codes," *Journal of Manufacturing Systems*, vol. 43, pp. 178–186, 2017.

20. X. Wang, S. Xu, S. Zhou, W. Xu, M. Leary, P. Choong, M. Qian, M. Brandt, and Y. M. Xie, "Topological design and additive manufacturing of porous metals for bone scaffolds and orthopaedic implants: A review," *Biomaterials*, vol. 83, pp. 127–141, 2016.
21. H. Bikas, P. Stavropoulos, and G. Chryssolouris, "Additive manufacturing methods and modelling approaches: A critical review," *The International Journal of Advanced Manufacturing Technology*, vol. 83, no. 1–4, pp. 389–405, 2016.

(Bio)Materials for AM

22. A. A. Zadpoor and J. Malda, "Additive manufacturing of biomaterials, tissues, and organs." *Annals of Biomedical Engineering*, vol. 45, pp. 1–11, 2017.
23. Y. Li, J. Zhou, P. Pavanram, M. Leeflang, L. Fockaert, B. Pouran, N. Tümer et al., "Additively manufactured biodegradable porous magnesium," *Acta Biomaterialia*, vol. 67, pp. 378–392, 2018.
24. Y. Li, H. Jahr, K. Lietaert, P. Pavanram, A. Yilmaz, L. Fockaert, M. Leeflang et al., "Additively manufactured biodegradable porous iron," *Acta Biomaterialia*, vol. 77, pp. 380–393, 2018.
25. R. Wauthle, J. Van Der Stok, S. A. Yavari, J. Van Humbeeck, J.-P. Kruth, A. A. Zadpoor, H. Weinans, M. Mulier, and J. Schrooten, "Additively manufactured porous tantalum implants," *Acta Biomaterialia*, vol. 14, pp. 217–225, 2015.
26. S. Bose, D. Ke, H. Sahasrabudhe, and A. Bandyopadhyay, "Additive manufacturing of biomaterials," *Progress in Materials Science*, vol. 93, pp. 45–111, 2018.
27. D. Bourell, J. P. Kruth, M. Leu, G. Levy, D. Rosen, A. M. Beese, and A. Clare, "Materials for additive manufacturing," *CIRP Annals*, vol. 66, no. 2, pp. 659–681, 2017.
28. F. Froes and B. Dutta, "The additive manufacturing (AM) of titanium alloys," in *Advanced Materials Research*, vol. 1019, pp. 19–25, 2014.

AM Post-treatment Processes and (Bio-)Functionalization

29. S. Amin Yavari, L. Loozen, F. L. Paganelli, S. Bakhshandeh, K. Lietaert, J. A. Groot, A. C. Fluit et al., "Antibacterial behavior of additively manufactured porous titanium with nanotubular surfaces releasing silver ions," *ACS Applied Materials & Interfaces*, vol. 8, no. 27, pp. 17080–17089, 2016.
30. S. Ahmadi, R. Kumar, E. Borisov, R. Petrov, S. Leeflang, Y. Li, N. Tümer et al., "From microstructural design to surface engineering: A tailored approach for improving fatigue life of additively manufactured meta-biomaterials," *Acta Biomaterialia*, vol. 83, pp. 153–166, 2019.

Mechanical Properties of AM Parts

31. A. A. Zadpoor, "Mechanics of additively manufactured biomaterials." *Journal of the Mechanical Behavior of Biomedical Materials*, vol. 70, pp. 1-6, 2017.

32. S. A. Yavari, R. Wauthlé, J. van der Stok, A. Riemslag, M. Janssen, M. Mulier, J.-P. Kruth, J. Schrooten, H. Weinans, and A. A. Zadpoor, "Fatigue behavior of porous biomaterials manufactured using selective laser melting," *Materials Science and Engineering: C*, vol. 33, no. 8, pp. 4849–4858, 2013.
33. G. Campoli, M. Borleffs, S. A. Yavari, R. Wauthle, H. Weinans, and A. A. Zadpoor, "Mechanical properties of open-cell metallic biomaterials manufactured using additive manufacturing," *Materials & Design*, vol. 49, pp. 957–965, 2013.
34. R. Hedayati, S. Ahmadi, K. Lietaert, B. Pouran, Y. Li, H. Weinans, C. Rans, and A. Zadpoor, "Isolated and modulated effects of topology and material type on the mechanical properties of additively manufactured porous biomaterials," *Journal of the Mechanical Behavior of Biomedical Materials*, vol. 79, pp. 254–263, 2018.
35. J. J. Lewandowski and M. Seifi, "Metal additive manufacturing: A review of mechanical properties," *Annual Review of Materials Research*, vol. 46, pp. 151–186, 2016.
36. H. Rafi, N. Karthik, H. Gong, T. L. Starr, and B. E. Stucker, "Microstructures and mechanical properties of Ti6Al4V parts fabricated by selective laser melting and electron beam melting," *Journal of Materials Engineering and Performance*, vol. 22, no. 12, pp. 3872–3883, 2013.

REFERENCES

1. W. Gao, Y. Zhang, D. Ramanujan, K. Ramani, Y. Chen, C. B. Williams, C. C. Wang, Y. C. Shin, S. Zhang, and P. D. Zavattieri, "The status, challenges, and future of additive manufacturing in engineering," *Computer-Aided Design*, vol. 69, pp. 65–89, 2015.
2. A. Zadpoor, "Frontiers of Additively Manufactured Metallic Materials." Multidisciplinary Digital Publishing Institute, 2018.
3. S. Bose, D. Ke, H. Sahasrabudhe, and A. Bandyopadhyay, "Additive manufacturing of biomaterials," *Progress in Materials Science*, vol. 93, pp. 45–111, 2018.
4. A. A. Zadpoor and J. Malda, "Additive manufacturing of biomaterials, tissues, and organs." Springer, vol. 45, pp. 1–11, 2017.
5. R. Huang, M. Riddle, D. Graziano, J. Warren, S. Das, S. Nimbalkar, J. Cresko, and E. Masanet, "Energy and emissions saving potential of additive manufacturing: The case of lightweight aircraft components," *Journal of Cleaner Production*, vol. 135, pp. 1559–1570, 2016.
6. L. E. Murr, "Frontiers of 3D printing/additive manufacturing: From human organs to aircraft fabrication," *Journal of Materials Science & Technology*, vol. 32, no. 10, pp. 987–995, 2016.
7. A. Uriondo, M. Esperon-Miguez, and S. Perinpanayagam, "The present and future of additive manufacturing in the aerospace sector: A review of important aspects," *Proceedings of the Institution of Mechanical Engineers, Part G: Journal of Aerospace Engineering*, vol. 229, no. 11, pp. 2132–2147, 2015.
8. J.-P. Kruth, M.-C. Leu, and T. Nakagawa, "Progress in additive manufacturing and rapid prototyping," *CIRP Annals*, vol. 47, no. 2, pp. 525–540, 1998.
9. M. K. Thompson, G. Moroni, T. Vaneker, G. Fadel, R. I. Campbell, I. Gibson, A. Bernard et al. "Design for Additive Manufacturing: Trends, opportunities, considerations, and constraints," *CIRP Annals*, vol. 65, no. 2, pp. 737–760, 2016.

10. T. Wohlers, T. Caffrey, R. I. Campbell, O. Diegel, and J. Kowen. "Wohlers Report 2018: 3D Printing and Additive Manufacturing State of the Industry;" Annual Worldwide Progress Report. *Wohlers Associates*, 2018.
11. P. K. Gokuldoss, S. Kolla, and J. Eckert, "Additive manufacturing processes: Selective laser melting, electron beam melting and binder jetting—Selection guidelines," *Materials*, vol. 10, no. 6, p. 672, 2017.
12. N. Hrabe and T. Quinn, "Effects of processing on microstructure and mechanical properties of a titanium alloy (Ti–6Al–4V) fabricated using electron beam melting (EBM), part 1: Distance from build plate and part size," *Materials Science and Engineering: A*, vol. 573, pp. 264–270, 2013.
13. S. L. Sing, J. An, W. Y. Yeong, and F. E. Wiria, "Laser and electron-beam powder-bed additive manufacturing of metallic implants: A review on processes, materials and designs," *Journal of Orthopaedic Research*, vol. 34, no. 3, pp. 369–385, 2016.
14. M. Annoni, H. Giberti, and M. Strano, "Feasibility study of an extrusion-based direct metal additive manufacturing technique," *Procedia Manufacturing*, vol. 5, pp. 916–927, 2016.
15. D. Ding, Z. Pan, D. Cuiuri, and H. Li, "Wire-feed additive manufacturing of metal components: Technologies, developments and future interests," *The International Journal of Advanced Manufacturing Technology*, vol. 81, no. 1–4, pp. 465–481, 2015.
16. B. N. Turner, R. Strong, and S. A. Gold, "A review of melt extrusion additive manufacturing processes: I. Process design and modeling," *Rapid Prototyping Journal*, vol. 20, no. 3, pp. 192–204, 2014.
17. B. Mueller, "Additive manufacturing technologies–Rapid prototyping to direct digital manufacturing," *Assembly Automation*, vol. 32, no. 2, 2012.
18. F. Froes and B. Dutta, "The additive manufacturing (AM) of titanium alloys," in *Advanced Materials Research*, 2014, vol. 1019, pp. 19–25.
19. A. Saboori, D. Gallo, S. Biamino, P. Fino, and M. Lombardi, "An overview of additive manufacturing of titanium components by directed energy deposition: Microstructure and mechanical properties," *Applied Sciences*, vol. 7, no. 9, p. 883, 2017.
20. N. Shamsaei, A. Yadollahi, L. Bian, and S. M. Thompson, "An overview of Direct Laser Deposition for additive manufacturing; Part II: Mechanical behavior, process parameter optimization and control," *Additive Manufacturing*, vol. 8, pp. 12–35, 2015.
21. W. E. Frazier, "Metal additive manufacturing: A review," *Journal of Materials Engineering and Performance*, vol. 23, no. 6, pp. 1917–1928, 2014.
22. D. Herzog, V. Seyda, E. Wycisk, and C. Emmelmann, "Additive manufacturing of metals," *Acta Materialia*, vol. 117, pp. 371–392, 2016.
23. B. P. Conner, G. P. Manogharan, A. N. Martof, L. M. Rodomsky, C. M. Rodomsky, D. C. Jordan, and J. W. Limperos, "Making sense of 3-D printing: Creating a map of additive manufacturing products and services," *Additive Manufacturing*, vol. 1, pp. 64–76, 2014.
24. S. Bremen, W. Meiners, and A. Diatlov, "Selective laser melting: A manufacturing technology for the future?," *Laser Technik Journal*, vol. 9, no. 2, pp. 33–38, 2012.
25. L. E. Murr, S. M. Gaytan, D. A. Ramirez, E. Martinez, J. Hernandez, K. N. Amato, P. W. Shindo, F. R. Medina, and R. B. Wicker, "Metal fabrication by additive manufacturing using laser and electron beam melting technologies," *Journal of Materials Science & Technology*, vol. 28, no. 1, pp. 1–14, 2012.
26. S. Ahmadi, R. Hedayati, R. A. K. Jain, Y. Li, S. Leeflang, and A. Zadpoor, "Effects of laser processing parameters on the mechanical properties, topology, and microstructure of additively manufactured porous metallic biomaterials: A vector-based approach," *Materials & Design*, vol. 134, pp. 234–243, 2017.
27. T. DebRoy, H. Wei, J. Zuback, T. Mukherjee, J. Elmer, J. Milewski, A. Beese, A. Wilson-Heid, A. De, and W. Zhang, "Additive manufacturing of metallic components–process, structure and properties," *Progress in Materials Science*, 2017.

28. H. Rafi, N. Karthik, H. Gong, T. L. Starr, and B. E. Stucker, "Microstructures and mechanical properties of Ti6Al4V parts fabricated by selective laser melting and electron beam melting," *Journal of Materials Engineering and Performance*, vol. 22, no. 12, pp. 3872–3883, 2013.
29. H. Y. Jung, S. J. Choi, K. G. Prashanth, M. Stoica, S. Scudino, S. Yi, U. Kühn, D. H. Kim, K. B. Kim, and J. Eckert, "Fabrication of Fe-based bulk metallic glass by selective laser melting: A parameter study," *Materials & Design*, vol. 86, pp. 703–708, 2015.
30. K. Prashanth and J. Eckert, "Formation of metastable cellular microstructures in selective laser melted alloys," *Journal of Alloys and Compounds*, vol. 707, pp. 27–34, 2017.
31. J. Suryawanshi, K. Prashanth, S. Scudino, J. Eckert, O. Prakash, and U. Ramamurty, "Simultaneous enhancements of strength and toughness in an Al-12Si alloy synthesized using selective laser melting," *Acta Materialia*, vol. 115, pp. 285–294, 2016.
32. D.-S. Shim, G.-Y. Baek, J.-S. Seo, G.-Y. Shin, K.-P. Kim, and K.-Y. Lee, "Effect of layer thickness setting on deposition characteristics in direct energy deposition (DED) process," *Optics & Laser Technology*, vol. 86, pp. 69–78, 2016.
33. F. Mazzucato, S. Tusacciu, M. Lai, S. Biamino, M. Lombardi, and A. Valente, "Monitoring approach to evaluate the performances of a new deposition nozzle solution for DED systems," *Technologies*, vol. 5, no. 2, p. 29, 2017.
34. B. Zheng, Y. Zhou, J. Smugeresky, J. Schoenung, and E. Lavernia, "Thermal behavior and microstructural evolution during laser deposition with laser-engineered net shaping: Part I. Numerical calculations," *Metallurgical and Materials Transactions A*, vol. 39, no. 9, pp. 2228–2236, 2008.
35. B. Zheng, Y. Zhou, J. Smugeresky, J. Schoenung, and E. Lavernia, "Thermal behavior and microstructure evolution during laser deposition with laser-engineered net shaping: Part II. Experimental investigation and discussion," *Metallurgical and Materials Transactions A*, vol. 39, no. 9, pp. 2237–2245, 2008.
36. M. Hedges and N. Calder, "Near net shape rapid manufacture & repair by LENS," in *Cost Effective Manufacture via Net-Shape Processing, Meeting Proceedings RTO-MP-AVT-139, Paper*, 2006, vol. 13.
37. C. Selcuk, "Laser metal deposition for powder metallurgy parts," *Powder Metallurgy*, vol. 54, no. 2, pp. 94–99, 2011.
38. S. Fafenrot, N. Grimmelsmann, M. Wortmann, and A. Ehrmann, "Three-dimensional (3D) printing of polymer-metal hybrid materials by fused deposition modeling," *Materials*, vol. 10, no. 10, p. 1199, 2017.
39. M. Merklein, D. Junker, A. Schaub, and F. Neubauer, "Hybrid additive manufacturing technologies–an analysis regarding potentials and applications," *Physics Procedia*, vol. 83, pp. 549–559, 2016.
40. X. Shi, S. Ma, C. Liu, Q. Wu, J. Lu, Y. Liu, and W. Shi, "Selective laser melting-wire arc additive manufacturing hybrid fabrication of Ti-6Al-4V alloy: Microstructure and mechanical properties," *Materials Science and Engineering: A*, vol. 684, pp. 196–204, 2017.
41. I. Campbell, D. Bourell, and I. Gibson, "Additive manufacturing: Rapid prototyping comes of age," *Rapid Prototyping Journal*, vol. 18, no. 4, pp. 255–258, 2012.
42. J. Karlsson, A. Snis, H. Engqvist, and J. Lausmaa, "Characterization and comparison of materials produced by Electron Beam Melting (EBM) of two different Ti–6Al–4V powder fractions," *Journal of Materials Processing Technology*, vol. 213, no. 12, pp. 2109–2118, 2013.
43. A. Strondl, O. Lyckfeldt, H. Brodin, and U. Ackelid, "Characterization and control of powder properties for additive manufacturing," *JOM*, vol. 67, no. 3, pp. 549–554, 2015.
44. X. Zhao, J. Chen, X. Lin, and W. Huang, "Study on microstructure and mechanical properties of laser rapid forming Inconel 718," *Materials Science and Engineering: A*, vol. 478, no. 1–2, pp. 119–124, 2008.

45. J. A. Slotwinski, E. J. Garboczi, P. E. Stutzman, C. F. Ferraris, S. S. Watson, and M. A. Peltz, "Characterization of metal powders used for additive manufacturing," *Journal of Research of the National Institute of Standards and Technology*, vol. 119, p. 460, 2014.
46. Y. Seki, S. Okamoto, H. Takigawa, and N. Kawai, "Effect of atomization variables on powder characteristics in the high-pressured water atomization process," *Metal Powder Report*, vol. 45, no. 1, pp. 38–40, 1990.
47. A. J. Pinkerton and L. Li, "Direct additive laser manufacturing using gas-and water-atomised H_{13} tool steel powders," *The International Journal of Advanced Manufacturing Technology*, vol. 25, no. 5–6, pp. 471–479, 2005.
48. I. Anderson, R. Figliola, and H. Morton, "Flow mechanisms in high pressure gas atomization," *Materials Science and Engineering: A*, vol. 148, no. 1, pp. 101–114, 1991.
49. R. J. Hebert, "Metallurgical aspects of powder bed metal additive manufacturing," *Journal of Materials Science*, vol. 51, no. 3, pp. 1165–1175, 2016.
50. L. E. Murr, E. Martinez, J. Hernandez, S. Collins, K. N. Amato, S. M. Gaytan, and P. W. Shindo, "Microstructures and properties of 17-4 PH stainless steel fabricated by selective laser melting," *Journal of Materials Research and Technology*, vol. 1, no. 3, pp. 167–177, 2012.
51. T. L. Starr, K. Rafi, B. Stucker, and C. M. Scherzer, "Controlling phase composition in selective laser melted stainless steels," *Power (W)*, vol. 195, p. 195, 2012.
52. H. Qi, M. Azer, and A. Ritter, "Studies of standard heat treatment effects on microstructure and mechanical properties of laser net shape manufactured Inconel 718," *Metallurgical and Materials Transactions A*, vol. 40, no. 10, pp. 2410–2422, 2009.
53. B. Champagne and R. Angers, "REP(Rotating Electrode Process) atomization mechanisms," *International Journal of Powder Metallurgy*, vol. 16, no. 3, pp. 125–128, 1984.
54. A. Ozols, H. Sirkin, and E. Vicente, "Segregation in Stellite powders produced by the plasma rotating electrode process," *Materials Science and Engineering: A*, vol. 262, no. 1–2, pp. 64–69, 1999.
55. M. I. Boulos, "New frontiers in thermal plasmas from space to nanomaterials," *Nuclear Engineering and Technology*, vol. 44, no. 1, pp. 1–8, 2012.
56. M. Entezarian, F. Allaire, P. Tsantrizos, and R. Drew, "Plasma atomization: A new process for the production of fine, spherical powders," *JOM*, vol. 48, no. 6, pp. 53–55, 1996.
57. C. Doblin, A. Chryss, and A. Monch, "Titanium powder from the TiRO™ process," in *Key Engineering Materials*, 2012, vol. 520, pp. 95–100.
58. M. Qian and F. H. Froes, *Titanium Powder Metallurgy: Science, Technology and Applications*. Butterworth-Heinemann, Waltham, MA, 2015.
59. B. Liu, R. Wildman, C. Tuck, I. Ashcroft, and R. Hague, "Investigation the effect of particle size distribution on processing parameters optimisation in selective laser melting process," *Additive Manufacturing Research Group, Loughborough University*, pp. 227–238, 2011.
60. A. Santomaso, P. Lazzaro, and P. Canu, "Powder flowability and density ratios: The impact of granules packing," *Chemical Engineering Science*, vol. 58, no. 13, pp. 2857–2874, 2003.
61. D. Bourell, J. P. Kruth, M. Leu, G. Levy, D. Rosen, A. M. Beese, and A. Clare, "Materials for additive manufacturing," *CIRP Annals*, vol. 66, no. 2, pp. 659–681, 2017.
62. R. Banerjee, P. Collins, A. Genc, and H. Fraser, "Direct laser deposition of in situ Ti–6Al–4V–TiB composites," *Materials Science and Engineering: A*, vol. 358, no. 1–2, pp. 343–349, 2003.
63. M. Fischer, D. Joguet, G. Robin, L. Peltier, and P. Laheurte, "In situ elaboration of a binary Ti–26Nb alloy by selective laser melting of elemental titanium and niobium mixed powders," *Materials Science and Engineering: C*, vol. 62, pp. 852–859, 2016.

64. P. Vora, K. Mumtaz, I. Todd, and N. Hopkinson, "AlSi12 in-situ alloy formation and residual stress reduction using anchorless selective laser melting," *Additive Manufacturing*, vol. 7, pp. 12–19, 2015.
65. J. Beuth, J. Fox, J. Gockel, C. Montgomery, R. Yang, H. Qiao, E. Soylemez et al., "Process mapping for qualification across multiple direct metal additive manufacturing processes," in *Proceedings of SFF Symposium*, Austin, TX, Aug, 2013, pp. 12–14.
66. J. Beuth and N. Klingbeil, "The role of process variables in laser-based direct metal solid freeform fabrication," *JOM*, vol. 53, no. 9, pp. 36–39, 2001.
67. J. Gockel and J. Beuth, "Understanding Ti-6Al-4V microstructure control in additive manufacturing via process maps," *Solid Freeform Fabrication Proceedings*, Austin, TX, Aug, pp. 12–14, 2013.
68. S. Lambrakos and K. Cooper, "An Algorithm for Inverse Modeling of Layer-by-Layer Deposition Processes," *Journal of Materials Engineering and Performance*, vol. 18, no. 3, pp. 221–230, 2009.
69. T. Vilaro, C. Colin, and J.-D. Bartout, "As-fabricated and heat-treated microstructures of the Ti-6Al-4V alloy processed by selective laser melting," *Metallurgical and Materials Transactions A*, vol. 42, no. 10, pp. 3190–3199, 2011.
70. A. B. Spierings and G. Levy, "Comparison of density of stainless steel 316L parts produced with selective laser melting using different powder grades," in *Proceedings of the Annual International Solid Freeform Fabrication Symposium*, 2009, pp. 342–353.
71. T. Brajlih, B. Valentan, J. Balic, and I. Drstvensek, "Speed and accuracy evaluation of additive manufacturing machines," *Rapid Prototyping Journal*, vol. 17, no. 1, pp. 64–75, 2011.
72. B. N. Turner and S. A. Gold, "A review of melt extrusion additive manufacturing processes: II. Materials, dimensional accuracy, and surface roughness," *Rapid Prototyping Journal*, vol. 21, no. 3, pp. 250–261, 2015.
73. H. Bikas, P. Stavropoulos, and G. Chryssolouris, "Additive manufacturing methods and modelling approaches: A critical review," *The International Journal of Advanced Manufacturing Technology*, vol. 83, no. 1–4, pp. 389–405, 2016.
74. G. Campoli, M. Borleffs, S. A. Yavari, R. Wauthle, H. Weinans, and A. A. Zadpoor, "Mechanical properties of open-cell metallic biomaterials manufactured using additive manufacturing," *Materials & Design*, vol. 49, pp. 957–965, 2013.
75. G. Strano, L. Hao, R. M. Everson, and K. E. Evans, "Surface roughness analysis, modelling and prediction in selective laser melting," *Journal of Materials Processing Technology*, vol. 213, no. 4, pp. 589–597, 2013.
76. K. Mumtaz and N. Hopkinson, "Top surface and side roughness of Inconel 625 parts processed using selective laser melting," *Rapid Prototyping Journal*, vol. 15, no. 2, pp. 96–103, 2009.
77. A. AB, "Case study: Additive manufacturing of aerospace brackets," *Advanced Materials & Processes*, p. 19, 2013.
78. E. Lyczkowska, P. Szymczyk, B. Dybala, and E. Chlebus, "Chemical polishing of scaffolds made of Ti–6Al–7Nb alloy by additive manufacturing," *Archives of Civil and Mechanical Engineering*, vol. 14, no. 4, pp. 586–594, 2014.
79. H. Tang, M. Qian, N. Liu, X. Zhang, G. Yang, and J. Wang, "Effect of powder reuse times on additive manufacturing of Ti-6Al-4V by selective electron beam melting," *JOM*, vol. 67, no. 3, pp. 555–563, 2015.
80. D. Gu and Y. Shen, "Balling phenomena in direct laser sintering of stainless steel powder: Metallurgical mechanisms and control methods," *Materials & Design*, vol. 30, no. 8, pp. 2903–2910, 2009.
81. J.-P. Kruth, L. Froyen, J. Van Vaerenbergh, P. Mercelis, M. Rombouts, and B. Lauwers, "Selective laser melting of iron-based powder," *Journal of Materials Processing Technology*, vol. 149, no. 1–3, pp. 616–622, 2004.

82. H. Niu and I. Chang, "Instability of scan tracks of selective laser sintering of high speed steel powder," *Scripta Materialia*, vol. 41, no. 11, pp. 1229–1234, 1999.
83. D. Buchbinder, W. Meiners, K. Wissenbach, and R. Poprawe, "Selective laser melting of aluminum die-cast alloy—Correlations between process parameters, solidification conditions, and resulting mechanical properties," *Journal of Laser Applications*, vol. 27, no. S2, p. S29205, 2015.
84. L. Thijs, K. Kempen, J.-P. Kruth, and J. Van Humbeeck, "Fine-structured aluminium products with controllable texture by selective laser melting of pre-alloyed AlSi10Mg powder," *Acta Materialia*, vol. 61, no. 5, pp. 1809–1819, 2013.
85. L. Murr, "Metallurgy of additive manufacturing: Examples from electron beam melting," *Additive Manufacturing*, vol. 5, pp. 40–53, 2015.
86. X. Tan, Y. Kok, Y. J. Tan, M. Descoins, D. Mangelinck, S. B. Tor, K. F. Leong, and C. K. Chua, "Graded microstructure and mechanical properties of additive manufactured Ti–6Al–4V via electron beam melting," *Acta Materialia*, vol. 97, pp. 1–16, 2015.
87. L. Thijs, F. Verhaeghe, T. Craeghs, J. Van Humbeeck, and J.-P. Kruth, "A study of the microstructural evolution during selective laser melting of Ti–6Al–4V," *Acta Materialia*, vol. 58, no. 9, pp. 3303–3312, 2010.
88. M. Seifi, A. Salem, J. Beuth, O. Harrysson, and J. J. Lewandowski, "Overview of materials qualification needs for metal additive manufacturing," *JOM*, vol. 68, no. 3, pp. 747–764, 2016.
89. J. Gockel, J. Beuth, and K. Taminger, "Integrated control of solidification microstructure and melt pool dimensions in electron beam wire feed additive manufacturing of Ti-6Al-4V," *Additive Manufacturing*, vol. 1, pp. 119–126, 2014.
90. A. R. Nassar, J. S. Keist, E. W. Reutzel, and T. J. Spurgeon, "Intra-layer closed-loop control of build plan during directed energy additive manufacturing of Ti–6Al–4V," *Additive Manufacturing*, vol. 6, pp. 39–52, 2015.
91. J. Alcisto, A. Enriquez, H. Garcia, S. Hinkson, T. Steelman, E. Silverman, P. Valdovino et al. "Tensile properties and microstructures of laser-formed Ti-6Al-4V," *Journal of Materials Engineering and Performance*, vol. 20, no. 2, pp. 203–212, 2011.
92. L. E. Murr, E. Martinez, S. Gaytan, D. Ramirez, B. Machado, P. Shindo, J. Martinez et al. "Microstructural architecture, microstructures, and mechanical properties for a nickel-base superalloy fabricated by electron beam melting," *Metallurgical and Materials Transactions A*, vol. 42, no. 11, pp. 3491–3508, 2011.
93. B. E. Carroll, T. A. Palmer, and A. M. Beese, "Anisotropic tensile behavior of Ti–6Al–4V components fabricated with directed energy deposition additive manufacturing," *Acta Materialia*, vol. 87, pp. 309–320, 2015.
94. C. Körner, H. Helmer, A. Bauereiß, and R. F. Singer, "Tailoring the grain structure of IN718 during selective electron beam melting," in *MATEC Web of Conferences*, 2014, vol. 14, p. 08001.
95. L. Thijs, M. L. M. Sistiaga, R. Wauthle, Q. Xie, J.-P. Kruth, and J. Van Humbeeck, "Strong morphological and crystallographic texture and resulting yield strength anisotropy in selective laser melted tantalum," *Acta Materialia*, vol. 61, no. 12, pp. 4657–4668, 2013.
96. S. Bontha, N. W. Klingbeil, P. A. Kobryn, and H. L. Fraser, "Thermal process maps for predicting solidification microstructure in laser fabrication of thin-wall structures," *Journal of Materials Processing Technology*, vol. 178, no. 1–3, pp. 135–142, 2006.
97. S. Bontha, N. W. Klingbeil, P. A. Kobryn, and H. L. Fraser, "Effects of process variables and size-scale on solidification microstructure in beam-based fabrication of bulky 3D structures," *Materials Science and Engineering: A*, vol. 513, pp. 311–318, 2009.
98. P. Collins, D. Brice, P. Samimi, I. Ghamarian, and H. Fraser, "Microstructural control of additively manufactured metallic materials," *Annual Review of Materials Research*, vol. 46, pp. 63–91, 2016.

99. R. Wauthle, B. Vrancken, B. Beynaerts, K. Jorissen, J. Schrooten, J.-P. Kruth, and J. Van Humbeeck, "Effects of build orientation and heat treatment on the microstructure and mechanical properties of selective laser melted Ti6Al4V lattice structures," *Additive Manufacturing*, vol. 5, pp. 77–84, 2015.
100. J.-P. Kruth, G. Levy, F. Klocke, and T. Childs, "Consolidation phenomena in laser and powder-bed based layered manufacturing," *CIRP Annals*, vol. 56, no. 2, pp. 730–759, 2007.
101. A. Zadpoor, "Design for additive bio-manufacturing: From patient-specific medical devices to rationally designed meta-biomaterials," *International Journal of Molecular Sciences*, vol. 18, no. 8, p. 1607, 2017.
102. A. A. Zadpoor, "Mechanical meta-materials," *Materials Horizons*, vol. 3, no. 5, pp. 371–381, 2016.
103. J. Berger, H. Wadley, and R. McMeeking, "Mechanical metamaterials at the theoretical limit of isotropic elastic stiffness," *Nature*, vol. 543, no. 7646, p. 533, 2017.
104. R. Hedayati, A. Leeflang, and A. Zadpoor, "Additively manufactured metallic pentamode meta-materials," *Applied Physics Letters*, vol. 110, no. 9, p. 091905, 2017.
105. X. Zheng, H. Lee, T. H. Weisgraber, M. Shusteff, J. DeOtte, E. B. Duoss, J. D. Kuntz et al. "Ultralight, ultrastiff mechanical metamaterials," *Science*, vol. 344, no. 6190, pp. 1373–1377, 2014.
106. Q. Wang, J. A. Jackson, Q. Ge, J. B. Hopkins, C. M. Spadaccini, and N. X. Fang, "Lightweight mechanical metamaterials with tunable negative thermal expansion," *Physical Review Letters*, vol. 117, no. 17, p. 175901, 2016.
107. S. A. Cummer, J. Christensen, and A. Alù, "Controlling sound with acoustic metamaterials," *Nature Reviews Materials*, vol. 1, no. 3, p. 16001, 2016.
108. H. M. Kolken, S. Janbaz, S. M. Leeflang, K. Lietaert, H. H. Weinans, and A. A. Zadpoor, "Rationally designed meta-implants: A combination of auxetic and conventional meta-biomaterials," *Materials Horizons*, vol. 5, no. 1, pp. 28–35, 2018.
109. P. Dérand, L.-E. Rännar, and J.-M. Hirsch, "Imaging, virtual planning, design, and production of patient-specific implants and clinical validation in craniomaxillofacial surgery," *Craniomaxillofacial Trauma & Reconstruction*, vol. 5, no. 3, p. 137, 2012.
110. A. L. Jardini, M. A. Larosa, R. Maciel Filho, C. A. de Carvalho Zavaglia, L. F. Bernardes, C. S. Lambert, D. R. Calderoni, and P. Kharmandayan, "Cranial reconstruction: 3D biomodel and custom-built implant created using additive manufacturing," *Journal of Cranio-Maxillofacial Surgery*, vol. 42, no. 8, pp. 1877–1884, 2014.
111. M. Salmi, J. Tuomi, K.-S. Paloheimo, R. Björkstrand, M. Paloheimo, J. Salo, R. Kontio, K. Mesimäki, and A. A. Mäkitie, "Patient-specific reconstruction with 3D modeling and DMLS additive manufacturing," *Rapid Prototyping Journal*, vol. 18, no. 3, pp. 209–214, 2012.
112. Tosca, "http://www.3ds.com/products-services/simulia/products/tosca/.".
113. Paretoworks, "http://www.sciartsoft.com/paretoworks/.".
114. T. D. Blacker, J. Robbins, S. J. Owen, M. A. Aguilovalentin, B. W. Clark, and T. E. Voth, "PLATO Platinum Topology Optimization," Sandia National Lab (SNL-NM), Albuquerque, NM (United States), 2015.
115. D. Brackett, I. Ashcroft, and R. Hague, "Topology optimization for additive manufacturing," in *Proceedings of the Solid Freeform Fabrication Symposium*, Austin, TX, 2011, vol. 1, pp. 348–362.
116. O. Sigmund, "A 99 line topology optimization code written in Matlab," *Structural and Multidisciplinary Optimization*, vol. 21, no. 2, pp. 120–127, 2001.
117. A. M. Mirzendehdel and K. Suresh, "A pareto-optimal approach to multimaterial topology optimization," *Journal of Mechanical Design*, vol. 137, no. 10, p. 101701, 2015.
118. A. M. Mirzendehdel and K. Suresh, "Support structure constrained topology optimization for additive manufacturing," *Computer-Aided Design*, vol. 81, pp. 1–13, 2016.

119. M. M. Francois, A. Sun, W. E. King, N. J. Henson, D. Tourret, C. A. Bronkhorst, N. N. Carlson et al., "Modeling of additive manufacturing processes for metals: Challenges and opportunities," *Current Opinion in Solid State and Materials Science*, vol. 21, no. LA-UR-16–24513, 2017.
120. Y. Saadlaoui, J.-L. Milan, J.-M. Rossi, and P. Chabrand, "Topology optimization and additive manufacturing: Comparison of conception methods using industrial codes," *Journal of Manufacturing Systems*, vol. 43, pp. 178–186, 2017.
121. P. Zhang, J. Toman, Y. Yu, E. Biyikli, M. Kirca, M. Chmielus, and A. C. To, "Efficient design-optimization of variable-density hexagonal cellular structure by additive manufacturing: Theory and validation," *Journal of Manufacturing Science and Engineering*, vol. 137, no. 2, p. 021004, 2015.
122. X. Wang, S. Xu, S. Zhou, W. Xu, M. Leary, P. Choong, M. Qian, M. Brandt, and Y. M. Xie, "Topological design and additive manufacturing of porous metals for bone scaffolds and orthopaedic implants: A review," *Biomaterials*, vol. 83, pp. 127–141, 2016.
123. F. Calignano, "Design optimization of supports for overhanging structures in aluminum and titanium alloys by selective laser melting," *Materials & Design*, vol. 64, pp. 203–213, 2014.
124. A. Hussein, L. Hao, C. Yan, R. Everson, and P. Young, "Advanced lattice support structures for metal additive manufacturing," *Journal of Materials Processing Technology*, vol. 213, no. 7, pp. 1019–1026, 2013.
125. F. Bobbert, K. Lietaert, A. A. Eftekhari, B. Pouran, S. Ahmadi, H. Weinans, and A. Zadpoor, "Additively manufactured metallic porous biomaterials based on minimal surfaces: A unique combination of topological, mechanical, and mass transport properties," *Acta Biomaterialia*, vol. 53, pp. 572–584, 2017.
126. F. Bobbert and A. Zadpoor, "Effects of bone substitute architecture and surface properties on cell response, angiogenesis, and structure of new bone," *Journal of Materials Chemistry B*, vol. 5, no. 31, pp. 6175–6192, 2017.
127. S. Van Bael, Y. C. Chai, S. Truscello, M. Moesen, G. Kerckhofs, H. Van Oosterwyck, J.-P. Kruth, and J. Schrooten, "The effect of pore geometry on the in vitro biological behavior of human periosteum-derived cells seeded on selective laser-melted Ti6Al4V bone scaffolds," *Acta Biomaterialia*, vol. 8, no. 7, pp. 2824–2834, 2012.
128. A. A. Zadpoor, "Bone tissue regeneration: The role of scaffold geometry," *Biomaterials Science*, vol. 3, no. 2, pp. 231–245, 2015.
129. D. W. Hutmacher, "Scaffolds in tissue engineering bone and cartilage," in *The Biomaterials: Silver Jubilee Compendium*, Elsevier, 2006, pp. 175–189.
130. X. Liu, S. Wu, K. W. Yeung, Y. Chan, T. Hu, Z. Xu, X. Liu, J. C. Chung, K. M. Cheung, and P. K. Chu, "Relationship between osseointegration and superelastic biomechanics in porous NiTi scaffolds," *Biomaterials*, vol. 32, no. 2, pp. 330–338, 2011.
131. J. van der Stok, H. Wang, S. Amin Yavari, M. Siebelt, M. Sandker, J. H. Waarsing et al. "Enhanced bone regeneration of cortical segmental bone defects using porous titanium scaffolds incorporated with colloidal gelatin gels for time-and dose-controlled delivery of dual growth factors," *Tissue Engineering Part A*, vol. 19, no. 23–24, pp. 2605–2614, 2013.
132. Z. G. Karaji, R. Hedayati, B. Pouran, I. Apachitei, and A. A. Zadpoor, "Effects of plasma electrolytic oxidation process on the mechanical properties of additively manufactured porous biomaterials," *Materials Science and Engineering: C*, vol. 76, pp. 406–416, 2017.
133. S. Amin Yavari, L. Loozen, F. L. Paganelli, S. Bakhshandeh, K. Lietaert, J. A. Groot, A. C. Fluit, C. Boel, J. Alblas, H. C. Vogely, H. Weinans, and A. A. Zadpoor, "Antibacterial behavior of additively manufactured porous titanium with nanotubular surfaces releasing silver ions," *ACS Applied Materials & Interfaces*, vol. 8, no. 27, pp. 17080–17089, 2016.

134. S. Bakhshandeh, Z. Gorgin Karaji, K. Lietaert, A. C. Fluit, C. E. Boel, H. C. Vogely, T. Vermonden et al., "Simultaneous delivery of multiple antibacterial agents from additively manufactured porous biomaterials to fully eradicate planktonic and adherent staphylococcus aureus," *ACS Applied Materials & Interfaces*, vol. 9, no. 31, pp. 25691–25699, 2017.
135. I. A. van Hengel, M. Riool, L. E. Fratila-Apachitei, J. Witte-Bouma, E. Farrell, A. A. Zadpoor, S. A. Zaat, and I. Apachitei, "Selective laser melting porous metallic implants with immobilized silver nanoparticles kill and prevent biofilm formation by methicillin-resistant Staphylococcus aureus," *Biomaterials*, vol. 140, pp. 1–15, 2017.
136. L. Thijs, J. Van Humbeeck, K. Kempen, E. Yasa, J. Kruth, and M. Rombouts, "Investigation on the inclusions in maraging steel produced by selective laser melting," *Innovative Developments in Virtual and Physical Prototyping*, pp. 297–304, 2011.
137. Y. Zhang, X. Cao, P. Wanjara, and M. Medraj, "Oxide films in laser additive manufactured Inconel 718," *Acta Materialia*, vol. 61, no. 17, pp. 6562–6576, 2013.
138. C. Ng, M. Savalani, H. Man, and I. Gibson, "Layer manufacturing of magnesium and its alloy structures for future applications," *Virtual and Physical Prototyping*, vol. 5, no. 1, pp. 13–19, 2010.
139. J. Yu, M. Rombouts, G. Maes, and F. Motmans, "Material properties of Ti6Al4 V parts produced by laser metal deposition," *Physics Procedia*, vol. 39, pp. 416–424, 2012.
140. A. T. Sidambe, "Biocompatibility of advanced manufactured titanium implants—A review," *Materials*, vol. 7, no. 12, pp. 8168–8188, 2014.
141. A. Bagheri, N. Shamsaei, and S. M. Thompson, "Microstructure and mechanical properties of Ti-6Al-4V Parts fabricated by laser engineered net shaping," in *ASME 2015 International Mechanical Engineering Congress and Exposition*, 2015, pp. V02AT02A005–V02AT02A005.
142. S. Ahmadi, G. Campoli, S. A. Yavari, B. Sajadi, R. Wauthlé, J. Schrooten, H. Weinans, and A. Zadpoor, "Mechanical behavior of regular open-cell porous biomaterials made of diamond lattice unit cells," *Journal of the Mechanical Behavior of Biomedical Materials*, vol. 34, pp. 106–115, 2014.
143. Z. Fan, "On the Young's moduli of Ti-6Al-4V alloys," *Scripta Metallurgica et Materialia*, vol. 29, no. 11, 1993.
144. Y. Lee and G. Welsch, "Young's modulus and damping of Ti-6Al-4V alloy as a function of heat treatment and oxygen concentration," *Materials Science and Engineering: A*, vol. 128, no. 1, pp. 77–89, 1990.
145. T. Ahmed and H. Rack, "Phase transformations during cooling in α+ β titanium alloys," *Materials Science and Engineering: A*, vol. 243, no. 1–2, pp. 206–211, 1998.
146. Y. Zhai, H. Galarraga, and D. A. Lados, "Microstructure, static properties, and fatigue crack growth mechanisms in Ti-6Al-4V fabricated by additive manufacturing: LENS and EBM," *Engineering Failure Analysis*, vol. 69, pp. 3–14, 2016.
147. X.-Y. Zhang, G. Fang, S. Leeflang, A. J. Böttger, A. A. Zadpoor, and J. Zhou, "Effect of subtransus heat treatment on the microstructure and mechanical properties of additively manufactured Ti-6Al-4V alloy," *Journal of Alloys and Compounds*, vol. 735, pp. 1562–1575, 2018.
148. D. Kuroda, M. Niinomi, M. Morinaga, Y. Kato, and T. Yashiro, "Design and mechanical properties of new β type titanium alloys for implant materials," *Materials Science and Engineering: A*, vol. 243, no. 1–2, pp. 244–249, 1998.
149. E. Chlebus, B. Kunicka, T. Kurzynowski, and B. Dybala, "Microstructure and mechanical behaviour of Ti–6Al–7Nb alloy produced by selective laser melting," *Materials Characterization*, vol. 62, no. 5, pp. 488–495, 2011.
150. L. Zhang, D. Klemm, J. Eckert, Y. Hao, and T. Sercombe, "Manufacture by selective laser melting and mechanical behavior of a biomedical Ti–24Nb–4Zr–8Sn alloy," *Scripta Materialia*, vol. 65, no. 1, pp. 21–24, 2011.

151. A. Barbas, A.-S. Bonnet, P. Lipinski, R. Pesci, and G. Dubois, "Development and mechanical characterization of porous titanium bone substitutes," *Journal of the Mechanical Behavior of Biomedical Materials*, vol. 9, pp. 34–44, 2012.
152. M. de Wild, R. Schumacher, K. Mayer, E. Schkommodau, D. Thoma, M. Bredell, A. Kruse Gujer, K. W. Grätz, and F. E. Weber, "Bone regeneration by the osteoconductivity of porous titanium implants manufactured by selective laser melting: A histological and micro computed tomography study in the rabbit," *Tissue Engineering Part A*, vol. 19, no. 23–24, pp. 2645–2654, 2013.
153. X. Liu, P. K. Chu, and C. Ding, "Surface modification of titanium, titanium alloys, and related materials for biomedical applications," *Materials Science and Engineering: R: Reports*, vol. 47, no. 3–4, pp. 49–121, 2004.
154. R. Wauthle, S. M. Ahmadi, S. A. Yavari, M. Mulier, A. A. Zadpoor, H. Weinans, J. Van Humbeeck, J.-P. Kruth, and J. Schrooten, "Revival of pure titanium for dynamically loaded porous implants using additive manufacturing," *Materials Science and Engineering: C*, vol. 54, pp. 94–100, 2015.
155. D. M. Brunette, P. Tengvall, M. Textor, and P. Thomsen, *Titanium in Medicine: Material Science, Surface Science, Engineering, Biological Responses and Medical Applications*. Springer Science & Business Media, Berlin, Germany, 2012.
156. A. L. Jardini, M. A. Larosa, C. A. de Carvalho Zavaglia, L. F. Bernardes, C. S. Lambert, P. Kharmandayan, D. Calderoni, and R. Maciel Filho, "Customised titanium implant fabricated in additive manufacturing for craniomaxillofacial surgery: This paper discusses the design and fabrication of a metallic implant for the reconstruction of a large cranial defect," *Virtual and Physical Prototyping*, vol. 9, no. 2, pp. 115–125, 2014.
157. J. A. Helsen and Y. Missirlis, *Biomaterials: A Tantalus Experience*. Springer Science & Business Media, Berlin, Germany, 2010.
158. T. Niendorf, S. Leuders, A. Riemer, H. A. Richard, T. Tröster, and D. Schwarze, "Highly anisotropic steel processed by selective laser melting," *Metallurgical and Materials Transactions B*, vol. 44, no. 4, pp. 794–796, 2013.
159. E. Yasa and J.-P. Kruth, "Microstructural investigation of selective laser melting 316L stainless steel parts exposed to laser re-melting," *Procedia Engineering*, vol. 19, pp. 389–395, 2011.
160. K. Abd-Elghany and D. Bourell, "Property evaluation of 304L stainless steel fabricated by selective laser melting," *Rapid Prototyping Journal*, vol. 18, no. 5, pp. 420–428, 2012.
161. G. Casalino, S. Campanelli, N. Contuzzi, and A. Ludovico, "Experimental investigation and statistical optimisation of the selective laser melting process of a maraging steel," *Optics & Laser Technology*, vol. 65, pp. 151–158, 2015.
162. K. Kempen, E. Yasa, L. Thijs, J.-P. Kruth, and J. Van Humbeeck, "Microstructure and mechanical properties of selective laser melted 18Ni-300 steel," *Physics Procedia*, vol. 12, pp. 255–263, 2011.
163. M. Garibaldi, I. Ashcroft, M. Simonelli, and R. Hague, "Metallurgy of high-silicon steel parts produced using Selective Laser Melting," *Acta Materialia*, vol. 110, pp. 207–216, 2016.
164. D. Cormier, O. Harrysson, and H. West, "Characterization of H13 steel produced via electron beam melting," *Rapid Prototyping Journal*, vol. 10, no. 1, pp. 35–41, 2004.
165. J. Suryawanshi, K. Prashanth, and U. Ramamurty, "Mechanical behavior of selective laser melted 316L stainless steel," *Materials Science and Engineering: A*, vol. 696, pp. 113–121, 2017.
166. C. Brice, R. Shenoy, M. Kral, and K. Buchannan, "Precipitation behavior of aluminum alloy 2139 fabricated using additive manufacturing," *Materials Science and Engineering: A*, vol. 648, pp. 9–14, 2015.

167. S. Lathabai, "Additive manufacturing of aluminium-based alloys and composites," in *Fundamentals of Aluminium Metallurgy*, Elsevier, pp. 47–92, 2018.
168. K. Bartkowiak, S. Ullrich, T. Frick, and M. Schmidt, "New developments of laser processing aluminium alloys via additive manufacturing technique," *Physics Procedia*, vol. 12, pp. 393–401, 2011.
169. B. Vandenbroucke and J.-P. Kruth, "Selective laser melting of biocompatible metals for rapid manufacturing of medical parts," *Rapid Prototyping Journal*, vol. 13, no. 4, pp. 196–203, 2007.
170. N. Xiang, X.-Z. Xin, J. Chen, and B. Wei, "Metal–ceramic bond strength of Co–Cr alloy fabricated by selective laser melting," *Journal of Dentistry*, vol. 40, no. 6, pp. 453–457, 2012.
171. X. Xin, N. Xiang, J. Chen, and B. Wei, "In vitro biocompatibility of Co–Cr alloy fabricated by selective laser melting or traditional casting techniques," *Materials Letters*, vol. 88, pp. 101–103, 2012.
172. F. A. España, V. K. Balla, S. Bose, and A. Bandyopadhyay, "Design and fabrication of CoCrMo alloy based novel structures for load bearing implants using laser engineered net shaping," *Materials Science and Engineering: C*, vol. 30, no. 1, pp. 50–57, 2010.
173. T. Koutsoukis, S. Zinelis, G. Eliades, K. Al-Wazzan, M. A. Rifaiy, and Y. S. Al Jabbari, "Selective laser melting technique of Co-Cr dental alloys: A review of structure and properties and comparative analysis with other available techniques," *Journal of Prosthodontics*, vol. 24, no. 4, pp. 303–312, 2015.
174. L. E. Murr, S. M. Gaytan, E. Martinez, F. Medina, and R. B. Wicker, "Next generation orthopaedic implants by additive manufacturing using electron beam melting," *International Journal of Biomaterials*, vol. 2012, 2012.
175. L. Murr, K. Amato, S. Li, Y. Tian, X. Cheng, S. Gaytan, E. Martinez, P. Shindo, F. Medina, and R. Wicker, "Microstructure and mechanical properties of open-cellular biomaterials prototypes for total knee replacement implants fabricated by electron beam melting," *Journal of the Mechanical Behavior of Biomedical Materials*, vol. 4, no. 7, pp. 1396–1411, 2011.
176. K. Amato, S. Gaytan, L. Murr, E. Martinez, P. Shindo, J. Hernandez, S. Collins, and F. Medina, "Microstructures and mechanical behavior of Inconel 718 fabricated by selective laser melting," *Acta Materialia*, vol. 60, no. 5, pp. 2229–2239, 2012.
177. S. Li, Q. Wei, Y. Shi, Z. Zhu, and D. Zhang, "Microstructure characteristics of Inconel 625 superalloy manufactured by selective laser melting," *Journal of Materials Science & Technology*, vol. 31, no. 9, pp. 946–952, 2015.
178. P. Nie, O. Ojo, and Z. Li, "Numerical modeling of microstructure evolution during laser additive manufacturing of a nickel-based superalloy," *Acta Materialia*, vol. 77, pp. 85–95, 2014.
179. L. L. Parimi, G. Ravi, D. Clark, and M. M. Attallah, "Microstructural and texture development in direct laser fabricated IN718," *Materials Characterization*, vol. 89, pp. 102–111, 2014.
180. R. Dehoff, M. Kirka, W. Sames, H. Bilheux, A. Tremsin, L. Lowe, and S. Babu, "Site specific control of crystallographic grain orientation through electron beam additive manufacturing," *Materials Science and Technology*, vol. 31, no. 8, pp. 931–938, 2015.
181. Y. Chen, Z. Xu, C. Smith, and J. Sankar, "Recent advances on the development of magnesium alloys for biodegradable implants," *Acta Biomaterialia*, vol. 10, no. 11, pp. 4561–4573, 2014.
182. H. Hermawan, "Biodegradable metals: State of the art," in *Biodegradable Metals*, Springer, Berlin, Germany, 2012, pp. 13–22.
183. Q. Chen and G. A. Thouas, "Metallic implant biomaterials," *Materials Science and Engineering: R: Reports*, vol. 87, pp. 1–57, 2015.

184. Y. Zheng, X. Gu, and F. Witte, "Biodegradable metals," *Materials Science and Engineering: R: Reports*, vol. 77, pp. 1–34, 2014.
185. R. Erbel, C. Di Mario, J. Bartunek, J. Bonnier, B. de Bruyne, F. R. Eberli, P. Erne et al., "Temporary scaffolding of coronary arteries with bioabsorbable magnesium stents: A prospective, non-randomised multicentre trial," *The Lancet*, vol. 369, no. 9576, pp. 1869–1875, 2007.
186. M. P. Staiger, A. M. Pietak, J. Huadmai, and G. Dias, "Magnesium and its alloys as orthopedic biomaterials: A review," *Biomaterials*, vol. 27, no. 9, pp. 1728–1734, 2006.
187. Y. Li, H. Jahr, K. Lietaert, P. Pavanram, A. Yilmaz, L. Fockaert, M. Leeflang et al., "Additively manufactured biodegradable porous iron," *Acta Biomaterialia*, vol. 77, pp. 380–393, 2018.
188. Y. Li, J. Zhou, P. Pavanram, M. Leeflang, L. Fockaert, B. Pouran, N. Tümer et al., "Additively manufactured biodegradable porous magnesium," *Acta Biomaterialia*, vol. 67, pp. 378–392, 2018.
189. M. Gieseke, C. Noelke, S. Kaierle, V. Wesling, and H. Haferkamp, "Selective laser melting of magnesium and magnesium alloys," in *Magnesium Technology 2013*, Springer, Cham, Switzerland, 2013, pp. 65–68.
190. B. Zberg, P. J. Uggowitzer, and J. F. Löffler, "MgZnCa glasses without clinically observable hydrogen evolution for biodegradable implants," *Nature Materials*, vol. 8, no. 11, p. 887, 2009.
191. B. Zberg, E. R. Arata, P. J. Uggowitzer, and J. F. Löffler, "Tensile properties of glassy MgZnCa wires and reliability analysis using Weibull statistics," *Acta Materialia*, vol. 57, no. 11, pp. 3223–3231, 2009.
192. M. Montani, A. G. Demir, E. Mostaed, M. Vedani, and B. Previtali, "Processability of pure Zn and pure Fe by SLM for biodegradable metallic implant manufacturing," *Rapid Prototyping Journal*, vol. 23, no. 3, pp. 514–523, 2017.
193. P. Wen, M. Voshage, L. Jauer, Y. Chen, Y. Qin, R. Poprawe, and J. H. Schleifenbaum, "Laser additive manufacturing of Zn metal parts for biodegradable applications: Processing, formation quality and mechanical properties," *Materials & Design*, 2018.
194. S. Dadbakhsh, M. Speirs, J. Van Humbeeck, and J.-P. Kruth, "Laser additive manufacturing of bulk and porous shape-memory NiTi alloys: From processes to potential biomedical applications," *MRS Bulletin*, vol. 41, no. 10, pp. 765–774, 2016.
195. M. H. Elahinia, M. Hashemi, M. Tabesh, and S. B. Bhaduri, "Manufacturing and processing of NiTi implants: A review," *Progress in Materials Science*, vol. 57, no. 5, pp. 911–946, 2012.
196. S. Dadbakhsh, B. Vrancken, J.-P. Kruth, J. Luyten, and J. Van Humbeeck, "Texture and anisotropy in selective laser melting of NiTi alloy," *Materials Science and Engineering: A*, vol. 650, pp. 225–232, 2016.
197. C. Haberland, M. Elahinia, J. M. Walker, H. Meier, and J. Frenzel, "On the development of high quality NiTi shape memory and pseudoelastic parts by additive manufacturing," *Smart Materials and Structures*, vol. 23, no. 10, p. 104002, 2014.
198. Z. Gorgin Karaji, M. Speirs, S. Dadbakhsh, J.-P. Kruth, H. Weinans, A. Zadpoor, and S. Amin Yavari, "Additively manufactured and surface biofunctionalized porous nitinol," *ACS Applied Materials & Interfaces*, vol. 9, no. 2, pp. 1293–1304, 2017.
199. R. Wauthle, J. Van Der Stok, S. A. Yavari, J. Van Humbeeck, J.-P. Kruth, A. A. Zadpoor, H. Weinans, M. Mulier, and J. Schrooten, "Additively manufactured porous tantalum implants," *Acta Biomaterialia*, vol. 14, pp. 217–225, 2015.
200. V. K. Balla, S. Bodhak, S. Bose, and A. Bandyopadhyay, "Porous tantalum structures for bone implants: Fabrication, mechanical and in vitro biological properties," *Acta Biomaterialia*, vol. 6, no. 8, pp. 3349–3359, 2010.

201. E. Marin, L. Fedrizzi, and L. Zagra, "Porous metallic structures for orthopaedic applications: A short review of materials and technologies," *European Orthopaedics and Traumatology*, vol. 1, no. 3–4, pp. 103–109, 2010.
202. V. K. Balla, S. Banerjee, S. Bose, and A. Bandyopadhyay, "Direct laser processing of a tantalum coating on titanium for bone replacement structures," *Acta Biomaterialia*, vol. 6, no. 6, pp. 2329–2334, 2010.
203. P. Fox, S. Pogson, C. Sutcliffe, and E. Jones, "Interface interactions between porous titanium/tantalum coatings, produced by Selective Laser Melting (SLM), on a cobalt–chromium alloy," *Surface and Coatings Technology*, vol. 202, no. 20, pp. 5001–5007, 2008.
204. M. Stiehler, M. Lind, T. Mygind, A. Baatrup, A. Dolatshahi-Pirouz, H. Li, M. Foss, F. Besenbacher, M. Kassem, and C. Bünger, "Morphology, proliferation, and osteogenic differentiation of mesenchymal stem cells cultured on titanium, tantalum, and chromium surfaces," *Journal of Biomedical Materials Research Part A: An Official Journal of The Society for Biomaterials, The Japanese Society for Biomaterials, and The Australian Society for Biomaterials and the Korean Society for Biomaterials*, vol. 86, no. 2, pp. 448–458, 2008.
205. V. K. Balla, S. Bose, N. M. Davies, and A. Bandyopadhyay, "Tantalum—A bioactive metal for implants," *JOM*, vol. 62, no. 7, pp. 61–64, 2010.
206. J. J. Lewandowski and M. Seifi, "Metal additive manufacturing: A review of mechanical properties," *Annual Review of Materials Research*, vol. 46, pp. 151–186, 2016.
207. P. Mercelis and J.-P. Kruth, "Residual stresses in selective laser sintering and selective laser melting," *Rapid Prototyping Journal*, vol. 12, no. 5, pp. 254–265, 2006.
208. G. K. Lewis and E. Schlienger, "Practical considerations and capabilities for laser assisted direct metal deposition," *Materials & Design*, vol. 21, no. 4, pp. 417–423, 2000.
209. R. Moat, A. Pinkerton, L. Li, P. Withers, and M. Preuss, "Residual stresses in laser direct metal deposited Waspaloy," *Materials Science and Engineering: A*, vol. 528, no. 6, pp. 2288–2298, 2011.
210. K. Dai and L. Shaw, "Distortion minimization of laser-processed components through control of laser scanning patterns," *Rapid Prototyping Journal*, vol. 8, no. 5, pp. 270–276, 2002.
211. P. Rangaswamy, M. Griffith, M. Prime, T. Holden, R. Rogge, J. Edwards, and R. Sebring, "Residual stresses in LENS® components using neutron diffraction and contour method," *Materials Science and Engineering: A*, vol. 399, no. 1–2, pp. 72–83, 2005.
212. E. Brandl and D. Greitemeier, "Microstructure of additive layer manufactured Ti–6Al–4V after exceptional post heat treatments," *Materials Letters*, vol. 81, pp. 84–87, 2012.
213. B. Song, S. Dong, Q. Liu, H. Liao, and C. Coddet, "Vacuum heat treatment of iron parts produced by selective laser melting: Microstructure, residual stress and tensile behavior," *Materials & Design (1980–2015)*, vol. 54, pp. 727–733, 2014.
214. W. J. Sames, F. List, S. Pannala, R. R. Dehoff, and S. S. Babu, "The metallurgy and processing science of metal additive manufacturing," *International Materials Reviews*, vol. 61, no. 5, pp. 315–360, 2016.
215. K. S. Chan, M. Koike, R. L. Mason, and T. Okabe, "Fatigue life of titanium alloys fabricated by additive layer manufacturing techniques for dental implants," *Metallurgical and Materials Transactions A*, vol. 44, no. 2, pp. 1010–1022, 2013.
216. W.-Q. Yan, T. Nakamura, M. Kobayashi, H.-M. Kim, F. Miyaji, and T. Kokubo, "Bonding of chemically treated titanium implants to bone," *Journal of Biomedical Materials Research: An Official Journal of The Society for Biomaterials and The Japanese Society for Biomaterials*, vol. 37, no. 2, pp. 267–275, 1997.

217. S. A. Yavari, J. van der Stok, Y. C. Chai, R. Wauthle, Z. T. Birgani, P. Habibovic, M. Mulier, J. Schrooten, H. Weinans, and A. A. Zadpoor, "Bone regeneration performance of surface-treated porous titanium," *Biomaterials*, vol. 35, no. 24, pp. 6172–6181, 2014.
218. I.-C. Brie, O. Soritau, N. Dirzu, C. Berce, A. Vulpoi, C. Popa, M. Todea et al., "Comparative in vitro study regarding the biocompatibility of titanium-base composites infiltrated with hydroxyapatite or silicatitanate," *Journal of Biological Engineering*, vol. 8, no. 1, p. 14, 2014.
219. S. F. S. Shirazi, S. Gharehkhani, M. Mehrali, H. Yarmand, H. S. C. Metselaar, N. A. Kadri, and N. A. A. Osman, "A review on powder-based additive manufacturing for tissue engineering: selective laser sintering and inkjet 3D printing," *Science and Technology of Advanced Materials*, vol. 16, no. 3, p. 033502, 2015.
220. P. Heinl, L. Müller, C. Körner, R. F. Singer, and F. A. Müller, "Cellular Ti–6Al–4V structures with interconnected macro porosity for bone implants fabricated by selective electron beam melting," *Acta Biomaterialia*, vol. 4, no. 5, pp. 1536–1544, 2008.
221. X. Li, Y.-F. Feng, C.-T. Wang, G.-C. Li, W. Lei, Z.-Y. Zhang, and L. Wang, "Evaluation of biological properties of electron beam melted Ti6Al4V implant with biomimetic coating in vitro and in vivo," *PLoS One*, vol. 7, no. 12, p. e52049, 2012.
222. S. A. Yavari, S. Ahmadi, J. van der Stok, R. Wauthlé, A. Riemslag, M. Janssen, J. Schrooten, H. Weinans, and A. A. Zadpoor, "Effects of bio-functionalizing surface treatments on the mechanical behavior of open porous titanium biomaterials," *Journal of the Mechanical Behavior of Biomedical Materials*, vol. 36, pp. 109–119, 2014.
223. G. Pyka, A. Burakowski, G. Kerckhofs, M. Moesen, S. Van Bael, J. Schrooten, and M. Wevers, "Surface modification of Ti6Al4V open porous structures produced by additive manufacturing," *Advanced Engineering Materials*, vol. 14, no. 6, pp. 363–370, 2012.
224. S. A. Yavari, R. Wauthlé, A. J. Böttger, J. Schrooten, H. Weinans, and A. A. Zadpoor, "Crystal structure and nanotopographical features on the surface of heat-treated and anodized porous titanium biomaterials produced using selective laser melting," *Applied Surface Science*, vol. 290, pp. 287–294, 2014.
225. L. N. Carter, M. M. Attallah, and R. C. Reed, "Laser powder bed fabrication of nickel-base superalloys: Influence of parameters; characterisation, quantification and mitigation of cracking," *Superalloys*, pp. 577–586, 2012.
226. P. Kobryn and S. Semiatin, "Mechanical properties of laser-deposited Ti-6Al-4V," in *Solid Freeform Fabrication Proceedings*, 2001, pp. 6–8.
227. S. Ahmadi, R. Kumar, E. Borisov, R. Petrov, S. Leeflang, Y. Li, N. Tümer et al. "From microstructural design to surface engineering: A tailored approach for improving fatigue life of additively manufactured meta-biomaterials," *Acta Biomaterialia*, vol. 83, pp. 153–166, 2019.
228. ISO/ASTM, "Standard terminology for additive manufacturing-coordinate systems and test methodologies, ASTM/ISO Standard 52921," International Organization for Standardization, 2013.
229. M. Simonelli, Y. Y. Tse, and C. Tuck, "Effect of the build orientation on the mechanical properties and fracture modes of SLM Ti–6Al–4V," *Materials Science and Engineering: A*, vol. 616, pp. 1–11, 2014.
230. A. A. Zadpoor, "Mechanics of additively manufactured biomaterials." Elsevier, vol. 70, pp. 1–6, 2017.
231. B. Vrancken, S. Buls, J.-P. Kruth, and J. V. Humbeeck, "Preheating of selective laser melted Ti6Al4V: Microstructure and Mechanical Properties," in *Proceedings of the 13th World Conference on Titanium*, 2016, pp. 1269–1277.
232. N. Guo and M. C. Leu, "Additive manufacturing: Technology, applications and research needs," *Frontiers of Mechanical Engineering*, vol. 8, no. 3, pp. 215–243, 2013.

233. S. M. Ahmadi, S. A. Yavari, R. Wauthle, B. Pouran, J. Schrooten, H. Weinans, and A. A. Zadpoor, "Additively manufactured open-cell porous biomaterials made from six different space-filling unit cells: The mechanical and morphological properties," *Materials*, vol. 8, no. 4, pp. 1871–1896, 2015.
234. A. Bandyopadhyay, F. Espana, V. K. Balla, S. Bose, Y. Ohgami, and N. M. Davies, "Influence of porosity on mechanical properties and in vivo response of Ti6Al4V implants," *Acta Biomaterialia*, vol. 6, no. 4, pp. 1640–1648, 2010.
235. C. Y. Lin, N. Kikuchi, and S. J. Hollister, "A novel method for biomaterial scaffold internal architecture design to match bone elastic properties with desired porosity," *Journal of Biomechanics*, vol. 37, no. 5, pp. 623–636, 2004.
236. A. A. Zadpoor and R. Hedayati, "Analytical relationships for prediction of the mechanical properties of additively manufactured porous biomaterials," *Journal of Biomedical Materials Research Part A*, vol. 104, no. 12, pp. 3164–3174, 2016.
237. L. J. Gibson, "Biomechanics of cellular solids," *Journal of Biomechanics*, vol. 38, no. 3, pp. 377–399, 2005.
238. L. J. Gibson and M. F. Ashby, *Cellular Solids: Structure and Properties*. Cambridge University Press, Cambrdige, 1999.
239. R. Hedayati, S. Ahmadi, K. Lietaert, B. Pouran, Y. Li, H. Weinans, C. Rans, and A. Zadpoor, "Isolated and modulated effects of topology and material type on the mechanical properties of additively manufactured porous biomaterials," *Journal of the Mechanical Behavior of Biomedical Materials*, vol. 79, pp. 254–263, 2018.
240. A. Bandyopadhyay, Bv. Krishna, W. Xue, and S. Bose, "Application of laser engineered net shaping (LENS) to manufacture porous and functionally graded structures for load bearing implants," *Journal of Materials Science: Materials in Medicine*, vol. 20, no. 1, p. 29, 2009.
241. B. V. Krishna, S. Bose, and A. Bandyopadhyay, "Low stiffness porous Ti structures for load-bearing implants," *Acta Biomaterialia*, vol. 3, no. 6, pp. 997–1006, 2007.
242. L. Mullen, R. C. Stamp, W. K. Brooks, E. Jones, and C. J. Sutcliffe, "Selective Laser Melting: A regular unit cell approach for the manufacture of porous, titanium, bone in-growth constructs, suitable for orthopedic applications," *Journal of Biomedical Materials Research Part B: Applied Biomaterials: An Official Journal of The Society for Biomaterials, The Japanese Society for Biomaterials, and The Australian Society for Biomaterials and the Korean Society for Biomaterials*, vol. 89, no. 2, pp. 325–334, 2009.
243. S. Ahmadi, R. Hedayati, Y. Li, K. Lietaert, N. Tümer, A. Fatemi, C. Rans, B. Pouran, H. Weinans, and A. Zadpoor, "Fatigue performance of additively manufactured meta-biomaterials: The effects of topology and material type," *Acta Biomaterialia*, vol. 65, pp. 292–304, 2018.
244. J. Kadkhodapour, H. Montazerian, A. C. Darabi, A. Anaraki, S. Ahmadi, A. Zadpoor, and S. Schmauder, "Failure mechanisms of additively manufactured porous biomaterials: Effects of porosity and type of unit cell," *Journal of the Mechanical Behavior of Biomedical Materials*, vol. 50, pp. 180–191, 2015.
245. R. Hedayati, M. Sadighi, M. Mohammadi-Aghdam, and A. Zadpoor, "Mechanical properties of regular porous biomaterials made from truncated cube repeating unit cells: Analytical solutions and computational models," *Materials Science and Engineering: C*, vol. 60, pp. 163–183, 2016.
246. R. Hedayati, M. Sadighi, M. Mohammadi-Aghdam, and A. Zadpoor, "Analytical relationships for the mechanical properties of additively manufactured porous biomaterials based on octahedral unit cells," *Applied Mathematical Modelling*, vol. 46, pp. 408–422, 2017.

247. R. Hedayati, M. Sadighi, M. Mohammadi-Aghdam, and A. Zadpoor, "Mechanics of additively manufactured porous biomaterials based on the rhombicuboctahedron unit cell," *Journal of the Mechanical Behavior of Biomedical Materials*, vol. 53, pp. 272–294, 2016.
248. R. Hedayati, M. Sadighi, M. Mohammadi-Aghdam, and A. Zadpoor, "Mechanical behavior of additively manufactured porous biomaterials made from truncated cuboctahedron unit cells," *International Journal of Mechanical Sciences*, vol. 106, pp. 19–38, 2016.
249. K. Genovese, S. Leeflang, and A. A. Zadpoor, "Microscopic full-field three-dimensional strain measurement during the mechanical testing of additively manufactured porous biomaterials," *Journal of the Mechanical Behavior of Biomedical Materials*, vol. 69, pp. 327–341, 2017.
250. ASTM International. "Standard practice for conducting constant amplitude axial fatigue tests of metallic materials," ASTM E 466-07: ASTM International, 2007.
251. E. Amsterdam and G. Kool, "High cycle fatigue of laser beam deposited Ti-6Al-4V and Inconel 718," in *ICAF 2009, Bridging the Gap between Theory and Operational Practice*, Springer, Dordrecht, the Netherlands, 2009, pp. 1261–1274.
252. N. W. Hrabe, P. Heinl, B. Flinn, C. Körner, and R. K. Bordia, "Compression-compression fatigue of selective electron beam melted cellular titanium (Ti-6Al-4V)," *Journal of Biomedical Materials Research Part B: Applied Biomaterials*, vol. 99, no. 2, pp. 313–320, 2011.
253. K. Rekedal and D. Liu, "Fatigue life of selective laser melted and hot isostatically pressed Ti-6Al-4v absent of surface machining," in *56th AIAA/ASCE/AHS/ASC Structures, Structural Dynamics, and Materials Conference*, 2015, p. 0894.
254. S. A. Yavari, R. Wauthlé, J. van der Stok, A. Riemslag, M. Janssen, M. Mulier, J.-P. Kruth, J. Schrooten, H. Weinans, and A. A. Zadpoor, "Fatigue behavior of porous biomaterials manufactured using selective laser melting," *Materials Science and Engineering: C*, vol. 33, no. 8, pp. 4849–4858, 2013.
255. E. Wycisk, A. Solbach, S. Siddique, D. Herzog, F. Walther, and C. Emmelmann, "Effects of defects in laser additive manufactured Ti-6Al-4V on fatigue properties," *Physics Procedia*, vol. 56, pp. 371–378, 2014.
256. A. Prabhu, T. Vincent, A. Chaudhary, W. Zhang, and S. Babu, "Effect of microstructure and defects on fatigue behaviour of directed energy deposited Ti–6Al–4V," *Science and Technology of Welding and Joining*, vol. 20, no. 8, pp. 659–669, 2015.
257. P. Li, D. Warner, A. Fatemi, and N. Phan, "Critical assessment of the fatigue performance of additively manufactured Ti–6Al–4V and perspective for future research," *International Journal of Fatigue*, vol. 85, pp. 130–143, 2016.
258. B. Van Hooreweder, Y. Apers, K. Lietaert, and J.-P. Kruth, "Improving the fatigue performance of porous metallic biomaterials produced by selective laser melting," *Acta Biomaterialia*, vol. 47, pp. 193–202, 2017.
259. H. Gong, K. Rafi, H. Gu, G. J. Ram, T. Starr, and B. Stucker, "Influence of defects on mechanical properties of Ti–6Al–4 V components produced by selective laser melting and electron beam melting," *Materials & Design*, vol. 86, pp. 545–554, 2015.
260. S. A. Yavari, S. Ahmadi, R. Wauthle, B. Pouran, J. Schrooten, H. Weinans, and A. Zadpoor, "Relationship between unit cell type and porosity and the fatigue behavior of selective laser melted meta-biomaterials," *Journal of the Mechanical Behavior of Biomedical Materials*, vol. 43, pp. 91–100, 2015.
261. J. de Krijger, C. Rans, B. Van Hooreweder, K. Lietaert, B. Pouran, and A. A. Zadpoor, "Effects of applied stress ratio on the fatigue behavior of additively manufactured porous biomaterials under compressive loading," *Journal of the Mechanical Behavior of Biomedical Materials*, vol. 70, pp. 7–16, 2017.

262. R. Hedayati, H. Hosseini-Toudeshky, M. Sadighi, M. Mohammadi-Aghdam, and A. Zadpoor, "Computational prediction of the fatigue behavior of additively manufactured porous metallic biomaterials," *International Journal of Fatigue*, vol. 84, pp. 67–79, 2016.
263. R. Hedayati, H. Hosseini-Toudeshky, M. Sadighi, M. Mohammadi-Aghdam, and A. Zadpoor, "Multiscale modeling of fatigue crack propagation in additively manufactured porous biomaterials," *International Journal of Fatigue*, vol. 113, pp. 416–427, 2018.
264. R. Hedayati, S. A. Yavari, and A. Zadpoor, "Fatigue crack propagation in additively manufactured porous biomaterials," *Materials Science and Engineering: C*, vol. 76, pp. 457–463, 2017.
265. G. Tapia and A. Elwany, "A review on process monitoring and control in metal-based additive manufacturing," *Journal of Manufacturing Science and Engineering*, vol. 136, no. 6, p. 060801, 2014.
266. P. Sun, Z. Z. Fang, Y. Zhang, and Y. Xia, "Review of the methods for production of spherical Ti and Ti alloy powder," *JOM*, vol. 69, no. 10, pp. 1853–1860, 2017.
267. B. Dutta and F. H. S. Froes, "The additive manufacturing (AM) of titanium alloys," in *Titanium Powder Metallurgy*, Elsevier, Amsterdam, the Netherlands, 2015, pp. 447–468.
268. G. Kasperovich and J. Hausmann, "Improvement of fatigue resistance and ductility of TiAl6V4 processed by selective laser melting," *Journal of Materials Processing Technology*, vol. 220, pp. 202–214, 2015.
269. V. Cain, L. Thijs, J. Van Humbeeck, B. Van Hooreweder, and R. Knutsen, "Crack propagation and fracture toughness of Ti6Al4V alloy produced by selective laser melting," *Additive Manufacturing*, vol. 5, pp. 68–76, 2015.
270. L. Facchini, E. Magalini, P. Robotti, A. Molinari, S. Höges, and K. Wissenbach, "Ductility of a Ti-6Al-4V alloy produced by selective laser melting of prealloyed powders," *Rapid Prototyping Journal*, vol. 16, no. 6, pp. 450–459, 2010.
271. D. A. Hollander, M. Von Walter, T. Wirtz, R. Sellei, B. Schmidt-Rohlfing, O. Paar, and H.-J. Erli, "Structural, mechanical and in vitro characterization of individually structured Ti–6Al–4V produced by direct laser forming," *Biomaterials*, vol. 27, no. 7, pp. 955–963, 2006.
272. B. Vrancken, L. Thijs, J.-P. Kruth, and J. Van Humbeeck, "Heat treatment of Ti6Al4V produced by selective laser melting: Microstructure and mechanical properties," *Journal of Alloys and Compounds*, vol. 541, pp. 177–185, 2012.
273. U. Ackelid and M. Svensson, "Additive manufacturing of dense metal parts by electron beam melting," in *Proceedings of the Materials Science and Technology Conference*, Pittsburgh, PA, 2009, vol. 2529.
274. P. Edwards, A. O'conner, and M. Ramulu, "Electron beam additive manufacturing of titanium components: Properties and performance," *Journal of Manufacturing Science and Engineering*, vol. 135, no. 6, p. 061016, 2013.
275. A. Mohammadhosseini, D. Fraser, S. Masood, and M. Jahedi, "Microstructure and mechanical properties of Ti–6Al–4V manufactured by electron beam melting process," *Materials Research Innovations*, vol. 17, no. supp. 2, pp. s106–s112, 2013.
276. L. Murr, E. Esquivel, S. Quinones, S. Gaytan, M. Lopez, E. Martinez, F. Medina et al., "Microstructures and mechanical properties of electron beam-rapid manufactured Ti–6Al–4V biomedical prototypes compared to wrought Ti–6Al–4V," *Materials Characterization*, vol. 60, no. 2, pp. 96–105, 2009.
277. H. K. Rafi, N. Karthik, T. L. Starr, and B. E. Stucker, "Mechanical property evaluation of Ti-6Al-4V parts made using electron beam melting," in *Proceedings of the Solid Freeform Fabrication Symposium*, 2012, pp. 526–535.

278. O. L. Rodriguez, P. G. Allison, W. R. Whittington, D. K. Francis, O. G. Rivera, K. Chou, X. Gong, T. Butler, and J. F. Burroughs, "Dynamic tensile behavior of electron beam additive manufactured Ti6Al4V," *Materials Science and Engineering: A*, vol. 641, pp. 323–327, 2015.
279. P. Wang, M. L. S. Nai, X. Tan, W. J. Sin, S. B. Tor, and J. Wei, "Anisotropic mechanical properties in a big-sized Ti-6Al-4V plate fabricated by electron beam melting," in *TMS 2016 145th Annual Meeting & Exhibition*, 2016, pp. 5–12.
280. G. Dinda, L. Song, and J. Mazumder, "Fabrication of Ti-6Al-4V scaffolds by direct metal deposition," *Metallurgical and Materials Transactions A*, vol. 39, no. 12, pp. 2914–2922, 2008.
281. D. Keicher, W. Miller, J. Smugeresky, and J. Romero, "Laser engineered net shaping(LENS): Beyond rapid prototyping to direct fabrication," *Hard Coatings Based on Borides, Carbides & Nitrides: Synthesis, Characterization & Applications*, pp. 369–377, 1998.
282. T. M. Mower and M. J. Long, "Mechanical behavior of additive manufactured, powder-bed laser-fused materials," *Materials Science and Engineering: A*, vol. 651, pp. 198–213, 2016.
283. C. Qiu, N. J. Adkins, and M. M. Attallah, "Microstructure and tensile properties of selectively laser-melted and of HIPed laser-melted Ti–6Al–4V," *Materials Science and Engineering: A*, vol. 578, pp. 230–239, 2013.
284. A. Christensen, R. Kircher, and A. Lippincott, "Qualification of electron beam melted (EBM) Ti6Al4V-ELI for orthopaedic applications," in *Medical Device Materials IV: Proceedings of the Materials and Processes for Medical Devices Conference*, 2008, pp. 48–53.
285. A. International, "ASTM F1472-14, Standard Specification for Wrought Titanium-6Aluminum-4Vanadium Alloy for Surgical Implant Applications (UNS R56400)," West Conshohocken, PA, 2014.
286. A. International, "ASTM F1108-14, Standard Specification for Titanium-6Aluminum-4Vanadium Alloy Castings for Surgical Implants (UNS R56406)," West Conshohocken, PA, 2014.
287. M. J. Donachie, "Titanium," *A Technical Guide, ASM International*, Materials Park, OH, 2000.
288. D. R. Carter and D. M. Spengler, "Mechanical properties and composition of cortical bone," *Clinical Orthopaedics and Related Research®*, no. 135, pp. 192–217, 1978.
289. M. J. Mirzaali, A. Bürki, J. Schwiedrzik, P. K. Zysset, and U. Wolfram, "Continuum damage interactions between tension and compression in osteonal bone," *Journal of the Mechanical Behavior of Biomedical Materials*, vol. 49, pp. 355–369, 2015.
290. M. J. Mirzaali, J. J. Schwiedrzik, S. Thaiwichai, J. P. Best, J. Michler, P. K. Zysset, and U. Wolfram, "Mechanical properties of cortical bone and their relationships with age, gender, composition and microindentation properties in the elderly," *Bone*, vol. 93, pp. 196–211, 2016.
291. S. A. Goldstein, "The mechanical properties of trabecular bone: Dependence on anatomic location and function," *Journal of Biomechanics*, vol. 20, no. 11–12, pp. 1055–1061, 1987.
292. M. J. Mirzaali, F. Libonati, D. Ferrario, L. Rinaudo, C. Messina, F. M. Ulivieri, B. M. Cesana, M. Strano, and L. Vergani, "Determinants of bone damage: An ex-vivo study on porcine vertebrae," *PLoS One*, vol. 13, no. 8, p. e0202210, 2018.
293. M. J. Mirzaali, V. Mussi, P. Vena, F. Libonati, L. Vergani, and M. Strano, "Mimicking the loading adaptation of bone microstructure with aluminum foams," *Materials & Design*, vol. 126, pp. 207–218, 2017.

294. E. Andrews, W. Sanders, and L. J. Gibson, "Compressive and tensile behaviour of aluminum foams," *Materials Science and Engineering: A*, vol. 270, no. 2, pp. 113–124, 1999.
295. T. Miyoshi, M. Itoh, S. Akiyama, and A. Kitahara, "ALPORAS aluminum foam: Production process, properties, and applications," *Advanced Engineering Materials*, vol. 2, no. 4, pp. 179–183, 2000.
296. M. Mirzaali Mazandarani, F. Libonati, P. Vena, V. Mussi, L. Vergani, and M. Strano, "Investigation of the effect of internal pores distribution on the elastic properties of closed-cell aluminum foam: A comparison with cancellous bone," *Procedia Structural Integrity*, pp. 1285–1294, 2016.
297. A. Baca, R. Konecná, G. Nicoletto, and L. Kunz, "Effect of surface roughness on the fatigue life of laser additive manufactured Ti6Al4V alloy," *Manufacturing Technology*, vol. 15, no. 4, pp. 498–502, 2015.
298. K. D. Rekedal, "Investigation of the High-Cycle Fatigue Life of Selective Laser Melted and Hot Isostatically Pressed Ti-6Al-4V," Air Force Institute of Technology Wright-Patterson AFB OH,Graduate School of Engineering and Management, 2015.

5 Deposition-Based and Solid-State Additive Manufacturing Technologies for Metals

Vamsi Krishna Balla

CONTENTS

5.1 INTRODUCTION

Initially, technologies to create three-dimensional (3D) components from computer-aided design (CAD) files have been termed as Rapid Prototyping technologies as these are primarily used to create prototypes of the parts with different materials, primarily plastics. However, there is paradigm shift from prototyping to direct manufacturing/production of 3D components and therefore these technologies have been improved over the last few decades and are being called as additive manufacturing (AM) technologies. Currently, the output of AM technologies include up to 20% final products and is estimated to increase to 50% by 2020 (The Economist 2011). While the invention of technologies is being argued to be a "Third Industrial Revolution" (The Economist 2012), huge investment and development efforts are required to fully realize their potential (Reeves and Hague 2013). The unique benefits of these agile manufacturing technologies include rapid production of components with efficient utilization of available resources, reverse engineering to develop functional components, new materials development such as light weight structures, complex integration of materials including assemblies with moving parts, functionally graded materials, etc.

Current AM technologies for metals/alloys are aimed at producing complex, unique geometries, tailored materials development and customization, and functionally graded materials development, which find applications in aerospace, defense, automotive, and biomedical industries with demanding requirements. Although several AM techniques have been developed for creating metallic objects, only deposition-based, solid-state, and some new AM techniques will be discussed in this chapter. These techniques can be categorized based on energy source, processing state (liquid or solid), and feedstock material as shown in Figure 5.1.

Processes that fuse feedstock material include Laser Engineered Net Shaping (LENS™; developed at Sandia National Laboratory, USA, and marketed by Optomec, USA), direct metal deposition (DMD; developed at Michigan University, USA and

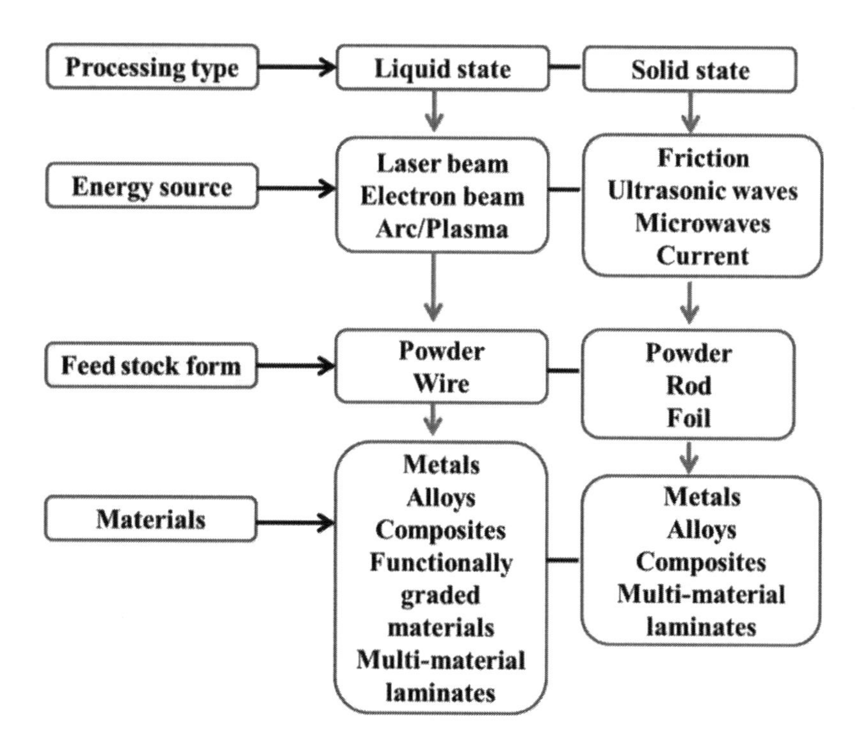

FIGURE 5.1 Classification of AM processes for metallic objects.

TABLE 5.1
Energy Sources Used for Fusion-Based AM Techniques

Characteristic	Laser	Electron Beam	Arc
Atmosphere	Inert	Vacuum	Inert
Energy density (W/mm^2)	10^6	10^8	Very high
Power efficiency (%)	Poor (2–5)	Good (15–20)	Excellent (>80)
Material utilization (%)	10–15 for powder ~100 for wire	~100	~100
Deposition rate	Medium	Medium	High
Unit size and cost	Bulky and expensive	Bulky and expensive	Compact and economical
Deposit quality	Good	Excellent	Good
Geometrical quality	Near net shape	Requires machining	Requires machining

Source: Karunakaran, K.P. et al., *Rapid Prototyp. J.*, 18, 264–280, 2012.

marketed by Precision Optical Manufacturing, Inc. USA), laser augmented manufacturing (developed by Aeromet, USA), directed light fabrication (DLF), and electron beam free form fabrication (EBF^3; developed at NASA Langley Research Center, USA). These processes use either lasers or an electron beam the as energy source to melt the metal during deposition. Other fusion-based processes which use arc-based energy sources are hybrid plasma deposition and milling (HPDM) and shape metal deposition (SMD; developed at Stanford and Carnegie-Mellon Universities, USA) where metal wire is used as feedstock. Table 5.1 compares the characteristics of these energy sources. Solid-state deposition processes include ultrasonic consolidation (UC; developed by Solidica, USA), electrochemical FABrication (EFAB; marketed by Microfabrica, Inc. developed at the University of Southern California, USA), and emerging technology, namely, friction freeform fabrication (FFF; developed at Indian Institute of Technology Madras, India). Both UC and FFF are considered as hybrid AM technologies as machining is required for each layer to give desired contour, where friction generated heat and plastic deformation is the source of bonding. EFAB is based on electrodeposition and primarily used to fabricate micron-scale devices.

5.2 CURRENT TECHNOLOGIES

5.2.1 Powder Deposition-Based Techniques

The most popular powder deposition-based AM technique uses lasers as energy source and no other type of energy sources have been reported yet. In this process, the metal powder is delivered to the melt pool using an inert gas such as argon and therefore use of electron beam energy sources is precluded as it requires high vacuum. There are four major versions of this process namely LENS™, DMD, laser augmented manufacturing, and DLF that share a common AM principle where a high power laser is used as energy source and metal powder as feedstock material. However, in LENS™ and DLF the deposition process is carried out in a glove box with controlled atmosphere and DMD process uses inert gas shroud to prevent oxidation of deposit with the process being carried out in a chamber without inert atmosphere. In all techniques, the deposition process begins with the creation of a small liquid metal pool on the substrate to which a predetermined amount of metal powder is delivered using inert gas as carrier. The powder melts in the liquid metal pool and the substrate (fixed to a CNC table) moves relative to the deposition head creating solidified metal track. Deposition of overlapping metal tracks completes a layer, and the deposition head along with the powder delivery nozzles move up by small distance (slice thickness) to deposit the next layer. The process continues for all layers producing near net shape metallic components represented by a 3D CAD model. The deposition path, distance between successive metal tracks, and slice thickness are usually created using customized software in each process. Typical processing steps and various components of LENS™ system are presented in Figure 5.2.

Latest LMPD systems are equipped with multiple powder feeders, multi-axis deposition and closed loop process control systems, which enable fabrication of near net shape metallic components with high surface finish, dimensional accuracy, microstructural uniformity, and compositional and/or structural gradients. These techniques are also being used for repair, remanufacturing, feature addition, claddin, and hardfacing of aerospace and engineering components. However, the unique capabilities

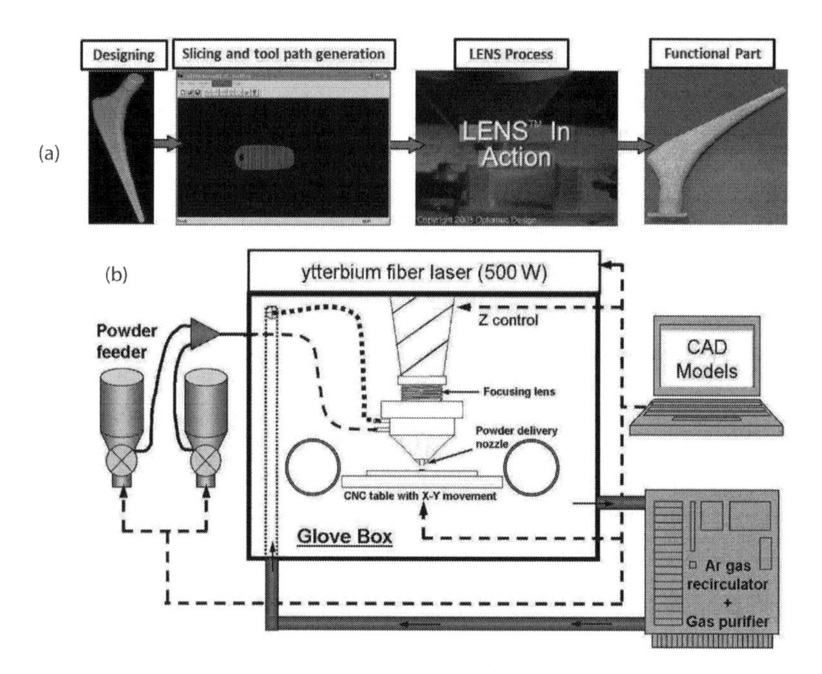

FIGURE 5.2 (a) Typical processing steps involved in laser metal powder deposition (LMPD); (b) Schematic of LENS™ system. (Reprinted from Das, M. et al., *Trans. Indian Ceram. Soc.*, 72, 169–174, 2013. With permission.)

of laser-based deposition techniques have been effectively exploited to produce new/designed materials such as compositionally graded materials, structurally graded materials, porous structures, and custom implants with tailored mechanical, physical, and chemical properties (Mazumder 2000, Mazumder et al. 2000, Shin et al. 2003).

The stability and hence the quality of deposits prepared by LMPD depends on physical phenomena of the process, which is dictated by absorbance of laser beam by metals, surface tension, and viscosity of the melt. The laser absorbance of metal powder is a very important factor to control heating and melting of the powder—too low absorbance requires high energy input or results in partial melting leading to porous deposits and excessively high could lead to evaporation of material during deposition. For example, net shape bulk alumina ceramic parts have been successfully fabricated at a laser power of 175 W (Balla et al. 2008) and silica-based lunar regolith parts at 50 W (Balla et al. 2012b), whereas the fabrication of fully dense metallic parts would require much higher laser power. The difference is primarily attributable to more effective laser absorbance of ceramic materials compared to highly conductive metals. The optimal processing window for laser processing of materials also depends on laser light absorptivities of the constituent elements in the materials. España et al. (2011) processed Al-12Si alloy using LENS™ where large difference in laser absorptivities of Al and Si posed severe difficulties in achieving sound and stable deposits.

Since the LMPD process relies on melting of metals, the surface tension/wettability of the liquid metals against substrate and/or previous deposits is very important for deposit stability during processing. Das (2003) reported that formation of an oxide layer on the powders due to contamination could lead to defects in the deposits such as balling and therefore the protective atmosphere should be carefully controlled using high purity inert gases. Additionally, the viscosity of liquid metal should be optimum to achieve good spreading of freshly deposited metal on previous layers/substrate. It is generally accepted that a high total energy input (combined effect of laser power, scan velocity, and powder feed rate) during deposition decreases the melt viscosity and aids spreading in the majority of metals and alloys. However, in multi-material deposition, the viscosity may increase with energy input if intermetallic compounds form during deposition. Another important consequence of melt viscosity is the balling effect in the LMPD processing. Very high melt viscosity (at low energy input) generates severe balling effect (Figure 5.3), and high energy input with very low melt viscosity results in melt spreading (España et al. 2011). It appears that precise control of melt pool temperatures and hence the melt viscosity by process parameter optimization is very critical to deposit new materials such as metal matrix composites, where constituent elements/compounds have different laser absorbance capacities.

In general, the surface finish in terms of roughness of the parts produced by LMPD processes is relatively higher than the parts fabricated using powder bed-based processes. The surface finish has been reported to be influenced by layer thickness, laser power, deposition speed, and powder feed rate. Gharbi et al. (2013) reported that a combination of deep melt pools and thin layers can reduce the surface roughness of Ti6Al4V alloy parts produced using DMD. The surface finish can also be improved with slow deposition speed, particularly the speed of wall/contour deposition (Mazumder et al. 2000). However, Kong et al. (2007) reported that the Inconel

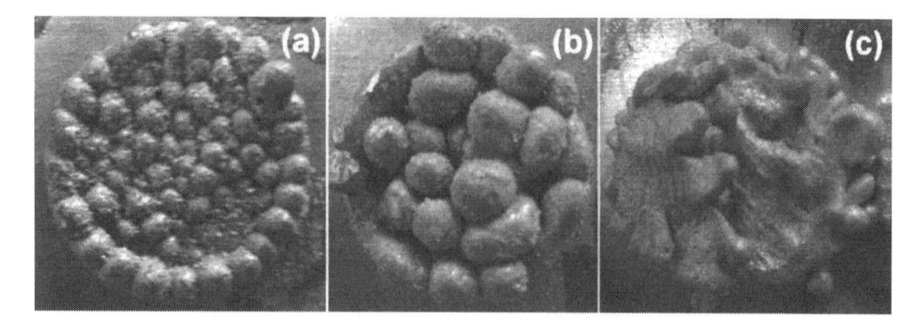

FIGURE 5.3 Laser deposited Al-12Si alloy. (a) Porous deposit at low energy input; (b) severe balling to due high melt viscosity; and (c) melt spreading due to low melt viscosity at excessively high energy input. (Reprinted from España, F.A. et al., *Philos. Mag.*, 91, 574–588, 2011. With permission.)

625 parts produced using finer powder size exhibited superior surface finish and deposition efficiency compared to coarse powder. Further, the surface finish of the DMD processed parts has been shown to improve by changing the position of powder entry into the melt pool (Zhu et al. 2012). Recently it was found that the use of pulsed lasers, instead of continuous lasers, help to improve the surface finish (Pinkerton and Li 2003). Deposition of Ni-based super alloy using DMD in pulsed mode resulted in average surface roughness of 2 μm (Xue et al. 2011). Reduced thermal gradients and Marangoni flows in the melt pool are thought to be responsible for forming smoother deposits in pulsed mode compared to continuous mode of lasers (Gharbi et al. 2014).

For stable deposition process, continuous and precise control over powder feed rate, laser power, and deposition speed is very essential as these dictate the melt pool size, thermal gradients, and cooling rates (Hofmeister et al. 1995). Therefore, real-time thermal imaging of the melt pool and closed loop feedback control for melt pool have been developed. The cooling rates during LMPD can vary between 10^3 and 10^8 K/s (Hofmeister et al. 2001, Zheng et al. 2008, Das et al. 2010) and can be controlled via process parameters enabling creation of tailored microstructures and properties. One critical application area of such controlled cooling rates is the processing of metallic glasses. Balla and Bandyopadhyay (2010) fabricated Fe-based bulk glass forming alloy components without losing amorphous structure of feedstock powder via high cooling rates achieved by maintaining low prior deposition temperature using short time delay between successive laser scans. It appears that LMPD techniques present a viable processing route to create amorphous components using existing bulk amorphous alloy powders. However, each deposit experiences several reheating cycles during deposition of fresh layer, leading to complex solidification and transformed microstructures (Balla and Bandyopadhyay 2010). In addition, rapid cooling rates are also responsible for locked in residual stresses leading to warpage, cracking, and deterioration of mechanical properties of final parts. The beneficial effect of rapid cooling rates during LMPD are fine grains, fine precipitates, absence of segregation, etc. Another inherent characteristic of this process is directional solidification due to preferential heat flow though the substrate, which results in some anisotropic properties. Further, the heat buildup with deposition of large number of layers could produce large variation in microstructures between the first layer to last layer of the part (Hofmeister et al. 2001, Wu 2007). A detailed review on laser-based additive manufacturing of metals can be found in Gu et al. (2012).

Development of materials with gradual change in composition using LMPD is regarded as the best approach to incorporate such variations in net shape components with tailored properties (Schwendner et al. 2001, Banerjee et al. 2002, 2003, Collins et al. 2003, Oruganti and Ghosh 2003). Compositionally graded coatings for biomedical and other applications have been successfully fabricated using LENS™ (Bandyopadhyay et al. 2007, Balla et al. 2007, 2009a, Krishna et al. 2008a, Dittrick et al. 2011). Balla et al. (2009b) created a thin ZrO_2 layer on zirconium via laser assisted oxidation by controlling the concentration of oxygen in the glove box of LENS™. These films have been shown to exhibit good wear resistance and biocompatibility. Similarly, fabrication of unitized acetabular shell structures with porous titanium on one side and compositionally graded TiO_2 on the other side have also been successfully demonstrated (Balla et al. 2009a). Unique capabilities of the LENS™ process in creating novel structures are reported in (Bandyopadhyay et al. 2009, España et al. 2010, DeVasConCellos et al. 2012, Das et al. 2013). Custom implants with the desired porosity level in the proximal region of the implant to enable bone ingrowth and a fully dense distal region to support a mechanical load fabricated using LENS™ are shown in Figure 5.4. Another unique possibility of fabricating two separate parts in a single step using this technique has also been reported by Espana et al. (2010). For example, a part with dense sleeve and porous

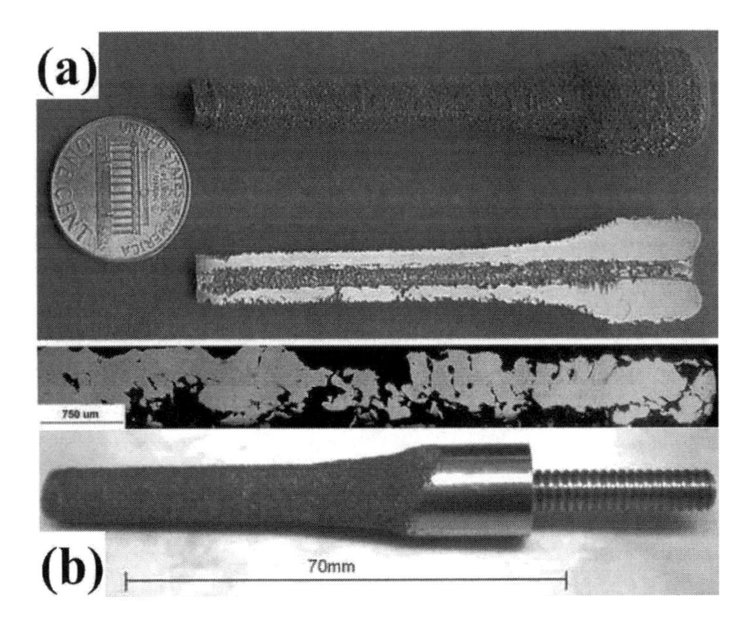

FIGURE 5.4 (a) LENS™ processed custom implants. Lower cross-sectional image shows the porosity in proximal region of the implant and (b) complete assembled implant. (Reprinted from DeVasConCellos, P. et al., *Vet. Comp. Orthopaedi. Traumatol.*, 25, 286–296, 2012. With permission.)

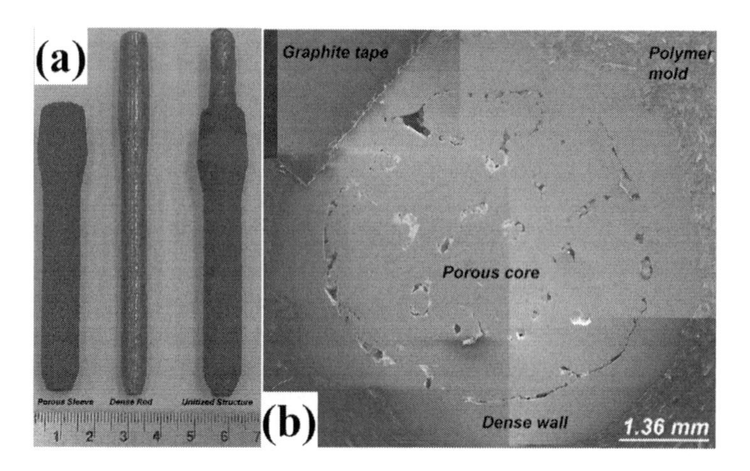

FIGURE 5.5 LENS™ processed components and materials (a) left: porous sleeve; middle: solid core; right: unitized structure fabricated in single step and (b) CoCrMo alloy structure with solid shell and porous core. (Reprinted from España, F.A. et al., *Mater. Sci. Eng. C*, 30, 50–57, 2019. With permission.)

core requires assembly of these two parts in convention manufacturing and the sharp interface could be a source of failure. However, manufacturing them in a single step using AM techniques (Figure 5.5) not only solve this issue, but also eliminate time consuming machining of interface surfaces required for assembly.

Extensive research has also been done in the area of creating porous structures using LMPD process (Krishna et al. 2007, 2008b, 2009, Xue et al. 2007, Bandyopadhyay et al. 2010, Balla et al. 2010b). A novel design concept has been proposed by Krishna et al. (2007) to create porous structures with desired pore characteristics and distribution as shown in Figure 5.6. It has been demonstrated that by controlling the extent of metal powder melting, via the appropriate combination of process parameters, the residual porosity in the deposited tracks can be tailored (Approach A in Figure 5.6). By utilizing the design flexibility of AM processes, porous structures with designed porosity characteristics (pore size, shape, and distribution) can be fabricated by changing the layer thickness and distance between two successive metal tracks as shown in Approach B. Three dimensionally interconnected porosity in the structures can be created by combining the above two approaches. LENS™ processed porous titanium samples with and without designed porosity have also been tested for their mechanical properties and deformation behavior (Balla et al. 2010c). It was found that regular arrangement of pores (tailored pore distribution) in designed porosity samples improve 0.2% proof strength to 485 MPa from 220 MPa in random porosity samples with comparable total porosity and pore size. This observation clearly demonstrates that a drop in mechanical strength of porous metals can be compensated by tailoring pore distribution. Balla et al. (2011) discovered that brittleness associated with porous metals processed using powder metallurgical routes can be eliminated in laser processed porous metals and is primarily due to differences in particle bonding in these processing routes. However, Bernard et al. reported that presence of 10% porosity decreases the rotating bending

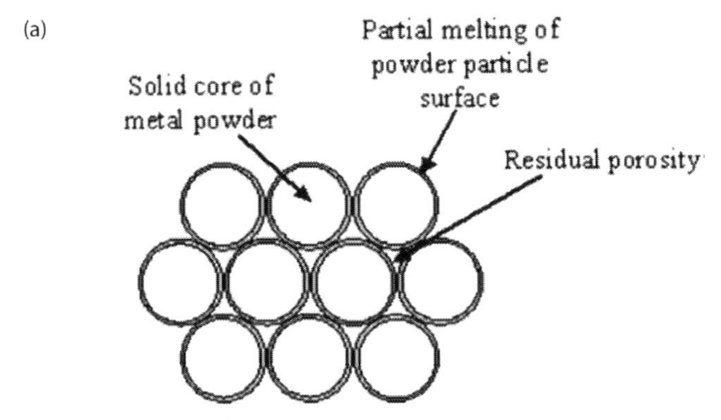

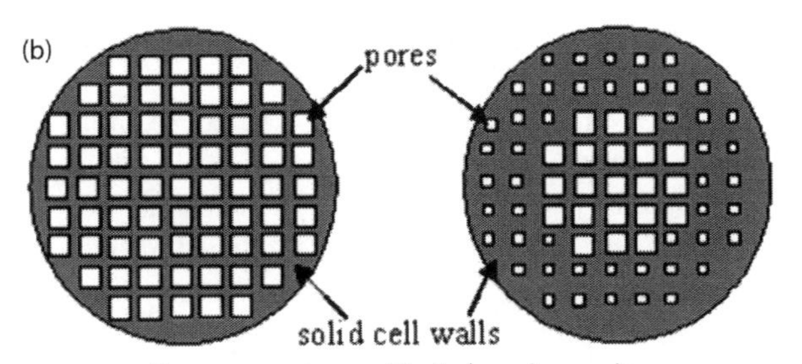

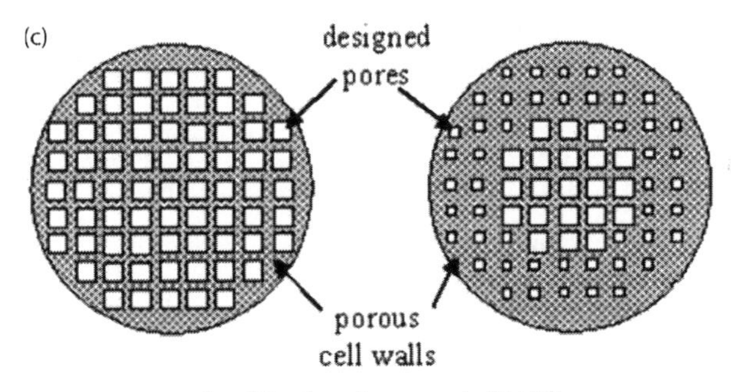

FIGURE 5.6 (a–c) Design approaches to create functional implants with tailored pore characteristics such as size, shape, and distribution. (Reprinted from Krishna, B.V. et al., *Acta Biomater.*, 3, 997–1006, 2007. With permission.)

fatigue strength of NiTi alloy by 54% (Bernard et al. 2011), while compression fatigue testing demonstrated that porous NiTi alloy samples (up to 20% porosity) processed using LENS™ are able to sustain stresses up to 1.4 times their yield strength without failure (Bernard et al. 2012). Several biocompatible coatings and composite coatings (Roy et al. 2008, 2012, Balla et al. 2010a, 2010d, 2012b, 2013, Bhat et al. 2011, Das et al. 2011, 2012) and bulk ceramics (Balla et al. 2008a, 2012a, Bernard et al. 2010) processed using LENS™ have been reported.

5.2.2 Wire Deposition-Based Processes

Additive manufacturing based on powder deposition are the most widely used and researched technologies for metals. These technologies demonstrated their capabilities to manufacture complex, but small components. However, powder deposition-based techniques suffer from low deposition rate and yield, high surface roughness, and residual gas porosity. For example, the deposition efficiency of powder-based AM techniques depends on melt pool area and problems associated with powder recycling, contamination, and storage are also high (Kukreja et al. 2012). As a result fabrication of large area structures using these techniques could become expensive. A majority of these issues can be obviated using alternative feedstock materials, and one such approach is the use of metal wire as feedstock.

Wire-based deposition for AM of components has been found very promising (Nurminen 2008, Heralić 2012) particularly for large components where dimensional accuracy is vital. Figure 5.7 shows the process of wire deposition-based additive manufacturing of metallic components. The process starts with creating of small melt pool on the substrate using appropriate energy source. Then the wire is fed to the melt pool at a controlled rate and is melted by a focused energy source. By moving the wire nozzle and energy source, relative to the substrate, along the desired path creates a thin metal bead. A complete layer is produced by depositing overlapping beads, and the process is repeated until a 3D component is created. Normally, the deposition is carried out in a controlled atmosphere. Post-processing such as grinding or machining may be performed depending on the final requirements.

Compared to powder feeding, the wire feeding for AM of metals offers several benefits. Significantly high deposition rates up to 1500 cm^3/hr have been reported for EBF^3 (Taminger and Hafley 2003, Seufzer and Taminger 2007). Similarly, laser-based wire deposition has been shown to provide high deposition rates (Syed and Li 2005, Syed et al. 2005, Nurminen 2008). Martina et al. reported a deposition rate of 1.8 kg/h with Ti6Al4V alloy wire using plasma wire deposition (Martina et al. 2012). Irrespective of energy source, wire feeding gave better surface finish, material

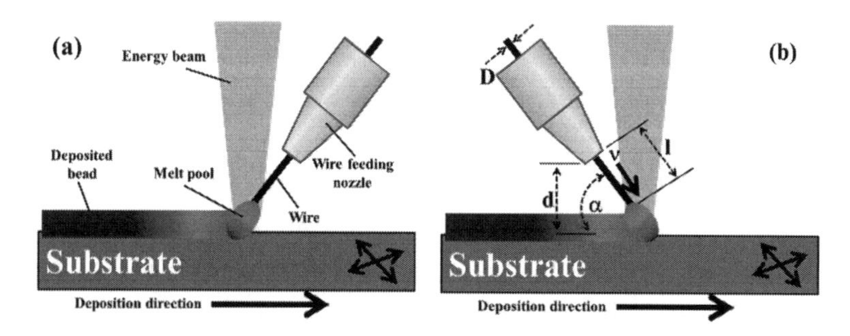

FIGURE 5.7 Schematic showing wire deposition-based additive manufacturing. (a) Front feeding with different components of the processing equipment and (b) rear feeding process and important geometrical process parameters. D: wire diameter; d: stand-off (too small 'd' lead to stubbing, and too high results in dripping); α: wire feed angle; v: wire feed rate; l: stick-out length. The wire tip position can be at leading edge, middle, or tailing edge of the melt pool.

quality (Ader et al. 2004), and usage efficiency (Syed and Li 2005, Syed et al. 2005, Nurminen 2008). Other benefits include low cost of wire preparation (Kim and Peng 2000), clean work environment due to almost 100% wire utilization, and minimal health hazards. However, wire-based deposition is very sensitive to several process parameters and should be carefully controlled. Therefore, process optimization and control is extremely important to achieve stable deposition. Important process parameters include type of energy source, energy input, wire feed rate and feeding position, wire tip position in the melt pool, and traverse speed (Figure 5.7). Major energy sources used for wire deposition-based AM are laser, electron beam, and electric arc. Among these, laser-based wire deposition has been extensively studied followed by the electron beam-based deposition process. Although the electric arc source is not as precise as laser and electron beam, recently, 3D net shape components with mesoscale features have been successfully fabricated using a miniature deposition process consisting of a micron-size wire and micron-tungsten inert gas welding system (Horii et al. 2009). Combination of wire and powder feeding has also been reported (Syed et al. 2006, 2007a, 2007b, Wang et al. 2006, 2007).

5.2.2.1 Laser-Based Metal Wire Deposition

Laser-based metal wire deposition has been widely used to deposit Ti and Ti6Al4V alloy, and their microstructural and mechanical properties evaluations have also been done (Kim and Peng 2000, Cao et al. 2008, Miranda et al. 2008, Hussein et al. 2008, Mok et al. 2008a, 2008b, Medrano et al. 2009, Brandl et al. 2010, 2011a, 2011b, 2011c, 2012, Baufeld et al. 2011, Abioye et al. 2013). Detailed microstructural analysis of single bead Ti6Al4V alloy deposited using laser and wire as feedstock material was studied by Brandl et al. (2011a). It appears that the laser power and deposition speed during wire deposition process had similar effect to that of powder deposition processes. The microstructural features such as prior β-grains were found to increase in size with laser power and decreased with deposition speed. While increasing the wire feed factor (deposition speed/wire feed rate) increased the feature size (Brandl et al. 2011a). The influence of these microstructural features on hardness of single beads was studied (Brandl et al. 2011b). In this study, they measured the hardness and bead dimensions and attempted to correlate with thermal history during deposition. The bead dimensions provided good qualitative information of thermal history but hardness mapping failed to provide good correlation. Large columnar grains spanning across many layers were formed (Brandl 2012). Post deposition heat treatment had stronger influence on hardness compared to process parameters. As-deposited Ti6Al4V alloy exhibited tensile yield strength in the range of 697–884 MPa and elongation between 5% and 12% depending on process parameters and post-deposition heat treatment (Brandl et al. 2011c). Importantly, the impurity levels of wire deposited Ti6Al4V alloy were below acceptable levels of aerospace material specifications (AMS 4911L) and mechanical properties meet AMS 4928 specifications (Brandl et al. 2011c). Example of deposits and parts prepared in (Brandl et al. 2011b) are shows in Figure 5.8.

Very recently, wire laser deposition has been employed to fabricate Ni-based superalloy, Inconel 625, and process parameters have been optimized to achieve sound beads (Abioye et al. 2013), wherein energy input and deposition volume per

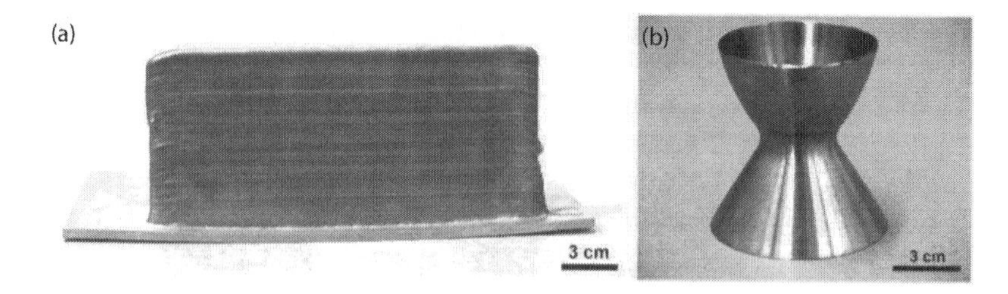

FIGURE 5.8 Archetypal thin wall deposit (a) and machined thruster (b) fabricated using Ti6Al4V wire deposition process. (Reprinted from Brandl, E. et al., *Surf. Coat. Technol.*, 206, 1130–1141, 2011b. With permission.)

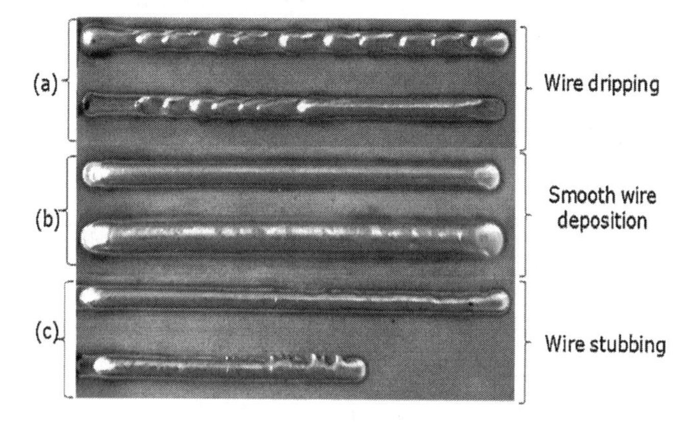

FIGURE 5.9 Laser wire deposited beads of Inconel 625. (Reprinted from Abioye, T.E. et al., *J. Mater. Process. Technol.*, 213, 2145–2151, 2013.)

unit track length are identified as key process parameters. As shown in Figure 5.9a wire dripping occurs when the deposition volume is very low, and when it becomes excessively high wire stubbing (Figure 5.9c) was observed. Smooth bead deposits with good dimensional stability can only be attained (Figure 5.9b) with parameters that provide smooth wire transfer during deposition (Abioye et al. 2013). The distance from the wire tip and the substrate (d in Figure 5.7) also have been reported to produce similar effects on deposited beads (Heralić 2012). Low dilution was achieved with high wire feed rate, high deposition speed, and low laser power (Abioye et al. 2013). Wire feeding direction (front feeding or rear feeding, Figure 5.7), feeding angle (α in Figure 5.7b), and the position of wire tip in the melt pool (leading edge, middle, or tailing edge) are also found to have a strong effect on overall quality of the deposit in terms of porosity, surface finish, and geometrical control (Syed and Li 2005). Feeding angle effect on bead roughness depended on wire feeding direction—high angles resulted in rough and smooth beads for front and rear feeding, respectively. For stable bead deposition, the wire tip position in the melt pool should always be away from the solidification start point and good quality deposits can be obtained with either front or rear feeding, but with a different set of process parameters (Syed and Li 2005).

From the above discussion, it is clear that the wire-based deposition process is sensitive to a large number of process parameters, and maintaining and controlling stable deposition is of utmost importance to achieve high quality parts. Therefore, continuous monitoring and control of wire deposition has been attempted by several authors (Heralic et al. 2008, 2010, 2012, Liu et al. 2014, Hagqvist et al. 2014). Hagqvist et al. (2014) proposed an innovative approach for controlling the laser metal wire deposition process via electrical resistance between wire and the melt pool. They demonstrated that this approach effectively controlled wire dripping and stubbing by automatic adjustment of stand-off distance (d in Figure 5.7b). The result of resistance measurement for online wire deposition control is shown in Figure 5.10. A 3D scanning-based system has also been used to control the stand-off distance thus achieving flat deposits (Heralic et al. 2012). The wire feed rate control based on deposits' 3D scanned data helped to compensate the deviations in deposit heights.

5.2.2.2 Electron Beam Free Form Fabrication

EBF^3 was developed at NASA Langley Research Center, USA and is capable of producing complex parts using a variety of metals and alloys. The process is very similar to the laser-based wire deposition process except that it is carried out in high vacuum (typically between 1×10^{-4} and 1×10^{-5} Torr) with an electron beam as energy source. Several advantages of EBF^3 over the laser-based deposition process have been reported (Stecker et al. 2006) including high power efficiency ($\geq$90%) and high coupling efficiency. Therefore, it is highly suitable for materials that reflect lasers such as aluminum and copper and is highly flexible in terms of achieving desired surface finish and feature size. The high vacuum environment of EBF^3 ensures clean deposits while loss of some elements from the melt pool is also unavoidable. In general, fine diameter wires are used for complex components with fine features, and for high deposition rates large diameter wires are preferred. Recent developments enabled deposition of compositionally graded components using dual wire feeders. Further, the EBF^3 process enables part fabrication in space as well (Taminger 2009). The surface finish of the parts produced using EBF^3 is also excellent as shown in Figure 5.11.

The EBF^3 process is controlled by several parameters, namely, beam power and beam pattern apart from other parameters shown in Figure 5.7 for laser-based deposition process. These parameters strongly influence the deposit quality, residual stresses, final chemical composition, etc. Matz and Eagar (2002) examined net shape fabrication of Alloy 718 using EBF^3. It was found that the spherical carbide precipitates size can be significantly reduced using the EBF^3 process and is attributed to rapid cooling rates. Similarly, detrimental Cr-carbides were suppressed during

FIGURE 5.10 Wire dipping without online controller (above) and smooth deposit produced using resistance measurement-based online control (below). (Reprinted from Hagqvist, P. et al., *Opt. Lasers Eng.*, 54, 62–67, 2014. With permission.)

FIGURE 5.11 Typical part produced using EBF^3. Note the macrostructure showing columnar grains oriented along the part axis, which demonstrate EBF^3 ability to produce smooth parts. (From Taminger, K., *Adv. Mater. Process.*, 11–12, 45, 2009.)

EBF^3 processing of 347 stainless steel leading to tensile properties comparable to that of wrought equivalent (Wanjara et al. 2007). Several authors reported the influence of the EBF^3 process parameters on microstructures and mechanical properties of aluminum alloys (Taminger and Hafley 2002, 2003, Taminger et al. 2006), which emphasize process optimization and control (Seufzer and Taminger 2007). Other issues that require close attention include loss of certain elements form the deposit (for example, Al from Ti6Al4V alloy), improvements in repeatability, residual stresses and distortion, gradient deposits, and tailored microstructures.

5.2.2.3 Arc-Based Wire Deposition Processes

Arc-based wire deposition processes are known as SMD and use a metal inert gas welding technique to produce dense components (Akula and Karunakaran 2006). The process was originally developed by Rolls-Royce. Typically, the process is controlled by commercial welding robots with dimensional accuracy and surface finish comparable to that of beam-based processes. Advantages of this technique over beam-based processes are relatively high deposition rate, power density at low cost, and ability to pulse the arc providing additional microstructural control. To date, the majority of weldable alloys have been deposited using the SMD technique, which include Ti alloys (Katou et al. 2007, Baufeld et al. 2009, 2010, Baufeld and Van der Biest 2009), steels (Skiba et al. 2009, 2011) and Ni-base alloy (Clark et al. 2008). One important challenge in this process is deposition of overhang structures due to lack of support to liquid metal pool. However, recently, electromagnetic confinement of liquid metal pool was found to increase the tilt angle by 10° (Bai et al. 2013). Typical Ti6Al4V alloy components produced using SMD are presented in Figure 5.12.

Clark et al. (2008) developed combustion outer casing with Alloy 718 using SMD based on their initial multi-pass deposits. However, they could not control the formation of laves and delta phases in alloy 718 during solidification. Ti6Al4V alloy samples fabricated by SMD exhibited tensile strength in the range of 929 and 1014 MPa, which are comparable to equivalent cast material (Baufeld et al. 2010). To address feature resolution of SMD, recently, micro-arc-based deposition processes have been developed (Horii et al. 2009, Jhavar et al. 2014). Net shape manufacturing of meso-scale parts using micro-tungsten inert gas welding was reported by Horii et al. (2009). Very recently, more energy efficient and cost effective deposition processes based on micro-plasma transferred arc reportedly produced tool steel deposits to repair dies and molds (Jhavar et al. 2014). The process has been demonstrated to achieve wall width of ~2 mm with deposition efficiency of 87% and deposition rate

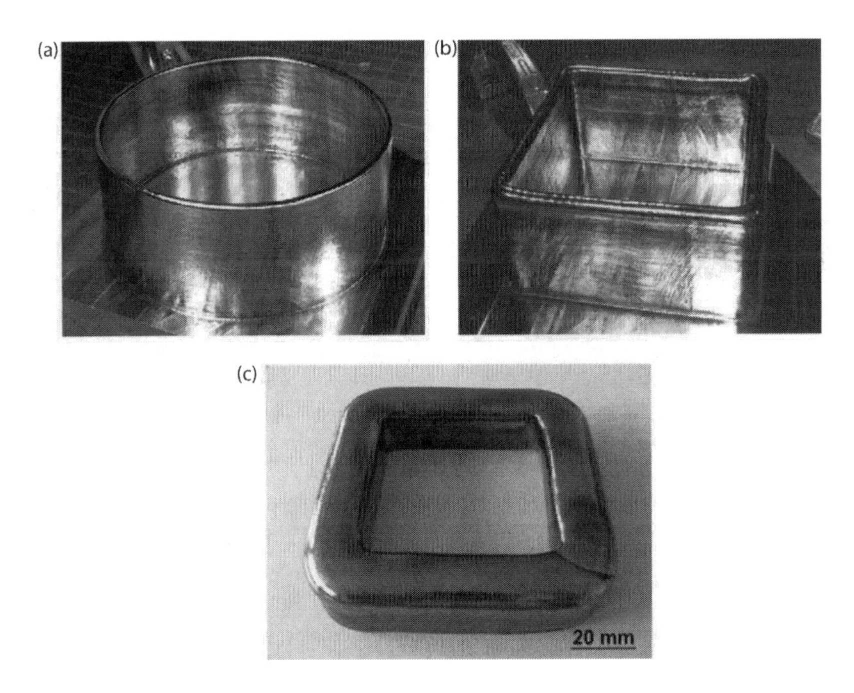

FIGURE 5.12 Tubular parts fabricated using SMD (a) and (b) thin wall components and (c) thick wall (20 mm) component. (Reprinted from Baufeld, B. et al., *Mater. Des.*, 31, S106–S111, 2010. With permission.)

of 42 g/h. The deposits were also metallurgically and physically sound without any defects. The properties of Ti6Al4V alloy fabricated using laser and arc beam deposition were found to be comparable (Brandl et al. 2010). Similar observations were also reported by Baufeld et al. (2011) where properties of the same alloy produced via laser beam-based deposition and SMD processes were compared. Other reports include fabrication of Ti6Al4V alloy using the wire arc AM process (Wang et al. 2013) and stainless steel powder consolidation using the electric arc (Rangesh and O'Neill 2011).

5.2.3 Solid-State Additive Manufacturing Processes

Solid-state additive manufacturing techniques have been developed to create complex 3D structures with metals that are difficult to process using fusion-based techniques such as LENS, DMD, SMD, etc. Additionally, solid-state processes enable processing of metallurgically incompatible metals, create laminated materials, and embedded structures. UC is the only solid-state additive manufacturing technology based on ultrasonic metal joining that is commercially available since 2000 from Solidica Inc., USA. UC is a hybrid additive manufacturing technique, and commercial UC machines consist of ultrasonic welding head (sonotrode), thin metal foil feeding system, and a CNC milling station. Like other AM processes, UC also uses custom software to generate layers and processing conditions. However, the layer thickness is decided based on available metal foil thickness. The UC process and bonding mechanism are presented in Figure 5.13.

The process begins with feeding thin metal foil (typically between 100 and 150 μm thick) which will be pressed against a base plate using normal load applied through sonotrode. The sonotrode vibrates transversely at 20 kHz under specified normal load and travel across the length of the part creating metallurgical bond. A layer will be created by a deposition series of foil strips side-by-side, and the final shape/contour of the layer will be achieved using CNC milling. Then compressed air is used to clean off the machining debris and the next layer of deposition starts. The CNC milling is usually performed after several layers have been deposited, and the process of deposition and milling continue until the 3D component is produced. The ultrasonic welding heads usually have a rough knurl surface which keeps the foil intact with the sonotrode head while the sonotrodes oscillate at high frequency. The ultrasonic oscillations of top foil against the bottom foil/base plate create frictional forces and break up oxide layers that bring the atomically clean metal surface together. The preheating and friction generated heat accelerate the atomic diffusion across the metal interfaces and a strong metallurgical bond forms under the influence of normal force. UC has been extensively used to fabricate different multi-material and multi-functional metal structures with embedded sensors, circuits (Kong 2005, George 2006, Janaki Ram et al. 2007a, Siggard 2007, Obielodan et al. 2010, Friel and Harris 2013). It has been demonstrated that metal matrix composites can also be fabricated using UC (Yang et al. 2007). Fabrication of a novel Al composite with a tailored coefficient of thermal expansion has also been attempted by incorporating shape memory alloy in an Al 3003 matrix using UC (Hahnlen and Dapino 2014).

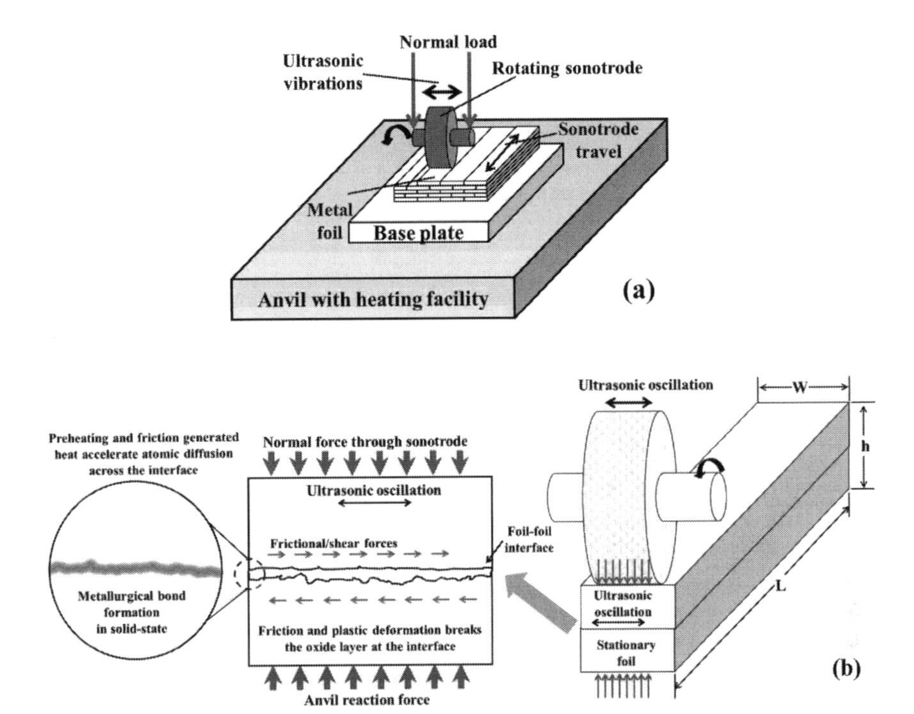

FIGURE 5.13 (a) Ultrasonic consolidation process and its components and (b) illustration of foil geometrical parameter (right) and formation of metallurgical bond between the foils during ultrasonic consolidation (left).

Important process parameters include normal force (500–2000 N), sonotrode texture (Ra between 4 and 15 μm), sonotrode amplitude (5–150 μm), and sonotrode travel speed (10–50 mms/), preheating temperature (93°C–150°C). Too low sonotrode amplitude and normal force produce very weak bonds, and very high values of these parameters could lead to excessive foil deformation and misalignment of the layers. Therefore, to achieve strong bonding and bulk components, the optimal choice of process parameters is very important for each material and part geometry (Soar and Dickens 2003, Kong et al. 2004). UC of dissimilar metals has been reported by Obielodan et al. (2011), and the influence of ultrasonic energy on material softening was studied by Langenecker (Langenecker 1966). A study by Gonzalez and Stucker (2012) demonstrated that linear weld density strongly influenced by process parameters and highest linear weld density of 95.89% was obtained at 1800 N normal load, 27 μm ultrasonic amplitude, 11 mm/s sonotrode travel speed, and a temperature of 204°C. Further, they emphasize the use of high power UC to achieve strong structures using high-strength materials. Recently, the sonotrode surface texture found to influence the bond strength in UC-processed Al alloy and surface roughness around 6 μm has been suggested (Li and Soar 2009). Increasing the surface roughness can potentially eliminate foil overlap and crinkling due to high ultrasonic energy transfer to the foil while improving the peel strength and linear weld density. It is also important to state here that increased surface roughness of sonotrode may transfer this roughness to deposited foils, which may affect bonding of subsequent foils. Excessively high and low roughness of sonotrode were found to result in low linear weld density (Friel et al. 2010). It appears that there exists optimal sonotrode topography that ensures strong and effective bonding during UC due to efficient energy transfer and inter-foil deformation. Similarly, the build geometry strongly influences the stability of the UC process (Gibert et al. 2010).

Understanding the fundamental mechanism of bond formation during UC is still an important, but challenging area of research. Plastic deformation is an essential part of UC which bring the two metals in intimate contact and break the surface oxide layer. Earlier studies show that bond formation during ultrasonic welding, after intimate contact is achieved, is due to mechanical interlocking, interfacial melting, and metal diffusion (Joshi 1971). It is extremely important to identify process parameters dependent dominant mechanism and which mechanisms enable formation of strong metallurgical bonding during UC. Experimental investigations on bond formation during UC of similar and dissimilar metals have been reported (Janaki Ram et al. 2007b, Yang et al. 2009). The results showed no evidence of above mechanisms, namely, mechanical loading, melting, and diffusion, which suggest that the bonding occurred purely in solid-state. It was concluded that removal of oxide layers and formation of intimate contact between the metal surfaces is responsible for bond formation (Yang et al. 2009). To understand the influence of process parameters on bonding, a term "total transmitted energy" (E_t) has been developed which primarily depends on normal force, sonotrode oscillating amplitude, and sonotrode travel speed (welding speed). Earlier studies show clear dependence of linear weld density on E_t (Kong et al. 2004, Janaki Ram et al. 2007c), where high E_t improved the bond formation. However, excessively high E_t could damage the previous bonds leading to a drop in linear weld density. Interestingly, the deterioration of bonds was found to be influenced by energy input during single cycle of ultrasonic vibration (E_0) not by E_t (Kong et al. 2004, Janaki Ram et al. 2007c). These energy terms were defined in (Yang et al. 2010), where process parameters such as welding speed, sonotrode amplitude, and normal force were correlated with energy terms and linear weld density. In this study, an analytical model also has been developed to estimate the linear weld density from energy input. In line with this study, Kelly et al. (2014) confirmed through experiments that the bonding in UC occurs in solid-state and is not due to thermal softening or melting. A model was developed to understand the influence of energy input on weld strength and linear correlation was observed. In another study, acoustic softening was found to reduce the yield strength of Al 1100 foils up to 82% and thermal softening was very minimum (Kelly et al. 2013). Interfacial microstructures of UC-processed Al 3003 alloy showed fine-scale microstructural modifications at the foil-foil interface and are due to local plastic deformation as a result of sonotrode texture (Dehoff and Babu 2010).

5.2.4 Electrodeposition-Based Additive Manufacturing

EFAB is one of the additive manufacturing technologies based on Instant Masking™ and the electrodeposition process that can effectively build miniature 3D metal structures with micro-scale resolution. The process is currently being marketed by Microfabrica, Inc. Although originally developed at the University of Southern California in late 1990s (Cohen et al. 1998) later developments (Cohen 1999, Reid and Webster 2006, Kruglick et al. 2006) enabled the process to fabricate functional components as small as 4 × 25 × 25 μm weighing 0.02 μg (Cohen et al 2010). This flexible process produces highly intricate metallic structures/devices of the order of millimeters to centimeters in size and is cost-effective for batches up to 1000 parts. This process can be considered as a hybrid additive manufacturing process where additive and subtractive steps are involved. In general, the process of making each layer consists of electrodeposition of selective pattern followed by blanket deposition and final mechanical planarization. Use of electrodeposition in EFAB enables extremely fine deposits, low residual stresses, no shrinkage, and fine features. Apart from part complexity, the EFAB process can create devices with moving parts that are pre-assembled during the fabrication process. Currently, the process geometrical capabilities include ≥4 μm thick layers having ±1.5 μm inter-layer alignment, 10–20 μm in-plane features with tolerances of ±2 μm, and ±1 μm for Z-axis and X-Y-axis, respectively. The surface finish of the devices fabricated using EFAB is typically around 0.15 μm and further improvements are also possible (Cohen et al. 2010).

The EFAB process is a micro additive manufacturing process and involves three basic steps to generate each layer, and these three steps are repeated until the complex 3D component is built (Vaezi et al. 2013). As with other additive manufacturing technologies, the EFAB process also relies on deposition of structural material (forming feature of final component/device) and sacrificial material (forming support structures), and both materials should be electrically conductive because these materials are deposited using the electrodeposition technique. The three sequential process steps for each layer include: (a) electrodeposition of sacrificial material, (b) structural material electrodeposition, and (c) mechanical planarization. The process starts with creation of "instant masks" that include cross sectional geometry of each layer using custom software (Layerize™) from a 3D CAD model of the final device—single part or assembly of multiple parts. Layerize™ generates: (i) 2D cross sections of each layer in a format compatible with commercial photomask pattern generators and (ii) automated EFAB process control file used for electrodeposition of structural and sacrificial materials. The photomask patterns produced using generated 2D cross sections are used to fabricate "instant masks" using a micromolding technique (Cohen 2002) and are used in the EFAB machine for selective deposition of materials in each layer. The EFAB process begins with selective electrodeposition of sacrificial material with the use of "instant masks." Figure 5.14 shows a typical EFAB process.

The first step involves electrodeposition of sacrificial material on a substrate at selected areas pre-determined by the "instant mask" of the first layer. This is achieved by pressing the substrate (cathode) against "instant mask" (mounted on anode) placed in an electrodeposition cell where the electrolyte occupies the openings in the masks. Then the electrodeposition process is initiated by passing an electric current through the cell electrodes leading to selective deposition of sacrificial material on the substrate at areas defined by the mask. After this, the "instant mask" along with the anode is removed leaving behind the deposited sacrificial material. In the second step, the structural material is electrodeposited

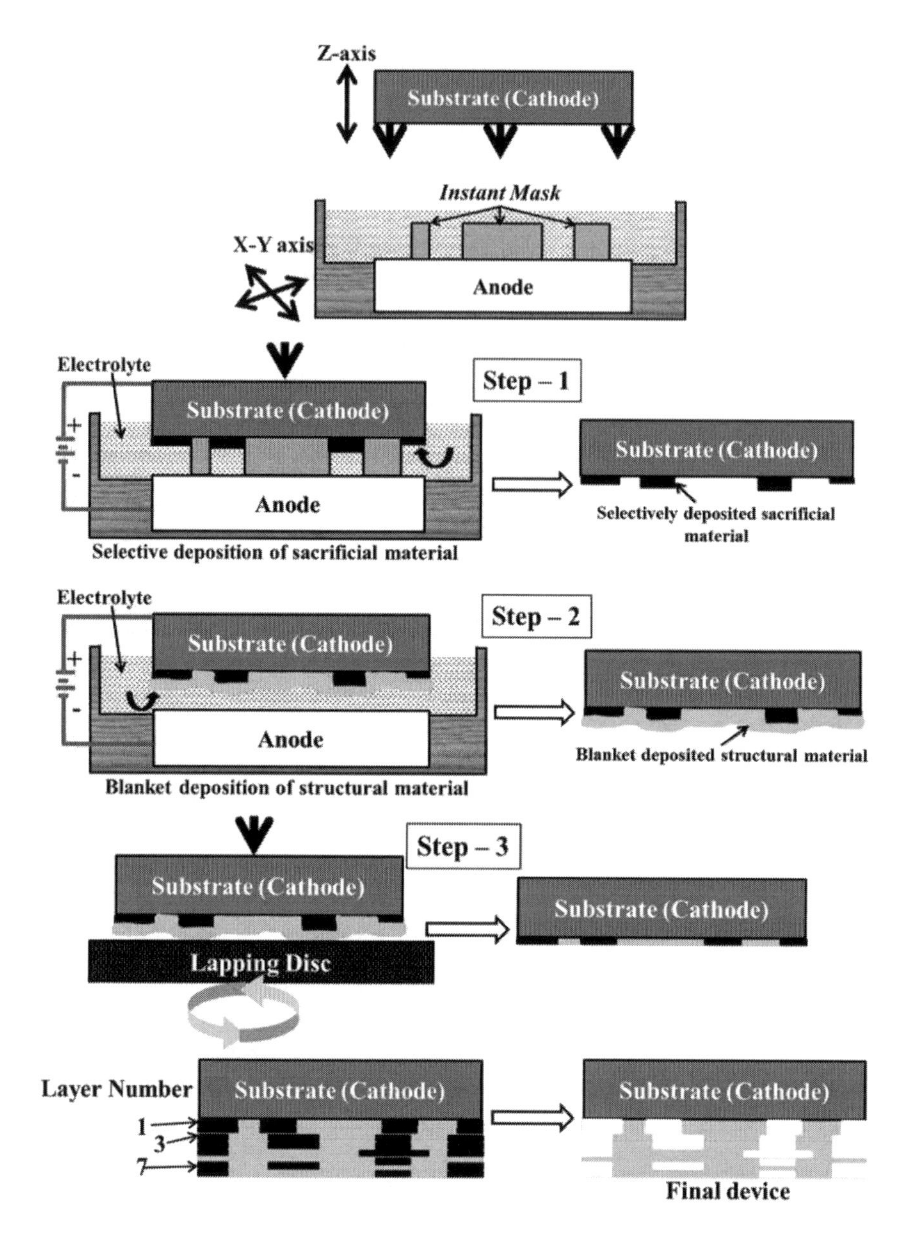

FIGURE 5.14 EFAB process flow.

non-selectively (blanket-deposited) covering the entire area including previously deposited materials and other open areas on the substrate. This process takes place in a separate electrodeposition cell with the appropriate electrolyte and anode. As a third step, the entire deposit is mechanically planarized using lapping plate until both materials are visible and a desired layer thickness with flatness and smoothness is achieved. Other reasons for planarization can be found in Cohen (2002).

Repetition of above three steps for all layers creates a final device embedded in the sacrificial material, which is then chemically etched producing desired structure as represented in the 3D CAD model.

In principle, any material that can be deposited using electrolytic/electroless deposition are good candidate materials for EFAB. Therefore, structures with many pure metals and alloys can be fabricated using EFAB. However, Microfabrica, Inc. developed some limited number of materials such as Ni-Co and rhodium. Ni-Co alloy processed using EFAB has been shown to exhibit good mechanical properties, corrosion resistance, and short-term biocompatibility properties (Cohen et al. 2010). Rhodium also achieved good mechanical properties in as-fabricated condition and the interlayer adhesion of EFAB structures was also found to be >20% of the bulk strength of the structural material (Cohen 2002). One important material consideration for sacrificial material is that it should be selectively etchable after EFAB process.

EFAB process is an enabling micro additive manufacturing technology with very strong future potential. However, like other processes, EFAB has some limitations which include throughput, part size, stair-step effect, and maximum number of layers. Compared to other additive manufacturing technologies, the build rates in Z direction (several hundred mirons/day) is significantly less for EFAB process. The process is limited to a maximum 50 number of layers which is again linked to build rates. Similarly, geometrically large devices (large volume) cannot be easily fabricated using EFAB. The stair-step effect poses some problems in certain devices with moving parts where the clearance between moving parts is smaller than minimum layer thickness. The effective removal of sacrificial materials require designed release holes. Fabrication of microdevices with moving parts and other elements has been successfully demonstrated by EFAB process (Cohen et al. 2010). Further developments in the area of new metals, alloys, and other sensing devices are also anticipated (Vaezi et al. 2013).

5.3 EMERGING ADDITIVE MANUFACTURING TECHNOLOGIES

5.3.1 Friction Freeform Fabrication

Very recently, Dilip et al. proposed FFF that uses friction surfacing—a solid-state surface deposition process, to deposit material layer-by-layer creating 3D metal structures (Dilip et al. 2013). In this technique, the process of depositing a single track of metal on a substrate is very similar to conventional friction surfacing. A consumable rod is rotated at high speed and is forced against a substrate with desired axial force generating frictional heat sufficient to plastically deform the consumable rod. Then, moving the substrate in a predetermined path creates deposition of the consumable rod onto the substrate forming a track. Following this procedure, parallel tracks can be deposited creating a layer, which is then machined using CNC machining to give desired slice/layer contour. The processing of deposition and CNC machining is repeated several times to complete the fabrication of 3D metallic structures. Typically, the track width is of the order of consumable rod diameter, but can be varied along with layer thickness depending on the process parameters. A typical FFF process is schematically shown in Figure 5.15.

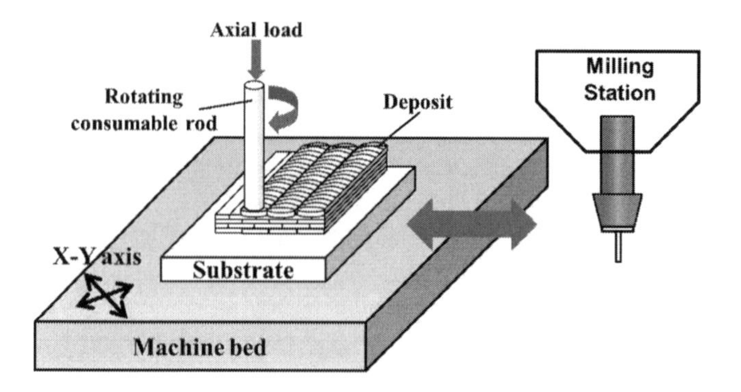

FIGURE 5.15 Friction free form fabrication process.

Different samples with dissimilar metals and structures with enclosed internal cavities have been successfully fabricated using FFF (Dilip et al. 2013). FFF metals exhibited excellent inter-track and inter-layer bonding, fine grained microstructures, and comparable mechanical properties with that of wrought equivalents (Dilip et al. 2013). FFF-processed Inconel 718 alloy was also shown to have good room temperature mechanical properties after direct aging (Dilip and Janaki Ram 2014). However, FFF appears to be detrimental to mechanical properties of heat treatable Al alloys due to precipitate coarsening (Dilip and Janaki Ram 2013). The precipitate coarsening is attributable to very high friction-generated temperatures and repeated heating and cooling cycles (Rafi et al. 2011, Puli and Janaki Ram 2012). Therefore, like fusion-based AM techniques, the evolution of microstructure during FFF of metals is also highly complex. Currently, this technology is being commercialized as MELD by Meld Manufacturing Corporation, Christiansburg, VA.

5.3.2 Hybrid Techniques

A new hybrid AM technique named HPDM has been proposed by Xiong et al. (2008). The HPDM unit consists of a plasma torch and CNC milling station. In this process, metal is deposited using a plasma arc and CNC milling creates layer contour. Important advantages of this process include high deposition rates, near-net shape manufacturing, and an economical energy source. Good surface finish and dimensional accuracy have been achieved using HPDM (Xiong et al. 2008). Another process based on plasma deposition known as electromagnetic compressed PDM has also been reported (Zhang et al. 2003). The feedstock material is a metallic powder that is fed into the molten metal pool created by plasma heating. The powder melts and moving the deposition head relative to the substrate creates a thin metal track. Overlapping the tracks creates one layer, and the process is repeated to fabricate a 3D structure. The major difference compared to HPDM is that the plasma is magnetically confined, and this may help in achieving better feature resolution than HPDM. It appears the overall process is similar to laser-based deposition technique, except the energy source. The deposit quality depends on the powder feed rate, scan velocity, and arc current (Zou et al. 2009).

Fusion-based AM techniques lack optimal balance of efficiency and accuracy with additional problems such as residual porosity, coarse columnar microstructure, and anisotropic properties (Dinda et al. 2009). To address some of these issues, a novel hybrid AM process was introduced by Zhang et al., which combine all advantages of additive manufacturing and microstructural benefits of metal deformation (Zhang et al. 2013). The basic principle of this hybrid AM process is the fusion deposition of metal followed by hot deformation of the same. To achieve this, a micro roller will be positioned behind the deposition head, and the distance between these two is one important process parameter the other being the rolling deformation. As a result of hot rolling, the deposit top surface is always flat and the microstructure changes from cast to wrought. The flat top surface of the deposit ensures stable and accurate deposition of subsequent layer.

The flatness and quality of deposit after rolling was found to depend on the distance between the deposition head and the roller, and the amount of deformation (Zhang et al. 2013). Too short a distance between energy source and the roller would lead to deposit surface peel off, possibly due to sticking at high temperature, while large distance cannot deform the deposit due to a drop in deposit temperature (require high pressure). Good deposit shape and dimensions can be achieved at optimal process parameters. The process refined the deposit microstructure leading to improved mechanical properties (Zhang et al. 2013). Accurate slice thickness control, tailorable deposit width, and economical large-scale part production are some of the other benefits of this process.

5.4 OPPORTUNITIES AND CHALLENGES

Having known the capabilities of various AM technologies available for manufacturing metallic components, it is too early to comment on their readiness to compete with conventional manufacturing at least in some important areas (Reeves and Hague 2013). For example fundamental understanding of the process and precise process control is an essential requirement to reduce or even eliminate variability and uncertainty in product properties. AM standards may be required to qualify their products for use in applications related to aerospace and military. Initially, these materials should meet the performance of materials manufactured via conventional routes. In this context, a large number of opportunities and serious challenges exist to realize full potential of these enabling technologies, which are discussed briefly in the following sections.

5.4.1 MATERIALS RELATED

AM technologies have been used to process several existing metals and alloys, but comprehensive correlation and understanding of processing-microstructure-property-performance relationships of these alloys is yet to be established. Such an understanding not only enables development of new materials, but also help designers consider AM-processed existing materials so that full potential of these technologies can be realized (Scott et al. 2012). Other requirements include availability, consistency and quality of feedstock materials (all materials not available in either powder/wire or foil form), complete properties and characteristics of feedstock materials (dictate process stability and quality of final product), creation and access to materials database that includes microstructures and properties of finished product, feed stock, recycled materials, etc. Further, the influence of post-processing operations on properties and performance also needs to be studied ("Roadmap Workshop on Measurement Science for Metal-Based Additive Manufacturing, Workshop Summary Report" May 2013). Important challenges related to AM materials are presented in Table 5.2.

5.4.2 Process Related

Path dependent attributes such as residual stresses, distortion, microstructure, related properties, and part geometry dependent temperature fields, and thermal gradients/history must be clearly understood for fusion-based technologies to minimize variability and uncertainty. For this purpose, process sensing such as melt pool size, temperature, and temperature variations, are required while these signals are used to control the process on-line. Similar types of processing monitoring and closed-loop control systems are required for other solid-state and emerging AM processes as well (Kinsella 2011). Currently, these facilities are partly available in only selected AM processing equipment, but they lack desired ability to control and require further improvements. For example, they cannot detect defects and correlate them with processing variations (such as powder, speed, deposited material thickness variations, etc.) in real-time and make

TABLE 5.2
Identified Challenges for AM

Area	Challenges
Materials	• Lack of processing-microstructure-property-performance relationships.
	• Understanding issues related to post-processing of AM parts.
	• Feedstock materials characterization, testing, and availability.
Process	• High-speed image process enabling real-time process diagnosis and control.
	• Models and devices for new real-time measurement capabilities composition, dimensions, stresses, and distortion.
	• Models that can effectively combine AM design, process, and materials.
Machine	• Non-availability of research versions of AM machines with high flexibility.
	• New sensors to monitor and measure process parameters such as temperature, stress, and their effective integration of above for robust feedback control.
	• Real-time measurement and control of microstructures, surface finish, etc.
	• Ability to produce net-shape components (currently near net-shape) with improved feature resolution, part size, and isotropic properties.
	• Lack of standards for AM processes, materials properties, defects, geometrical parameters, test procedures, and samples.

Source: Roadmap Workshop on Measurement Science for Metal-Based Additive Manufacturing, Workshop Summary Report, May 2013.

necessary corrections. Development of non-destructive evaluation systems that can detect defects and provide feedback to control the same in real-time are also required to improve the product quality (Bourell et al. 2009).

Multi-scale and multi-level modeling, simulation, and analysis are needed to understand the physical phenomenon operating during processing and predict final microstructures, residual stresses, properties, and surface quality (Frazier 2010). It is intuitive to expect that the accuracy of developed models rely on comprehensive and fundamental understanding of AM processes and materials. Moreover, AM machine capabilities in terms of in-situ sensing, monitoring, and control process are also essential to develop reliable and accurate models based on information such as temperature fields, dimensions, and composition. The modeling efforts provide more options to tailor materials properties to suit desired end use while providing information on essential requirements to achieve these such as sensing, measuring, monitoring, and control systems.

5.4.3 Machine Related

Production type AM machines may be qualified per government qualification procedures thus improving overall repeatability (Kinsella 2011). The most important issue with current AM machines is their flexibility, i.e., that they come with restricted ability to create/test custom processing parameters and materials. Therefore, development of new materials and processing routes is hampered. It is opined that the AM machine should be grouped into production types and development types, the former types may be customized for production where new material development is not required. However, AM machines for developmental activities must have greater flexibility to change process parameters, custom materials, and composites. Other immediate requirements appear to be improved overall product quality (dimensional accuracy, surface finish, etc.), production rate, process efficiency, and cost competiveness of AM equipment. Some AM processing requires controlled atmosphere, and their overall process efficiency is typically less than conventional processing. Standards are another key area which help making parts with identical properties and geometrical quality using different AM techniques.

The quality and performance of parts produced by AM depends on inherent physical phenomenon during processing, which rely on manufacturing paths. Therefore, these must be considered as priority and may be right from the beginning of designing the components. This approach requires concentrated efforts in the area of design for additive manufacturing (Ponche et al 2014). Design for additive manufacturing enables effective designing of components after considering unique capabilities and limitations of AM processes (Vayrea et al. 2012). For example, build orientation has been recognized to strongly effect mechanical properties in different directions of the parts, which must be considered in the process of designing a component. Similarly, rate of acceleration and deceleration during deposition usually leads to variations in deposition height between contour and away from it. Therefore, designing a part without sharp corners may eliminate this problem. Due to the agile nature of AM techniques, designers can consider complex geometries that can improve the performance and efficiency during service.

5.5 SUMMARY AND FUTURE DIRECTIONS

Concentrated efforts during the past two decades made some of the AM technologies matured and are able to produce components that are of high quality and superior properties compared to conventionally process materials. Some AM technologies offer outstanding benefits in manufacturing components such as micro-scale devices with moving parts. Other demonstrated capabilities of AM technologies for metals include creation of novel compositional variations across the sample, multi-materials such as composites/alloys, structurally graded materials (such as porous metals) with tailored mechanical, physical, chemical, and multi-functional properties. However, fusion-based AM processes are highly complex problems to understand and model primarily due to multi-factorial effects. Solid-state processes also have similar complexities related to material and heat flow, bonding mechanisms, and properties. Several inherent aspects of processes controlling the stability of AM processes, process and microstructural control, process optimization, and machine capabilities still require significant improvements.

The major barriers for widespread utilization and development of AM are relatively immature technology (compared to conventional manufacturing technologies), limited number of available materials, cost effectiveness, and lack of confidence among various industries as these processes are not robust enough to create components with high repeatability, accuracy, and properties. The future potential of these enabling technologies depends on how effectively we can overcome these barriers. Further, the future research focus will remain on development of novel and unique material with designed properties, feedstock materials providing more flexibility, process modeling, simulation and control, materials and property database, and standards.

SUGGESTED FURTHER READING

Bintao, W., Zengxi, P., Donghong, D., Dominic, C., Huijun, L., Jing, X., John, N. 2018. A review of the wire arc additive manufacturing of metals: Properties, defects and quality improvement. *Journal of Manufacturing Processes* 35:127–139.

Cohen, A., Chen, R., Frodis, U., Wu, M.-T., Folk, C. 2010. Microscale metal additive manufacturing of multi-component medical devices. *Rapid Prototyping Journal* 16(3):209–215.

Deb Roy, T., Wei, H.L., Zuback, J.S., Mukherjee, T., Elmer, J.W., Milewski, J.O., Beese, A.M., Wilson-Heid, A., De, A., Zhang, W. 2018. Additive manufacturing of metallic components—Process, structure and properties. *Progress in Materials Science* 92:112–224.

Dilip, J.J.S., Babu, S., Rajan, S.V., Rafi, K.H., Janaki Ram, G.D., Stucker, B.E. 2013. Use of friction surfacing for additive manufacturing. *Materials and Manufacturing Processes* 28(2):189–194.

Dirk, H., Vanessa, S., Eric, W., Claus, E. 2016. Additive manufacturing of metals. *Acta Materialia* 117:371–392.

Friel, R.J., Harris, R.A. 2013. Ultrasonic additive manufacturing A hybrid production process for novel functional products. *Procedia CIRP* 6:35–40.

Gu, D.D., Meiners, W., Wissenbach, K., Poprawe, R. 2012. Laser additive manufacturing of metallic components: Materials, processes and mechanisms. *International Materials Reviews* 57(3):133–164.

Janaki Ram, G.D., Robinson, C., Yang, Y., Stucker, B. 2007a. Use of ultrasonic consolidation for fabrication of multi-material structures. *Rapid Prototyping Journal* 13:226–235.

John, J. L., Mohsen, S. 2016. Metal additive manufacturing: A review of mechanical properties. *Annual Review of Materials Research* 46:151–186.

Karunakaran, K.P., Bernard, A., Suryakumar, S., Dembinski, L., Taillandier, G. 2012. Rapid manufacturing of metallic objects. *Rapid Prototyping Journal* 18(4):264–280.

Michele, C., Xin, L., Miguel, C., Wei, L., Yuxiang, Z., Weidong, H. 2017. Numerical simulation and experimental calibration of additive manufacturing by blown powder technology. Part I: Thermal analysis. *Rapid Prototyping Journal* 23(2):448–463.

Panagis, F., Alexios, P., Panagiotis, S. 2018. On thermal modeling of additive manufacturing processes. *CIRP Journal of Manufacturing Science and Technology* 20:66–83.

Patrick, P., Zicheng, Z., Richard, B., James, M. 2018. A framework for mapping design for additive manufacturing knowledge for industrial and product design. *Journal of Engineering Design* 29(6):291–326.

Sames, W.J., List, F.A., Pannala, S., Dehoff, R.R., Babu, S.S. 2016. The metallurgy and processing science of metal additive manufacturing. *International Materials Reviews* 61(5):315–360.

Samperi, M.T. 2014. Development of Design Guidelines for Metal Additive Manufacturing and Process Selection, Master Thesis, The Pennsylvania State University. https://etda.libraries.psu.edu/files/final_submissions/9524

Sebastian, H., Lars, P., Jens, E. 2016. (Re)Design for additive manufacturing. *Procedia CIRP* 50:246–251.

Vaezi, M., Seitz, H., Yang, S. 2013. A review on 3D micro-additive manufacturing technologies. *International Journal Advanced Manufacturing Technology* 67:1721–1754.

William, E.F. 2014. Metal additive manufacturing: A review. *Journal of Materials Engineering and Performance* 23(6):1917–1928.

QUESTIONS

1. Differentiate deposition-based and powder-bed-based AM technologies.
2. Why solid-state AM technologies have been developed?
3. Which energy source you select for fusion-based AM of metals and why?
4. LENS and DMD processes avoid oxidation of deposits. But which process is economical in achieving this?
5. Describe the importance of laser absorption in LMPD processes.
6. How surface finish of a component can be controlled in powder deposition-based AM technologies?
7. Why control of melt pool size during LMPD is important?
8. Describe the challenges associated with development of new materials such as metal matrix composites and compositionally graded materials using powder deposition-based AM technologies.
9. Discuss unique advantages of wire deposition-based processes over powder-based processed.
10. List important process variable in wire deposition processes and describe approaches to control quality of deposited beads.
11. Distinguish electron beam free form fabrication from other wire deposition-based processes.
12. Describe ultrasonic consolidation process and its advantages.
13. What is the role of sonotrode texture and normal load in UC process?

14. Describe the mechanism of bond formation in the UC process.
15. List disadvantages of the UC process and describe the challenges associated with processing high-strength metals/alloys.
16. Why EFAB is not popular and what is the fundamental limitation of this process?
17. Briefly describe the EFAB process.
18. Compare and contrast the UC process and friction free form fabrication.
19. Hybrid AM technologies—how they are better than standard AM technologies?
20. What are AM process related challenges and how can they can addressed?
21. Describe the importance of feedstock materials and its characteristics in achieving high-quality AM parts.
22. What is the biggest and immediate challenge in these AM processes to be addressed and why?

REFERENCES

"A third industrial revolution." *The Economist*, April 21, 2012. Accessed August 20, 2018. http://www.economist.com/node/21552901.

Abioye, T.E., Folkes, J., Clare, A.T. 2013. A parametric study of Inconel 625 wire laser deposition. *Journal of Materials Processing Technology* 213:2145–2151.

Ader, C., Brosemer, M., Freyer, C., Fricke, H., Hennings, D., Klocke, F., Kühne, V., Meiners, W. et al. 2003. In *Solid Freeform Fabrication Symposium Proceedings* 26–30.

Akula, S., Karunakaran, K.P. 2006. Hybrid adaptive layer manufacturing: An intelligent art of direct metal rapid tooling process. *Robotics and Computer-Integrated Manufacturing* 22:113–123.

Bai, X.W., Zhang, H.O., Wang, G.L. 2013. Electromagnetically confined weld-based additive manufacturing. *Procedia CIRP* 6:515–520.

Balla, V.K., Bandyopadhyay, A. 2010. Laser processing of Fe based bulk amorphous alloy. *Surface & Coatings Technology* 205(7):2661–2667.

Balla, V.K., Bandyopadhyay, P.P., Bose, S., Bandyopadhyay, A. 2007. Compositionally graded Yttria stabilized zirconia coating on stainless steel using laser engineered net shaping (LENS™). *Scripta Materialia* 57(9):861–864.

Balla, V.K., Banerjee, S., Bose, S., Bandyopadhyay, A. 2010a. Direct laser processing of tantalum coating on Ti for bone replacement structures. *Acta Biomaterialia* 6(6):2329–2334.

Balla, V.K., Bhat, A., Bose, S., Bandyopadhyay, A. 2012a. Laser processed TiN reinforced Ti6Al4V alloy composite coatings. *Journal of Mechanical Behavior of Biomedical Materials* 6:9–20.

Balla, V.K., Bodhak, S., Bose, S., Bandyopadhyay, A. 2010b. Porous tantalum structures for bone implants: Fabrication, mechanical and *in vitro* biological properties. *Acta Biomaterialia* 6(8):3349–3359.

Balla, V.K., Bose, S., Bandyopadhyay, A. 2010c. Understanding compressive deformation in porous titanium. *Philosophical Magazine* 90(22):3081–3094.

Balla, V.K., Bose, S., Bandyopadhyay, A. 2010d. Microstructure and wear properties of laser deposited WC-12%Co composites. *Materials Science and Engineering A* 527(24–25):6677–6682.

Balla, V.K., Das, M., Bose, D., Ram, G.D.J., Manna, I. 2013. Laser surface modification of 316L stainless steel with bioactive hydroxyapatite. *Materials Science & Engineering C* 33(8):4594–4598.

Balla, V.K., DeVasConCellos, P., Xue, W., Bose, S., Bandyopadhyay, A. 2009a. Fabrication of compositionally and structurally graded Ti-TiO2 structures using laser engineered net shaping (LENS). *Acta Biomaterialia* 5(5):1831–1837.

Balla, V.K., Martinez, S., Rogoza, B.T., Livingston, C., Venkateswaran, D., Bose, S., Bandyopadhyay, A. 2011. Quasi-static torsional deformation behavior of porous Ti–6Al–4V alloy. *Materials Science and Engineering C* 31(5):945–949.

Balla, V.K., Roberson, L.B., O'Connor, G.W., Trigwell, S., Bose, S. Bandyopadhyay, S. 2012b. First demonstration on direct laser fabrication of lunar regolith parts. *Rapid Prototyping Journal* 18(6):451–457.

Balla, V.K., Xue, W., Bose, S., Bandyopadhyay, A. 2009b. Laser assisted Zr/ZrO2 coating on Ti for load-bearing implants. *Acta Biomaterialia* 5(7):2800–2809.

Balla, V.K., Bose, S., Bandyopadhyay, A. 2008. Processing of bulk alumina ceramics using laser engineered net shaping. *International Journal of Applied Ceramic Technology* 5(3):234–242.

Bandyopadhyay, A., España, F.A., Balla, V.K., Bose, S., Ohgami, Y., Davies, N.M. 2010. Influence of porosity on mechanical properties and *in vivo* response of Ti6Al4V implants. *Acta Biomaterialia* 6(4):1640–1648.

Bandyopadhyay, A., Krishna, B.V., Xue, W., Bose, S. 2009. Application of Laser Engineered Net Shaping (LENS) to manufacture porous and functionally graded structures for load bearing implants. *Journal of Materials Science - Materials in Medicine* 20(S1):S29–S34.

Bandyopadhyay, P.P., Balla, V.K., Bose, S., Bandyopadhyay, A. 2007. Compositionally graded aluminum oxide coatings on stainless steel using laser processing. *Journal of American Ceramic Society* 90(7):1989–1991.

Banerjee, R., Collins, P.C., Bhattacharyya, D., Banerjee, S., Fraser, H.L. 2003. Microstructural evolution in laser deposited compositionally graded α/β titanium-vanadium alloys. *Acta Materialia* 51(11):3277–3292.

Banerjee, R., Collins, P.C., Fraser, H.L. 2002. Phase evolution in laser-deposited titanium-chromium alloys. *Metallurgical and Materials Transactions A* 33A(7):2129–2138.

Baufeld, B., Brandl, E., Biest, O. 2011. Wire based additive layer manufacturing: Comparison of microstructure and mechanical properties of Ti–6Al–4V components fabricated by laser-beam deposition and shaped metal deposition. *Journal of Materials Processing Technology* 211:1146–1158.

Baufeld, B., Van der Biest, O. 2009. Mechanical properties of Ti–6Al–4V specimens produced by shaped metal deposition. *Science and Technology of Advanced Materials* 10(1):015008.

Baufeld, B., Van der Biest, O., Gault, R. 2009. Microstructure of Ti–6Al–4V specimens produced by shaped metal deposition. *International Journal of Materials Research* 100(11):1536–1542.

Baufeld, B., Van der Biest, O., Gault, R. 2010. Additive manufacturing of Ti–6Al–4V components by shaped metal deposition: Microstructure and mechanical properties. *Materials and Design* 31:S106–S111.

Bernard, S., Balla, V.K., Bose, S., Bandyopadhyay, A. 2011. Rotating bending fatigue response of laser processed porous NiTi alloy. *Materials Science and Engineering C* 31(4):815–820.

Bernard, S., Balla, V.K., Bose, S., Bandyopadhyay, A. 2012. Compression fatigue behavior of laser processed porous NiTi alloy. *Journal of Mechanical Behavior of Biomedical Materials* 13:62–68.

Bernard, S.A., Balla, V.K., Bose, S., Bandyopadhyay, A. 2010. Direct laser processing of bulk lead zirconate titanate ceramics. *Materials Science and Engineering B* 172(1):85–88.

Bhat, A., Balla, V.K., Bysakh, S., Basu, D., Bose, S., Bandyopadhyay, A. 2011. Carbon nanotube reinforced Cu-10Sn alloy composites: Mechanical and thermal properties. *Materials Science and Engineering A* 528(22–23):6727–6732.

Bourell, D.L., Leu, M.C., Rosen, D.W. 2009. Roadmap for additive manufacturing: Identifying the future of freeform processing, In *SFF Symposium*, Austin, TX.

Brandl, E., Baufeld, B., Leyens, C., Gault, R. 2010. Additive manufactured Ti6Al4V using welding wire: Comparison of laser and arc beam deposition and evaluation with respect to aerospace material specifications. *Physics Procedia* 5:595–606.

Brandl, E., Michailov, V., Viehweger, B., Leyens, C. 2011a. Deposition of Ti–6Al–4V using laser and wire, part I: Microstructural properties of single beads. *Surface & Coatings Technology* 206:1120–1129.

Brandl, E., Michailov, V., Viehweger, B., Leyens, C. 2011b. Deposition of Ti–6Al–4V using laser and wire, part II: Hardness and dimensions of single beads. *Surface & Coatings Technology* 206:1130–1141.

Brandl, E., Palm, F., Michailov, V., Viehweger, B., Leyens, C. 2011c. Mechanical properties of additive manufactured titanium (Ti–6Al–4V) blocks deposited by a solid-state laser and wire. *Materials and Design* 32:4665–4675.

Brandl, E., Schoberth, A., Leyens, C. 2012. Morphology, microstructure, and hardness of titanium (Ti-6Al-4V) blocks deposited by wire-feed additive layer manufacturing (ALM). *Materials Science and Engineering A* 532:295–307.

Cao, X., Jahazi, M., Fournier, J., Alain, M. 2008. Optimization of bead spacing during laser cladding of ZE41A–T5 magnesium alloy castings. *J Mater Process Technol* 205:322–331.

Clark, D., Bache, M.R., Whittaker, M.T. 2008. Shaped metal deposition of a nickel alloy for aero engine applications. *Journal of Materials Processing Technology* 203:439–448.

Cohen, A. 1999. 3-D micromachining by electrochemical fabrication. *Micromachine Devices* March, 6–7.

Cohen, A., Chen, R., Frodis, U., Wu, M.-T., Folk, C. 2010. Microscale metal additive manufacturing of multi-component medical devices. *Rapid Prototyping Journal* 16(3):209-–215.

Cohen, A., Zhang, G., Tseng, F.-G., Mansfield, F., Frodis, U., Will, P. 1998. EFAB: Batch production of functional, fully-dense metal parts with micron-scale features. In *Solid Freeform Fabrication Symposium Proceedings*, The University of Texas, Austin, TX, 161–168.

Cohen, A.L. 2002. Electrochemical fabrication (EFAB™). In *The MEMS Handbook*, Mohamed Gad-el-Hak (ed). Boca Raton, FL, CRC Press LLC.

Collins, P.C., Banerjee, R., Banerjee, S., Fraser, H.L. 2003. Laser deposition of compositionally graded titanium–vanadium and titanium–molybdenum alloys. *Materials Science and Engineering A* A352(1–2):118–128.

Das, M., Balla, V.K., Basu, D., Bose, S., Bandyopadhyay, A. 2010. Laser processing of SiC-particle-reinforced coating on titanium. *Scripta Materialia* 63(4):438–441.

Das, M., Balla, V.K., Basu, D., Manna, I., Kumar, T.S.S., Bandyopadhyay, A. 2012. Laser processing of in-situ synthesized TiB-TiN reinforced Ti6Al4V alloy composite coatings. *Scripta Materialia* 66(8):578–581.

Das, M., Balla, V.K., Kumar, T.S.S., Manna, I. 2013. Fabrication of biomedical implants using laser engineered net shaping (LENS™). *Transactions of the Indian Ceramic Society* 72(3):169–174.

Das, M., Bysakh, S., Basu, D., Kumar, T.S.S., Balla, V.K., Bose, S., Bandyopadhyay, A. 2011. Microstructure, mechanical and wear properties of laser processed SiC particle reinforced coatings on titanium. *Surface & Coatings Technology* 205(19):4366–4373.

Das, S. 2003. Physical aspects of process control in selective laser sintering of metals. *Advanced Engineering Materials* 5(10):701–711.

Dehoff, R.R., Babu, S.S. 2010. Characterization of interfacial microstructures in 3003 aluminum alloy blocks fabricated by ultrasonic additive manufacturing. *Acta Materialia* 58:4305–4315.

DeVasConCellos, P., Balla, V.K., Bose, S., Bandyopadhyay, A., Fugazzi, R., Dernell, W.S. 2012. Patient specific implants for amputation prostheses: Design, manufacture and analysis. *Veterinary and Comparative Orthopaedics and Traumatology* 25(4):286–296.

Dilip, J.J.S., Babu, S., Rajan, S.V., Rafi, K.H., Janaki Ram, G.D., Stucker, B.E. 2013. Use of friction surfacing for additive manufacturing. *Materials and Manufacturing Processes* 28(2):189–194.

Dilip, J.J.S., Janaki Ram, G.D. 2013. Microstructure evolution in aluminum alloy AA 2014 during multi-layer friction deposition. *Materials Characterization* 86:146–151.

Dilip, J.J.S., Janaki Ram, G.D. 2014. Friction freeform fabrication of superalloy inconel 718: Prospects and problems. *Metallurgical and Materials Transactions B* 45B:182–192.

Dinda, G.P., Dasgupta, A.K., Mazumder, J. 2009. Laser aided direct metal deposition of Inconel 625 superalloy: Microstructural evolution and thermal stability. *Materials Science and Engineering A* 509(1–2):98–104.

Dittrick, S., Balla, V.K., Davies, N.M., Bose, S., Bandyopadhyay, A. 2011. *In vitro* wear rate and Co Ion release of compositionally and structurally graded CoCrMo-Ti6Al4V structures. *Materials Science and Engineering C* 31(4):809–814.

España, F.A., Balla, V.K., Bandyopadhyay, A. 2011. Laser processing of bulk Al-12Si alloy: Influence of microstructure on thermal properties. *Philosophical Magazine* 91(4):574–588.

España, F.A., Balla, V.K., Bose, S., Bandyopadhyay, A. 2010. Design and fabrication of CoCrMo based novel structures for load bearing implants using laser engineered net shaping. *Materials Science and Engineering C* 30(1):50–57.

Frazier, W.E. 2010. Direct digital manufacturing of metallic components: Vision and roadmap. *Paper read at Direct Digital Manufacturing of Metallic Components: Affordable, Durable, and Structurally Efficient Airframes*, Solomons Island, MD.

Friel, R.J., Harris, R.A. 2013. Ultrasonic additive manufacturing A hybrid production process for novel functional products. *Procedia CIRP* 6:35–40.

Friel, R.J., Johnson, K.E., Dickens, P.M., Harris R.A. 2010. The effect of interface topography for ultrasonic consolidation of aluminium. *Materials Science and Engineering A* 527:4474–4483.

George, J.L. 2006. Utilization of ultrasonic consolidation in fabricating satellite decking. MS dissertation, Utah State University.

Gharbi, M., Peyre, P., Gorny, C., Carin, M., Morville, S., Le Masson, P., Carron, D., Fabbro, R., 2013. Influence of various process conditions on surface finishes induced by the direct metal deposition laser technique on a Ti64 alloy. *Journal of Materials Processing Technology* 213:791–800.

Gharbi, M., Peyre, P., Gorny, C., Carin, M., Morville, S., Masson, P.L., Carron, D., Fabbro, R. 2014. Influence of a pulsed laser regime on surface finish induced by the direct metal deposition process on a Ti64 alloy. *Journal of Materials Processing Technology* 214:485–495.

Gibert, J.M., Austin, E.M. Fadel, G. 2010. Effect of height to width ratio on the dynamics of ultrasonic consolidation. *Rapid Prototyping Journal* 16(4):284–294.

Gonzalez, R., Stucker, B. 2012. Experimental determination of optimum parameters for stainless steel 316L annealed ultrasonic consolidation. *Rapid Prototyping Journal* 18(2):172–183.

Gu, D.D., Meiners, W., Wissenbach, K., Poprawe, R. 2012. Laser additive manufacturing of metallic components: Materials, processes and mechanisms. *International Materials Reviews* 57(3):133–164.

Hagqvist, P., Heralić, A., Christiansson, A.-K., Lennartson, B. 2014. Resistance measurements for control of laser metal wire deposition. *Optics and Lasers in Engineering* 54:62–67.

Hahnlen, R., Dapino, M.J. 2014. NiTi-Al interface strength in ultrasonic additive manufacturing composites. *Composites: Part B* 59:101–108.

Heralić, A. 2012 Monitoring and control of robotized laser metal-wire deposition. PhD dissertation, Chalmers University of Technology.

Heralic, A., Christiansson, A.-K., Lennartson, B. 2012. Height control of laser metal-wire deposition based on iterative learning control and 3D scanning. *Optics and Lasers in Engineering* 50:1230–1241.

Heralic, A., Christiansson, A.-K., Ottosson, M., Lennartson, B. 2010. Increased stability in laser metal wire deposition through feedback from optical measurements. *Optics and Lasers in Engineering* 48(4):478–485.

Heralic, A., Ottosson, M., Hurtig, K., Christiansson, A.-K. 2008. Visual feedback for operator interaction in robotized laser metal deposition. In *Proceedings of the 22nd International Conference on Surface Modification Technologies*, 297–304.

Hofmeister, W., Griffith, M., Ensz, M., Smugeresky, J. 2001. Solidification in direct metal deposition by LENS processing. *JOM* 53(9):30–34.

Hofmeister, W., Wert, M., Smugeresky, J., Philliber, J.A., Griffith, M., Ensz, M. 1995. Investigating solidification with the laser-engineered net shaping (LENS™) process. *JOM* 51(7). Accessed August 20, 2018. http://www.tms.org/pubs/journals/JOM/9907/Hofmeister/Hofmeister-9907.html

Horii, T., Kirihara, S., Miyamoto, Y., 2009. Freeform fabrication of superalloy objects by 3D micro welding. *Materials & Design* 30:1093–1097.

Hussein, N., Segal, J., Mc Cartney, D., Pashby, I. 2008. Microstructure formation in Waspaloy multilayer builds following direct metal deposition with laser and wire. *Mater Sci Eng A* 497(1–2):260–269.

Janaki Ram, G.D., Robinson, C., Yang, Y., Stucker, B. 2007a. Use of ultrasonic consolidation for fabrication of multi-material structures. *Rapid Prototyping Journal* 13:226–235.

Janaki Ram, G.D., Yang, Y., Nylander, C., Aydelotte, B., Stucker, B.E., Adams, B.L. 2007b. Interface microstructures and bond formation in ultrasonic consolidation. In *Proceedings of the 18th Solid Freeform Fabrication Symposium*, Austin, TX.

Janaki Ram, G.D., Yang, Y., Stucker, B.E. 2007c. Effect of process parameters on bond formation during ultrasonic consolidation of aluminum alloy 3003. *Journal of Manufacturing Systems* 25:221–238.

Jhavar, S., Jain, N.K., Paul, C.P. 2014. Development of micro-plasma transferred arc (-PTA) wire deposition process for additive layer manufacturing applications. *Journal of Materials Processing Technology* 214:1102–1110.

Joshi, K.C., 1971. The formation of ultrasonic bonds between metals. *Welding Journal* 50:840–848.

Karunakaran, K.P., Bernard, A., Suryakumar, S., Dembinski, L., Taillandier, G. 2012. Rapid manufacturing of metallic objects. *Rapid Prototyping Journal* 18(4):264–280.

Katou, M., Oh, J., Miyamoto, Y., Matsuura, K., Kudoh, M. 2007. Freeform fabrication of titanium metal and intermetallic alloys by three-dimensional micro welding. *Materials and Design* 28:2093–2098.

Kelly, G.S., Advania, S.G., Gillespie Jr. J.W., Bogetti, T.A. 2013. A model to characterize acoustic softening during ultrasonic consolidation. *Journal of Materials Processing Technology* 213:1835–1845.

Kelly, G.S., Just Jr. M.S., Advani, S.G., Gillespie Jr. J.W. 2014. Energy and bond strength development during ultrasonic consolidation. *Journal of Materials Processing Technology* 214:1665–1672.

Kim, J.-D., Peng, Y., 2000. Plunging method for Nd: YAG laser cladding with wire feeding. *Optics and Lasers in Engineering* 33:299–309.

Kinsella, M. 2011. *Additive Manufacturing Workshop: Results and Plans.* Washington, DC, Air Force Research Laboratory.

Kong, C., Soar, R., Dickens, P., 2003. Characterization of aluminum alloy 6061 for the ultrasonic consolidation process. *Materials Science and Engineering A* 363:99–106.

Kong, C., Soar, R., Dickens, P., 2004. Optimum process parameters for ultrasonic consolidation of 3003 aluminum. *Journal of Materials Processing Technology* 146:181–187.

Kong, C.Y. 2005. Investigation of ultrasonic consolidation for embedding active/passive fibers in aluminum matrices. PhD dissertation, Loughborough University.

Kong, C.Y., Carroll, P.A., Brown, P., Scudamore, R.J. 2007. "The effect of average powder particle size on deposition efficiency, deposit height and surface roughness in the direct metal laser deposition process" (Paper presented at 14th International Conference on Joining of Materials, Helsingør, Denmark, 29 April - 2 May, 2007).

Krishna, B.V., Bose, S., Bandyopadhyay, A. 2007. Low stiffness porous Ti structures for load bearing implants. *Acta Biomaterialia* 3(6):997–1006.

Krishna, B.V., Bose, S., Bandyopadhyay, A. 2009. Fabrication of porous NiTi shape memory alloy structures using laser engineered net shaping. *Journal of Biomedical Materials Research Part B—Applied Biomaterials* 89B(2):481–490.

Krishna, B.V., Xue, W., Bose, S., Bandyopadhyay, A. 2008a. Functionally graded Co-Cr-Mo coating on Ti-6Al-4V alloy structures. *Acta Biomaterialia* 4(3):697–706.

Krishna, B. V., Xue, W., Bose, S., Bandyopadhyay, S. 2008b. Engineered porous metals for implants. *JOM* 60(5):45–48.

Kruglick, E., Cohen, A. and Bang, C. 2006. EFAB technology and applications. In *The MEMS Handbook*, Gad-El-Hak, M. (ed.), 2nd ed., CRC Press, Boca Raton, FL, 6-1–6-20.

Kukreja, L.M., Kaul, R., Paul, C.P., Ganesh, P., Rao, B.T. 2012. Emerging laser materials processing techniques for future industrial applications. In *Laser Assisted Fabrication of Materials*, ed. Manna, I., Majumder, J. 423–478. Berlin, Germany, Springer-Verlag.

Langenecker, B. 1966. Effects of ultrasound on deformation characteristics of metals. *IEEE Transactions on Sonics and Ultrasonics* 13(1):1–8.

Li, D., Soar, R. 2009. Influence of sonotrode texture on the performance of an ultrasonic consolidation machine and the interfacial bond strength. *Journal of Materials Processing Technology* 209:1627–1634.

Liu, S., Liu, W., Harooni, M., Ma, J., Kovacevic, R. 2014. Real-time monitoring of laser hot-wire cladding of Inconel 625. *Optics & Laser Technology* 62:124–134.

Martina, F., Mehnen, J., Williams, S.W., Colegrove, P., Wang, F., 2012. Investigation of the benefits of plasma deposition for the additive layer manufacture of Ti–6Al–4V. *Journal of Materials Processing Technology* 212:1377–1386.

Matz, J.E., Eagar, T.W. 2002. Carbide formation in Alloy 718 and electron-beam solid free-form fabrication. *Metallurgical and Materials Transactions* 33A:2559–2567.

Mazumder, J. 2000. Crystal ball view of direct-metal deposition. *JOM* 52(12):28–29.

Mazumder, J., Dutta, D., Kikuchi, N., Ghosh, A. 2000. Closed loop direct metal deposition: Art to Part. *Optics and Lasers in Engineering* 34(4–6):397–414.

Medrano, A., Folkes, J., Segal, J., Pashby, I. 2009. Fibre laser metal deposition with wire: Parameters study and temperature monitoring system. In *Proceedings of the SPIE* 7131:713122.

Miranda, R., Lopes, G., Quintino, L., Rodrigues, J., Williams, S. 2008. Rapid prototyping with high power fiber lasers. *Materials & Design* 29:2072–2075.

Mok, S., Bi, G., Folkes, J., Pashby, I. 2008a. Deposition of Ti6Al4V using a high power diode laser and wire, Part I: Investigation on the process characteristics. *Surface & Coatings Technology* 202:3933–3939.

Mok, S., Bi, G., Folkes, J., Pashby, I., Segal, J. 2008b. Deposition of Ti6Al4V using a high power diode laser and wire, Part II: Investigation on the mechanical properties. *Surface & Coatings Technology* 202:4613–4619.

Nurminen, J. 2008. Hot-wire laser cladding: Process, materials and their properties. PhD dissertation, Tampere University of Technology.

Obielodan, J.O., Ceylan, A., Murr, L.E., Stucker, B.E. 2010. Multi-material bonding in ultrasonic consolidation. *Rapid Prototyping Journal* 16(3):180–188.

Obielodan, J.O., Stucker, B.E., Martinez, J.L., Hernandez, D.H., Ramirez, D.A., Murr, L.E., 2011. Optimization of the shear strengths of ultrasonically consolidated Ti/Al 3003 dual-materials structures. *Journal of Materials Processing Technology* 211(6):988–995.

Oruganti, R.K., Ghosh, A.K. 2003. Fabrication and creep properties of superalloy-zirconia composites. *Metallurgical and Materials Transactions A* 34A(11):2643–2653.

Pinkerton, A.J., Li, L., 2003. The effect of laser pulse width on multiple-layer 316L steel clad microstructure and surface finish. *Applied Surface Science* 208–209:405–410.

Ponche, R., Kerbrat, O., Mognol, P., Hascoet, J.-Y. 2014. A novel methodology of design for additive manufacturing applied to additive laser manufacturing process. *Robotics and Computer-Integrated Manufacturing* 30:389–398.

Puli, R., Janaki Ram, G.D. 2012. Dynamic recrystallization in friction surfaced austenitic stainless steel coatings. *Materials Characterization* 74:49–54.

Rafi, H.K., Balasubramaniam, K., Phanikumar, G., Prasad Rao, K. 2011. Thermal profiling using infrared thermography in friction surfacing. *Metallurgical and Materials Transactions* 42:3425–3429.

Rangesh, A., O'Neill, W. 2011. Rapid prototyping by consolidation of stainless steel powder using an electrical arc. *Rapid Prototyping Journal* 17(4) 280–287.

Reeves, P., Hague, R. 2013. Additive manufacturing or 3D printing—you decide!. *The Royal Academy of Engineering Magazine* 6:39–45.

Reid, J., Webster, R. 2006. A 55 GHz band pass filter realized with integrated TEM transmission lines. In *IEEE MTT-S International Microwave Symposium Digest*, San Francisco, CA, 132–135.

"Roadmap Workshop on Measurement Science for Metal-Based Additive Manufacturing, Workshop Summary Report", May 2013. Accessed August 20, 2018. http://events.energetics.com/NIST-AdditiveMfgWorkshop/pdfs/NISTAdd_Mfg_report_FINAL.pdf

Roy, M., Balla, V.K., Bandyopadhyay, A., Bose, S. 2012. MgO doped tantalum coating on Ti: Microstructural study and biocompatibility evaluation. *ACS Applied Materials & Interfaces* 4(2):577–580.

Roy, M., Krishna, B.V., Bandyopadhyay, A., Bose S. 2008. Laser processing of bioactive tricalcium phosphate coating on titanium for load bearing implant. *Acta Biomaterialia* 4(2):324–333.

Schwendner, K.I., Banerjee, R., Collins, P.C., Brice, C.A., Fraser, H.L. 2001. Direct laser deposition of alloys from elemental powder blends. *Scripta Materialia* 45(10):1123–1129.

Scott, J., Gupta, N., Weber, C., Newsome, S., Wohlers, T., Caffrey, T. 2012. Additive manufacturing: Status and opportunities. *Occasional Papers in Science and Technology Policy*, March 2012. Accessed May 05, 2014. https://www.ida.org/stpi/occasionalpapers/papers/AM3D_33012_Final.pdf

Seufzer, W.J., Taminger, K.M. 2007. Control methods for the electron beam free form fabrication process. In *Solid Freeform Fabrication Symposium Proceedings* 13–21.

Shin, K-H., Natu, H., Dutta, D., Mazumder, J. 2003. A method for the design and fabrication of heterogeneous objects. *Materials and Design* 24(5):339–353.

Siggard, E.J. 2007. Investigative research into the structure embedding of electrical and mechanical systems using ultrasonic consolidation (UC). MS dissertation, Utah State University.

Skiba, T., Baufeld, B., Van der Biest, O. 2009. Microstructure and mechanical properties of stainless steel component manufactured by shaped metal deposition. *ISIJ International* 49:1588–1591.

Skiba, T., Baufeld, B., Van der Biest, O. 2011. Shaped metal deposition of 300M steel. *Proceedings of the Institution of Mechanical Engineers, Part B: Journal of Engineering Manufacture* 225(6):831–839.

Stecker, S., Lachenberg, K.W., Wang, H., Salo, R.C. 2006. Advanced electron beam free form fabrication methods & technology. In *Professional Program & Poster Session*, Atlanta, GA, Session 2: Electron beam welding, New York, American Welding Society, 35–46.

Syed, W.U.H., Li, L. 2005. Effects of wire feeding direction and location in multiple layer diode laser direct metal deposition. *Applied Surface Science* 248:518–524

Syed, W.U.H., Pinkerton, A.J., Li, L. 2006. Combining wire and coaxial powder feeding in laser direct metal deposition for rapid prototyping. *Applied Surface Science* 252:4803–4808.

Syed, W.U.H., Pinkerton, A.J., Li, L., 2005. A comparative study of wire feeding and powder feeding in direct diode laser deposition for rapid prototyping. *Applied Surface Science* 247:268–276.

Syed, W.U.H., Pinkerton, A.J., Liu, Z., Li, L. 2007a. Coincident wire and powder deposition by laser to form compositionally graded material. *Surface & Coatings Technology* 201:7083–7091.

Syed, W.U.H., Pinkerton, A.J., Liu, Z., Li, L. 2007b. Single-step laser deposition of functionally graded coating by dual 'wire–powder' or 'powder–powder' feeding-A comparative study. *Applied Surface Science* 253:7926–7931.

Taminger, K. 2009. Electron beam freeform fabrication. *Advanced Materials & Processes* 11–12:45.

Taminger, K.M.B., Hafley, R.A. 2002. Characterization of 2219 aluminum produced by electron beam freeform fabrication. In *Proceedings of 13th SFF Symposium* 482–489.

Taminger, K.M.B., Hafley, R.A. 2003. Electron beam freeform fabrication: A rapid metal deposition process. In *Proceedings of the 3rd Annual Automotive Composites Conference*, Troy, MI, Society of Plastic Engineers, September 9–10, CD-ROM.

Taminger, K.M.B., Hafley, R.A., Domack, M.S. 2006. Evolution and control of 2219 aluminum microstructural features through electron beam freeform fabrication. *Materials Science Forum* 519–521:1297–1304.

"The printed world." *The Economist*, February 10, 2011. Accessed August 20, 2018. http://www.economist.com/node/18114221.

Vaezi, M., Seitz, H., Yang, S. 2013. A review on 3D micro-additive manufacturing technologies. *The International Journal of Advanced Manufacturing Technology* 67:1721–1754.

Vayrea, B., Vignata, F., Villeneuvea, F. 2012. Designing for additive manufacturing. *Procedia CIRP* 3:632–637.

Wang, F., Mei, J., Jiang, H., Wu, X. 2007. Laser fabrication of Ti6Al4V/TiC composites using simultaneous powder and wire feed. *Materials Science and Engineering* 445–446:461–466;

Wang, F., Mei, J., Wu, X. 2006. Microstructure study of direct laser fabricated Ti alloys using powder and wire. *Applied Surface Science* 253:1424–1430.

Wang, F., Williams, S., Colegrove, P., Antonysamy, A.A. 2013. Microstructure and mechanical properties of wire and arc additive manufactured Ti-6Al-4V. *Metallurgical and Materials Transactions A* 44A:968–977.

Wanjara, P., Brochu, M., Jahazi, M. 2007. Electron beam freeforming of stainless steel using solid wire feed. *Materials and Design* 28:2278–2286.

Wu, X. 2007. A review of laser fabrication of metallic engineering components and of materials. *Materials Science and Technology* 23(6):631–640.

Xiong, X., Haiou, Z., Guilan, W. 2008. A new method of direct metal prototyping: Hybrid plasma deposition and milling. *Rapid Prototyping Journal* 14(1):53–56.

Xue, L., Li, Y., Wang, S., 2011. Direct manufacturing of net-shape functional components/test pieces for aerospace, automotive and other applications. *Journal of Laser Applications* 23(4):042004.

Xue, W., Krishna, B.V., Bandyopadhyay, A., Bose, S. 2007. Processing and biocompatibility evaluation of laser processed porous titanium. *Acta Biomaterialia* 3(6):1007–1018.

Yang, Y., Janaki Ram, G.D., Stucker, B. 2007. An experimental determination of optimum processing parameters for Al/SiC metal matrix composites made using ultrasonic consolidation. *Journal of Engineering Materials and Technology* 129:538–549.

Yang, Y., Janaki Ram, G.D., Stucker, B.E. 2009. Bond formation and fiber embedment during ultrasonic consolidation. *Journal of Materials Processing Technology* 209:4915–4924.

Yang, Y., Janaki Ram, G.D., Stucker, B.E. 2010. An analytical energy model for metal foil deposition in ultrasonic consolidation. *Rapid Prototyping Journal* 16(1):20–28.

Zhang, H., Wang, X., Wang, G., Zhang, Y. 2013. Hybrid direct manufacturing method of metallic parts using deposition and micro continuous rolling. *Rapid Prototyping Journal* 19(6):387–394.

Zhang, H., Xu, J., Wang, G. 2003. Fundamental study on plasma deposition manufacturing. *Surface & Coatings Technology* 171:112–118.

Zheng, B., Zhou, Y., Smugeresky, J.E., Schoenung, J.M. Lavernia, E.J. 2008. Thermal behavior and microstructural evolution during laser deposition with laser-engineered net shaping: Part I. Numerical calculations. *Metallurgical and Materials Transactions A* 39A(9):2228–2236.

Zhu, G., Li, D., Zhang, A., Pi, G., Tang, Y. 2012. The influence of laser and powder defocusing characteristics on the surface quality in laser direct metal deposition. *Optics & Laser Technology* 44:349–356.

Zou, H., Zhang, H., Wang, G., Li, J. 2009. Rapid manufacturing of FGM components by using electromagnetic compressed plasma deposition. *Progress In Electromagnetics Research Symposium Proceedings*, Moscow, Russia, August 18–21, 1953–1956.

6 Additive Manufacturing of Ceramics

Susmita Bose, Naboneeta Sarkar, Sahar Vahabzadeh, Dongxu Ke, and Amit Bandyopadhyay

CONTENTS

6.1 INTRODUCTION

The concept of solid freeform fabrication or 3D printing (3DP) was first introduced by Chuck Hull in 1986 through the stereolithography method [1]. Since then, several processes such as selective laser sintering (SLS), fused deposition modeling (FDM) [2,3], inkjet printing [4–6], and laser engineered net shaping (LENS) have been introduced to fabricate metals, ceramics, polymers, and composites. Various approaches of additive manufacturing (AM) techniques used to fabricate 3D ceramic structures can be classified as—(1) laser-assisted sintering (e.g., selective laser sintering and laser engineered net shaping), (2) extrusion (e.g., fused deposition of ceramics), (3) polymerization (e.g., stereolithography), and (4) direct writing (e.g., inkjet 3DP) processes [7–9]. In all AM techniques, a 3D model is created by computer-aided design (CAD) program, and then is converted to a Standard Tessellation Language (.stl) file. The 3D object is then sliced into two-dimensional (2D) cross-sections, and fabrication of the part is started from the base in alternating layers. Manufacturing is continued layer-by-layer until the entire part is fabricated. Compared to traditional manufacturing processes, the primary advantage of AM is its ability to create complex designs with high accuracy in dimensions and structural features without the need for part-specific tooling [10]. Besides, almost all AM techniques are faster than traditional processes without the need for further surface finishing. However, not all the AM techniques are useful for fabrication of similar ceramic structures due to some of the inherent limitations of those techniques [11]. In this chapter, the methodology of these techniques is explained in detail and various ceramic structures processed by each technique for different applications are introduced as examples for readers to understand the process better.

6.2 STEREOLITHOGRAPHY

6.2.1 History and Methodology

Stereolithography or SLA, first developed by Chuck Hull in 1986, is one of the most prevalent additive manufacturing technique. In this process, an ultraviolet (UV) light or laser is used to fabricate a solid structure from photopolymerizable monomers [12]. It is a layer-by-layer process, similar to other AM techniques, and the finished structure requires post-processing or curing by heat or UV-light treatment to polymerize any residual monomers. Figure 6.1 represents a simple schematic of

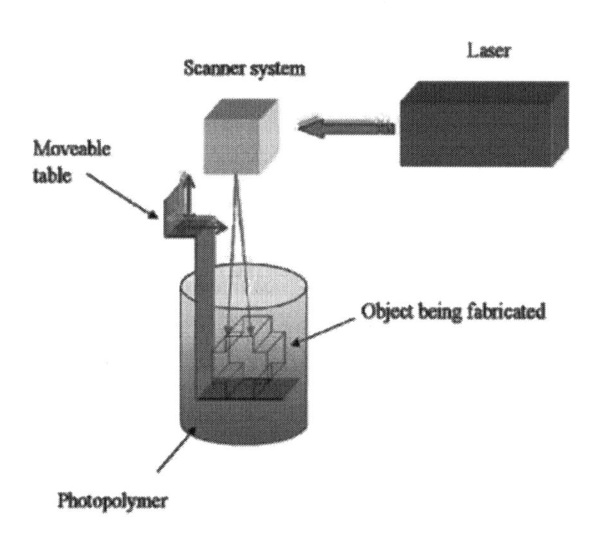

FIGURE 6.1 Schematic illustration of the stereolithography process.

the SLA process. SLA has the capability to fabricate parts with a very fine surface quality and micrometer scale resolution [13]. Generally, SLA is applied to manufacture polymeric structures; however, further modifications allowed SLA to be used for processing of ceramic materials. Ceramic stereolithography is an extended method for SLA [14]. This technique involves photopolymerization of a slurry containing ~40 vol.% ceramic powder, a monomer solution, a photo-initiator, and dispersants [14]. Ceramic SLA has grown importance and being used to fabricate dental and bone-tissue engineering scaffolds, sensors, piezoelectrics, microelectromechanical systems, as well as some structural parts [15–20].

Flexibility, efficiency, and high dimensional accuracy of ceramic structures with different geometries are the main advantages of SLA. On the other hand, drawbacks of SLA includes large non-uniform shrinkage during sintering, need for supporting material during manufacturing, difficulty of loading bioactive agents, and inability to process multi-material structure. Commercially available epoxy-based or acrylate-based resins are not biocompatible or biodegradable [21–22]. Besides, in order to prepare the photopolymerizable resin, toxic solvents are required. Therefore, preparation of biomedical implants is typically carried out using an indirect approach. Levy et al. prepared a slurry-based suspension using hydroxyapatite and liquid photocurable acrylic resin [23]. The porous hydroxyapatite part was processed by laser curing, followed by removal of resin via heat treatment. The ceramic powder ratio, chemistry and concentration of photocurable polymer, and UV power are important parameters for successive printing in SLA. Hinczewski et al. reported the effects of dispersant and diluent concentrations on the viscosity and rheological behavior of highly loaded alumina particles in photocurable acrylate monomer solutions [24]. The effects of exposure conditions, powder characteristics, reactive system, and cured depth and width on the SLA process were also reported by Chartier et al. [13]. Different types of polymers along with ceramic powder compositions that have been used in SLA are listed in Table 6.1 for various applications including photonic crystals, piezoelectrics, cellular ceramic structures, and bone scaffolds.

To ensure successful part production, the ceramic suspension must possess suitable rheological behavior, long term stability, and desirable viscosity. Ceramic particles must be homogeneously dispersed in the photocurable resin. Additionally, the ceramic suspension must retain its stability and viscosity during the manufacturing process. Earlier, suspension viscosity had to be similar to conventional SLA resins (<3000 mPa·s) [14], however, current SLA techniques can work with a viscosity in the range of 10–5000 mPa·s [14]. In addition, the ceramic suspension should be photoactive with high cure depth and low cure width to achieve high efficiency and resolution during the AM operation [13,15,24]. Furthermore, the cured ceramic green part must have a high density to prevent cracking, warpage, or significant shrinkage after polymer removal [14,24].

Ceramic powders have negligible solubility in polymer solutions. To achieve a homogeneous dispersion of ceramic powders in polymer solution with high ceramic loading, *dispersants* are needed. Dispersants are widely used to disperse ceramic particles in a non or low-polar organic solvent through electrostatic and steric repulsion forces [16,24]. Many dispersants have been introduced to improve ceramic powder loading and homogeneity of suspension for the SLA process.

TABLE 6.1
Ceramic—Photocurable Polymer Systems in SLA

Ceramic Powder	Photocurable Polymer	References
Alumina (Al_2O_3)	Di-ethoxylated bisphenol a dimethacrylate (diacryl 101)	[24]
	Hexanediol diacrylate (HDDA)	[25]
	Diacryl 101 and HDDA	[26]
	Acrylamide	[14]
	Acrylic and silicon acrylate	[27]
	Acrylate	[18]
	Zirconate+3% Irgacure 184	[28]
Silica (SiO_2)	Acrylate	[18]
	Acrylamide	[14]
	Acrylic and silicone acrylate	[29]
Lead zirconate titanate (PZT)	Acrylates (Diacryl 101 and HDDA) and epoxy-acrylates (SOMOS 6100)	[19,20]
Hydroxyapatite (HA)	SL5180 resin (Huntsman)	[18]
Barium Titanate ($BaTiO_3$)	Hexanediol diacrylate (HDDA)	[30]
Titanium oxide (TiO_2)	Epoxy resin	[31]

Using quaternary ammonium acetate as dispersant allows 50 vol.% of alumina loading in hexanediol diacrylate (HDDA); however, same amount of alumina loading in HDDA without dispersant results in a stiff paste-like colloidal gel [25]. Triton X-100 is another dispersant used in a barium titanate-HDDA (BT-HDDA) resin system. 6 wt.% triton X-100 is the optimized concentration for 30 vol.% barium titanate-HDDA suspension [30]. In recent studies, pre-ceramic polymers such as silicon oxycarbide in liquid form are often used for photopolymerization instead of homogeneously dispersed ceramic suspensions. High-temperature sintering and ceramization results in the transformation of pre-ceramic polymers to polymer derived ceramic components. Various metal alkoxides are often incorporated within pre-ceramic polymer composition or structure to produce high-performance ceramic parts with desired functionalities. Figure 6.2c depicts complex polymer derived SiC and SiOC ceramic structures fabricated using SLA with photopolymerization of the pre-ceramic polymer [32].

Viscosity is another important factor that decides the rheological behavior of the suspension. The viscosity of an SLA suspension can be calculated using a modified Krieger-Dougherty equation [14,33]. It is usually larger than pure photocurable resin. Thus, decreasing the viscosity in the range of 2–5 Pa.s is vital to assure satisfactory layer recoating and successful ceramic stereolithography [26]. Suspension with low viscosity can be achieved by using an appropriate polymer, dispersant, and diluent [26,34]. Not only the initial viscosity, but also the change in viscosity with regard to shear rate plays a significant role during the SLA fabrication. With increased shear rate, the viscosity may decrease (shear-thinning), increase (shear-thickening), or remain unchanged (shear immunity). Shear-thinning is desired in

FIGURE 6.2 (a) From left to right: different CSL structures that can be employed as molds for turbine airfoils (fused silica) [48], as microwave-guides (zirconia) [16], and as bone implants (hydroxyapatite) [49]. (b) Alumina drill bits (left) and mesoporous tricalcium phosphate structures (right) at above 99.5% of theoretical density manufactured using CSL. (Courtesy of Homa & Co. KG, Vienna, Austria.) (c) SiC parts prepared by SLA on preceramic polymers. Comparison of printed and sintered samples (bottom). (From de Hazan, Y., and D. Penner, *J. Eur. Ceram. Soc.*, 37, 5205–5212, 2017.)

the standard SLA process as it allows the tape casting of layers prior to UV treatment [34]. The Viscosity of the suspensions decreases with an increase in temperature. However, the temperature should not be higher than the polymerization temperature as it can cause undesired polymerization before UV scanning.

In SLA, cured depth and width are important parameters that determine the rate and accuracy of the process. Ceramic powder characteristics such as particle size and refractive index, the monomer properties such as minimal energy for polymerization and refractive index, and suspension characteristics such as the volume fraction of ceramic powder, interparticle distance, UV energy, and wavelength of irradiation affects the cured depth of SLA. Cured depth (C_d) is the thickness of the gelled resin [24,35]. It can be theoretically calculated using the Beer-Lambert law:

$$C_d = D_p \ln\left(\frac{E}{E_c}\right), \quad [26]$$

where D_p, E_c, and E are penetration depth, minimal energy for polymerization of the monomer, and provided energy, respectively. D_p depends on the volume fraction of ceramic powder, particle size, and the refractive index difference between the UV curable solution and ceramic powder [14,35,36,37,38].

A major concern in ceramic SLA is a light scattering phenomena due to the presence of ceramic particles in the suspension. This light scattering not only affects the light penetration to the suspension but also broadens the cured width. Cured width (W_c) should be low enough to ensure high resolution and quality of the SLA processed parts. It has been associated with ceramic particle size and volume fraction, light exposure power, the density of energy, photoinitiator concentration, powder concentration, particle diameter, and the difference of refractive index between ceramic powder and monomer. A linear relationship was found between the mean particle diameter and cured width, with slopes depending on the density of energy. Also cured width corresponds to a power law with powder concentration, with an exponent close to –1.

6.2.2 SLA Processed Ceramics

SLA has been implemented in ceramic manufacturing to fabricate parts with complex geometry for a variety of applications such as casting molds, photonic crystals, sensors, tissue engineering scaffolds, and dental implants. Compared to a traditional multi-layer ceramic shell, SLA processed alumina/silica molds are more efficient and cost-effective for investment casting [39–42]. SLA is also used to process piezoelectric ceramics such as lead zirconate titanate [Pb(Zr, Ti)O_3, (PZT)] and barium titanate [$BaTiO_3$, (BT)] due to its flexibility, efficiency, and accuracy [19,20,30,43]. In addition, complex 3D photonic crystals for telecommunication domain applications such as antennas, filters, and resonators have also been manufactured using ceramic loaded SLA [16,31,44–47].

Researchers also employed SLA to fabricate porous ceramics for various applications owing to its ability to generate high precision part with dense struts. Kirihara et al. synthesized samples with controlled porosity, with a total porosity of 75 vol.% [50]. Yttria-stabilized zirconia (YSZ) lattice for anode electrodes in solid oxide fuel cells was fabricated with a lattice constant of 250 μm and a thick strut of 100 μm length. In another study, Al_2O_3 parts with a diamond structure were synthesized and properties were modified for the application of photonic crystals. Recently, SLA has been employed to fabricate synthetic bone scaffolds and implants using various bioceramics, such as hydroxyapatite, tricalcium phosphate, zirconia and bioglass for tissue engineering application [51–54]. Bian et al. reported successful fabrication of osteochondral scaffold using β-tricalcium phosphate (TCP). The green part was subjected to debinding and sintering. Freeze casting was carried out to simulate the cartilage phase by cross-linking in type I collagen suspension. The hydroxyapatite (HA) part with a dendrite structure and graded porosity is fabricated for tissue engineering applications [55]. Chu et al. synthesized HA porous implant with orthogonal pores and designed porosity of 40 vol.%. These implants exhibited high compressive strength of 30 MPa and good *in vivo* biocompatibility [49].

A novel improvisation of SLA, named as nano-stereolithography technique, utilizes two-photon equipment for the cross-linking of resin with a resolution of 200 nm [56]. Pham et al. fabricated a complex micrometer-shaped SiCN ceramic utilizing a polyvinyl silazane preceramic polymer. The polymer was first pyrolyzed at 600°C in nitrogen environment, and then silica nanoparticles were incorporated to obtain a reduced and isotropic shrinkage [57]. The future development of this process may focus on improved resolution, lower processing time as well as enhanced repeatability. Major challenges related to SLA are presence of residual monomers, inhomogeneous dispersion of ceramic particles, light curing process, and cracking during debinding. To obtain higher green strength in SLA fabricated parts, non-aqueous suspensions such as acrylamide or resin are mostly used. Chen et al. reported higher

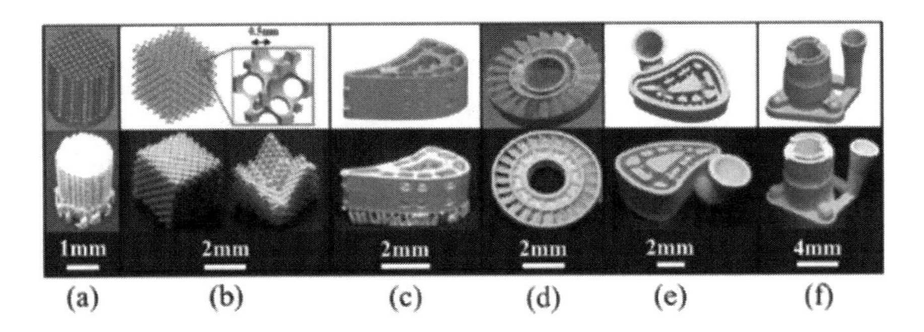

FIGURE 6.3 Advanced ceramic parts fabricated using SiO_2 via SLA [19]: (a) porous bioceramic scaffold; (b) photonic crystals; (c) hollow turbine blade; (d) impeller; (e)–(f) investment casting molds.

green strength and lower viscosity by improvising the aqueous binder by imparting silica sol instead of deionized water. The SiO_2 ceramic parts of different applications are shown in Figure 6.3.

6.3 SELECTIVE LASER SINTERING OF CERAMICS

6.3.1 History and Methodology

SLS is an additive manufacturing method in which a pulsed laser is used to fuse fine particles together [58]. SLS is the first commercialized powder bed-based manufacturing method [27]. The first commercial SLS machine was introduced in 1992 by DTM Corporation. With the future development of SLS technology, part accuracy, temperature uniformity, build speed, process repeatability, feature definition, and surface finish are improved; but the basic processing features and system configuration remain mostly unchanged [58].

Prior to laser sintering, the powder needs to be preheated just below the melting point or glass transition temperature to minimize thermal distortion and facilitate fusion of a new layer to the previous layer [60,61]. During the process, the chamber should be sealed with nitrogen gas to avoid oxidation and degradation of the powder. The focused laser beam is then directed onto the powder bed to form the pattern according to the CAD design. Meanwhile, surrounding powder remains loose and serves as a support for the subsequent layers [60]. After sintering each layer, the build platform is lowered and the roller spreads the next layer of powder to the build bed. The process is repeated until manufacturing of the designed part is complete. Finally, the SLS processed part should be kept long enough in the chamber to cool down. This prevents any degradation of powder and shape deformation due to the presence of oxygen and uneven thermal contraction. Figure 6.4 shows a schematic of the SLS process.

Unlike metals or polymers processing, manufacturing of ceramics by SLS involves several challenges mainly due to the high melting point of ceramics. As a result, higher laser energy and longer cooling times are required, which reduces the

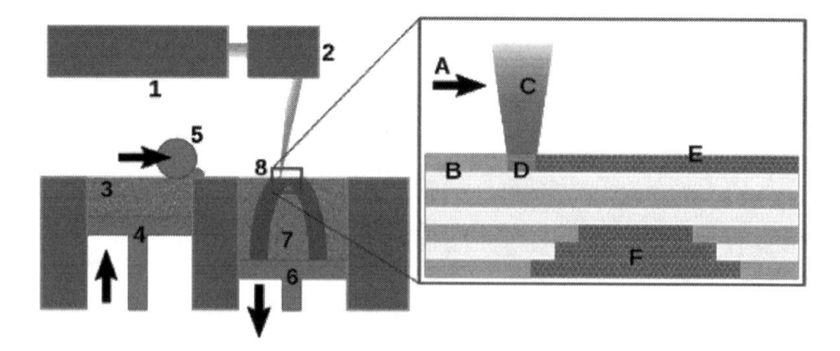

FIGURE 6.4 Selective laser sintering process (1) laser; (2) scanner system; (3) powder delivery system; (4) powder delivery piston; (5) roller; (6) fabrication piston; (7) fabrication powder bed; and (8) object being fabricated (see inset)—(A) laser scanning direction; (B) sintered powder particles (brown state); (C) laser beam; (D) laser sintering; (E) pre-placed powder bed (green state); and (F) unsintered material in previous layers. (Courtesy of Materialgeeza/Creative Commons.)

efficiency and cost-effectiveness of this process. For example, Hagedorn et al. reported an SLS process to manufacture approximately 100% dense Al_2O_3-ZrO_2-Y_2O_3 parts at the eutectic composition. The electrical heating energy was 7 kW and the powder was preheated to higher than 1600°C using a secondary laser to reach the ceramic melting temperature and obtain a crack-free part [60]. In general, ceramics can be processed by either direct or indirect SLS. For the direct SLS method, ceramic powders are sintered or melted directly; however, for indirect SLS an organic phase is melted, which acts as a binder to manufacture green ceramic parts [61].

6.3.2 Direct Selective Laser Sintering

Direct selective laser sintering of ceramics can be further divided into two subcategories: (1) powder and (2) slurry-based direct laser sintering. In powder-based direct SLS, a roller spreads a layer of powder from feed bed to build bed. This method allows processing of 3D products with physical, mechanical, and chemical properties different from original properties of those powder compositions [62]. However, low packing density in powder bed is not desired as it may lead to low sintered density and crack formation due to thermal stresses [61]. In addition, sometimes the laser energy must be high enough to directly sinter the ceramic powder with high-temperature resistance. Friedel et al. utilized a binary system as feedstock to process polymer-derived ceramic parts with complex geometries, as shown in Figure 6.5 [63]. Silicon infiltration after pyrolysis resulted in the formation of dense part and enhanced the bending strength from 17 to 220 MPa.

Shishkovsky et al. reported the manufacturing process of a dense aluminum and zirconium mixture using SLS. This high-speed laser sintering creates ceramics with high density and uniform distribution of stable phases. It showed a great potential to be used as thermal and electrical insulators and wear-resistant coating for the application of solid oxide fuel cells, crucibles, heating elements, and biomedical

FIGURE 6.5 SiOC/SiC turbine wheel produced: (a) after SLS; (b) after firing; (c) after infiltration with silicon. (From Friedel, T. et al., *J. Eur. Ceram Soc.*, 25, 193–197, 2005.)

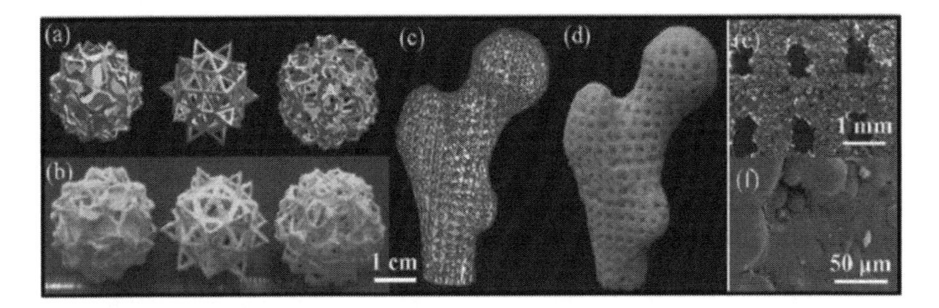

FIGURE 6.6 (a) Digitally designed complex 3D cellular models; (b) final CP-PHBV porous structures made with SLS; (c) 3D model of a human bone reconstructed by computed tomography (CT) and processed into a porous scaffold with cubic units; (d) final CP-PHBV scaffold made with SLS; (e) SEM image of one layer of the fabricated scaffold; and (f) cytocompatibility tests: morphology of SaOS-2 cells cultured for seven days on the scaffold. (From Duan, B., and Wang, M., *J. Royal Soc. Interface*, 7, S615–S629, 2010.)

tools [62]. Yttria-zirconia has also been processed by powder-based direct SLS. However, the preparation of powder with appropriate characteristics suitable for the SLS process is still challenging [64]. Figure 6.6 illustrates an example of ceramic-glass biomedical scaffolds made from calcium phosphate-poly hydroxy butyrate-co hydroxy valerate [65]. The sintered scaffold was subjected to surface modification using gelatin and heparin, which significantly improved the wettability of the scaffold. Incorporation of growth factor (rhBMP-2) within this scaffold resulted in enhanced ALP activity and osteogenic differentiation marker expression. Table 6.2 summarizes some of the ceramic compositions processed by powder-based direct SLS, along with their potential applications.

Slurry-based direct SLS, however, starts from a homogeneous slurry as feeding layers. Unlike powder-based SLS, slurry suspensions are commonly fed to the build platform by a doctor blade or spray deposition followed by subsequent drying and laser processing [61]. The high density of the final product is one of the most important advantages of this method. This is mainly due to the possibility of using smaller ceramic particles in slurry [71,72].

TABLE 6.2
Ceramic Materials Processed by Powder-Based Direct SLS and Their Potential Applications

Ceramic Composition	Potential Applications	References
Alumina (Al_2O_3) and zirconia (ZrO_2)	Automotive, aerospace, or biomedical sector and implant	[60,62]
Yttria (Y_2O_3) and Zirconia (ZrO_2)	Ceramic shell molds and solid oxide fuel cells	[64]
Calcium phosphate (CP)/ hydroxybutyrate-co hydroxy valerate	Implant materials in tissue engineering	[65]
Nano-hydroxyapatite	Bone tissue engineering scaffolds	[66,67]
Cu-Ni and ZrB_2	Electrical discharge machining (EDM) electrodes	[68]
45S5 bioactive glass	Bone tissue engineering scaffolds	[69]
PZT (PbO, ZrO_2, and TiO_2)	Piezoelectric material	[70]

Tian et al. have reported a slurry-based SLS for making 3D porcelain parts with a sintered density of ~86%; however, the mechanical strength of those parts were low due to the microstructural inhomogeneities and thermal cracking [61,73]. Recently, Liu reported a novel slurry-based SLS using HA, silica sol, and sodium tripolyphosphate (STPP) as the suspension for biomedical scaffolds preparation. Sintering at 1300°C resulted in scaffolds with a porosity of ~14% and compressive strength of 43 MPa. The scaffolds also demonstrated high resolution and dimensional accuracy, which indicated the great potential of applying this method for bone tissue engineering scaffolds manufacturing [74]. This approach have also been used in other ceramic-ceramic composite systems involving alumina, silica and hydroxyapatite ceramics [75–77]. Table 6.3 summarizes the details of some of the slurry-based direct SLS systems.

6.3.3 Indirect Selective Laser Sintering for Ceramics

In the case of indirect selective laser sintering, ceramic powders coated with an organic phase are used as the feedstock material. Due to the lower melting point compared to that of ceramics, polymers act as a binder to fuse ceramic particles using the laser beam. They also consolidate ceramic layers. Similar to direct SLS, indirect SLS is categorized into powder-based and slurry-based processes [78]. Powder-based indirect SLS makes it possible for using conventional SLS equipment to produce ceramic parts with high melting points [61,79–82]. While selecting a polymer for indirect SLS process, semi-crystalline polymers are generally preferred as a binder due to their higher density compared to amorphous polymers [61,78]. However, powder-based indirect SLS with semi-crystalline polymers as binder might show high volume shrinkage during densification, which may lead to distortion of the part.

Various polymers have been used with ceramic particles for powder-based indirect SLS. Alumina parts by indirect SLS were prepared using epoxy resin E06 and polyvinyl

TABLE 6.3
Slurry Based Direct Selective Laser Sintering Systems

Suspension Composition	Layer Feeding Method	Application	References
SiO_2 powder and silica sol	Doctor blade	Manufacturing ceramic shell mold with high permeability	[71]
3 mol.% yttria stabilized ZrO_2	Spray deposition	Ceramic shell mold	[72]
Al_2O_3/SiO_2 powder mixtures	Doctor blade	Ceramic dental components	[77]
HA, silica sol, and sodium tripolyphosphate (STPP)	Doctor blade	Bone tissue engineering scaffolds	[74]
HA and silica	Doctor blade	Implant devices for biological and abiological interfaces	[77]

alcohol (PVA) as an organic binder. Cold isostatic pressing (CIP) was performed after the laser sintering to eliminate the pores and increase the density of green ceramics. The final SLS alumina parts achieved a relative density of more than 92% after sintering [79]. Another alumina indirect SLS processing was introduced with polystyrene as a binder. Interestingly, the binder was not just blended with alumina by simple stir mixing. An *in situ* polymerization method was used for mixing ceramic powder and binder together. This study demonstrated the processing feasibility of different 3D geometries via polystyrene-coated alumina powders, but only parts with small dimension did not show crack formation [80]. In a process optimization study, three operating parameters (laser remelting, warm isostatic pressing [WIP], and ceramic suspension infiltration) were manipulated to increase the green density of the indirect SLS parts using alumina-polyamide composite powder. Remelting was found to be effective in improving poor green density of the part, which is caused due to the unwanted "dross formation" while high laser energies were applied during SLS processing. WIP tended to reduce the final dimensional shrinkage of the parts, but not to increase the final density. Finally, ceramic suspension increased the final density up to 71% without introducing unwanted cracks [81].

Similar to the slurry-based direct selective laser sintering, the slurry-based indirect SLS was applied to make up for the low density of ceramic parts manufactured via powder-based indirect SLS. However, the part building efficiency was decreased a lot for this method because of the additional drying step. In one study, airbrush spray was utilized for uniform spreading of the slurry on the substrate. This technique was successful for producing homogeneous slurry layers suitable for the laser sintering process and the green density of the part was increased by this process [82]. A similar approach was used to demonstrate high density (~94%) complex Al_2O_3 parts as shown in Figure 6.7 [83]. In another study, high density (~92%) YSZ parts were built using SLS combined with warm isostatic pressing and shown in Figure 6.7 [84]. Initially, the sintered density of the SLS parts was only 32%, which was improved by PI and WIP to produce crack free, dense, and complex parts. Some more indirect selective laser sinter information is also presented in Table 6.4.

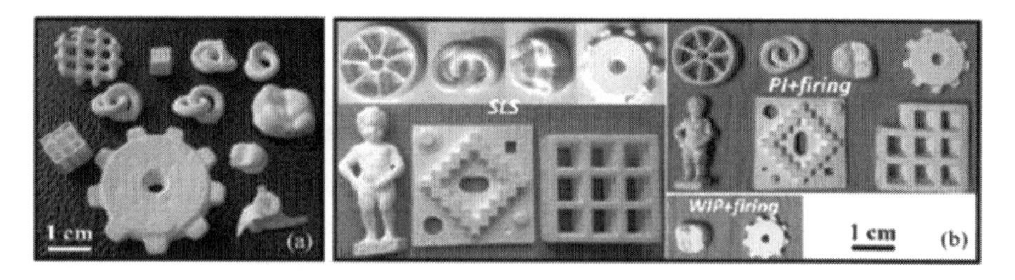

FIGURE 6.7 Complex ceramic parts produced by SLS: (a) Al_2O_3 parts with the assistance of quasi-isostatic pressing and final firing. (From Shahzad, K. et al., *Ceram. Int.*, 38, 1241–1247, 2012.) (b) 3YSZ parts after SLS and after combination with pressure infiltration (PI)/ warm isostatic pressing (WIP) and final firing. (From Shahzad, K. et al., *J. Eur. Ceram Soc.*, 34, 81–89, 2014.)

TABLE 6.4
Ceramic Composition, Binder and Feeding Layer Phase of Some Reported Indirect SLS Processing

Ceramic Composition	Polymer Binder Composition	Feeding Layer Phase	References
3 mol% Y_2O_3–ZrO_2	Isotactic polypropylene (PP)	Powder	[59]
Castable glass (SiO_2–Al_2O_3–P_2O_5–CaO–CaF_2)	Acrylic binder	Powder	[85]
13–93 bioactive glass	Stearic acid polymeric binder	Powder	[86]
α-Al_2O_3 powder	Epoxy resin and polyvinyl alcohol	Powder	[80]
α-Al_2O_3 powder	Polystyrene by dispersion polymerization	Powder	[81]
Al_2O_3 powder	Hydrolyzed PVA	Slurry	[73]
Alumina	PA composite	Powder	[82]
Si_3N_4 and Al_2O_3 powders	Self-made copolymer, which polymerized by MMA and BMA	Powder	[87]
Tricalcium phosphate (β-TCP)	Epoxy resin and nylon	Powder	[88]

Selective laser sintering is an additive manufacture process which is excellent for complex geometry because no additional support structure is needed during the process. Hence, this method has been used in many fields based on their specific requirements. It has been adopted to produce PZT ceramics from precursor powders. The property of the part was manipulated to match the requirements of some medical ultrasonic equipment such as hydrostatic charge and voltage [89]. Bone tissue engineering scaffolds were also prepared by selective laser sintering. Bioceramics, such as hydroxyapatite and tricalcium phosphate, were manufactured by SLS with high processing accuracy and biocompatibility, which is excellent for bone regeneration [66,67,88,90,91].

There are three main challenges for the SLS processing of ceramics:

1. The density of parts is usually modest, which may result in poor mechanical strength
2. Due to the high processing temperature, the cooling cycle is an important issue. An inappropriate cooling might cause warpage in the part
3. Ceramic parts in large dimension are difficult to manufacture using SLS.

The future development will move forward to control the processing parameters and overcome these drawbacks.

6.4 INKJET THREE-DIMENSIONAL PRINTING FOR CERAMICS

6.4.1 History and Methodology

3DP is a powder-bed based ceramic inkjet printing process. In 1990, Sachs et al. of Massachusetts Institute of Technology (MIT) first implemented 3D printing in ceramic manufacturing by using alumina and silicon carbide particles and colloidal silica as a binder [6,92]. In this process, the binder is deposited through an inkjet print-head onto a selected region of the powder layer. Prior to printing, the powder should be packed in the powder feed bed. A roller then spreads a layer of powder with a predetermined thickness to the powder build bed. Usually, a few primary layers are spread to build a foundation layer as a support for the actual part. According to the CAD file, a regular inkjet print-head selectively sprays the binder to the build powder layer [93]. Binder, water or organic-based, starts a hydraulic setting reaction or binds the particles together [94–96]. A new powder layer is spread on the build bed and the process continues until the final part is printed. After this, loosely adhered powders are removed to obtain the green part, which is later sintered to produce the dense final part. Figure 6.8 shows the schematic of the 3DP process [93].

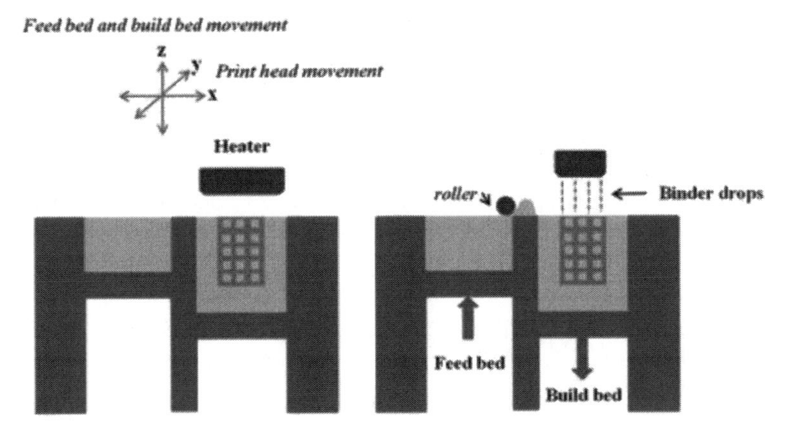

FIGURE 6.8 Schematic of inkjet 3DP processing. (From Bose, S. et al., *Mater. Today*, 16, 496–504, 2013.)

In this method, binder and powder characteristics should be determined prior to printing. A variety of binder and powder properties such as powder packing density, particle size, powder flowability, powder wettability, layer thickness, binder drop volume, binder saturation, and powder wettability, as well as drying time and heating rate, play crucial roles toward the success of this method. Powder packing density can be defined as the relative density of the powder after spreading into the build bed. Binder drop volume is the volume of binder deposited from one nozzle in each drop. Binder saturation is calculated according to the following equation:

$$S = \frac{V_b}{\left(1 - \frac{PR}{100}\%\right) \times a \times b \times LT},$$

where S is the relative binder saturation, V_b is the binder drop volume (pL), PR is the relative bulk density of the powder bed, $a \times b$ is the lateral pixel area of one droplet on the substrate (μm^2), and LT is the thickness of the spread layer (μm) [97]. Tarafder et al. reported that high binder saturation can cause binder spreading over multiple layers of powder as well as bumping appearance in build layers; whereas, low binder saturation causes the layer displacement and/or fragile parts [98]. Powder flowability is another critical parameter, which is regulated by particle size, size distribution, shape, and surface roughness. Large particle size improves flowability, however, densification and sinterability of the ceramic part is compromised due to a smaller surface area. On the other hand, the use of fine particles can cause agglomeration. Flowability is determined according to the Hausner ratio, H, using the following equation:

$$H = \frac{\rho_{Tap}}{\rho_{Bulk}},$$

where ρ_{Tap} and ρ_{Bulk} are the tapped and freely settled bulk densities of the powder, respectively [99]. Similarly, powder wettability is related to particle chemistry and surface energy. Low wettability causes weak integration between particle and binder; however, high wettability results in binder spreading among different layers [100,101]. The powder must be dried properly before the binder deposition. In addition, the binder should possess desired rheological properties in order to successfully eject through the print head without clogging.

The main advantages of the inkjet 3DP are the simplicity of the technology and flexibility in part geometry. Inkjet 3DP is categorized as one of the low-cost additive manufacturing methods. In addition, it does not require any external platform or support, and the powder bed supports the structure during the printing [102]. Furthermore, this method does not demand liquid with modified viscosity or photopolymerizable material [103]. However, inkjet 3DP processed parts suffer from a considerable amount of porosity because of the high friction between particles, lack of external compression force to provide better packing, and random agglomeration [102,104]. Although porosities are desirable for tissue engineering scaffold application, this limits the application of 3DP to produce dense high-performance ceramics. In addition, the polymeric binders used in this process are often toxic and harmful for specific applications such as drug delivery and tissue engineering [105]. Biocompatible binders/printing solution available in the market is not suitable for printing all ceramic powders. As a result, this method has a limitation for printing biocompatible ceramics. Another major disadvantage of 3DP lies in the post-processing steps. Removal of loosely adhered powders or depowdering is a critical step, which causes cracks in the part due to the low green density of printed ceramic [106]. The residual non-bound powders can also be found in the final part, which not only creates undesirable surface roughness, but also lowers the resolution of the complex geometry part. 3D-printed ceramics and glass-ceramics need to be sintered to enhance the densification and mechanical properties [107,108]. However, sintering often causes shrinkage of the scaffolds, which may result in distortions and cracking in the final part.

6.4.2 Inkjet 3DP Processed Ceramics

Originally, 3DP was developed to produce a wide variety of ceramics, metals, polymers, or composite parts. However, at present ExOne and 3D Systems are the only manufacturers of commercialized ceramic 3D printers. In 1990, Sachs et al. first implemented 3D printing in ceramic manufacturing by using alumina and silicon carbide particles and colloidal silica as a binder [92]. Since then, several ceramic materials have been printed using inkjet 3DP for high-temperature applications, electronic devices, tissue engineering, and drug delivery.

Ti_3SiC_2-based ceramics were processed using a combination of 3DP and liquid silicon infiltration. TiC powder and dextrin were used as printing feedstock and binder, respectively. Liquid silicon filtration was performed in an Ar atmosphere at 1600°C–1700°C followed by annealing at 1400°C. The obtained Ti_3SiC_2-$TiSi_2$-SiC composite at 1700°C demonstrated relatively high bending strength of 293 MPa and Vickers hardness of 7.2 GPa [109]. In another study, SiOC polymer derived ceramics have been processed using 3DP of a pre-ceramic polymer. In addition to the specific properties such as luminescence and piezoresistivity, PDCs also have unique oxidation resistance and thermomechanical properties. In this study, a polymethylsilsesquioxane pre-ceramic polymer powder (MK) was used. Two different approaches were applied to print the parts: (1) MK powder was mixed with isopropanol and zirconium acetylacetonate (ZrAcAc) was used as a catalyst. This mixture was used after drying as the base powder precursor and isopropanol was used as the printing media. (2) Tin-octoate (TinOc) was dissolved in 1-hexanol and hexyl acetate mixture and isopropanol was used as the printing media [110].

Silicon nitride (Si_3N_4) is widely used for high-temperature applications due to low coefficient of thermal expansion, good mechanical properties, and resistance to thermal shock. For some applications, a porous structure with precise control over the pore size and pore distribution demands a unique fabrication method. A 3D structure of Si_3N_4 using a powder-based printer was printed using Lu_2O_3 and dextrin in water. After drying and crushing, printing was performed using a water-based binder. By this method, structure with 68% porosity and low fracture toughness of 0.3 MPa $m^{1/2}$ and Vickers hardness of 0.4 GPa was achieved [111]. In a recent study, 3D printing technology was applied to fabricate several meters long building components using lunar soil for the proposed colonization of moon [112].

3DP technology has shown promising applications in fabricating bioceramic scaffolds for the tissue engineering field, which require porous architecture and less precision in surface finish. A poly(lactic-co-glycolic acid) (PLGA) tissue engineering scaffold was fabricated with a homogeneously distributed pore channels [113]. After a co-culture experiment using hepatocytes and nonparenchymal cells, the scaffolds exhibited good cell attachment in both static and dynamic conditions. HA and TCP have been widely employed in 3DP to produce bone tissue engineering scaffolds. HA scaffolds have been printed using water-based solutions as a binder. Tetracalcium phosphate, dicalcium phosphate and TCP have also been fabricated using citric acid as binder. A 13–93 bioactive glass/HA composite with complex porosity structure was also processed via 3DP using 7:1 water:glycerol as a printing solution, as shown in Figure 6.9 [114].

Due to its precise control over pore size, connectivity and geometry, inkjet 3DP is widely used to fabricate scaffolds for localized and controlled delivery of drugs and growth factors. Several drugs and growth factors such as vancomycin, ofloxacin,

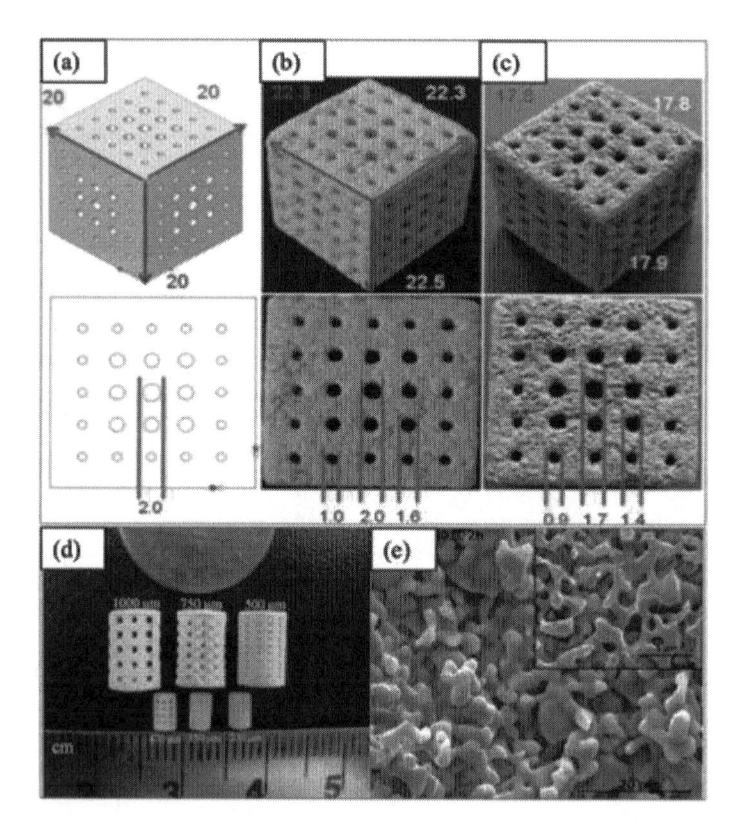

FIGURE 6.9 (a) Computer model, (b) photograph of the 3D-printed green body, and (c) sintered glass/HAp composite structure after heating to 750°C at 2 K/min). Labels indicate dimensions in mm. (From Winkel, A. et al., *J. Am. Ceram. Soc.*, 95, 3387–3393, 2012.) (d) Photograph and (e) FESEM images of TCP scaffolds by 3D printing. (From Tarafder, S. et al., *J. Tissue Eng. Regen. Med.*, 7, 631–641, 2013.)

tetracycline hydrochloride, and bone morphogenic protein have been locally delivered using 3DP scaffolds [115,116]. Tarafder et al. studied the alendronate release kinetics from 3D-printed TCP. Drug release kinetics was controlled by polycaprolactone coating and the scaffolds demonstrated enhanced osteogenesis *in vitro* and *in vivo* [117]. Another study by Tarafder et al. exhibited a successful fabrication of 3D-printed scaffolds by incorporating magnesium and silicon as dopants. These doped scaffolds showed significantly higher bone formation, compared to control TCP scaffolds, as shown in by Figure 6.10 [118].

3D-printed scaffolds were fabricated utilizing natural polymers, such as starch combined with an aqueous binder followed by mixing with gelatin and dextran [119]. These scaffolds exhibited low mechanical properties and poor structural integrity.

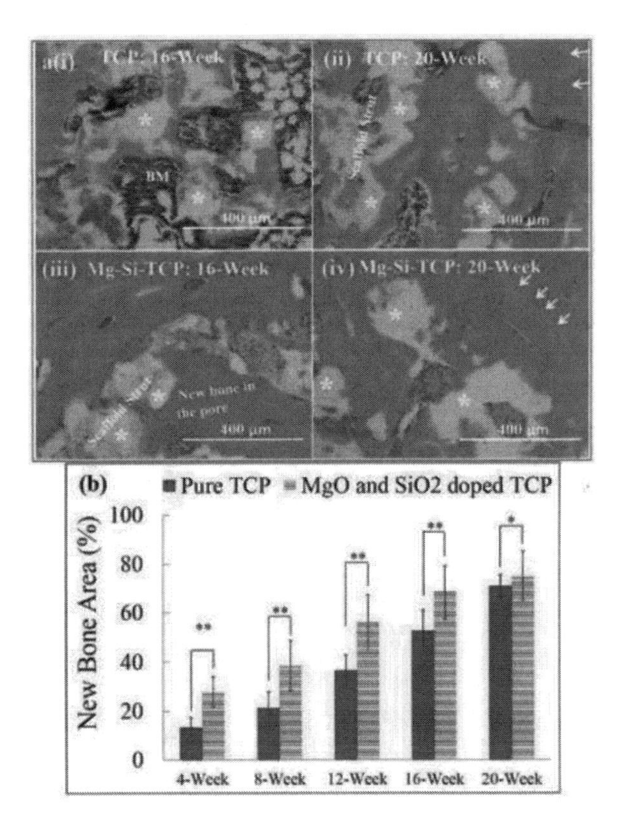

FIGURE 6.10 (a) Photomicrograph of 3DP pure (i & ii) and Mg-Si doped TCP (iii & iv) scaffolds showing the development of new bone formation inside the interconnected macro pores of the 3DP scaffolds after 16 (i & iii) and 20 (ii & iv) weeks in rat distal femur model. Hematoxylin and Eosin (H&E) staining of transverse section. BM = Bone marrow; arrows indicate the interface between scaffold and host bone; star (*) indicates acellular regions derived from the scaffold. Color description: Black = Bone marrow; Grey = New/old bone; White = acellular regions derive from scaffold and (b) Histomorphometric analysis of bone area fraction (total newly formed bone area/total area,%) from 800 μm width and 800 μm height H&E stained tissue sections ($^{**}p < 0.05$, $^{*}p > 0.05$, $n = 8$). (From Bose, S. et al., *Ann. Biomed. Eng.*, 45, 261–272, 2017.)

The addition of poly(L-lactide)-poly(ε-caprolactone) (PLLA-PCL) polymeric solution improved the mechanical properties. In a recent study, $CaSiO_3$-based bioceramic parts with 64 vol% porosity were printed using pre-ceramic polymer (silicon resin) as a binder system [120]. Scaffolds showed good cellular viability and no cytotoxic effects. The 3DP technique was also used to improve the binding of ceramic particles by mixing polymer and ceramic particles [121]. Later, the polymer phase is removed after sintering.

As mentioned before, post-processing of 3D-printed ceramics is essential to achieve proper handleability and desired mechanical strength, specifically for bone tissue engineering application [122]. To improve the mechanical properties of porous 3D-printed scaffolds, the effects of sintering condition, dopant addition, and polymer infiltration [123,124] on their final properties are well investigated. It was found that infiltration of bismethacrylated oligolactide macromer (DLM-1) increases the compressive strength of Tetracalcium phosphate/β-TCP from 0.7 to 76.1 MPa as shown in Figure 6.11a [107]. Compressive stiffness of starch increased from 11.15 to

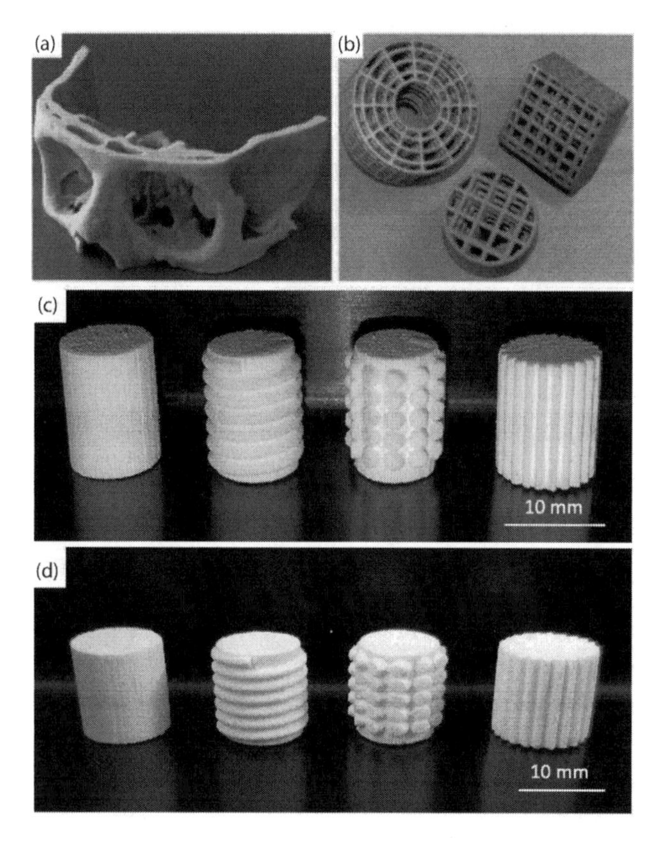

FIGURE 6.11 Exemplary ceramic structures, fabricated with indirect printing techniques: (a) bone implant; (From Khalyfa, A. et al., *J. Mater. Sci. Mater. Med.*, 18, 909–916, 2007.) (b) macrocellular structures with variable ligament lengths; (c) dense TCP scaffolds for bone tissue engineering application (before sintering); and (d) (after sintering). (Courtesy of Bose, S., Vu, A., and Bandyopadhyay, A., Unpublished data, Washington State University, 2019.)

55.19 MPa through PCL/PLLA infiltration [125]. Silica/zinc oxide incorporation in TCP resulted in an increase of compressive strength from 5.5 to 10.2 MPa [126]. Figure 6.11c and d demonstrates dense TCP scaffolds before and after sintering, respectively. These scaffolds are fabricated using synthesized β-TCP powder for bone tissue engineering application.

6.5 FUSED DEPOSITION MODELING

6.5.1 History and Methodology

FDM is one of the most widely used 3D printing techniques, which was first commercialized by Stratasys Inc. (Eden Prairie, MN) in 1990 [127]. FDM is very commonly used among researchers due to its low cost, simplicity, and flexible unit size. In an FDM process, a filament of a semi-solid thermoplastic polymer is fed into a liquefier by two rollers followed by heating and extrusion through a nozzle. The extruded part finally gets deposited on a platform. The heaters in liquefier increase the temperature of the polymer filament above but close to the melting point to easily extrude the filament through the nozzle. Based on the CAD file, the nozzle moves in an X-Y direction and "roads" or "rasters" get deposited [128,129]. Solidification starts from the outer surface of the roads and then continues radially to the core [130]. After deposition of the first layer, the platform moves downward and the second layer gets deposited on the first one. This process continues until the entire part is built.

Figure 6.12 shows a simple schematic representation of the FDM process. Fused deposition of ceramics (FDC) is a modified FDM process and was first developed in 1995 at Rutgers University [131]. In order to implement FDM in ceramic manufacturing, brittle ceramic powders are dispersed and mixed with a thermoplastic polymer mixture (40–65 vol.%) to prepare the flexible feedstock filament [132]. The semi-solid thermoplastic polymer mixture, which performs as the ceramic powder carrier, includes a binder, plasticizer, and dispersant [133]. Pre-treatment of ceramic powder with organic dispersant/surfactant is also crucial to obtain an extrudable material [134].

Filament fabrication is one of the main obstacles in FDC as it is one of the most time-consuming steps [133]. The filament should be compositionally homogeneous and agglomerate-free to prevent clogging of the FDC nozzle. In addition, to ensure the accuracy in deposition and final size of the part, the high dimensional tolerance of the filament is required [135]. In the case of brittle materials, optimization of filament composition and processing parameters is required to prevent filament buckling during the extrusion [134]. The stiffness of the filament, binder chemistry, and viscosity of the ceramic-polymer mixture are the principal parameters in FDC that determine the success of the process [136]. In addition to the filament fabrication, there are various other process parameters, which control the homogeneity of the final part in FDC. The flow rate of the process can be controlled by the rate of feedstock loading to the heated liquefier. Molten material temperature and deposition speed should match the cooling and solidification rate to prevent any discontinuity of the structure [130]. In addition, road width, thickness, the gap between the roads, and the angle between layers, control size, shape, and volume of the pores [129].

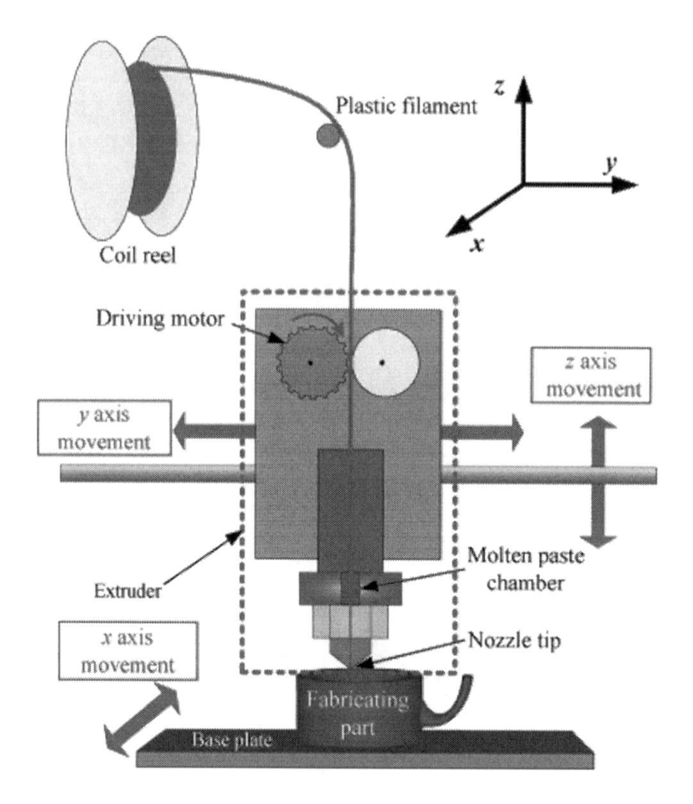

FIGURE 6.12 Schematic of FDM. (From Jin, Y. et al., *Addit. Manuf.*, 8, 142–148, 2015.)

Inappropriate vector spacing or road width, insufficient bonding between the roads (interlamellar) or build layers (interlaminar), and inaccurate filling between vectors and perimeter (sub-perimeter voids) can cause defects within the printed parts [136].

To achieve the final densified part, post-processing of the printed ceramic part is carried out, which includes binder removal and sintering. Removing a large amount of binder and organic components is a major concern in FDC [130]. This often involves relaxation of viscoelastic stresses, which subsequently causes a large shrinkage [137]. Shrinkage can be minimized by using a high solids loading during the part processing. Sintering is the final post-processing step, which can cause more shrinkage and warpage due to the thermal stress and relaxation in the binder. During the deposition, a high coefficient of thermal expansion of the binder causes tensile stresses in each building layer, which results in shrinkage while cooling down. Additionally, due to the constraint in bottom-built layers, these tensile stresses also initiate warping. These residual stresses arise because of the viscoelasticity of the binder, and relaxation is possible if enough time or temperature is given to the system [135].

6.5.2 FDC Processed Ceramics

FDC was patented in 1998 to process 3D ceramic structures including novel ceramic materials, ceramic/polymer composites, oriented/radial piezoelectrics, and photonic band-gap structures [139–144]. FDC has been used as a unique processing tool for piezoelectric materials fabrication. Compared to traditional methods, FDC allows the fabrication of composites with complex internal structure and symmetry. Bandyopadhyay et al. processed 3D PZT ceramics with a ladder and honeycomb structures through direct and indirect FDC processes, respectively [145]. PZT-Epoxy composites with a solid phase volume fraction of 0.3 were processed by FDC, and the effect of deposition angle on piezoelectric properties of PZT was investigated [146]. Piezoelectric material with ladder structure was fabricated for transducers application, which showed good uniformity and repeatability of structure with a road diameter of 300 μm [147]. FDC was also used for composites of polymers/piezoelectric ceramics such as PZT and lead-magnesium niobite as the feedstock materials (Figure 6.13) [148]. Multi-layered PZT sensor components were also manufactured using an FDM system with four extruders [149] where each layer of these multiple ceramic parts contains two types of soft and hard PZT ceramics.

Si_3N_4 ceramic parts with comparable mechanical properties have also been processed using FDC at different conditions [132,135]. Dense structural Si_3N_4 parts were produced using 55–60 vol.% ceramic loading. Oleyl alcohol and a commercially available investment casting wax were used as surfactant and binder, respectively. The effects of build process parameters on final part homogeneity and isotropy were studied using different angles of 0°, 90°, and ±45° between the roads and the long axis of the bar in alternating layers. It was found in the first two cases, significant warping occurred in samples; however, using the +45° or −45° of vector angles resulted in no warping. Binder removal resulted in ~1% shrinkage while major shrinkage of 16%–18% happened during the sintering. During binder removal, shrinkage was significantly higher in the direction of road vectors than across the

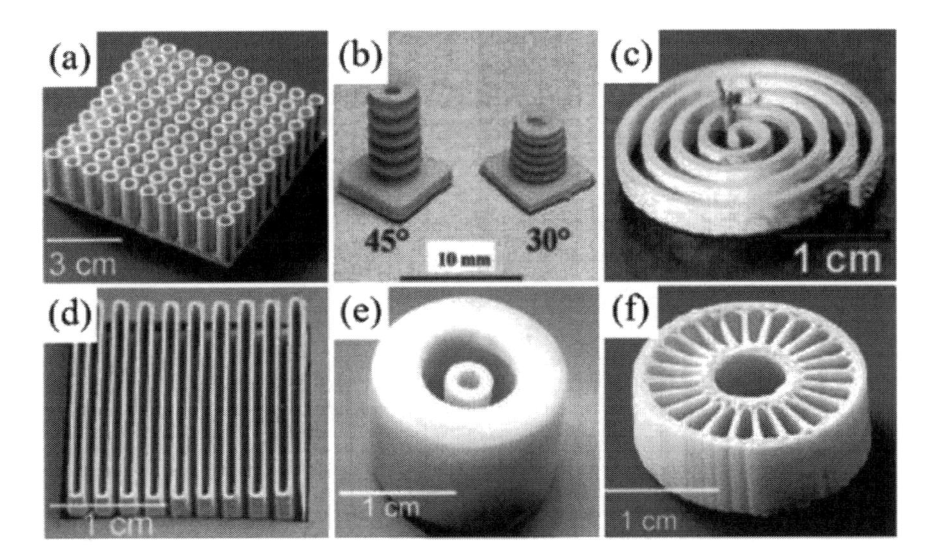

FIGURE 6.13 Various types of transducers made by FDM: (a) tube-array; (b) bellows; (c) spiral; (d) curved transducer; (e) telescoping; and (f) radial actuators. (From Allahverdi, M. et al., *J. Eur. Ceram. Soc.*, 21, 1485–1490, 2001.)

roads in parts using 0° and 90° of building orientation [135]. In another work, MgO-doped alumina with honeycomb structure was also processed using indirect FDM. The ceramic slurry was developed using 1-butanol and Darvan 821 as antifoaming agent and dispersant, respectively. The slurry was then infiltrated to an FDM-processed polymer mold, followed by drying, binder removal, and final sintering at 1600°C. Using the same technique, an alumina bone graft with horse knuckle structure was also processed [151]. Bandyopadhyay et al. processed *in situ* Al/Al_2O_3 composites with a controlled microstructure using FDC-processed silica structures and subsequent Al infiltration at 1150°ΔC [152].

Novel applications for FDC include fabrication of 3D woodpile lattice structures. The rod shape of the ceramic-polymer filament used in FDC is favorable for manufacturing lattice structures. Fused deposition of multi-materials is a subset of FDM, which is used to fabricate 3D alumina photonic bandgap structure [153]. This approach allows the fabrication of complex designs with periodic structure. Compared to traditional methods where structures are fabricated by making bulk pieces followed by etching and/or machining to achieve the desired dielectric properties, fused deposition of multi-materials allows better control over the structure. Alumina feedstock was prepared by mixing the 60 and 62 vol.% of alumina powder in stearic acid solution and toluene and ECG-9 was utilized as the surfactant and thermoplastic binder, respectively. Investment casting wax type polymer was used as the support for alumina and deposition was performed in a layer-by-layer manner of alumina and the wax. During binder burnout, zirconia setter powder was used as a support for the alumina to prevent bending of the rods. Calcium phosphate periodic woodpile structures were also processed with spatial resolutions less than 100 μm using FDC (Figure 6.14a–c) [154]. These lattice structures exhibit controlled pore morphologies and multi-scale hierarchical architectures, which can provide tailored mechanical, biological, and geometrical properties. PCL-HA and alumina bandgap woodpile structures are shown in Figure 16.14f, respectively [155].

HA and TCP structures have been also fabricated using direct and indirect FDC. In the direct process, HA powder was first coated by stearic acid as a surfactant and then mixed with binder solution including polyolefin-based binder, polyethylene wax, hydrocarbon resin tackifier, and polybutylene plasticizer. The HA scaffolds were fabricated using filament with 55 vol.% ceramic. Binder burnout was performed at 550°C, followed by final sintering at 1100°C [156]. Using the indirect process, various designs of wax molds were fabricated and infiltrated with slurries containing food grade TCP powder to fabricate TCP structures with different pore size and volume [157]. FDM is a promising additive manufacturing tool to manufacture porous bioceramic part and tissue engineering scaffolds. Processing of PCL and PCL-HA scaffolds with varying pore architectures was carried out using FDC [158]. PCL scaffolds showed *in vitro* fibroblast and periosteal cell proliferation, differentiation, and ECM production. Polymer selection is a critical step during the fabrication of ceramic-polymer composites for biomedical applications, as they must possess nontoxicity and excellent extrudable properties. Biocompatible composite of TCP/polypropylene was synthesized using vegetable oil as a plasticizer and VESTOWAX SH 105 pallets (Crenova, NJ) as viscosity modulator, including 20.5 vol.% TCP in the filament. Results showed that the fused deposition process allows precise control of

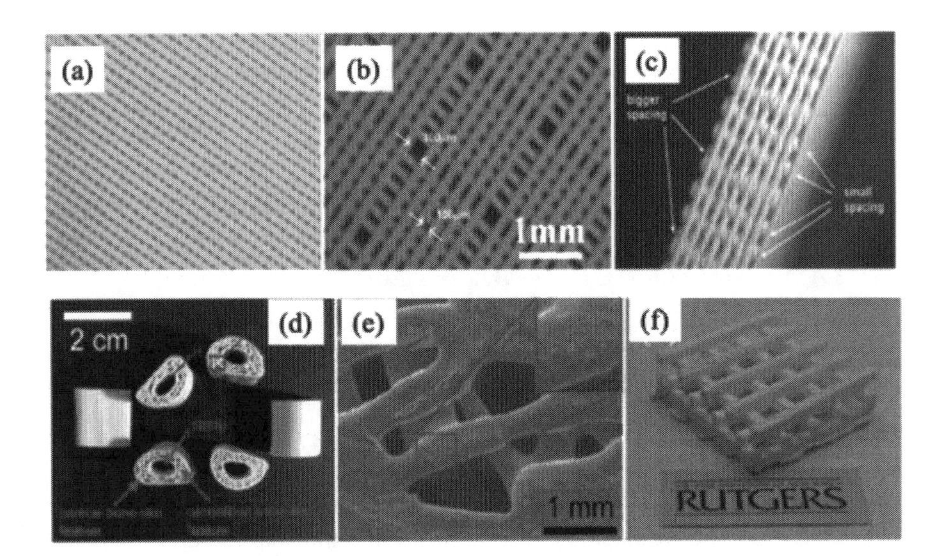

FIGURE 6.14 (a) Fine lattices with regular spacing; (b) lattices with graded spacing structures; and (c) side view of the lattices. (From Yang, H. et al., *J. Biomed. Mater. Res. B: Appl. Biomater.*, 79, 116–121, 2006.) (d) Bioceramic scaffold: photo of the artificial bone scaffolds made from PCL-HA using CT-guided FDM; (e) SEM image of the structure. (From Xu, N. et al., *ACS Appl. Mater. Inter.*, 6, 14952–14963, 2014.) and (f) photonic bandgap (PBG): unit cell structure of alumina fabricated using FDM. (From Chen, Y. et al., *Microw. Opt. Techn. Let.*, 30, 305–307, 2001.)

pore volume, and increasing the pore volume from 36%, to 40% and 52% decreases the strength gradually from 12.7 MPa to ~10 MPa [160]. Osteogenic properties of CaP/PCL scaffolds prepared by FDC were tested using mesenchymal progenitor cells. CaP/PCL filaments were fabricated through melt extrusion of CaP/PCL pellets containing 25 wt.% CaP. Results demonstrated a circular and centripetally-directed arrangement of the cells in pores, whereas multiple focal adhesions of elongated cells were observed in angles [159]. In a clinical study, PCL plug scaffolds were implanted in five patients, which showed excellent integration with the surrounding calvarial bone after 12 months [156]. Additionally, FDA-approved FDM-processed PCL scaffolds, such as interwoven PCL mesh (Osteomesh™) and 3D implants (Osteoplug™), are utilized for craniomaxillofacial applications. Implantation of FDC processed Osteoplug™ scaffolds in 12 patients with a chronic subdural hematoma showed no adverse effects and good osseointegration into the surrounding calvarial bone [161].

Biphasic construct consisting of a PCL cartilage scaffold and a PCL-TCP osseous matrix was fabricated by FDC [162]. These scaffolds showed *in vivo* bone regeneration and remodeling, cartilaginous repair, functional tissue restoration, and low occurrence of fibrocartilage in an osteochondral defect in pigs. Computer tomography-guided FDM was utilized to fabricate a PCL/HA and 3D artificial load-bearing PCL bone graft. The results exhibited suitable mechanical properties, excellent biocompatibility, and *in vivo* new bone formation ability [150]. A recent study has utilized FDM approach to build a polylactic acid–nanohydroxyapatite

(PLA/nanoHA) composites in a low-cost 3D printer. The composite consists of 5% and 15% nanoHA, which were uniformly distributed within the PLA matrix and the final part showed good mechanical properties [163]. PCL scaffolds by FDM were also investigated in comparison with polyurethane sponges for the engineering of adipose tissue. The scaffolds showed enhanced *in vivo* angiogenesis, fibrous tissue formation, and adipogenesis after 2 and 4 weeks of implantation in nude mice [164].

6.6 LAMINATED OBJECT MANUFACTURING OF CERAMICS

6.6.1 History and Methodology

Laminated object manufacturing (LOM) was first introduced by Helysis Corporation in 1986 to manufacture 3D components from sheets of paper, plastic, or metal [165]. They patented the process in 1987 and commercialized in 1991. In this method, thin sheets of materials are laser cut and stacked on top of another to manufacture a three-dimensional object. The system, as shown in Figure 6.15 includes a work table, which can move vertically, a feeder which is a continuous roll of the material, and an X-Y plotter. At first, the feeder sends the laminations/sheets over the build platform on the work table. The bottom side of the lamination layer is generally pre-coated with a heat-sensitive adhesive. Next, a hot lamination roller is used to melt the adhesive so that the thin sheets get attached to the bottom layer. The X-Y plotter uses a laser beam to cut the outline of the part at each layer. The excess part is also cut into “cubes” to facilitate its removal after manufacturing [166]. During the building process, this excess part remains in the building block to support the structure [167]. After completion of each layer, the build platform moves downward by a depth of sheet thickness. This process continues until the entire part is made. Once the part is manufactured, “decubing” carried out to remove any excess material [166,168].

Initially, LOM was developed to process papers. Later, this technique has been advanced to produce 3D ceramic parts from green ceramic tapes. Prior to fabrication

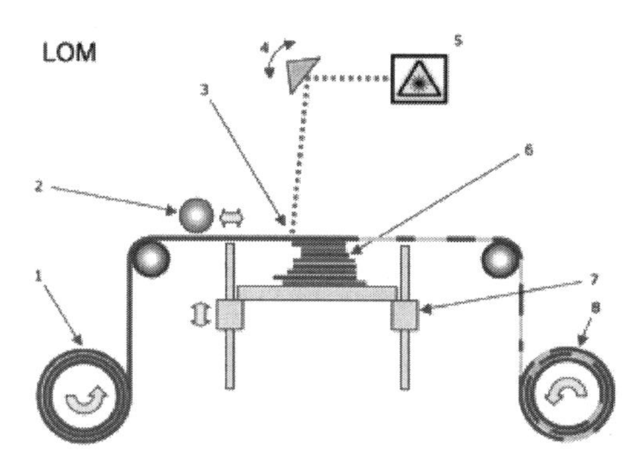

FIGURE 6.15 Laminated object manufacturing: principle drawing. (1) supply roll; (2) heated laminated roll; (3) laser cutting beam; (4) prism steering device; (5) laser; (6) laminated shape; (7) movable table; and (8) waste roll (with cutout shapes). (Courtesy of LaurensvanLieshout/CreativeCommons.)

of the ceramic roll, a suspension including the ceramic precursor, plasticizer, binder, and dispersant needs to be prepared. The ceramic suspension must possess suitable rheological properties as they determine the possibility of crack formation during the drying process, the green density and homogeneity of the final part, and overall success of the process [169]. The viscosity and shear thinning of suspensions are determined using the Herschel-Bulkley model [170]. The dispersant content has a significant effect on the viscosity of suspension [171]. Presence of a high amount of dispersant is not desirable because it increases the viscosity significantly, which not only reduces the green density but also enhances the risk of shrinkage. Besides, suspension with a viscosity lower than 20 Pa.s results in higher fluidibility, while higher viscosity causes formation of a paste, both of which are not suitable for the LOM process [172].

In LOM, the laser line energy is calculated using the following equation:

$$E_L = \frac{P_L}{V_L},$$

where P_L and V_L are laser power and scanning speed, respectively. E_L is considered to be a critical factor in the cutting and decubing process. High E_L results in undesirable cutting of the lower sheets and oxidation of the organic components in preceramic papers, whereas low E_L causes the insufficient cutting of the papers. Therefore, optimization of E_L helps to control the cutting depth and facilitates the decubing process [168]. The low-temperature manufacturing in the LOM process results in fewer distortions and deformations in the printed part due to the reduced thermal stress. Compared to the other AM techniques, it also has lower materials, machine, and process cost. Moreover, LOM parts exhibit moderate flexural strength at relatively high porosities [173]. The disadvantages include delamination, poor quality interfaces, interfacial porosities, and anisotropic mechanical properties. Various experiments have been carried out to improve the LOM process by eliminating the disadvantages. Bigger rollers have been used for better lamination, whereas small rollers result in quicker manufacturing time. Other than shortening the processing time, researchers have also focused on waste removal strategies, lamination, and cutting speed.

6.6.2 LOM Processed Ceramics

First-generation ceramic manufacturing by LOM was demonstrated utilizing Al_2O_3 and ZrO_2 green tapes [174]. Following this, other researchers were also able to fabricate successful 3D ceramic structural components using the LOM process. This includes, SiC, Si/SiC composite [175], ZrO_2, ZrO_2/Al_2O_3 composite [176], TiC/Ni composite [177], LiO_2–ZrO_2–SiO_2–Al_2O_3 (LZSA) glass-ceramic composite [178], functional ceramics such as PZT [179] and HA for bone graft substitutes [180].

LOM was used to fabricate Si-SiC composites. The pre-ceramic papers were prepared through pyrolysis of a commercially available filter paper made of cellulose fibers. The adhesive tape containing phenolic resin, polyvinyl butyral, benzyl butyl phthalate, and ethanol was used to bond the pyrolyzed paper. Followed by the lamination step, a second pyrolysis step was carried out at 800°C. Finally, Si infiltration was performed in vacuum at 1500°C in vacuum [175]. Rapid Köthen aqueous handsheet-forming process is utilized to process pre-ceramic papers comprising 76.8 wt% SiC powder, 20 wt% cellulose pulp, 3.2 wt% retention agent, and binders. A thermosetting polymeric adhesive with softening point of 60°C was used to coat the sheets. LOM was performed using a 25 W CO_2 laser at a wavelength of 10.6 μm [168].

LOM was carried out to manufacture Al_2O_3 parts using 7 wt% polyvinyl butyral and polyvinyl acetate as binder and adhesive, respectively. Binder was removed at 240°C–300°C and samples were sintered at 1580°C. Although binder removal did not cause any damage, sintering and cooling processes led to distortion and cracking [181]. Figure 6.16a and b show lightweight Al_2O_3 and SiC ceramic components, which were fabricated from pre-ceramic sheets using the LOM process.

Dense Si_3N_4 parts were processed using the LOM technique (Figure 6.16c) [182]. This part exhibited a post-sintering average final density of 97% with a volume shrinkage of 40%. The microstructures and mechanical properties such as Young's modulus,

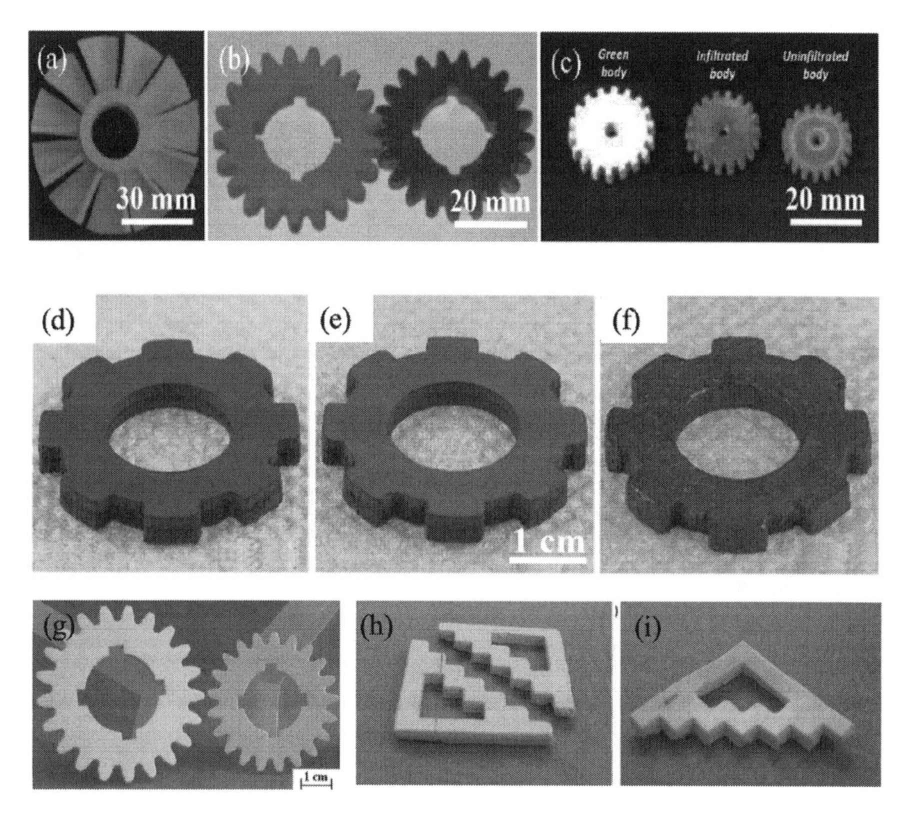

FIGURE 6.16 LOM fabricated (a) Al_2O_3 turbine rotor, (b) SiC gear wheels, (c) complex-shaped Si_3N_4 parts, (d) 3D gear with MAX phase Ti_3SiC_2 green sample, (e) after sintering in Ar, (f) after silicon infiltration, (g) LOM-processed LZSA glass-ceramics: gear wheel structure (left: green sample, right: sintered sample), (h) stair-like green sample, and (i) stair-like sintered sample. (From Gomes, C.M. et al., *J. Mater. Process Tech.*, 206, 194–201, 2008; Gomes, C.M. et al., *J. Am. Ceram. Soc.*, 92, 1186–1191, 2009.)

flexural strength, and fracture toughness measured for the final parts were shown to be comparable to the Si_3N_4 parts prepared by conventional methods such as reaction-bonding, slip casting, and pressureless sintering. In a more recent study, the LOM of ceramics has been demonstrated with $M_{n+1}AX_n$ (where M is a transition metal, A is a group element [mostly IIIA and IVA], and X is Carbon and/or Nitrogen) (MAX) phase Ti_3SiC_2 composites [183]. Again, a defect-free gear with a linear shrinkage of less than 3% relative to the green state was obtained, as shown in Figure 6.16d–f.

LOM has also been used to manufacture LiO_2–ZrO_2–SiO_2–Al_2O_3 (LZSA) glass-ceramics. Suspensions were prepared by dispersing glass powder in distilled water with ammonium polyacrylate being the dispersant, followed by mixing with polyvinyl alcohol, polyethylene glycol, and a blend of modified fatty and alkoxylated compound as the binder, plasticizer, and antifoaming agent, respectively. A glass ceramic roll was fabricated using a tape caster. A continuous-wave CO_2 laser with the power of 16.8W was used during the LOM process. Figure 6.16g–i shows the manufactured parts with a homogeneous distribution of the porosity within the microstructure [165,169]. Although the LOM technique promises to produce complex structures, application of this technique in ceramic manufacturing has been always limited to simple gear-like parts. Most of the abovementioned studies have been carried out years back and not much technological advancements have been reported in recent years. At present, there is no commercial vendor available to produce LOM based ceramic structures. However, a cheaper version of the same concept using sticky paper and knife is still practiced in Mcor (Dunleer, Ireland) to manufacture 3D parts from CAD files.

6.7 DIRECTED ENERGY DEPOSITION

6.7.1 History and Methodology

Directed energy deposition is a metal or ceramic AM technique where powders are injected at the focal point of a laser or electron beam to melt and create a melt pool that is used to build parts. LENS™ is a commercially available directed energy deposition-based AM process. In this method, a focused laser beam is used to melt the metal substrate first [184]. As shown in Figure 6.17, the focused beam produces a small molten area on the substrate, and metal powder is injected through a gas stream into the molten part. By moving the laser beam away, the molten area cools down and solidifies rapidly, and forms a strongly bonded solid material to the substrate. In this method, the cooling rate depends on processing parameters such as transverse velocity and laser output energy [185]. After deposition of the first layer, the laser head moves upward and a second layer gets deposited. This process continues until the entire part is fabricated. Success in this method depends on the interaction between the laser beam and the powder [186]. Generally, laser beam power can be increased by decreasing the laser beam diameter and the scan speed or increasing the specific laser energy input as shown in the following equation:

$$I = \frac{P}{vD},$$

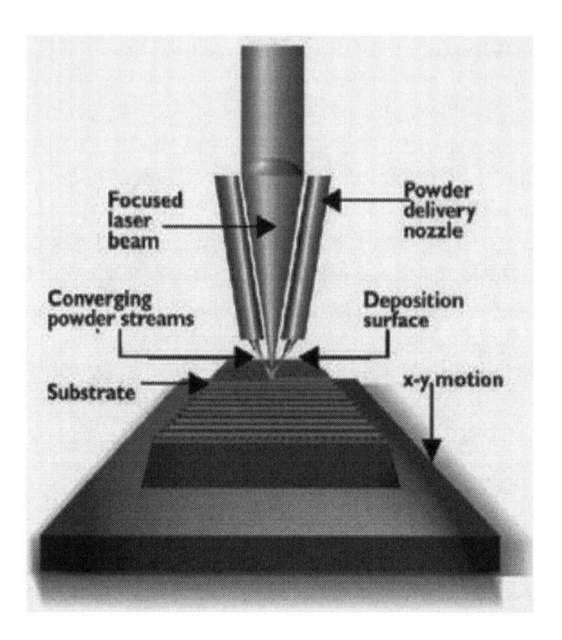

FIGURE 6.17 Schematic of LENS™ processing. (From Bandyopadhyay, A. et al., *J. Mater. Sci.: Mater. Med.*, 20, 29, 2009.)

where I is the laser energy input, P is the power of laser beam, v is the scan speed, and D is the laser beam diameter [187].

A small heating zone and high cooling rate result in a fine microstructure of deposited material [185]. Controlling the microstructure in this method is easily possible by changing the cooling rate [188,189]. During solidification, thermal gradient (G), and solidification velocity (R) determine the cooling rate $\left(\frac{\partial T}{\partial t}\right)$ [190]:

$$R = \frac{1}{G}\frac{\partial T}{\partial t}.$$

In addition, $\frac{\partial T}{\partial t}$ is related to the processing parameters according to the following equation [191]:

$$\frac{\partial T}{\partial t} \propto k\left(T - T_0\right)^2 / \left(\frac{P}{V}\right),$$

where k is the thermal conductivity of the material, T_0 is the temperature of the substrate, P is the power of the laser beam, and V is the transverse speed.

Unlike other AM processes, LENS™ allows the fabrication of fully dense structures [192]. Another advantage of this method is the possibility of gradient deposition of various materials in a single component [193,194]. Presence of thermal stresses is one of the main drawbacks of this method. Due to the nature of LENS™, a small volume of material is heated at a certain time, resulting in an intense temperature gradient within the sample in each layer. In addition, there are thermal stresses/strains between the solid deposited at the first layer and the solidifying second layer. Due to heat transfer from the top layer, there is a thermal expansion in the solidified layer and, as a result, compressive and tensile stresses develop in the solidified first layer and solidifying top layer, respectively. Depending on processing parameters and the magnitude of the stresses, distortion and possible failure may occur due to delamination or cracking [184]. Not all the ceramic materials can be processed using the LENS™ technique due to high thermal stresses [195]. Post-processing of LENS™ fabricated structures is needed in many cases. In addition, if the substrate is not a part of the final product, it should be removed [190].

6.7.2 LENS™ Processed Ceramics

The LENS™ process can fabricate complex prototypes in a near-net shape, leading to time and machining cost savings. Various ceramics with different shapes have been processed using LENS™. Al_2O_3–YSZ ceramics with fine-grained microstructure were processed via LENS™ [191]. The parts exhibited comparable mechanical properties with those synthesized by traditional processing. Though the surface precision and dimensional accuracy were not satisfactory, the eutectic spacing of the lamellar colony reached 100 nm due to the rapid melting process. In order to prevent thermal stress-induced crack formation and distortions, researchers have preheated the ceramic powder [192,193]. Results from another study reported that simultaneous crystallization during the solidification process resulted in a fine-grained microstructure and grain boundary strengthening, which improved mechanical properties of the part. Ceramic manufacturing in LENS™ requires a preheating temperature above 1000°C due to the high melting temperature of ceramics [194]. However, a temperature close to the melting point will create a large molten pool size, which results in undesirable outflow into the surrounding powder and poor dimensional accuracy. High-temperature preheating (~1700°C) with a CO_2 laser reduced the occurrence of thermal-gradient induced cracks. Parts received were fully dense and demonstrated flexural strength of more than 500 MPa without post-processing. Thermal gradients and poor-quality surface finish were still present and only parts with less than 3 mm height could be processed with top-down laser preheating. Pure ZrO_2 and Al_2O_3 ceramic powder were used to process ceramic parts with fine and homogeneous microstructure for dental graft applications [195]. In addition, other preheating techniques such as the bottom-up inducting approach and diode laser selective preheating were also reported. These approaches resulted in processing larger parts with a lower thermal gradient.

LENS™ was used to fabricate a crack-free functionally graded TiC/Ti composite. Using two separate powder feeders with controllable rotational speed, the functionally graded composite was processed. This was achieved by regulating the rotational speed of one material's powder feeder from zero to maximum, and the second material's feeder from maximum to zero, as the number of layers increased. Figure 6.18a shows the light microscope image of the structure. The bottom layer consists of α-Ti with a small amount of TiC. TiC concentration increases from the bottom layer to the top with a dendritic or equiaxed particulate structure. The TiC concentration at the top was 95 vol.%. Figure 6.18b–g shows the SEM microstructure of the sample at different layers [196]. Compositionally graded

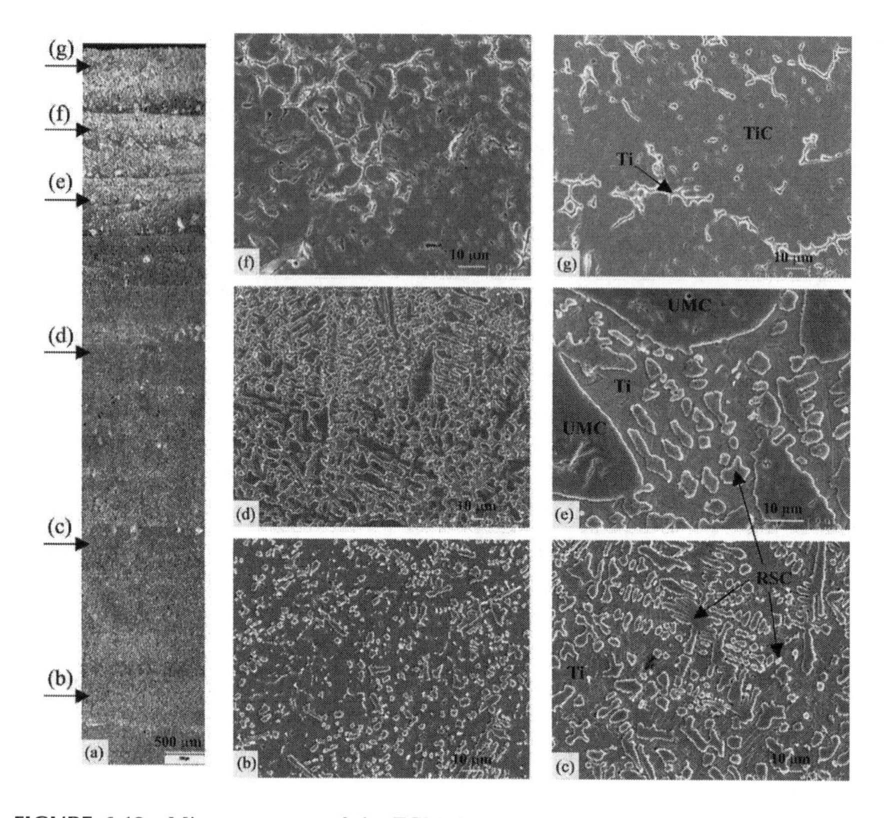

FIGURE 6.18 Microstructures of the FGM deposit: (a) Light optical microscopy photomicrograph of the deposit and (b–g) SEM photomicrographs with increasing TiC contents in different locations (UMC and RSC denote unmelted and resolidified TiC carbide respectively). (From Liu, W., and DuPont, J.N., *Scripta. Mater.*, 48, 1337–1342, 2003.)

aluminum oxide coating on stainless steel substrate with crack-free microstructure and high hardness was also fabricated using two optimized power levels of 400 and 500 W. Bond coating of Ni-20 wt.% Cr was deposited at first layer. Powder feed rate decreased from 13 g/min to 0 over the first three layers, while the aluminum oxide feed rate increased from 0 to 14 g/min and remained constant for four more layers [197].

LENS™ has also been used to produce particle-reinforced metallic matrix composites (MMCs) [185,198]. MMCs cannot be easily processed by conventional processes such as powder metallurgy techniques or casting. In conventional powder metallurgical processes, the product geometry is limited due to the complexity of the process. On the other hand, casting causes the particle segregation and undesirable interfacial reactions. Compared to these methods, LENS™ is a time, energy, and cost-effective approach for processing of MMCs [185]; however, use of very fine and irregular particles is still a challenge in fabrication of MMCs by LENS™. Ni-coated TiC particles with a size up to 45 μm as reinforcement were used in Ni-based IN625 superalloy. The process was performed using the laser with a power of 650 W in an Ar environment to prevent oxidation. During the process, Ni coating remained on the surface of the TiC, and the strength of the matrix increased significantly due to the presence of the reinforcement. In addition, the Ni coating resulted in better flowability of powders and interaction between the laser beam and TiC particles [199]. Ni-coated TiC particles in the Ti6V4Al matrix MMCs was also processed and shown in Figure 6.19a. Formation of the intermetallic phase of Ti-Ni resulted in a significant increase in strength, but a decrease in toughness [185].

LENS™ has been used to fabricate ceramic coatings on metallic structures for biomedical applications. The high crystallinity of coatings, controllable thickness,

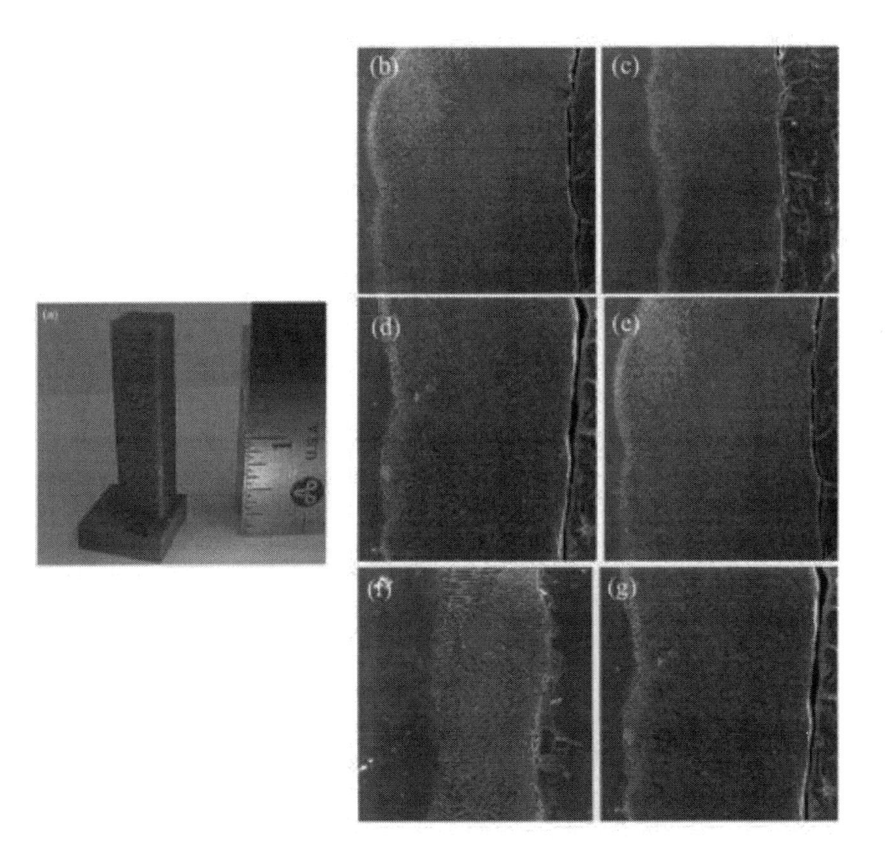

FIGURE 6.19 (a) Laser-deposited Ti6Al4V + 10 wt.% TiC/Ni cubic samples with LENS. (From Zheng, B. et al., *Metall. Mater. Trans. A*, 39, 1196–1205, 2008.) (b) speed of 15 mm s^{-1} with powder feed rate of 13 g min^{-1} and 500 W, (c) at a scan speed of 15 mm s^{-1} with powder feed rate of 13 g min^{-1} and 400 W, (d) 500 W power with a powder feed rate of 13 g min^{-1} and scan speed of 15 mm s^{-1}, (e) 500 W power with a powder feed rate of 13 g min^{-1} and scan speed of 10 mm s^{-1}, (f) 500 W power at a scan speed of 10 mm s^{-1} with a powder feed rate of 9 g min^{-1}, and (g) 500 W power at a scan speed of 10 mm s^{-1} with a powder feed rate of 13 g min^{-1}. (From Roy, M. et al., *Acta Biomater.*, 7, 866–873, 2011.)

and good adherence between the coatings and substrate are some the advantages of using LENS™ compared to other methods such as sol-gel, dip coating, biomimetic coating, and plasma spraying where one or more of the mentioned characteristics cannot be achieved. Laser beam power, laser scan speed, and powder feed rate control the coating thickness [186,200]. It was shown that increasing the power from 400 to 500 W resulted in an increase in coating thickness from 250 to 400 μm (Figure 6.19b and c). As shown in Figure 6.19d and e, keeping the laser power at 500 W and decreasing the scan speed increased the coating thickness. Increasing the powder feeding rate from 9 to 13 g/min also increased the coating thickness from 220 to 375 μm (Figure 6.19e and f). In addition to coating thickness, microstructure and TCP content in coating varied by powder feed rate, laser beam power, and scan speed. Ti grains had columnar and equiaxed structures at the bottom and top of the coating, respectively. An increase in powder feeding rate increased the TCP content in coatings; whereas, an increase in scan speed decreased the TCP volume fraction [186]. In addition, although the HA and Ti interaction was minimum, some $CaTiO_3$ was formed in this coating. Comparing the single and multi-layer coatings, HA layer was formed on top of the multi-layer coatings, representing a significant drop heat flow and resulting phase transformation from molten Ti to HA [200]. Processing of functionally graded Ti-TiO_2 structures was carried out using LENS™. Compositionally graded TiO_2 coating with 50% TiO_2 on the top surface of porous Ti increased the hardness and the wettability of Ti implants which enhanced their wear resistance and cell-material interactions [201]. Similarly, a functionally graded yttria-stabilized zirconia coating on stainless steel was also achieved. The composition varied from 100% bond coat on top of the stainless steel substrate to 100% YSZ topcoat at the third layer. Coatings were prepared at a laser power of 250 W and a scan speed of 40 mm/sec. The gradation in composition was achieved by altering the YSZ powder feed rate from 0 to 14 g/min and bond coat powder feed rate from 13 g/min to 0 over the first three layers. The microstructural analysis revealed the columnar grain-oriented structure including few segmented cracks along with the coating thickness, which made the obtained structure suitable for thermal barrier applications [202]. In a recent study by Bandyopadhyay et al., LENS™-processed implants were modified by introducing TiO_2 nanotubes to improve cytocompatibility. The result showed LENS™-induced porosity enhanced *in vivo* bone tissue integration, compared to dense Ti implants, as shown in Figure 6.20 [203].

LENS™ has been used to fabricate dense and crack-free Al_2O_3 parts. Using a laser power of 125 W, a scan speed of 10 mm/s, and powder feed rate of 14 g/min, bulk Al_2O_3 parts with a density of 94% were achieved. A smooth surface of the parts, as shown in Figure 6.21a, indicates that the Al_2O_3 particles were completely melted and resolidified during the process [184]. Dense and porous PZT can also be fabricated by LENS™. Using the laser power of 150 W, a scan speed of 5 mm/s, and powder feed rate of 1.3 g/min, crack-free and dense PZT with reasonable dielectric properties was achieved (Figure 6.21b). In addition, at very low energy density, the porous structure was obtained by incomplete melting and the presence of unbound particles [204].

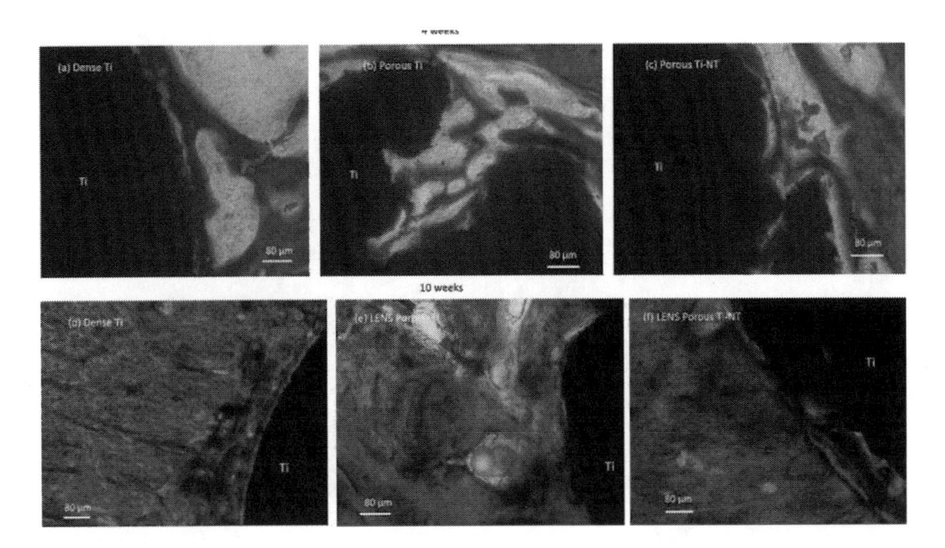

FIGURE 6.20 Photomicrograph showing the histology images after 4 weeks (a, b, c) and 10 weeks (d, e, f) where signs of osteoid-like new bone formation could be seen in darker shade within the bone area surrounding the Ti implant. Modified Masson Goldner's trichrome staining method was used. (From Bandyopadhyay, A. et al., *Ann. Biomed. Eng.*, 45, 249–260, 2017.)

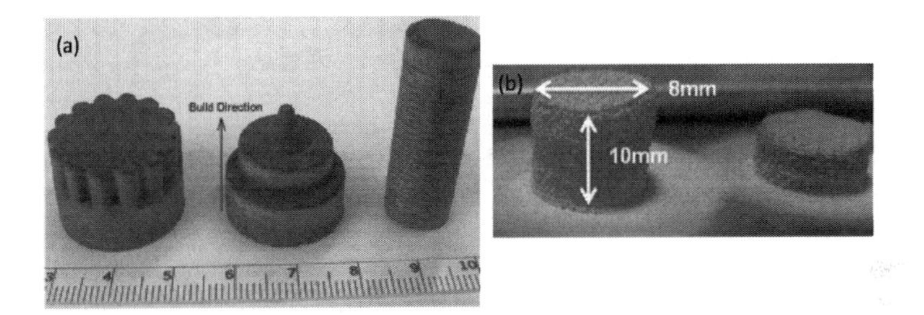

FIGURE 6.21 LENS™ processed structures: (a) Al_2O_3 and (b) PZT. (From Aggarangsi, P. and Beuth, J.L., Localized preheating approaches for reducing residual stress in additive manufacturing, *Proceedings of SFF Symposium*, Austin, 709–720, 2000.)

6.8 ROBOCASTING

Ceramic robocasting is an additive manufacturing process, which relies on the extrusion and layer-by-layer deposition of ceramic slurry through a syringe that follows a computer-aided design to form 3D structure. This AM technique emerged in the mid-1990s and first developed at Sandia National Laboratories, USA. Figure 6.22 illustrates a simple schematic for the ceramic robocasting process. Similar to other slurry-based AM process, the ceramic suspension for robocasting needs to possess suitable rheological properties for successful printing. To minimize cracking, the slurry composition should have relatively

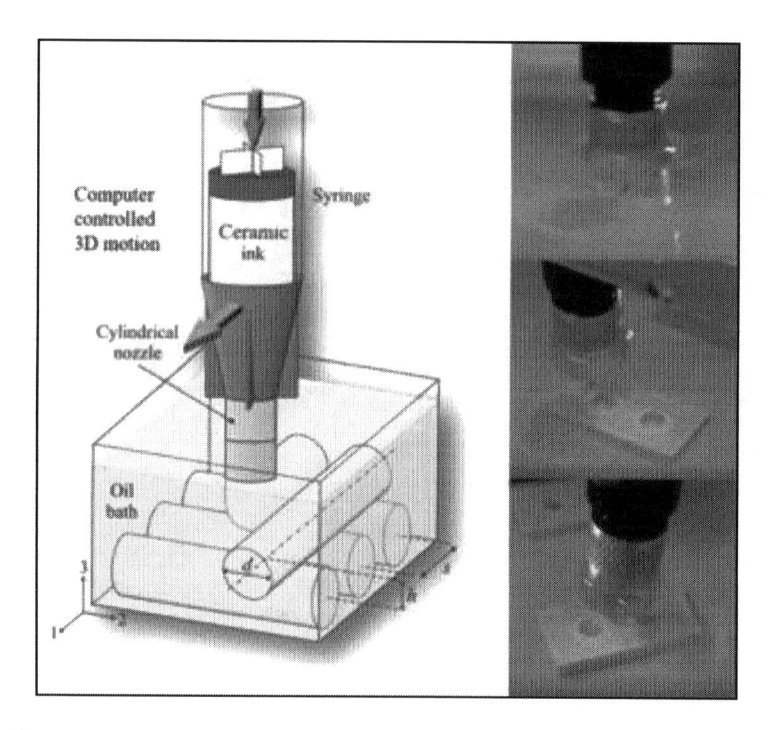

FIGURE 6.22 Schematic of robocasting process. (From Martínez-Vázquez, F.J. et al., *Acta Biomater.*, 6, 4361–4368, 2010.)

high solid ceramic powder loading (>40 vol%) and low amount of organic solvent (<3 vol%). Once the part is extruded, it needs to undergo post-processing, which involves debinding process to remove the organic additives and subsequent sintering to densify the final part.

Robocasting was used to fabricate parts using aqueous-mullite suspensions with an organic polyelectrolyte as a dispersant. Interparticle forces were employed to characterize the sedimentation and viscometry data. The pH and counter-ion addition of the suspension were controlled to optimize mullite suspensions for use in the robocasting process (Figure 6.23). The robocasted part was sintered at 1650°ΔC and resulted parts demonstrated greater than 96% density [206]. More recently, a kind of concentrated, aqueous colloidal ceramic slurry consisting of SiC, Al_2O_3, and Y_2O_3 was used to build parts with complex geometry via robocasting. After printing and drying, green parts were fired at 1700°C in argon by spark plasma sintering (SPS). The sintered structures exhibited an average grain size around 1–2 μm and above 97% of theoretical density [207].

With the development of this new technology, robocasting has also been applied to construct medical devices and scaffolds for tissue engineering. Calcium phosphate is one of the most common ceramics used for robocasting because of its excellent biocompatibility. TCP has been exploited to produce scaffolds using robocasting

FIGURE 6.23 SiC grid structures made by robocasting. (From Cai, K. et al., *J. Am. Ceram. Soc.*, 95, 2660, 2012.)

for orthopedic applications [208]. The particle size and morphology of TCP were optimized in order to prepare suspension suitable for robocasting. It turned out that TCP powders with reduced particle size and low specific surface area were more appropriate for slurry preparation. Moreover, optimal heat treatment and microstructure analysis resulted in the fabrication of TCP scaffolds with tailored performance which can be utilized for bone tissue engineering applications [209]. In another study, polylactide or polycaprolactone scaffolds with 70 wt% hydroxyapatite (HA) content and bioglass were fabricated using robocasting process. The mechanical properties of these scaffolds did not reduce significantly after submerging in simulated body fluid for 20 days. It was strongly dependent on the ratio of the organic and inorganic phase and could be controlled based on different applications [210]. HA scaffolds with multi-scale porosity to mimic the natural porous architecture of human bone has also been demonstrated. The scaffolds presented in Figure 6.24a and b resulted in excellent bone growth [211]. Information about ceramic suspension utilized for robocasting is given in Table 6.5.

The main challenge of robocasting includes optimization of ceramic slurry viscosity to achieve a suitable fluid-like consistency. High solid loading and minimal organic additives are desired to prevent drying crack and shrinkage. Additionally, robocasted part also suffers from poor surface finish, which can be minimized by using finer nozzle, lower layer height and high-quality slurries. With the recent advancement in multi-ceramic AM, a multi-material dispensing system, which will extrude more than one material, may fit into one of the foreseeable future trends in this technique.

FIGURE 6.24 (a, b) Optical micrographs of hydroxyapatite (HA) scaffolds prepared by robocasting, with four distinct periodic arrays. (From J.L. Simon, et al., *J. Biomed. Mater. Res. A*, 83, 747–758, 2007.) (c) Typical alumina lattice truss structure created by robocasting. (From Schlordt, T. et al., *J. Ceram. Sci. Technol.*, 3, 81, 2012.) (d) Alumina crucibles and labware produced by Robocasting. (Courtesy of Joe Cesarano, Robocasting Enterprises.)

TABLE 6.5
Ceramic Slurry for Robocasting

Ceramic Particle	Other Compositions	Relative Density	References
Mullite	Polyelectrolyte	>96% dense	[209]
Tricalcium phosphate	Darvan® C dispersant, hydroxypropyl methylcellulose, and polyethylenimine	<92% dense	[209,213]
HA, Bioglass 4S35, and metallic alloys (6P53B)	Polylactide or polycaprolactone (PCL)	Not mentioned	[210]
Hydroxyapatite	Darvan® C dispersant, hydroxypropyl methylcellulose, and polyethylenimine (PEI)	Not mentioned	[213,214]
Nanobioactive glass (nBG)	Chitosan (Chit)	Not mentioned	[215]
β-SiC, Al_2O_3, and Y_2O_3	H-PEI (high molecular weight) (polyethylenimine) and L-PEI	>97% dense	[207]
HA	PCL and CNTs	Not mentioned	[216]
Bioglass 45S5	Carboxymethyl cellulose (CMC)	Not mentioned	[217]

6.9 SUMMARY

In this chapter, different additive manufacturing processes for ceramics are discussed in detail. Table 6.6 summarizes the processing steps for these methods, as well as some of their advantages and disadvantages. Though ceramic processing using AM techniques started in the early 1990s, its widespread applications are still sparse. The primary reason is the extensive process optimization necessary from one powder to another powder due to changes in particle size, shape, and surface energy. Moreover, large ceramic parts are almost impossible to densify due to cracking or delamination. Finally, no dedicated machine is available even today that can be used to print only ceramics. However, the need for ceramic-based additive manufacturing techniques are growing, and we hope that more research and development work will focus on ceramic AM techniques in the coming days. Apart from green freeform fabrication, direct densification of ceramic structures via laser-based AM is also quite exciting and holds a lot of promise. Finally, AM approaches toward the fabrication of ceramic-loaded composites will probably see more real-world applications in the near future.

TABLE 6.6
AM Techniques for Ceramic Materials

Technique	Processing Steps	Advantages and Disadvantages
Stereolithography	• Mix ceramic particle with photocurable monomers. • Apply UV light to cure monomers based on the design. • Debind the parts from the build platform, and densify the parts by sintering.	Advantages: • SLA is a flexible and efficient manufacturing tool. • SLA is capable of producing high resolution parts with good dimensional tolerances. Disadvantages: • Support structure is needed when producing parts with complex geometry. • Maintaining uniform ceramic loading in the bath can be challenging.
Selective Laser Sintering	• Preheat the ceramic powder bed. • Fill the chamber with nitrogen gas to avoid oxidation. Then use a CO_2 laser to scan. • Cool down the system for part removal.	Advantages: • Excellent for complex geometry. • No support material is needed. Disadvantages: • Binder removal from larger parts can be challenging.
Inkjet 3 Dimensional Printing (3DP)	• Spreading the powder on build bed using a roller. • Selective spraying of binder/printing solution on the built layer. • Drying, depowdering, and sintering of part.	Advantages: • No need for support. • Possible for a wide range of materials. Disadvantages: • Difficult depowdering process. • Manufacturing larger parts can be challenging.

(Continued)

TABLE 6.6 (*Continued*)
AM Techniques for Ceramic Materials

Technique	Processing Steps	Advantages and Disadvantages
Fused Deposition of Ceramics (FDC)	• Mixing the ceramic powder with a polymer carrier. • Deposition of strands on the substrate using a liquefier and a nozzle. • Polymer removal and sintering.	Advantages: • High ceramic loading without trapping in voids. • Complete bonding between layers. Disadvantages: • Filament fabrication with optimized polymer/ceramic system is challenging. • Binder removal step is slow and critical for final sintered part.
Laminate Object Manufacturing (LOM)	• Preceramic paper preparation. • Cutting the paper using a laser beam. • Decubing of final part and sintering.	Advantages: • Processing dense structure with control on microstructural features. Disadvantages: • Need for post-processing, including decubing, surface finishing, and sintering.
Laser Engineered Net Shaping (LENS™)	• Melting the substrate using a laser. • Injection of ceramic powders to the molten area. • Cooling the structure by moving the laser away from the part.	Advantages: • Processing dense structure with control on microstructural features. • Control over the composition. • Gradient deposition of materials. Disadvantages: • Poor resolution and surface finish.
Robocasting	• Prepare ceramic slurry. • Deposit on a heated build platform. • Drying and sintering.	Advantages: • Dedicated process for ceramic 3D Printing. Disadvantages: • Difficult to make slurry with various ceramic compositions that are stable and suitable for robocasting.

REFERENCES

1. C.W. Hull, (1991), Apparatus for production of three-dimensional objects by stereolithography, US Patent # 4,575,330.
2. S. Bose, S. Sugiura, A. Bandyopadhyay, Processing of controlled porosity ceramic structures via fused deposition process. *Scripta Mater* 1999, 41(9), 1009–1014.
3. M.K. Agarwala, A. Bandyopadhyay, R. van Weeren, P. Whalen, A. Safari and S.C. Danforth, Fused deposition of ceramics: Rapid fabrication of structural ceramic components. *Ceramic Bulletin* 1996, 11, 60–65.
4. S. Bose, M. Roy, A. Bandyopadhyay, Recent advances in bone tissue engineering scaffolds. *Trends Biotechnol* 2012, 30(10), 546–554.

5. S. Bose, G. Fielding, S. Tarafder, A. Bandyopadhyay, Trace element doping in calcium phosphate ceramics to understand osteogenesis and angiogenesis. *Trends Biotechnol* 2013, 31(10), 594–605.
6. S. Bose, S. Vahabzadeh, A. Bandyopadhyay, Bone tissue engineering using 3D printing. *Mater Today* 2013, 16(12), 496–504.
7. S. Bose, D. Ke, H. Sahasrabudhe, A. Bandyopadhyay, Additive manufacturing of biomaterials. *Prog Mater Sci* 2018, 93, 45–111.
8. S. Bose, S.F. Robertson, A. Bandyopadhyay, Surface modification of biomaterials and biomedical devices using additive manufacturing. *Acta Biomater* 2018, 66, 6–22.
9. S. Bose, S.F. Robertson, A. Bandyopadhyay, 3D printing of bone implants and replacements. *Am Sci* 2018, 106, 112–119.
10. J. Darsell, S. Bose, H.L. Hosick, A. Bandyopadhyay, From CT scans to ceramic bone grafts. *J Am Ceram Soc* 2003, 86(7), 1076–1080.
11. S.A.M. Tofail, E.P. Koumoulos, A. Bandyopadhyay, S. Bose, L. O'Donoghue, C. Charitidis, Additive manufacturing: Scientific and technological challenges, market uptake and opportunities. *Mater Today* 2018, 21(1), 22–37.
12. F.P.W. Melchels, J. Feijen, D.W. Grijpma. A review on stereolithography and its applications in biomedical engineering. *Biomaterials* 2010, 31, 24, 6121–6130.
13. T. Chartier, C. Chaput, F. Doreau, M. Loiseau, Stereolithography of structural complex ceramic parts. *J Mater Sci* 2002, 37, 3141–3147.
14. M.L. Griffith, J.W. Halloran, Ultraviolet curable ceramic suspensions for stereolithography of ceramics. *Manuf Sci Eng* 1994, 68–72, 529–534.
15. S. Kanehira, S. Kirihara, Y. Miyamoto, Fabrication of TiO_2-SiO_2 photonic crystals with diamond structure. *J Am Ceram Soc* 2005, 88, 1461–1464.
16. T. Chartier, C. Duterte, N. Delhote et al., Fabrication of millimeter wave components via ceramic stereo- and microstereolithography processes. *J Am Ceram Soc* 2008, 91, 2469–2474.
17. R. Felzmann, S. Gruber, G. Mitteramskogler et al., Lithography-based additive manufacturing of cellular ceramic structures. *Adv Eng Mater* 2012, 14, 1052–1058.
18. J.Y. Kim, J.W. Lee, S.-J. Lee, E.K. Park, S.-Y. Kim, D.-W. Cho, Development of a bone scaffold using HA nanopowder and micro-stereolithography technology. *Microelectron Eng* 2007, 84, 1762–1765.
19. O. Dufaud, S. Corbel, Stereolithography of PZT ceramic suspensions. *Rapid Prototyp J* 2002, 8, 83–90.
20. O. Dufaud, P. Marchal, S. Corbel, Rheological properties of PZT suspensions for stereolithography. *J Eur Ceram Soc* 2002, 22, 2081–2092.
21. L. Elomaa, S. Teixeira, R. Hakala et al., Preparation of poly(e-caprolactone)- based tissue engineering scaffolds by stereolithography. *Acta Biomater* 2011, 7(11), 3850–3856.
22. J. Jansen, F.P.W. Melchels, D.W. Grijpma et al., Fumaric acid monoethyl esterfunctionalized poly(D, L-lactide)/N-vinyl- 2-pyrrolidone resins for the preparation of tissue engineering scaffolds by stereolithography. *Biomacromolecules* 2008, 10(2), 214–220.
23. R.A. Levy, T.M. Chu, J.W. Halloran et al., CT-generated porous hydroxyapatite orbital floor prosthesis as a prototype bioimplant. *J Neuroophthalmol* 1999, 19(2), 1522–1525.
24. C. Hinczewski, S. Corbel, T. Chartier, Ceramic suspensions suitable for stereolithography. *J Eur Ceram Soc* 1998, 18, 583–590.
25. G.A. Brady, J.W. Halloran, Stereolithography of ceramic suspensions. *Rapid Prototyp J* 1997, 3, 61–65.
26. C. Hinczewski, S. Corbel, T. Chartier, Stereolithography for the fabrication of ceramic three-dimensional parts. *Rapid Prototyp J* 1998, 4, 104–111.
27. A. Greco, A. Licciulli, A. Maffezzoli, Stereolithography of ceramic suspensions. *J Mater Sci* 2001, 36, 99–105.

28. A. Licciulli, C. Esposito Corcione, A. Greco, V. Amicarelli, A. Maffezzoli, Laser stereolithography of ZrO_2 toughened Al_2O_3. *J Eur Ceram Soc* 2005, 25, 1581–1589.
29. C.E. Corcione, A. Greco, F. Montagna, A. Licciulli, A. Maffezzoli, Silica moulds built by stereolithography. *J Mater Sci* 2005, 40, 4899–4904.
30. J.H. Jang, S. Wang, S.M. Pilgrim, W.A. Schulze, Preparation and characterization of barium titanate suspensions for stereolithography. *J Am Ceram Soc* 2000, 83, 1804–1806.
31. S. Kirihara, Y. Miyamoto, K. Takenaga, M.W. Takeda, K. Kajiyama, Fabrication of electromagnetic crystals with a complete diamond structure by stereolithography. *Solid State Commun* 2002, 121, 435–439.
32. Y. de Hazan, D. Penner, SiC and SiOC ceramic articles produced by stereolithography of acrylate modified polycarbosilane systems, *J Eur Ceram Soc* 2017, 37(16), 5205–5212.
33. I.M. Krieger, T.J. Dougherty, A mechanism for non-Newtonian flow in suspensions of rigid spheres. *Trans Soc Rheol.* 1959, 1957–1977, 3, 137–152.
34. S. Song, M. Park, J. Lee, and J. Yun, A study on the rheological and mechanical properties of photo-curable ceramic/polymer composites with different silane coupling agents for SLA 3D printing technology. *Nanomaterials* 2018, 8(2), 93.
35. M.L. Griffith, J.W. Halloran, Freeform fabrication of ceramics via stereolithography. *J Am Ceram Soc* 1996, 79, 2601–2068.
36. P.F. Jacobs, and D.T. Reid, Rapid prototyping & manufacturing: Fundamentals of stereolithography. Dearborn, MI: Society of Manufacturing Engineers in cooperation with the Computer and Automated Systems Association of SME, 1992.
37. W. Zimbeck, M. Pope, R.W. Rice, Microstructures and strengths of metals and ceramics made by photopolymer-based rapid prototyping. *Solid Free Fabr Proc* 1996, 411–418.
38. M.L. Griffith, T-M. Chu, W. Wagner, W.H. John, Ceramic stereolithography for investment casting and biomedical applications. *Proc Solid Free Fabr Symp* Austin, TX 1995, 31–38.
39. W.Z. Zhou, D. Li, Z.W. Chen, S. Chen, Direct fabrication of an integral ceramic mould by stereolithography. *Proc Inst Mech Eng Part B J Eng Manuf* 2010, 224, 237–243.
40. H. Wu, D. Li, N. Guo, Fabrication of integral ceramic mold for investment casting of hollow turbine blade based on stereolithography. *Rapid Prototyp J* 2009, 15, 232–237.
41. H. Wu, D. Li, Y. Tang, B. Sun, D. Xu, Rapid fabrication of alumina-based ceramic cores for gas turbine blades by stereolithography and gelcasting. *J Mater Process Tech* 2009, 209, 5886–5891.
42. C.E. Corcione, F. Montagna, A. Greco, A. Licciulli, A. Maffezzoli, Free form fabrication of silica moulds for aluminium casting by stereolithography. *Rapid Prototyp J* 2006, 12, 184–188.
43. O. Dufaud, S. Corbel, Oxygen diffusion in ceramic suspensions for stereolithography. *Chem Eng J* 2003, 92, 55–62.
44. A. Buerkle, K. Brakora, K. Sarabandi, Fabrication of a DRA array using ceramic stereolithography. *Antennas Wirel Propag Lett* 2006, 5, 479–482.
45. N. Delhote, D. Baillargeat, S. Verdeyme, C. Delage, C. Chaput, Innovative shielded high dielectric resonator made of alumina by layer-by-layer stereolithography. *IEEE Microw Wirel Compon Lett* 2007, 17, 433–435.
46. K.F. Brakora, J. Halloran, K. Sarabandi, Design of 3-D monolithic MMW antennas using ceramic stereolithography. *IEEE Trans Antennas Propag* 2007, 55, 790–797.
47. N. Delhote, D. Baillargeat, S. Verdeyme, C. Delage, C. Chaput, Ceramic layer-by-layer stereolithography for the manufacturing of 3-D millimeter-wave filters. *IEEE Trans Microw Theory Tech* 2007, 55, 548–554.
48. C.J. Bae, J.W. Halloran, Influence of residual monomer on cracking in ceramics fabricated by stereolithography. *Int J Appl Ceram Technol* 2011, 8, 1289–1295.

49. T.M.G. Chu, D.G. Orton, S.J. Hollister, S.E. Feinberg, J.W. Halloran, Mechanical and in vivo performance of hydroxyapatite implants with controlled architectures. *Biomaterials* 2002, 23, 1283–1293.
50. S. Kirihara, Creation of functional ceramics structures by using stereolithographic 3D printing. *Trans JWRI* 2014, 43(1), 5–10.
51. F. Scalera, C.E. Corcione, F. Montagna, A. Sannino, A. Maffezzoli, Development and characterization of UV curable epoxy/hydroxyapatite suspensions for stereolithography applied to bone tissue engineering. *Ceram Int* 2014, 40(10), 15455–15462.
52. D. Du, T. Asaoka, T. Ushida, K.S. Furukawa, Fabrication and perfusion culture of anatomically shaped artificial bone using stereolithography. *Biofabrication* 2014, 6(4), 045002.
53. D.P. Sarment, K. Al-Shammari, C.E. Kazor, Stereolithographic surgical templates for placement of dental implants in complex cases. *Int J Periodontics Restorative Dent* 2003, 23(3).
54. Q. Lian, W. Sui, X. Wu, F. Yang, S. Yang, Additive manufacturing of ZrO_2 ceramic dental bridges by stereolithography. *Rapid Prototyp J* 2018, 24(1), 114–119.
55. W. Bian, D. Li, Q. Lian, X. Li, W. Zhang, K. Wang, Z. Jin et al., Fabrication of a bio-inspired beta-tricalcium phosphate/collagen scaffold based on ceramic stereolithography and gel casting for osteochondral tissue engineering. *Rapid Prototyp J* 2012, 18(1), 68–80.
56. R.M. Felfel, D. Gupta, A.Z. Zabidi, A. Prosser, C.A. Scotchford, V. Sottile, D.M. Grant, Performance of multiphase scaffolds for bone repair based on two-photon polymerized poly (d, l-lactide-co-ε-caprolactone), recombinamers hydrogel and nano-HA. *Mater Design* 2018, 160, 455–467.
57. T.A. Pham, D.-P. Kim, T.-W. Lim, S.-H. Park, D.-Y. Yang, and K.-S. Lee, Three-dimensional SiCN ceramic microstructures via nano-stereolithography of inorganic polymer photoresists. *Adv Funct Mater* 2006, 16(9), 1235–1241.
58. I. Gibson, D.W. Rosen, B. Stucker, *Additive Manufacturing Technologies*. Boston, MA: Springer, 2010.
59. D.T. Pham, R.S. Gault, A comparison of rapid prototyping technologies. *Int J Mach Tools Manuf* 1998, 38, 1257–1287.
60. Y.-C. Hagedorn, N. Balachandran, W. Meiners, K. Wissenbach, R. Poprawe, SLM of net-shaped high strength ceramics: New opportunities for producing dental restorations. *Proc Solid Free Fabr Symp* 2011, 536–546.
61. A.N. Chen, J.M. Wu, K. Liu, J.Y. Chen, H. Xiao, P. Chen, C.H. Li, Y.S. Shi, High-performance ceramic parts with complex shape prepared by selective laser sintering: A review. *Adv Appl Ceram* 2018, 117(2), 100–117.
62. I. Shishkovsky, I. Yadroitsev, P. Bertrand, I. Smurov, Alumina–zirconium ceramics synthesis by selective laser sintering/melting. *Appl Surf Sci* 2007, 254, 966–970.
63. T. Friedel, N. Travitzky, F. Niebling, M. Scheffler, P. Greil, Fabrication of polymer derived ceramic parts by selective laser curing, *J Eur Ceram Soc* 2005, 25(2–3), 193–197.
64. P. Bertrand, F. Bayle, C. Combe, P. Goeuriot, I. Smurov, Ceramic components manufacturing by selective laser sintering. *Appl Surf Sci* 2007, 254, 989–992.
65. B. Duan, M. Wang. Customized Ca–P/PHBV nanocomposite scaffolds for bone tissue engineering: Design, fabrication, surface modification and sustained release of growth factor. *J Royal Soc Interface* 2010, 7(suppl_5), S615–S629.
66. C. Shuai, C. Gao, Y. Nie, H. Hu, Y. Zhou, S. Peng, Structure and properties of nano-hydroxypatite scaffolds for bone tissue engineering with a selective laser sintering system. *Nanotechnology* 2011, 22(28), 5703.
67. C. Shuai, C. Gao, Y. Nie, H. Hu, H. Qu, S. Peng, Structural design and experimental analysis of a selective laser sintering system with nano-hydroxyapatite powder. *J Biomed Nanotechnol* 2010, 6, 370–374.

68. T. Czelusniak, F.L. Amorim, A. Lohrengel, C.F. Higa, Development and application of copper–nickel zirconium diboride as EDM electrodes manufactured by selective laser sintering. *Int J Adv Manuf Technol* 2014, 72, 905–917.
69. J. Liu, H. Hu, P. Li, C. Shuai, S. Peng, Fabrication and characterization of porous 45S5 glass scaffolds via direct selective laser sintering. *Mater Manuf Process* 2013, 28(6), 610–615.
70. E.Y. Tarasova, G.V. Kryukova, A.L. Petrov, I.V. Shishkovsky, Structure and properties of porous PZT ceramics synthesized by selective laser sintering method. *SPIE Proc* 2000 3933, 502–504.
71. H.-C. Yen, A new slurry-based shaping process for fabricating ceramic green part by selective laser scanning the gelled layer. *J Eur Ceram Soc* 2012, 32, 3123–3128.
72. F. Klocke, C. Derichs, C. Ader, A. Demmer, Investigations on laser sintering of ceramic slurries. *Prod Eng* 2007, 1, 279–284.
73. X. Tian, D. Li, J.G. Heinrich, Rapid prototyping of porcelain products by layer-wise slurry deposition (LSD) and direct laser sintering. *Rapid Prototyp J* 2012, 18, 362–373.
74. F.-H. Liu, Synthesis of biomedical composite scaffolds by laser sintering: Mechanical properties and in vitro bioactivity evaluation. *Appl Surf Sci* 2014, 297, 1–8.
75. H.C. Yen, H.H. Tang, Study on direct fabrication of ceramic shell mold with slurry-based ceramic laser fusion and ceramic laser sintering. *Int J Adv Manuf Technol* 2012, 60, 1009–1015.
76. A. Gahler, J.G. Heinrich, J. Günster, Direct laser sintering of Al_2O_3-SiO_2 dental ceramic components by layer-wise slurry deposition. *J Am Ceram Soc* 2006, 89, 3076–3080.
77. F.-H. Liu, Y.-K. Shen, J.-L. Lee, Selective laser sintering of a hydroxyapatite-silica scaffold on cultured MG63 osteoblasts in vitro. *Int J Precis Eng Manuf* 2012, 13, 439–444.
78. D. Drummer, D. Rietzel, F. Kühnlein, Development of a characterization approach for the sintering behavior of new thermoplastics for selective laser sintering. *Phys Procedia* 2010, 5, 533–542.
79. K. Liu, Y. Shi, C. Li, L. Hao, J. Liu, Q. Wei, Indirect selective laser sintering of epoxy resin-Al_2O_3 ceramic powders combined with cold isostatic pressing. *Ceram Int* 2014, 40, 7099–7106.
80. L. Cardon, J. Deckers, A. Verberckmoes, K. Ragaert, L. Delva, K. Shahzad et al., Polystyrene-coated alumina powder via dispersion polymerization for indirect selective laser sintering applications. *J Appl Polym Sci* 2013, 128, 2121–2128.
81. J. Deckers, J.-P. Kruth, K, Shahzad, J. Vleugels. Density improvement of alumina parts produced through selective laser sintering of alumina-polyamide composite powder. *CIRP Ann Manuf Technol* 2012, 61, 211–214.
82. A.M. Waetjen, D.A. Polsakiewicz, I. Kuhl, R. Telle, H. Fischer, Slurry deposition by airbrush for selective laser sintering of ceramic components. *J Eur Ceram Soc* 2009, 29, 1–6.
83. K. Shahzad, J. Deckers, S. Boury, B. Neirinck, J.P. Kruth, J. Vleugels, Preparation and indirect selective laser sintering of alumina/PA microspheres. *Ceram Int* 2012, 38(2), 1241–1247.
84. K. Shahzad, J. Deckers, Z. Zhang, J.P. Kruth, J. Vleugels et al., Additive manufacturing of zirconia parts by indirect selective laser sintering. *J Eur Ceram Soc* 2014, 34(1), 81–89.
85. R.D. Goodridge, D.J. Wood, C. Ohtsuki, K.W. Dalgarno, Biological evaluation of an apatite–mullite glass-ceramic produced via selective laser sintering. *Acta Biomater* 2007, 3, 221–231.
86. K.C.R. Kolan, M.C. Leu, G.E. Hilmas, R.F. Brown, M. Velez, Fabrication of 13–93 bioactive glass scaffolds for bone tissue engineering using indirect selective laser sintering. *Biofabrication* 2011, 3, 025004.
87. J. Zhao, W.B. Cao, J.T. Li, Z. Han, Y.H. Li, C.C. Ge, Selective laser sintering of Si_3N_4 and Al_2O_3 ceramic powders ceramic powders. *Key Eng Mater* 2008, 368–372, 858–861.

88. B. Ma, L. Lin, X. Huang, Q. Hu, M. Fang, Bone tissue engineering using beta-tricalcium phosphate scaffolds fabricated via selective laser sintering. *Knowl Enterp Intell Strateg Prod Des Manuf Manag* 2006, 207, 710–716.
89. D.M. Gureev, R.V. Ruzhechko, I.V. Shishkovskii, Selective laser sintering of PZT ceramic powders. *Tech Phys Lett* 2000, 26, 262–264.
90. D. Liu, J. Zhuang, C. Shuai, S. Peng, Mechanical properties' improvement of a tricalcium phosphate scaffold with poly-l-lactic acid in selective laser sintering. *Biofabrication* 2013, 5, 025005.
91. C. Shuai, J. Zhuang, H. Hu, S. Peng, D. Liu, J. Liu, In vitro bioactivity and degradability of β-tricalcium phosphate porous scaffold fabricated via selective laser sintering: Properties of β-TCP Porous Scaffold. *Biotechnol Appl Biochem* 2013, 60, 266–273.
92. E. Sachs, M. Cima, J. Cornie, D. Brancazio, J. Bredt, A. Curodeau, T. Fan, S. Khanuja, A. Lauder, J. Lee, S. Michaels, Three-dimensional printing: The physics and implications of additive manufacturing. *CIRP Ann* 1993, 42(1), 257–260.
93. S. Bose, S. Vahabzadeh, A. Bandyopadhyay, Bone tissue engineering using 3D printing, *Mater Today* 2013, 16, 496–504.
94. P.H. Warnke, H. Seitz, F. Warnke, S.T. Becker, S. Sivananthan, E. Sherry, Q. Liu, J. Wiltfang, T. Douglas, Ceramic scaffolds produced by computer-assisted 3D printing and sintering: Characterization and biocompatibility investigations. *J Biomed Mater Res* 2010, 93B, 212–217.
95. E. Vorndran, M. Klarner, U. Klammert, L.M. Grover, S. Patel, J.E. Barralet, U. Gbureck, 3D powder printing of β-tricalcium phosphate ceramics using different strategies. *Adv Eng Mater* 2008, 10, B67–B71.
96. B. Leukers, H. Gülkan S.H. Irsen, S. Milz, C. Tille, H. Seitz, M. Schieker, Biocompatibility of ceramic scaffolds for bone replacement made by 3D printing. *Mater Wiss Werkstofftech* 2005, 36, 781–787.
97. A. Zocca, C.M. Gomes, E. Bernardo, R. Müller, J. Günster, P. Colombo, LAS glass–ceramic scaffolds by three-dimensional printing. *J Eur Ceram Soc* 2013, 33, 1525–1533.
98. S. Tarafder, V.K. Balla, N.M. Davies, A. Bandyopadhyay, S. Bose, Microwave-sintered 3D printed tricalcium phosphate scaffolds for bone tissue engineering. *J Tissue Eng Regen Med* 2013, 7, 631–641.
99. H. Hausner, Powder characteristics and their effect on powder processing. *Powder Technol.* 1981, 30, 3–8.
100. S.A. Uhland, R.K. Holman, S. Morissette, M.J. Cima, and E.M. Sachs, Strength of green ceramics with low binder content. *J Am Ceram Soc* 2001, 84, 2809–2818.
101. S. Amirkhani, R. Bagheri, A. Zehtab Yazdi, Effect of pore geometry and loading direction on deformation mechanism of rapid prototyped scaffolds. *Acta Mater* 2012, 60, 2778–2789.
102. S. Maleksaeedi, H. Eng, F.E. Wiria, T.M.H. Ha, Z. He, Property enhancement of 3D-printed alumina ceramics using vacuum infiltration. *J Mater Process Tech* 2014, 214, 1301–1306.
103. B.C. Gross, J.L. Erkal, S.Y. Lockwood, C. Chen, D.M. Spence, Evaluation of 3D printing and its potential impact on biotechnology and the chemical sciences. *Anal Chem* 2014, 86, 3240–3253.
104. J. Suwanprateeb, R. Sanngam, T. Panyathanmaporn, Influence of raw powder preparation routes on properties of hydroxyapatite fabricated by 3D printing technique. *Mater Sci Eng C* 2010, 30, 610–617.
105. S.F.S. Shirazi, S. Gharehkhani, M. Mehrali, H. Yarmand, H.S.C. Metselaar, N.A. Kadri, N.A.A. Osman, A review on powder-based additive manufacturing for tissue engineering: selective laser sintering and inkjet 3D printing. *Sci Technol Adv Mater* 2015, 16(3), 033502.

106. A. Doraiswamy, T.M. Dunaway, J.J Wilker, R.J.J. Narayan. *Biomed Mater Res Part B Appl Biomater* 2009, 89B, 28–35.
107. A. Khalyfa, S. Vogt, J. Weisser, G. Grimm, A. Rechtenbach, W. Meyer, M. Schnabelrauch, Development of a new calcium phosphate powder-binder system for the 3D printing of patient specific implants. *J Mater Sci Mater Med* 2007, 18, 909–916.
108. J. Suwanprateeb, R. Sanngam, W. Suvannapruk, T. Panyathanmaporn, Mechanical and in vitro performance of apatite-wollastonite glass ceramic reinforced hydroxyapatite composite fabricated by 3D-printing. *J Mater Sci Mater Med* 2009, 20, 1281–1289.
109. B. Nan, X. Yin, L. Zhang, L. Cheng, Three-dimensional printing of Ti_3SiC_2-based ceramics. *J Am Ceram Soc* 2011, 94(4), 969–972.
110. A. Zocca, C.M. Gomes, A. Staude, E. Bernardo, J. Günster, P. Colombo, SiOC ceramics with ordered porosity by 3D-printing of a preceramic polymer. *J Mater Res* 2013, 28, 2243–2252.
111. X. Li, L. Zhang, X. Yin, Microstructure and mechanical properties of three porous Si_3N_4 ceramics fabricated by different techniques. *Mat Sci Eng A-Struct* 2012, 549, 43–49.
112. G. Cesaretti, E. Dini, X. De Kestelier, V. Colla, L. Pambaguian, Building components for an outpost on the Lunar soil by means of a novel 3D printing technology. *Acta Astronautica* 2014, 93, 430–450.
113. S.S. Kim, H. Utsunomiya, J.A. Koski et al., Survival and function of hepatocytes on a novel three-dimensional synthetic biodegradable polymer scaffold with an intrinsic network of channels. *Ann Surg* 1998, 228(1), 8–13.
114. A. Winkel et al., Sintering of 3D-printed glass/HAp composites. *J Am Ceram Soc* 2012, 95(11), 3387–3393.
115. R. Gauvin et al., Microfabrication of complex porous tissue engineering scaffolds using 3D projection stereolithography. *Biomaterials* 2012, 33, 3824–3834.
116. E. Vorndran et al., Simultaneous immobilization of bioactives during 3D powder printing of bioceramic drug-release matrices. *Adv Funct Mater* 2010, 20, 1585–1591.
117. S. Tarafder, S. Bose, Polycaprolactone-Coated 3D printed tricalcium phosphate scaffolds for bone tissue engineering. *ACS Appl Mater Inter* 2014, 6, 9955–9965.
118. S. Bose, S. Tarafder, A. Bandyopadhyay, Effect of chemistry on osteogenesis and angiogenesis towards bone tissue engineering using 3D printed scaffolds. *Ann Biomed Eng* 2017, 45(1), 261–272.
119. C.X.F. Lam, X.M. Mo, S.H. Teoh et al., Scaffold development using 3D printing with a starch-based polymer. *Mater Sci Eng C Mater Biol Appl* 2002, 20(1–2), 49–56.
120. A. Zocca, H. Elsayed, E. Bernardo, C. Gomes, M. Lopez-Heredia, C. Knabe, P. Colombo, J. Günster, 3D-printed silicate porous bioceramics using a non-sacrificial preceramic polymer binder. *Biofabrication* 2015, 7(2), 025008.
121. Y. Shanjani, J.N. De Croos, R.M. Pilliar et al., Solid freeform fabrication and characterization of porous calcium polyphosphate structures for tissue engineering purposes. *J Biomed Mater Res B Appl Biomater* 2010, 93(2), 510–519.
122. N. Sarkar, S. Bose, Liposome-encapsulated curcumin-loaded 3D printed scaffold for bone tissue engineering. *ACS Appl Mater Interfaces* 2019, 11(19), 17184–17192.
123. S. Bose, G. Fielding, SiO_2 and ZnO Dopants in 3D printed TCP scaffolds enhances osteogenesis and angiogenesis in vivo. *Acta Biomater* 2013, 9, 9137–9148.
124. S. Tarafder, N.M. Davies, A. Bandyopadhyay, S. Bose, 3D Printed tricalcium phosphate scaffolds: Effect of SrO and MgO doping on in vivo osteogenesis in rat distal femoral defect model. *Biomater Sci* 2013, 1, 1250–1259.
125. C.X.F. Lam, X.M. Mo, S.H. Teoh, D.W. Hutmacher, Scaffold development using 3D printing with a starch-based polymer. *Mater Sci Eng C* 2002, 20, 49–56.

126. G.A. Fielding, A. Bandyopadhyay, S. Bose, Effects of silica and zinc oxide doping on mechanical and biological properties of 3D printed tricalcium phosphate tissue engineering scaffolds. *Dent Mater* 2012, 28, 113–122.
127. X. Yan, G. Peng, A review of rapid prototyping technologies and systems. *Comput Aided Des* 1996, 28(4), 307–318.
128. E.A. Griffin, S. McMillin, Selective laser sintering and fused deposition modeling processes for functional ceramic parts. *Solid Freeform Fabrication Proceedings*, University of Texas, 1995, 25–30.
129. S.J. Kalita, S. Bose, H.L. Hosick, A. Bandyopadhyay, Development of controlled porosity polymer-ceramic composite scaffolds via fused deposition modeling. *Mater Sci Eng C* 2003, 23, 611–620.
130. J.A. Lewis, J.E. Smay, J. Stuecker, J. Cesaran, Direct ink writing of three-dimensional ceramic structures. *J Am Ceram Soc* 2006, 89(12), 3599–3609.
131. S. Danforth, Fused deposition of ceramics: A new technique for the rapid fabrication of ceramic components. *Mater Technol* 1995, 10(7–8), 144–146.
132. M.K. Agarwala et al. Structural ceramics by fused deposition of ceramics, ibid., Reference #1, 1995, 6, 1–8.
133. L. Shor et al. Fabrication of three-dimensional polycaprolactone/hydroxyapatite tissue scaffolds and osteoblast-scaffold interactions in vitro. *Biomaterials* 2007, 28, 5291–5297.
134. A. Safari, M. Allahverdi, E.K. Akdogan, Solid freeform fabrication of piezoelectric sensors and actuators. *J Mater Sci* 2006, 41, 177–198.
135. S. Iyer et al., Microstructural characterization and mechanical properties of Si_3N_4 formed by fused deposition of ceramics. *Int J Appl Ceram Technol* 2008, 5(2), 127–137.
136. S. Onagoruwa, A. Bose, A. Bandyopadhyay, Fused deposition of ceramics (FDC) and composites. *Proc SFF*, Texas 2001, 224–231.
137. A. Bellini, Fused deposition of ceramics: A comprehensive experimental, analytical and computational study of material behavior. *Fabrication Process and Equipment Design*, PhD. Thesis, Drexel University, 2002.
138. Y. Jin et al., Quantitative analysis of surface profile in fused deposition modelling. *Addit Manuf* 2015, 8, 142–148.
139. M. Agarwala, A. Bandyopadhyay, S.C. Danforth, V.R. Jamalabad, R.N. Langrana, W.R. Priedeman, A. Safari, R. van Weeren (1999), Solid freeform fabrication methods. US Patent # 5,900,207.
140. M. Agarwala, A. Bandyopadhyay, S.C. Danforth, V.R. Jamalabad, N. Langrana, A. Safari, R. van Weeren (1998), Solid freeform fabrication methods. US Patent # 5,738,817.
141. M. Agarwala, A. Bandyopadhyay, S.C. Danforth, V.F. Janas, R.K. Panda, A. Safari (1999), Novel ceramic composites and methods for producing same. US Patent # 6,004,500.
142. A. Bandyopadhyay, S.C. Danforth, T. McNulty, R.K. Panda, A. Safari (2000), Radial ceramic piezoelectric composites. US Patent # 6,049,160.
143. A. Bandyopadhyay, S.C. Danforth, V.F. Janas, R.K. Panda, A. Safari (1998), Oriented piezoelectric ceramics and ceramic polymer composites. US Patent # 5,796,207.
144. J. Ballato, A. Bandyopadhyay, S.C. Danforth, A. Safari, R. van Weeren (1999), Process for forming of photonic band-gap structures. US Patent # 5,997,795.
145. A. Bandyopadhyay, R.K. Panda, V.F. Janas, M.K. Agarwala, S.C. Danforth, A. Safari, Processing of piezocomposites by fused deposition technique. *J Am Ceram Soc* 1997, 80, 1366–1372.
146. S. Turcu, B. Jadidian, S.C. Danforth, A. Safari, Piezoelectric properties of novel oriented ceramic-polymer composites with 2-2 and 3-3 connectivity. *J Electroc* 2002, 9, 165–171.

147. R.K. Panda, Novel piezoelectric ceramics by solid freeform fabrication. PhD Thesis, Rutgers University, New Brunswick, NJ, 1998.
148. M. Allahverdi, S. Danforth, M. Jafari, A. Safari, Processing of advanced electroceramic components by fused deposition technique. *J Eur Ceram Soc* 2001, 21(10–11), 1485–1490.
149. M. Jafari, W. Han, F. Mohammadi, A. Safari, S. Danforth, N. Langrana, A novel system for fused deposition of advanced multiple ceramics. *Rapid Prototyp J* 2000, 6(3), 161–175.
150. N. Xu et al., 3D artificial bones for bone repair prepared by computed tomography-guided fused deposition modeling for bone repair. *ACS Appl Mater Inter* 2014, 6(17), 14952–14963.
151. J. Darsell, S. Bose, H. Hosick, A. Bandyopadhyay, From CT scans to ceramic bone grafts. *J Am Ceram Soc* 2003, 86(7), 1076–1080.
152. A. Bandyopadhyay, K. Das, J. Marusich, S. Onagoruwa, Application of fused deposition in controlled microstructure metal-ceramic composites. *Rapid Prototyping J* 2006, 12(3), 121–128.
153. M.E. Pilleux et al., 3-D photonic bandgap structures in the microwave regime by fused deposition of multimaterials. *Rapid Prototyping J* 2002, 8, 46–52.
154. H. Yang, S. Yang, X. Chi, J.R. Evans, Fine ceramic lattices prepared by extrusion freeforming. *J Biomed Mater Res B: Appl Biomater* 2006, 79(1), 116–121.
155. Y. Chen, D. Bartzos, Y. Lu, E. Niver, M. Pilleux, M. Allahverdi, S. Danforth, A. Safari, Simulation, fabrication, and characterization of 3-D alumina photonic bandgap structures. *Microw Opt Techn Let* 2001, 30(5), 305–307.
156. T.F. McNulty, F. Mohammadi, A. Bandyopadhyay, D.J. Shanefield, S.C. Danforth, A. Safari, Development of binder formulation for fused deposition of ceramics. *Rapid Prototyping J* 1998, 4(4), 144–150.
157. S. Bose, J. Darsell, M. Kintner, H. Hosick, A. Bandyopadhyay, Pore size and pore volume effects on calcium phosphate based ceramics. *Mat Sci Eng C* 2003, 23, 479–486.
158. D.W. Hutmacher, T. Schantz, I. Zein, K.W. Ng, S.H. Teoh, K.C. Tan, Mechanical properties and cell cultural response of polycaprolactone scaffolds designed and fabricated via fused deposition modeling. *J Biomed Mater Res* 2001, 55(2), 203–216.
159. J.T. Schantz et al., Osteogenic differentiation of mesenchymal progenitor cells in computer designed fibrin-polymer-ceramic scaffolds manufactured by fused deposition modeling. *J Mater Sci Mater Med* 2005, 16, 807–819.
160. J.T. Schantz, T.C. Lim, C. Ning et al., Cranioplasty after trephination using a novel biodegradable burr hole cover: Technical case report. *Neurosurgery* 2006, 58(1suppl): ONS-E176; discussion ONS-E176.
161. S.W. Low, Y.J. Ng, T.T. Yeo et al., Use of Osteoplug polycaprolactone implants as novel burr-hole covers. *Singapore Med J* 2009, 50(8), 777–780.
162. S.T. Ho, D.W. Hutmacher, A.K. Ekaputra et al., The evaluation of a biphasic osteochondral implant coupled with an electrospun membrane in a large animal model. *Tissue Eng Part A* 2010, 16(4), 1123–1141.
163. C. Esposito Corcione, F. Gervaso, F. Scalera, F. Montagna, A. Sannino, A. Maffezzoli, The feasibility of printing polylactic acid–nanohydroxyapatite composites using a low-cost fused deposition modeling 3D printer. *J Appl Polym Sci* 2017, 134(13).
164. P.S. Wiggenhauser, D.F. Müller, F.P.W. Melchels et al., Engineering of vascularized adipose constructs. *Cell Tissue Res* 2012, 347(3), 747–757.
165. C.M. Gomes, A.P.N. Oliveira, D. Hotza, N. Travitzky, P. Greil, LZSA glass-ceramic laminates: Fabrication and mechanical properties. *J Mater Process Tech* 2008, 206, 194–201.
166. K. Schindler, A. Roosen, Manufacture of 3D structures by cold low pressure lamination of ceramic green tapes. *J Eur Ceram Soc* 2009, 29, 899–904.

167. I. Cho, K. Lee, W. Choi, Y.A. Song, Development of a new sheet deposition type rapid prototyping system. *Int J Mach Tools Manuf* 2000, 40, 1813–1829.
168. H. Windsheimer, N. Travitzky, A. Hofenauer, P. Greil, Laminated object manufacturing of preceramic-paper-derived Si-SiC composites. *Adv Mater* 2007, 19, 4515–4519.
169. C.M. Gomes, C.R. Rambo, A.P. Novaes de Oliveira, D. Hotza, D. Gouvea, Colloidal processing of glass: Ceramics for laminated object manufacturing. *J Am Ceram Soc* 2009, 92(6) 1186–1191.
170. C.W. Macosko, *Rheology: Principles, Measurements and Applications.* VCH Publishers, New York, 1993.
171. J.S. Reed, *Principles of Ceramics Processing*, 2nd ed. Wiley Interscience Publications, New York (1988).
172. R.E. Mistler, E.R. Twiname, *Tape Casting: Theory and Practice.* The American Ceramic Society, Westerville, OH (2000).
173. B. Mueller, D. Kochan, Laminated object manufacturing for rapid tooling and patternmaking in foundry industry. *Comput Ind* 1999, 39, 47–53.
174. C. Griffin, J. Daufenbach, S. McMillin, Desktop manufacturing: LOM vs. pressing. *Am Ceram Soc Bull* 1994, 73(8), 109–113.
175. L. Weisensel, N. Travitzky, H. Sieber, P. Greil, Laminated object manufacturing (LOM) of SiSiC composites. *Adv Eng Mater* 2004, 6(11), 899–903.
176. E. Griffin, D. Mumm, D. Marshall, Rapid prototyping of functional ceramic composites. *Am Ceram Soc Bull* 1996, 75(7), 65–68.
177. Y. Zhang, J. Han, X. Zhang, X. He, Z. Li, S. Du, Rapid prototyping and combustion synthesis of TiC/Ni functionally gradient materials. *Mat Sci Eng A* 2001, 299(1–2), 218–224.
178. C. Gomes, N. Travitzky, P. Greil, W. Acchar, H. Birol, A. Pedro Novaes de Oliveira, D. Hotza, Laminated object manufacturing of LZSA glass ceramics. *Rapid Prototyp J* 2011, 17(6), 424–428.
179. M.J. Pan, A. Leung, C. Wu, B. Bender, Optimizing the performance of telescoping actuators through rapid prototyping and finite element modeling. *Ceram Trans* 2004, 150, 53–62.
180. C. Steidle, D. Klosterman, R. Chartoff, G. Graves, N. Osborne, Automated fabrication of custom bone implants using rapid prototyping. *44th International SAMPE Symposium and Exhibition* 1999, 1866–1877.
181. Y. Zhang, X. He, S. Du, J. Zhang, Al_2O_3 Ceramics preparation by LOM (Laminated Object Manufacturing). *Int J Adv Manuf Technil* 2001, 17, 531–534.
182. S. J. Rodrigues, R. P. Chartoff, D. A. Klosterman, M. Agarwala, N. Hecht, Solid freeform fabrication of functional silicon nitride ceramics by laminated object manufacturing. *International Solid Freeform Fabrication Symposium*, 2000.
183. M. Krinitcyn, Z. Fu, J. Harris, K. Kostikov, G.A. Pribytkov, P. Greil, N. Travitzky, Laminated Object Manufacturing of in-situ synthesized MAX-phase composites. *Ceram Trans* 2017, 43(12), 9241–9245.
184. A. Bandyopadhyay et al., Application of laser engineered net shaping (LENS) to manufacture porous and functionally graded structures for load bearing implants. *J Mater Sci Mater Med* 2009, 20(1), 29.
185. B. Zheng, J.E. Smugereky, Y. Zhou, D. Baker, E.J. Lavernia, Microstructure and properties of laser-deposited $Ti6Al_4V$ metal matrix composites using Ni-coated powder. *Metall Mater Trans A* 2008, 39(5), 1196–1205.
186. M. Roy, V.K. Balla, A. Bandyopadhyay, S. Bose, Laser processing of bioactive tricalcium phosphate coating on titanium for load-bearing implants, *Acta Biomater* 2008, 4, 324–333.
187. J. Mazumder, A. Schifferer, J. Choi, Direct materials deposition: Designed macro and microstructure. *Mater Res Innovations* 1999, 3, 118–131.

188. S.P. Harimkar, A.N. Samant, A.A. Khangar, B.D. Narendra, Prediction of solidification microstructures during laser dressing of alumina-based grinding wheel material. *J Phys D Appl Phys*. 2006, 39, 1642–1649.
189. M. Das, V.K. Balla, T.S.S. Kumar, I. Manna, Fabrication of biomedical implants using laser engineered net shaping (LENS™). *Trans Ind Ceram Soc* 2013, 72(3), 169–174.
190. I. Gibson, D.W. Rosen, B. Stucker, Additive manufacturing technologies rapid prototyping to direct digital manufacturing. *Additive Manufacturing Technologies*. Springer, Boston, MA, 299–332 (2009).
191. F. Niu, D. Wu, G. Ma, J. Wang, M. Guo, B. Zhang, Nanosized microstructure of Al_2O_3–$ZrO_2(Y_2O_3)$ eutectics fabricated by laser engineered net shaping. *Scripta Mater* 2015, 95(1), 39–41.
192. J. Wilkes, Y.-C. Hagedorn, W. Meiners, K. Wissenbach, Additive manufacturing of ZrO_2-Al_2O_3 ceramic components by selective laser melting. *Rapid Prototyp J* 2013, 19(1), 51–57.
193. P. Aggarangsi, J.L. Beuth, Localized preheating approaches for reducing residual stress in additive manufacturing. *Proc SFF Symp* Austin 2006, 709–720.
194. J. Wilkes, Y.-C. Hagedorn, S. Ocylok, W. Meiners, K. Wissenbach, Rapid manufacturing of ceramic parts by selective laser melting. *Ceramic Engineering and Science Proceedings* 2010, 137.
195. Y. Hagedorn, N. Balachandran, W. Meiners, K. Wissenbach, R. Poprawe, SLM of net-shaped high strength ceramics: New opportunities for producing dental restorations, *Proceedings of the Solid Freeform Fabrication Symposium*, Austin (August), 2011, 8–10.
196. W. Liu, J.N. DuPont, Fabrication of functionally graded TiC/Ti composites by laser engineered net shaping. *Scripta Mater* 2003, 48, 1337–1342.
197. P.P. Bandyopadhyay, V.K. Balla, S. Bose, A. Bandyopadhyay, Compositionally graded aluminum oxide coatings on stainless steel using laser processing. *J Am Ceram Soc* 2007, 90(7), 1989–1991.
198. W. Liu, J.N. Dupont, Fabrication of carbide-particle-reinforced titanium aluminide-matrix composites by laser-engineered net shaping. *Metall Mater Trans A* 2004, 35 A, 1133–1140.
199. B. Zheng, T. Topping, J.E. Smugeresky, Y. Zhou, A. Biswas, D. Baker, E.J. Lavernia, The influence of Ni-Coated TiC on laser-deposited IN625 Metal matrix composites. *Metall Mater Trans A* 2010, 41 A, 568–573.
200. M. Roy, V.K. Balla, A. Bandyopadhyay, S. Bose, Compositionally graded hydroxyapatite/tricalcium phosphate coating on Ti by laser and induction plasma. *Acta Biomater.* 2011, 7(2), 866–873.
201. V.K. Balla, P.D. DeVasConCellos, W. Xue, S. Bose, A. Bandyopadhyay, Fabrication of compositionally and structurally graded Ti–TiO_2 structures using laser engineered net shaping (LENS). *Acta Biomater* 2009, 5, 1831–1837.
202. V.K. Balla, P.P. Bandyopadhyay, S. Bose, A. Bandyopadhyay, Compositionally graded yttria-stabilized zirconia coating on stainless steel using laser engineered net shaping (LENS™). *Scripta Mater* 2007, 57, 861–864.
203. A. Bandyopadhyay et al., In vivo response of laser processed porous titanium implants for load-bearing implants. *Ann Biomed Eng* 2017, 45(1), 249–260.
204. S.A. Bernard, V.K. Balla, S. Bose, A. Bandyopadhyay, Direct laser processing of bulk lead zirconate titanate ceramics. *Mater Sci Eng B Adv* 2010, 72, 85–88.
205. F.J. Martínez-Vázquez et al., Improving the compressive strength of bioceramic robocast scaffolds by polymer infiltration. *Acta Biomater* 2010, 6.11, 4361–4368.

206. J.N. Stuecker, J. Cesarano, D.A. Hirschfeld, Control of the viscous behavior of highly concentrated mullite suspensions for robocasting. *J Mater Process Tech* 2003, 142, 318–325.
207. K. Cai, B. Román-Manso, J.E. Smay, J. Zhou, M.I. Osendi, M. Belmonte et al., Geometrically complex silicon carbide structures fabricated by robocasting. *J Am Ceram Soc* 2012, 95, 2660–2666.
208. S. Raymond, Y. Maazouz, E.B. Montufar, R.A. Perez, B. González, J. Konka, J. Kaiser, M.P. Ginebra, Accelerated hardening of nanotextured 3D-plotted self-setting calcium phosphate inks. *Acta Biomater* 2018, 75, 451–462.
209. P. Miranda, E. Saiz, K. Gryn, A.P. Tomsia, Sintering and robocasting of β-tricalcium phosphate scaffolds for orthopaedic applications. *Acta Biomater* 2006, 2, 457–466.
210. J. Russias, E. Saiz, S. Deville, K. Gryn, G. Liu, R.K. Nalla et al., Fabrication and in vitro characterization of three-dimensional organic/inorganic scaffolds by robocasting. *J Biomed Mater Res A* 2007, 83A, 434–445.
211. J.L. Simon, S. Michna, J.A. Lewis, E.D. Rekow, V.P. Thompson, J.E. Smay, A. Yampolsky, J.R. Parsons, J.L. Ricci, In vivo bone response to 3D periodic hydroxyapatite scaffolds assembled by direct ink writing. *J Biomed Materi Res A* 2007, 83(3), 747–758.
212. T. Schlordt, F. Keppner, N. Travitzky, P. Greil, Robocasting of alumina lattice truss structures. *J Ceram Sci Technol* 2012, 3, 81.
213. P. Miranda, A. Pajares, E. Saiz, A.P. Tomsia, F. Guiberteau, Mechanical properties of calcium phosphate scaffolds fabricated by robocasting. *J Biomed Mater Res A* 2008, 85A, 218–227.
214. P. Miranda, A. Pajares, E. Saiz, A.P. Tomsia, F. Guiberteau, Fracture modes under uniaxial compression in hydroxyapatite scaffolds fabricated by robocasting. *J Biomed Mater Res A* 2007, 83A, 646–655.
215. B. Dorj, J.H. Park, H.W. Kim, Robocasting chitosan/nanobioactive glass dual-pore structured scaffolds for bone engineering. *Mater Lett* 2012, 73, 119–122.
216. B. Dorj, J.E. Won, J.H. Kim, S.J. Choi, U.S. Shin, H.W. Kim, Robocasting nanocomposite scaffolds of poly(caprolactone)/hydroxyapatite incorporating modified carbon nanotubes for hard tissue reconstruction. *J Biomed Mater Res A* 2013, 101A, 1670–1681.
217. S. Eqtesadi, A. Motealleh, P. Miranda, A. Pajares, A. Lemos, J.M.F. Ferreira, Robocasting of 45S5 bioactive glass scaffolds for bone tissue engineering. *J Eur Ceram Soc* 2014, 34, 107–118.

7 Designing for Additive Manufacturing

Timothy W. Simpson

CONTENTS

7.1 MOTIVATION

The excitement for additive manufacturing (AM) stems primarily from the unprecedented freedom it gives designers and engineers to create complex components and multi-material structures that were prohibitively expensive or impossible to make via conventional manufacturing techniques. This freedom can be used for a variety of benefits, including designing lightweight components for aerospace applications, creating lattice or cellular structures in medical implants that promote osseointegration and speed patient recovery, and consolidating multi-component assemblies to reduce assembly and inventory costs. Innovative designs can also increase material utilization and improve the buy-to-fly ratio of components that undergo extensive machining, and AM can even enable material substitutions that extend the life of the component. For instance, a material that was too expensive (or difficult) to machine into an end-use part may become economically viable when using AM.

None of these benefits can be achieved with additive manufacturing unless one knows how to design for AM. Designing for AM encompasses not only designing the part to take advantage of AM, but also designing how the part is going to be built to overcome the inherent limitations in the selected AM process. Given the current state of AM, if you simply design a part and "throw it over the wall" to be made with AM, then you may be surprised by the result. The build orientation and layout, process parameters, and post-processing needs vary for each AM process, and they can dramatically impact the performance (and cost) of your AM part. Sending a computer-aided design (CAD) file to a vendor or to your in-house manufacturing engineers and specifying only the material and tolerances for a part is no longer enough. Designers face many new tradeoffs with AM, and it is more important than ever that you think not only about the part that you are designing, but also about the AM process by which the part is going to be built layer-by-layer.

In the next section, the ten steps commonly encountered in the AM workflow are discussed to ensure that designers are aware of the tradeoffs inherent in AM and the different AM processes. Designers should be involved with many of these steps because they may necessitate redesign of the part to maintain a viable and cost-effective AM solution. Following the discussion of the workflow, topology optimization of an automotive component is detailed in Section 7.3, illustrating the possibilities as well as the challenges often encountered when designing a new part for AM. In Section 7.4, the tradeoffs and implications of replicating a part with AM, adapting a part for AM, and optimizing a part for AM are discussed. Closing remarks are offered in Section 7.5.

7.2 ADDITIVE MANUFACTURING (AM) WORKFLOW

When designing for AM, understanding the workflow to turn a 3D solid model into a part, be it a prototype or an end-use component, is critical. The ten steps commonly encountered when using any AM process are summarized in Figure 7.1. Details on each of these steps are discussed in Sections 7.2.1 through 7.2.10.

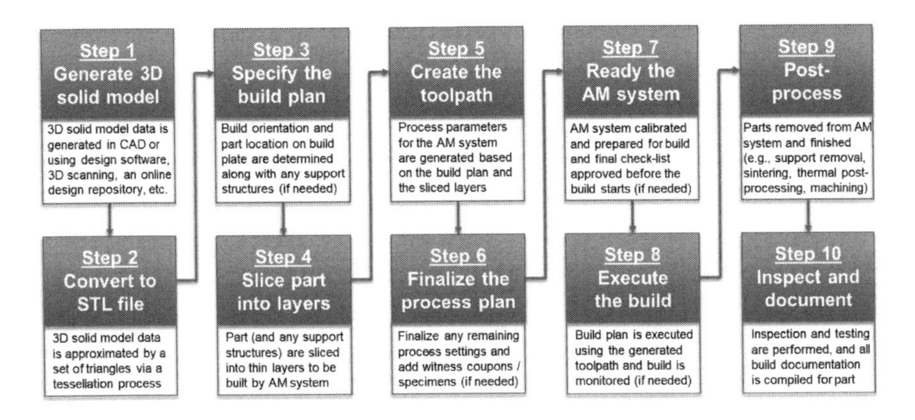

FIGURE 7.1 Ten steps commonly encountered for any AM process.

7.2.1 Step 1: Generating a 3D Solid Model

By definition, AM begins with 3D model data [1], which is typically generated with CAD software like SolidWorks,[1] CATIA,[2] Creo,[3] NX,[4] Fusion360,[5] or Onshape,[6] among others. The 3D solid model of a blower wheel in Figure 7.2 was developed in SolidWorks by Corey Dickman, a research engineer in Penn State CIMP-3D.[7] It will serve as an example to demonstrate the steps of the AM workflow and key tradeoffs that will often be encountered when designing for AM.

Note that in some cases, 3D model data may not be readily available, especially for legacy parts that were manufactured from 2D drawings. If you are trying to replicate an existing part with AM and do not have the 3D model data, then you may need to

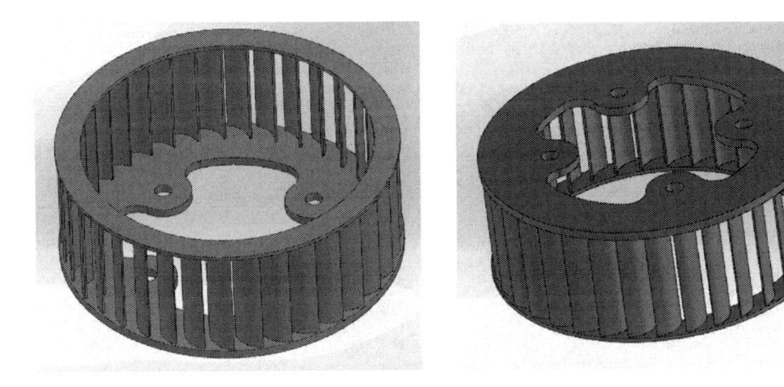

FIGURE 7.2 Top and bottom view of 3D solid model for blower wheel example.

1 https://www.solidworks.com/
2 https://www.3ds.com/products-services/catia/
3 https://www.ptc.com/en/products/cad/creo
4 https://www.plm.automation.siemens.com/global/en/products/nx/
5 https://www.autodesk.com/products/fusion-360/overview
6 https://www.onshape.com/
7 http://www.cimp-3d.org

measure (or 3D scan) the part and recreate it as a 3D solid model in CAD. Alternatively, you may be able to find a 3D model of the part (or a close replica) in an online design repository like GrabCAD[8] or Thingiverse,[9] which you can then modify to suit your needs. If you are designing a new component to be made with AM, then generative design tools like topology optimization may inspire new geometries to consider for AM; the example in Section 7.3 illustrates how this is done from start to finish.

7.2.2 Step 2: Converting to an STL File

Once you have 3D model data for your part, then you export it as an STL file, a file format initially developed for STereoLithography, which can be read into any build planning software. The STL file format consists of a triangulated mesh that approximates the boundaries and surfaces of the 3D solid model. This approximation of the 3D solid model into triangles occurs through the process of tessellation. The resulting STL file is simply a list of vertices for each triangle and its associated normal vector, which indicates which face points "outward" versus "inward."

The resolution of the tessellation process is typically controlled by two parameters: (1) surface deviation and (2) angular deviation. As shown in Figure 7.3, the surface deviation tolerance defines the maximum allowable difference between the tessellation

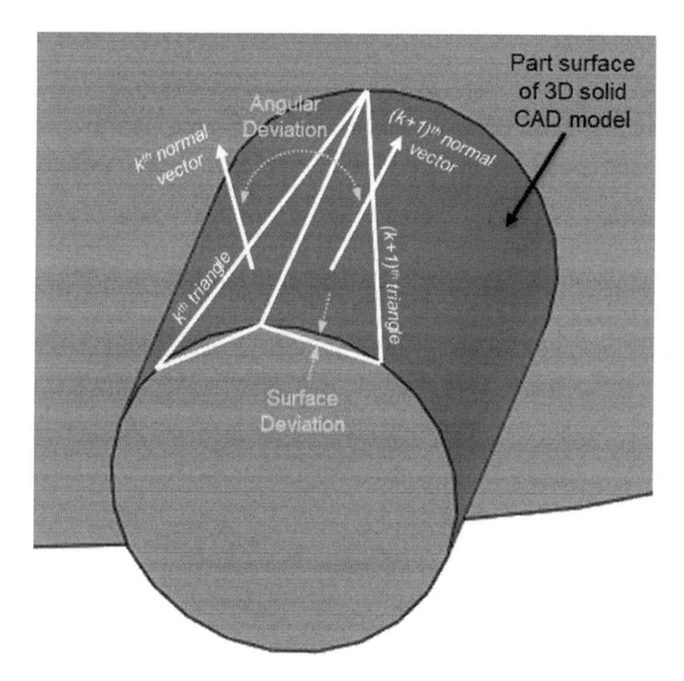

FIGURE 7.3 Angular and surface deviations control resolution of tessellated geometry in a STL file.

[8] https://grabcad.com/
[9] https://www.thingiverse.com/

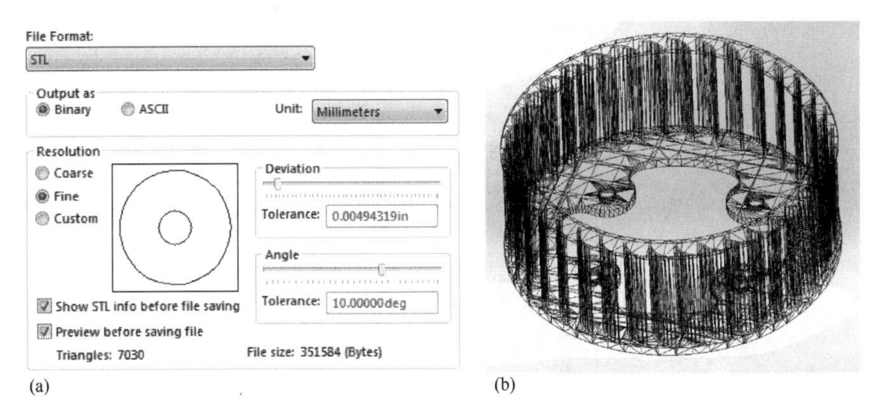

FIGURE 7.4 Example of STL file resolution settings and corresponding tessellated geometry. (a) Setting surface deviation and angle tolerances for STL file and (b) corresponding tessellated geometry.

and the actual part surface; the angular deviation tolerance specifies the maximum angle between adjacent normal vectors. Combined, these two settings control the accuracy and smoothness of the corresponding geometry in the exported STL file. Figure 7.4 shows an example of how these settings are specified within SolidWorks and the corresponding STL file for the default "Fine" resolution settings. Using the smallest tolerances yields the highest resolution, but this is often not practical due to the large number of triangles that are created: more tessellation errors can occur, and the large file size becomes too cumbersome to manipulate. Designers need to strike a good balance between accuracy/resolution and file size when converting CAD models into STL files.

To reduce the file size, STL files can be saved in binary format instead of ASCII format. ASCII files are readable by humans, but binary files are about 1/6th the size of the same file saved in ASCII format (typically 1:5.65 when saving as binary). STL files do not contain units (e.g., mm, inches), and there is no material information stored in a STL file. Despite these limitations, STL files have become the "de facto" standard in the AM industry; however, new AM file formats are in development (e.g., AMF [2], 3MF[10]), and some AM vendors (e.g., 3DXpert by 3D Systems[11]) are starting to skip the STL file conversion all together. This helps designers avoid tessellation and approximation errors and other errors (e.g., inverted normals) that can be introduced during STL file conversion, but layering effects will still need to be reconciled (see Section 7.2.4). Designers spend a lot of time perfecting their 3D solid models in CAD, and STL file conversion (and repair) remains a painful, yet necessary step, of the current AM workflow.

7.2.3 Step 3: Specifying the Build Plan

After converting your 3D solid model into an STL file, you begin the build planning process. Build planning involves determining the orientation of the part for the build, the

[10] https://3mf.io/specification/

[11] https://www.3dsystems.com/software/3dxpert

location of the part on the build plate (or build platform), etc. Specifying the part orientation is important because it drives many aspects of the AM build process, including:

- *Build time*: driven primarily by build height (i.e., a taller part requires more layers, which usually means longer build time) as well as deposition rates and other process parameters.
- *Dimensional integrity*: driven by layering effects (i.e., stair-casing), surface orientation (e.g., vertical surfaces will have different surface roughness than horizontal surfaces, downward-facing surfaces will have a different finish from upward-facing surfaces), size of unsupported features and overhanging geometries, tessellation effects from STL file conversion, etc.
- *Material usage*: dictated largely by the size of the part, but powder bed fusion, binder jetting, sheet lamination, and vat photo-polymerization will require additional material to fill up the entire build envelope (i.e., build volume) that encloses the part during the AM process.
- *Thermal history*: the heating (and cooling) of material as it is deposited layer-by-layer drives the microstructure evolution, which determines the material properties and the mechanical behavior of the part; this can lead to anisotropy (which can be exaggerated by the layering) in some AM processes.

Part orientation also affects the need for support structures in many AM processes. In material extrusion processes, for instance, support structures are needed to prevent overhanging features from sagging due to the effects of gravity. In metal laser-based powder bed fusion, the support structures help "anchor" overhanging structures to the build plate so that they do not distort or curl up (like a potato chip) and jam the recoater blade. Figure 7.5 shows how this manifests for the blower wheel example. Ideally, you would want to lay the part flat as shown in Figure 7.5a to minimize the build height, which reduces the build time and the amount of metal powder you need when fabricated using laser-based powder bed fusion. However, the underside of the top surface requires support structures (see Figure 7.5b) to ensure that it does not warp during the build. A better orientation is shown in Figure 7.5c. This orientation takes longer to build and requires more powder to fill the build volume, but there are no supports between the fins that need to be removed after it is built.

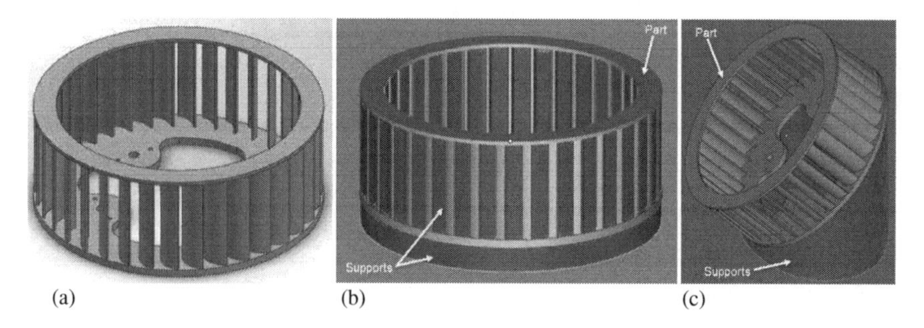

FIGURE 7.5 Support structures needed for different build orientations of the blower wheel. (a) Horizontal orientation builds fastest, (b) horizontal orientation needs support, and (c) better build orientation.

The critical overhang angle—the minimum overhanging angle that can be built without any support structures—is an important thing to know when designing for AM as it drives when (and where) you will need support structures, which impacts build time, material usage, and post-processing effort. The overhang angle is typically measured from the horizontal surface, and the critical value varies by process and by material. Generally speaking, overhangs that are greater than 45 degrees will not need supports, and overhangs less than 30 degrees will require supports. Overhangs between 30° and 45° degrees may build without supports, but may have poor surface finish, and companies will often print test parts to see what their system is capable of doing (see the example from Proto Labs,[12] for instance). There are a multitude of support structure options in commercial (e.g., 3DXpert,[13] Netfabb,[14] Magics[15]) and open source (e.g., Cura,[16] Slic3r[17]) build planning software. Research is also underway to optimize support structure design (e.g., [3]), and new thermo-mechanical process simulation tools such as Amphyon,[18] ANSYS,[19] Atlas3D,[20] ESI,[21] Magics,[22] Netfabb,[23] Simufact,[24] etc. are available to estimate when and where supports are needed in laser-based powder bed fusion and other AM processes.

Once the orientation of the part has been specified, it needs to be positioned on the build plate. The part's location on the build plate is driven by many factors, including the part's orientation, production quantity, nesting capacity, and many other process-specific factors (e.g., air cross-flow in a laser-based powder bed fusion system). The 3D printing knowledgebase on 3D Hubs[25] provides summaries of many of these considerations for each AM process, including AM design guidelines, and many hardware (e.g., 3D Systems,[26] EOS,[27] Renishaw,[28] Xact Metal[29]), software (e.g., Materialise[30]), and service (e.g., Xometry[31]) providers are starting to share their own design guidelines for metal AM given the high cost of failed builds.

---

12 https://www.additivemanufacturing.media/blog/post/7-helpful-numbers-quantify-design-rules-for-am
13 https://www.3dsystems.com/events/20-minutes-3d-systems-3dxpert-support-generation
14 https://knowledge.autodesk.com/support/netfabb/learn-explore/caas/CloudHelp/cloudhelp/2017/ENU/NETF/files/GUID-A89C5423-5662-4F71-A491-E2B56F406DCC-htm.html
15 https://www.materialise.com/en/software/magics/modules/metal-support-generation-module
16 https://ultimaker.com/en/resources/manuals/software
17 https://manual.slic3r.org/expert-mode/support-material
18 https://altairhyperworks.com/partner/amphyon
19 https://www.ansys.com/products/structures/additive-manufacturing
20 https://atlas3d.xyz/
21 https://www.esi-group.com/software-solutions/virtual-manufacturing/additive-manufacturing
22 https://www.materialise.com/en/software/magics/modules/simulation-module
23 https://www.autodesk.com/products/netfabb/overview
24 http://www.mscsoftware.com/Simufact-Additive/Introduction
25 https://www.3dhubs.com/knowledge-base
26 https://www.3dsystems.com/sites/default/files/2018-06/3DSystems-DirectMetalPrinting_DesignGuide_EN_Final.pdf
27 https://cdn1.scrvt.com/eos/public/ab4f0542d66453fc/5f889ab7e3f72bd3d44b22205ba8b68b/EOS-Basic-Design-Rules_Additive-Manufacturing_EN.pdf
28 https://www.renishaw.com/en/design-for-metal-am-a-beginners-guide--42652
29 https://www.xactmetal.com/design-guide.pdf
30 https://www.materialise.com/en/manufacturing/materials/design-guidelines
31 https://www.xometry.com/3d-printing-metal

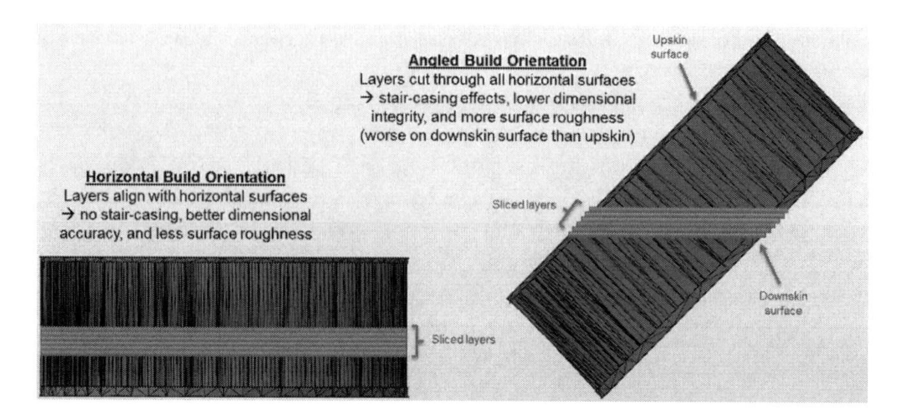

FIGURE 7.6 Potential implications of slicing part in different build orientations.

7.2.4 Step 4: Slicing the Part into Layers

Once the part's orientation and location on the build plate have been determined, the part (and any support structures) are "sliced" into thin layers by the build planning software. The layer thickness is specific to each material and each machine, and it is another important thing to know when designing for AM. Even though the layers are relatively small (20–80 μm in the case of laser-based powder bed fusion), they can lead to "stair-stepping" or "stair-casing" effects in the final part depending on the build orientation. While this may not be of much concern when prototyping with AM, it will impact the part's dimensional integrity, i.e., how closely the "as built" dimensions match the "as designed" dimensions of your part. For end-use parts, the "stair-casing" effects will often necessitate post-processing to achieve the specified tolerances, and designers will need to add machining allowances to the part to account for this. Of course, designers can reorient the part to avoid this "stair-stepping" effect on critical surfaces or features of the finished part. Figure 7.6 illustrates the effects of slicing the blower wheel in the horizontal and angled build orientations shown in Figure 7.5. Even though the horizontal build orientation offers better dimensional integrity and smoother horizontal surfaces (as well as a faster build time), the need for support structures in between the fins makes this orientation impractical.

7.2.5 Step 5: Creating the Toolpath

After slicing the part into layers, the build planning software creates a toolpath, which tells the AM system where and how to add material in each layer. For material extrusion systems, the nozzle temperature, speed, and deposition rate are specified to control the amount of extrudate that is deposited in each "road," and the infill pattern dictates how tightly the roads are spaced. For a laser beam-based system, the toolpath specifies the scan pattern and hatch spacing for the beam to follow as well as the power, scan speed, and spot size along with several hundred other process parameters. Examples of laser scan patterns for the blower wheel in the horizontal build orientation are shown in Figure 7.7. Zooming in on Layer 70 as an example (see Figure 7.7a), we can see how the laser scan pattern uses two contour passes and

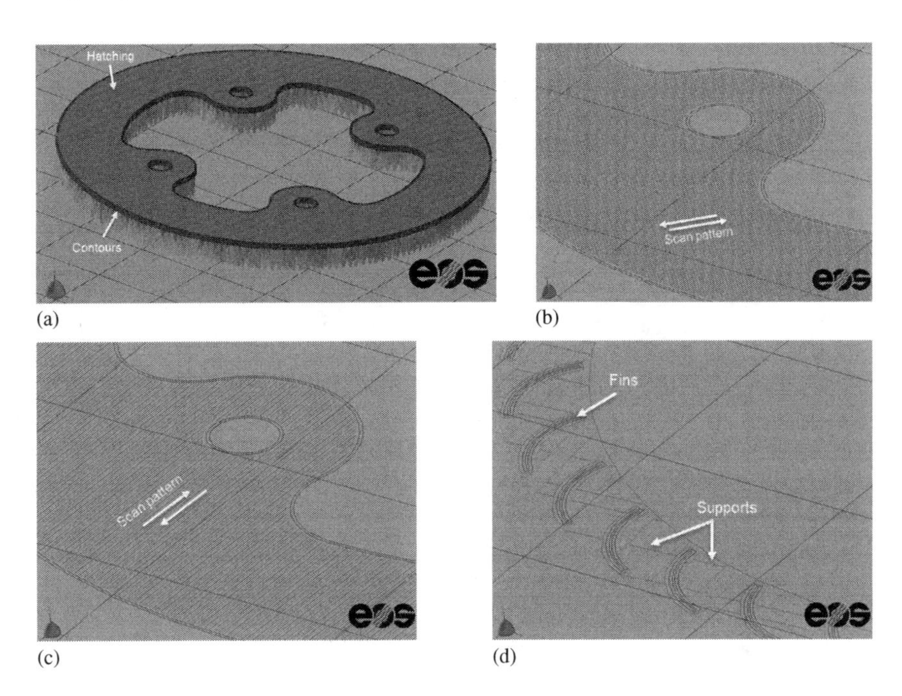

FIGURE 7.7 Examples of laser scan patterns for blower wheel in horizontal build orientation. (a) Layer 70 of the blower wheel in horizontal build orientation, (b) close up view of laser scan pattern in Layer 70, (c) close up view of laser scan pattern in Layer 71, and (d) close up view of scan pattern for fins in Layer 110.

a diagonal scan pattern to "hatch" the interior of the part in Figure 7.7b. This scan pattern alternates layer-by-layer as evidenced by the orientation of the scan pattern in Figure 7.7c, which shows the next layer in the build. Finally, Figure 7.7d shows how the scan pattern for the fins is much tighter than the scan pattern for the support structures. This happens because the support structures do not need to be fully dense like the material in the solid part since they will be removed after the part is built (and less scanning means a faster build).

Process parameters will vary by machine and material and may not be the same throughout the build (e.g., the process parameters for building an upskin surface may differ from those for a downskin surface in laser-based powder bed fusion). Magics, 3DXpert, QuantAM,[32] and other build planning software are giving designers more and more control over process parameters because the scan pattern/toolpath controls the thermal history (i.e., heating and cooling) that the material and the AM part experiences during the build. The thermal history that drives the microstructure that grows (and evolves), which dictates the material properties and mechanical strength of the part [4]. While many of these process parameters are "fixed" for a given material on a given machine, the designer needs to be aware of them and understand how the toolpath and

[32] https://www.renishaw.com/en/quantam-build-preparation-software--35455

process parameters not only affect build time, material usage, and cost, but also the microstructure, the mechanical properties, and the dimensional integrity of the part.

7.2.6 Step 6: Finalizing the Process Plan

The toolpath is part of the overall process plan, which instructs the machine on how to build the AM part. Other process parameter settings may need to be specified to heat the build plate or the build chamber during (or after) a layer, control the inert gas flow in a powder bed fusion system or directed energy deposition system, or switch materials when a machine is capable of printing support structures from an alternative material (e.g., in a material jetting system or in a material extrusion system with a dual head extruder), for example. Different machines require different settings for different process parameters, and default settings are built into the build planning software that comes with (or is recommended for) the AM system. Many CAD packages are also starting to provide "libraries" of AM machines to allow you to pick a specific AM system and start evaluating a build plan and layout within the CAD environment (e.g., 3DXpert[33]).

When making end-use parts, designers may also want to consider including some test specimens (e.g., hardness, tensile, fatigue, impact, corrosion), witness coupons, and chemistry/metallurgical samples to the build plate. For certified parts (e.g., medical devices, aerospace components) or production builds, witness coupons and test specimens are often made in conjunction with the part to help monitor the build quality and gather data that ensure that the process is operating within its limits [5]. Moreover, these test specimens can be destructively tested and analyzed to ensure that the right microstructure and mechanical properties are being obtained in each build.

7.2.7 Step 7: Readying the AM System

Once the process plan is established, the AM system is set up and readied for the build. Material is loaded into the machine, the build plate (i.e., platform) is leveled, and the system is calibrated (e.g., laser position and spot size, extruder head home location). Auxiliary equipment is set up and readied at this time as well. The time and effort to do this will vary by machine, and higher-end machines will automate many of these steps for you (e.g., automatic build plate leveling, cleaning of the nozzle or jets). Moreover, many AM operators and technicians will go through a check list to verify that everything is ready for the build (and avoid skipping steps). Once the AM system is ready to operate, the operator can finally hit "go" and start the build.

7.2.8 Step 8: Executing the Build

As the build proceeds, material (or binder in the case of binder jetting) is added layer-by-layer, forming a solid 3D object. As an example, the blower wheel was built in the horizontal and angled build orientations shown in Figure 7.5b and c, respectively, on an EOS M280, a laser-based powder bed fusion system using AlSi10Mg. The two resulting parts are shown in Figure 7.8, and a video of the build is available on YouTube.[34]

[33] https://www.3dsystems.com/software/3dxpert

[34] https://www.youtube.com/watch?v=FX72ZeF5MEE

FIGURE 7.8 Resulting blower wheel for angled and horizontal build orientations.

As a part is building, many AM systems are now equipped with *in situ* sensing systems to help monitor the build. Sensing could be something as simple as a digital camera that takes pictures of each layer to sophisticated photodiodes, infrared cameras, or high-speed imaging systems that are measuring laser emissions, melt pool temperature, and spatter, among other process variables and emissions in metal AM processes (e.g., [6,7]). All of these sensing systems generate data (sometimes as much as several GBs/layer) that can be used to help qualify the process and potentially certify the part. Some *in situ* monitoring systems are even capable of closed-loop control (e.g., Sciaky's Interlayer Realtime Imaging & Sensing System on their wire-fed electron beam directed energy deposition system[35]). These data are part of the "digital thread" that newer manufacturing technologies like AM are enabling, and this is an active area of research. As data become more readily available, designers will want to use this information to improve their designs using machine learning and artificial intelligence.

7.2.9 Step 9: Post-processing the Part

Once the build is complete, parts often have to undergo post-processing before they are ready for use. For instance, binder jetting systems (e.g., ExOne,[36] Voxeljet[37]) and metal material extrusion systems (e.g., Desktop Metal Studio System,[38] MarkForged Metal X[39]), require debinding and sintering to solidify the "green" part that comes out of the machine. Vat photopolymerization systems (e.g., 3D Systems,[40] Carbon,[41] EnvisionTec,[42] Formlabs[43]) often require a final curing step to harden the polymer to achieve its final cured state. In some cases, the part may print in 30 minutes or less, but the curing step can take 8 hours before the part is ready for use.

---

[35] http://www.sciaky.com/additive-manufacturing/iriss-closed-loop-control
[36] https://www.exone.com/
[37] https://www.voxeljet.com/
[38] https://www.desktopmetal.com/products/studio/
[39] https://markforged.com/metal-x/
[40] https://www.3dsystems.com/3d-printers/plastic#stereolithography-printers-sla
[41] https://www.carbon3d.com/
[42] https://envisiontec.com/
[43] https://formlabs.com/

With material extrusion, material jetting, and vat photopolymerization, support structures will need to be removed manually or with a support cleaning system if the supports be dissolved in water,[44] and surface treatments may be applied to improve the surface finish and enhance the mechanical properties (e.g., acetone vapor smoothing [8]). In laser-based metal powder bed fusion, unfused powder needs to be removed from the build chamber, and then parts need to be stress relieved, cut off the build plate, heat treated, and machined/finished to meet the specified part tolerances, increasing the cost and lead-time of the additively manufactured part [9].

Depending on the extent of the post-processing, this step can double or triple the time and cost for an end-use AM part based on current industry practices. Thus, designers need to carefully consider the need for, and implications of, post-processing when designing their parts for AM and specifying how they should be built (e.g., build orientation and layout). For instance, a topology optimized design or a novel lattice structure may save weight and build time, but if it requires an excessive amount of time to remove support structures from the part, for instance, then it may not be a viable solution.

7.2.10 Step 10: Inspecting the Part and Compiling Documentation

Non-destructive inspection, testing, and build documentation are also important steps in the AM process, especially when qualifying materials or machines or certifying parts. During the build planning process, engineers may add tensile bars, witness coupons, chemistry specimens, etc. to the build plate for mechanical (i.e., destructive) testing or microstructural/chemical analysis after the build. This helps provide additional data that the specified process parameters fabricated the AM part as intended and that the AM system operated as expected. Non-destructive inspections can be done with white light inspection, penetrant testing, digital radiography, and even x-ray computed tomography (CT) scanning (e.g., [10,11]). In fact, x-ray CT scanning is becoming increasingly popular for non-destructive evaluation of AM parts given the rich data that it provides for both external and internal features. All of this testing and inspection data are then compiled with 3D solid model, the build plan, the process plan, any *in situ* build data that were collected to document the part and how it was built. It is important to compile and maintain all of this information, especially for parts that need to be qualified or certified for use.

7.3 DESIGN FOR AM: TOPOLOGY OPTIMIZATION EXAMPLE

To illustrate the AM workflow in detail, let us consider the design of a titanium upright for a student Formula race car for the Society of Automotive Engineering (SAE). This is an annual competition wherein student teams design, build, and race a small Formula-style race car. In 2013, Vincent Maranan under the supervision of Dr. Todd Palmer, a professor in materials science and engineering at Penn State, redesigned the upright that connects the frame and suspension to each axle and wheel (see Figure 7.9). The previous design entailed welding small titanium plates together

[44] http://www.dimensionsca.com/

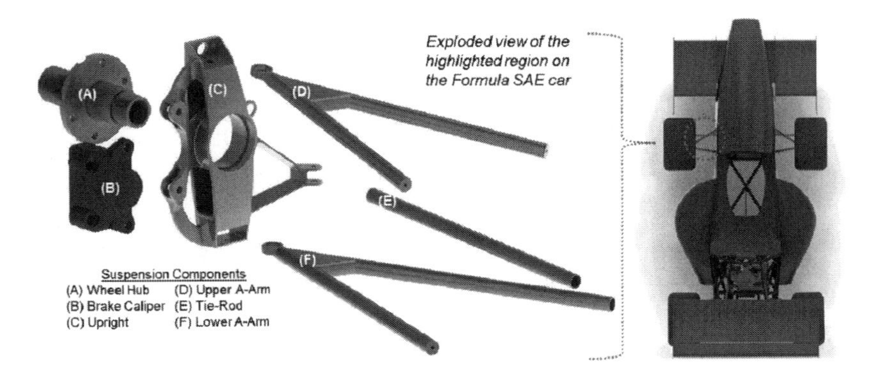

FIGURE 7.9 Upright and components in suspension on Formula SAE race car.

to create a 3D part, and Vincent had the opportunity to additively manufacture a new upright for the Penn State Formula SAE team using laser-based powder bed fusion and titanium (Ti-6Al-4V) powder. To accomplish this, he combined CAD, topology optimization, and finite element modeling to help optimize the upright for the new suspension geometry and loading conditions [12]. The following sections describe the workflow and decisions that were made when designing for AM.

7.3.1 Step 1: Generating a 3D Solid Model

As shown in Figure 7.9, the upright (C) connects the wheel hub (A) and brake caliper (B) to the suspension arms (D) and (F) and tie rod (E) on the race car. The upright shown in the figure is the additively manufactured part that was designed by Vincent using topology optimization. Topology optimization entails defining the design space, loading conditions, and objectives that are then solved by a computer algorithm to find the most lightweight structure (see, e.g., [13,14]). In other words, the designer defines the problem that the computer then solves to find the best structure. Figure 7.10 shows what the design space and sample results look like for different loading conditions. As seen in Figure 7.10a, the design space defines the volume wherein the computer can add (or subtract) material during optimization. Voids designate regions were material cannot be placed during the optimization (e.g., where the hub passes through the upright). Connection points and mating interfaces (e.g., connecting rods, mounting points for brake caliper) are then designated within the design space, and loads are added to the structure at these connections. For more details on the problem formulation, loading conditions, design constraints, etc., see [12].

Once the design space has been defined, different loading conditions are specified for forward and rearward braking, inside and outside cornering, etc., and combinations thereof. The loading conditions were larger than in prior years; hence, the objective was to achieve the same weight as the previous design while maximizing stiffness and maintaining a factor of safety of 3 or more. A free topology optimization algorithm, TopOpt[45], was used to solve the problem once it was formulated.

[45] https://www.topopt.mek.dtu.dk/

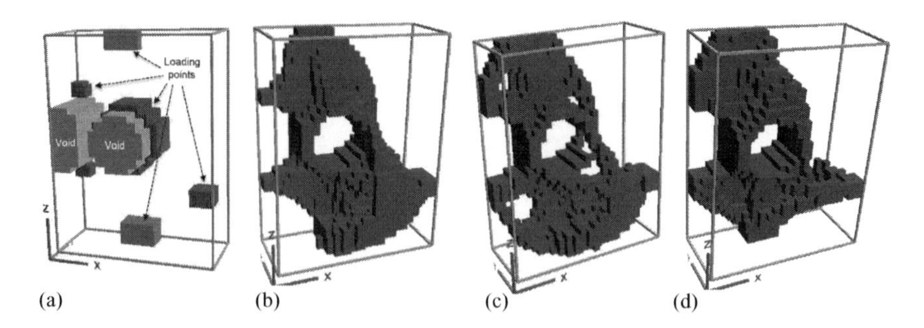

FIGURE 7.10 Topology optimization formulation and sample results. (a) Design space, (b) outside cornering result, (c) max braking result, and (d) combined result.

Results for the outside cornering and maximum braking conditions can be seen in Figure 7.10b and c, respectively, and the result of the combined loading case is shown in Figure 7.10d. The complete set of results can be found in [12] along with the problem formulations for each loading condition.

While it is possible to additively manufacture the results from a topology optimization algorithm, they are rarely meant to be final. The process of converting a "jagged" topology optimized result like those seen in Figure 7.10 into a "smooth" CAD model can be both labor- and time-intensive; however, many design software tools are trying to facilitate this step (e.g., 3-Matic,[46] Frustrum,[47] LiveParts,[48] Optistruct,[49] Paramatters[50]). Here, the results from the combined load case in Figure 7.10d served as inspiration for the CAD model that Vincent created manually in SolidWorks.

SolidWorks' optimization tools were used to help determine the best size of different features in the CAD model. Figure 7.11 shows the problem formulation for

Variable View | Table View | Results View

Run | Optimization | Total active scenarios: 210

Variables

Brake Caliper Mount Thickness	Range with Step	Min:	0.1in	Max:	0.4in	Step:	0.05in
Top Fillet	Range with Step	Min:	0.1in	Max:	0.5in	Step:	0.1in
Bottom Brake Caliper Mount Height	Range with Step	Min:	0.5in	Max:	1in	Step:	0.1in
Click here to add Variables							

Constraints

Stress1	is less than	Max:	45000 psi	Max Forward Braking - No R
Click here to add Constraints				

Goals

Mass1	Minimize

FIGURE 7.11 Example dimensional optimization used to size features on the upright.

[46] https://www.materialise.com/en/software/3-matic
[47] https://www.frustum.com/
[48] https://www.desktopmetal.com/products/software/live-parts/
[49] https://altairuniversity.com/learning-library/tutorial-hyperworks-hook-export-to-cad-with-ossmooth/
[50] https://paramatters.com/

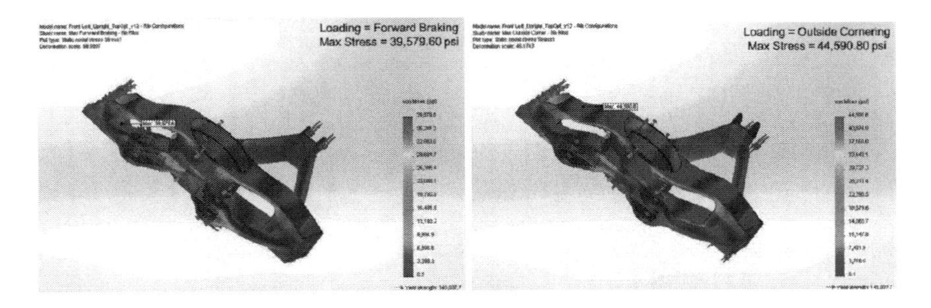

FIGURE 7.12 Sample finite element results for two load cases for the upright.

an optimization to help size three features of the upright. Specifically, the thickness of the brake caliper mount and top fillet are allowed to vary between 0.1–0.4 in and 0.1–0.5 in, respectively, while the height of the brake caliper mount could vary from 0.5–1.0 in. The maximum stress is set to be less than 45,000 psi using the outside cornering load case, which is one of the highest loading conditions [12]. The goal (i.e., objective) in this optimization problem is to minimize the mass of the structure while satisfying the stress constraint. The results from the optimization are used to more precisely size the geometry and features of the final CAD model for the upright.

Finite element analysis (FEA) was performed in parallel to the optimization using ANSYS[51] to confirm that the maximum stress stayed within the specified limits. ANSYS was also used to verify that the factor of safety remained at 3 or higher for the upright in each case. The FEA results from forward braking and outside cornering load cases can be seen in Figure 7.12a and b, respectively. The complete set of FEA results can be found in [12].

By combining topology optimization with CAD and FEA, a final 3D solid model was developed for the upright. CAD renderings of the upright can be seen in Figure 7.13. Note how the voids, clearances, and connection points are consistent with those designated in Figure 7.10, and the geometry mimics many aspects of the combined load case results (see Figure 7.10d), albeit a much "smoother" geometry that is ready to proceed through the rest of the AM workflow.

7.3.2 Step 2: Converting to an STL File

The next step in the AM workflow is to convert the CAD model of the upright into an STL file. When exporting a 3D solid model as an STL file from SolidWorks, the user has two default settings for the deviation and angle tolerances: (1) Coarse and (2) Fine. Alternatively, the user can move the slide bars to specify custom settings for these two tolerances (see Figure 7.4). Table 7.1 summarizes the results from the two default settings as well as two custom options. In the table, the "Finer" settings correspond to typical settings that lead to a reasonably accurate STL file (e.g., a deviation tolerance of 0.001 inches between the CAD file and the STL file), while the "Finest" settings correspond to the smallest values (and high resolution) for both settings.

---

[51] https://www.ansys.com/products/structures

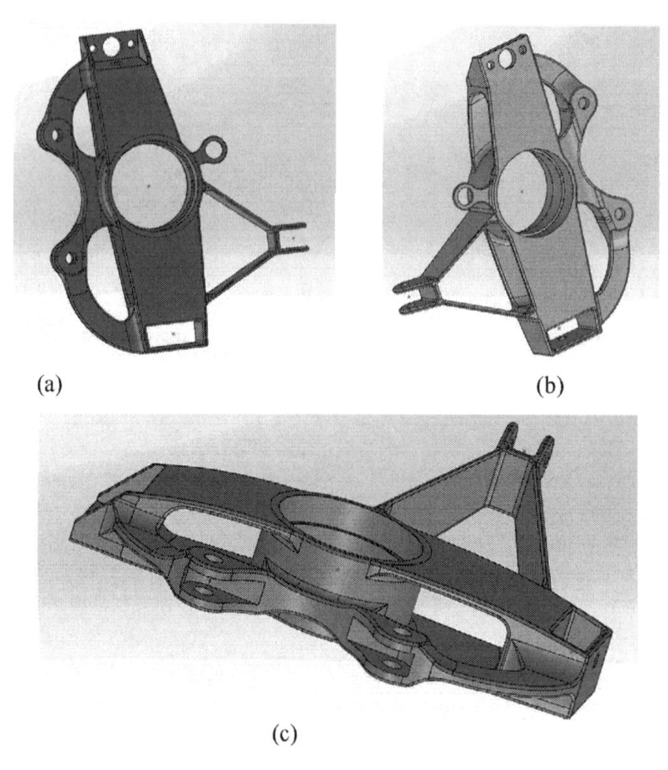

FIGURE 7.13 Final 3D solid model of the upright. (a) Front view, (b) rear view, and (c) side view.

TABLE 7.1
Summary of STL File Resolution Options

STL File Resolution	Deviation Tolerance	Angle Tolerance	No. of Triangles	File Size (Estimated)	File Size (Actual)
Coarse (Default)	0.01413 in	30°	16,134	806,784 Bytes	0.78 MB
Fine (Default)	0.00545 in	10°	62,580	3,129,084 Bytes	3.03 MB
Finer (Custom)	0.00100 in	7.5°	137,168	6,858,585 Bytes	6.66 MB
Finest (Custom)	0.00059 in	0.5°	15,751,336	787,566,884 Bytes	666.87 MB

The corresponding number of triangles and file size for each of these tolerance settings is listed in Table 7.1. While there is an appreciable jump in file size from the Coarse to the Fine (and Finer) resolution, the file size for the Finest resolution is two orders of magnitude larger. This may yield a more accurate STL file; however, working with this many triangles is tedious and burdensome to say the least. Conversion using the Fine and Finer settings took less than 1 sec each; conversion using the Finest resolution settings approached 2 min, and SolidWorks slowed to a crawl when manipulating this large STL file on a desktop computer.

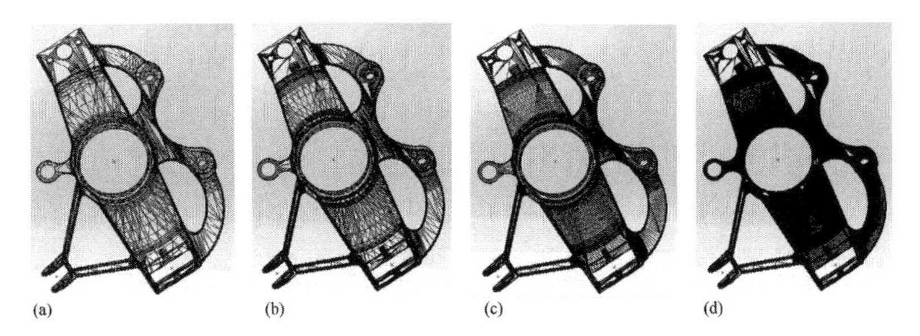

FIGURE 7.14 STL results for different file resolutions. (a) STL results with "Coarse" resolution, (b) STL results for "Fine" resolution, (c) STL results for "Finer" resolution, and (d) STL results for "Finest" resolution.

The corresponding STL representations for the four resolutions listed in Table 7.1 are shown in Figure 7.14. The impact of the different file resolution settings are immediately evident by the density of the triangles in each image. The "Coarse" resolution in Figure 7.14a has noticeably fewer triangles compared to the "Finer" and "Finest" resolutions in Figure 7.14c and d, respectively, and it is easy to see why the file size is so large for the "Finest" resolution. In the end, the "Finer" resolution provides a good balance between resolution and file size. It has over twice as many triangles as the "Fine" resolution (and 8x as many as the "Coarse" one), yet its file size is easy to handle, and the geometry is easy to manipulate on the screen. Also, given that the laser spot size on a powder bed fusion system is typically 50–100 μm, having a deviation tolerance less than 0.001 in (25.4 μm) does not make much sense since this is already half of the resolution that the AM system can achieve during fabrication.

7.3.3 Step 3: Specifying the Build Plan

Using the STL file with the Finer resolution settings in Figure 7.14c, the next step is to specify the build plan, which involves orienting the part and locating it on the build plate. An AM service bureau made the upright for Vincent, using Materialise Magics to set up the build plan. The selected build orientation is shown in Figure 7.15 with the upright in light gray, and support structures added underneath all overhanging surfaces. The supports are generated automatically, but the designer can modify (i.e., design) each support structure in Magics if desired. Figure 7.16 shows an example of the options available for one of the "Block" type support structures used to build the upright. Alternatively, designers can design their support structures in CAD along with their part (if they know the build orientation), exporting them as STL files along with the CAD model and then orienting them with the part during this step.

7.3.4 Step 4: Slicing the Part into Layers

For the given build orientation, the part and supports were sliced into 30 μm layers, which is the default layer thickness for Ti-6Al-4V on the EOS M270 used to build

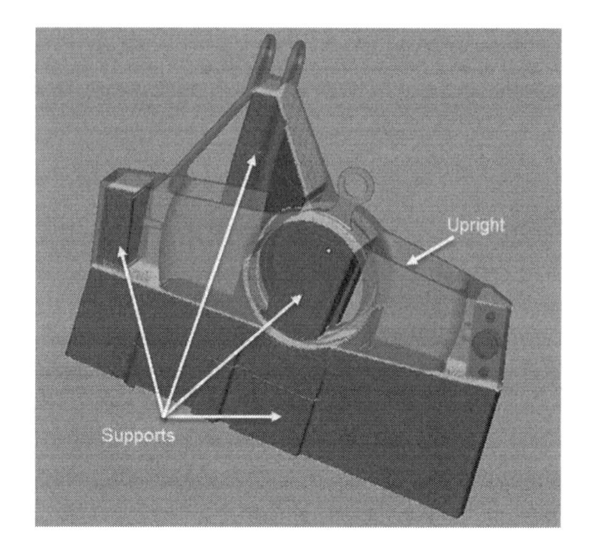

FIGURE 7.15 Build orientation and support structures for the upright.

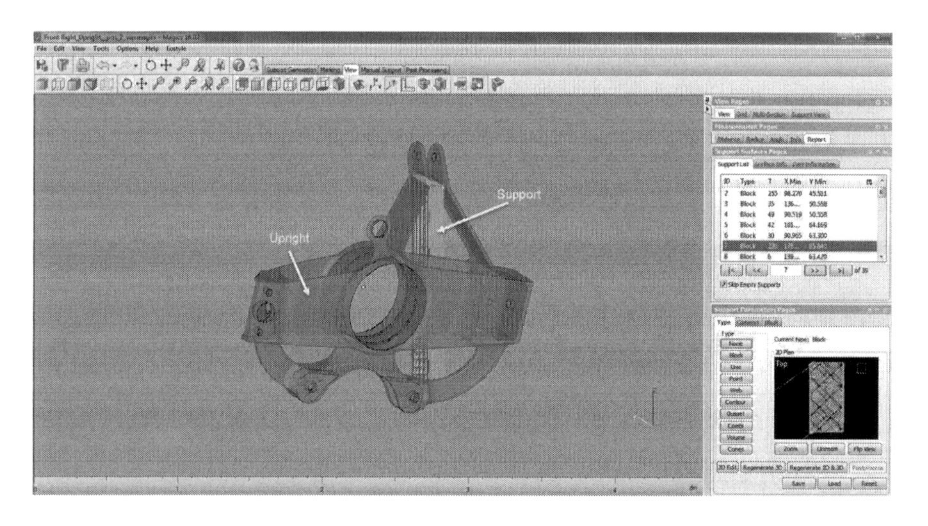

FIGURE 7.16 Example of "Block" type support structure used for the upright.

the upright. Figure 7.17 shows a preview of five layers at 4.44 mm into the build. The dense "hatches" correspond to the underside of one of the bolt holes to secure the brake caliper while the sparse "hatches" show where the laser solidifies the support structures. The tight hatching for the part ensures that the part will be fully dense; the sparse hatching for the supports leads to thin walls and hollow sections per the "Block" type support that was shown in Figure 7.16.

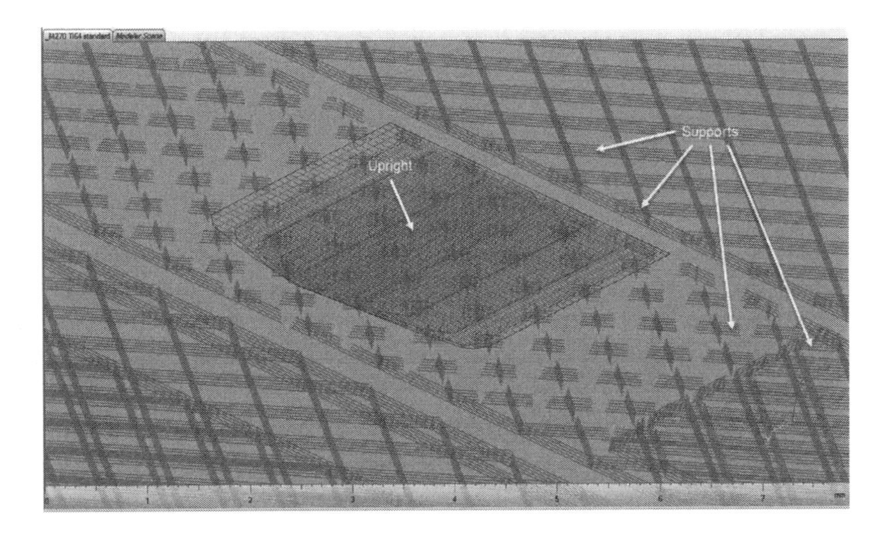

FIGURE 7.17 Example of scan pattern for the upright.

7.3.5 Step 5: Creating the Toolpath

The laser scan pattern and corresponding process parameters for Ti-6Al-4V are automatically generated after the part is sliced into layers. For this laser powder bed fusion system, PSW, EOS's proprietary process software, controls the process parameter settings for the machine.[52] Figure 7.18 shows an example of the laser scanning a portion of the upright during the build. Note how the laser moves so fast

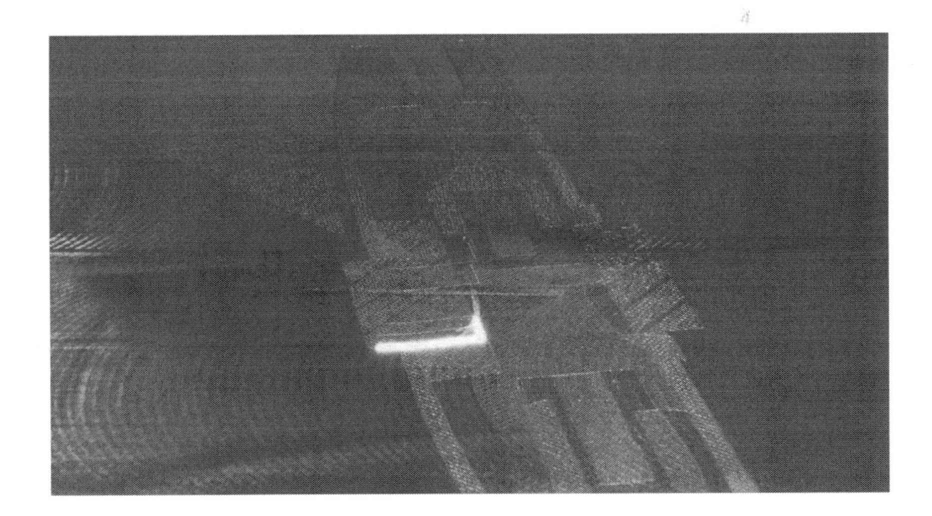

FIGURE 7.18 Example of laser scanning a portion of the upright during the build.

[52] https://www.eos.info/systems_solutions/software_data-preparation

(up to 7.0 m/s) that it looks like it is a line in the picture, but it is not. The EOS M270 uses a single 200W Yb-fiber laser with a spot size ranging from 100–500 μm depending on the process parameter settings used for a given material.

7.3.6 Step 6: Finalizing the Process Plan

In this case, no additional parts, tensile bars, witness coupons, or chemical/metallurgical samples were added to the build plate to reduce build time and material consumption. Theoretically, 2–4 uprights could have been printed in the specified build orientation; however, only one was made at a time in case the part failed during the build (e.g., warping or distortion leading to a recoater jam). Although it was not available at the time, thermo-mechanical process simulation could have been used after finalizing the process plan to ensure that the build was not likely to fail. A process simulation was conducted afterward for the upright; results are available on YouTube.[53]

7.3.7 Step 7: Readying the AM System

Once the process plan was finalized, the build plate was installed and leveled, and Ti-6Al-4V powder feedstock was loaded into the machine. Designers need to remember that the amount of powder needed for the build is not simply the mass of the part; the entire build volume needs to be filled when the part is being built on a powder bed fusion system. In this build orientation, that equates to a build volume of 250 mm × 250 mm × ~175 mm or ~11,000 cm^3, which is nearly 50 kg of titanium (at 4.41 g/cm^3) if it was fully dense. In reality, the powder density is less than the fully dense material, and therefore more powder must be loaded into the machine. In this case, nearly 70 kg of powder is needed to fill the machine, costing nearly $48,000 if using virgin Ti-6Al-4V powder (Ti-6Al-4V was $680/kg in 2013; the costs has dropped to under $300/kg in 2018). Granted, powders can be reused and recycled for a period of time (e.g., [15]), but this staggering number reinforces the importance of trying to minimize build height and use process simulation to reduce the likelihood of failures when designing for AM, especially laser-based powder bed fusion.

7.3.8 Step 8: Executing the Build

Once the machine is readied and enough inert gas is available for the build (i.e., Argon in the case of Ti-6Al-4V), the build can begin, fabricating the part layer-by-layer. Figure 7.19 shows the upright and its support structures in the build chamber after powder removal. The total build time was 54 hrs and 37 min of which 24 hrs and 14 min was needed to fabricate the upright. The remaining 30 hrs and 23 min (55% of the total build time) was required to fabricate the support structures, which eventually need to be removed from the part before use.

[53] https://youtu.be/dtJ3BnCIpoE

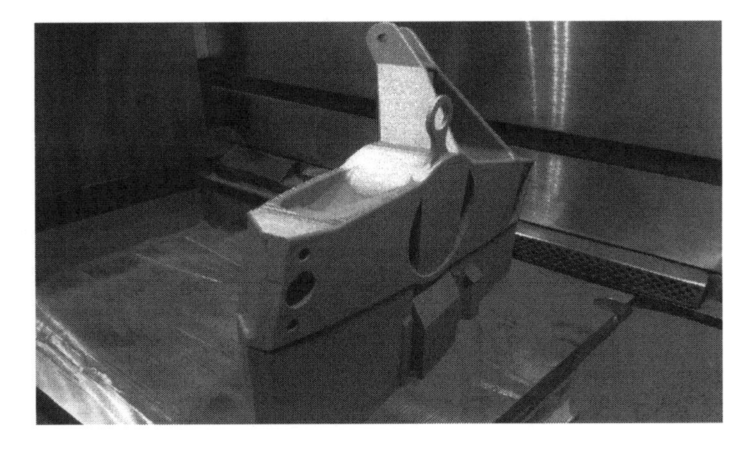

FIGURE 7.19 Image of upright and support structures on the build plate after powder removal.

7.3.9 Step 9: Post-processing the Part

After the build is complete and the build plate and the part have cooled down, post-processing begins. Figure 7.20 shows examples of some of the post-processing steps required when using laser-based powder bed fusion. The unfused powder is removed

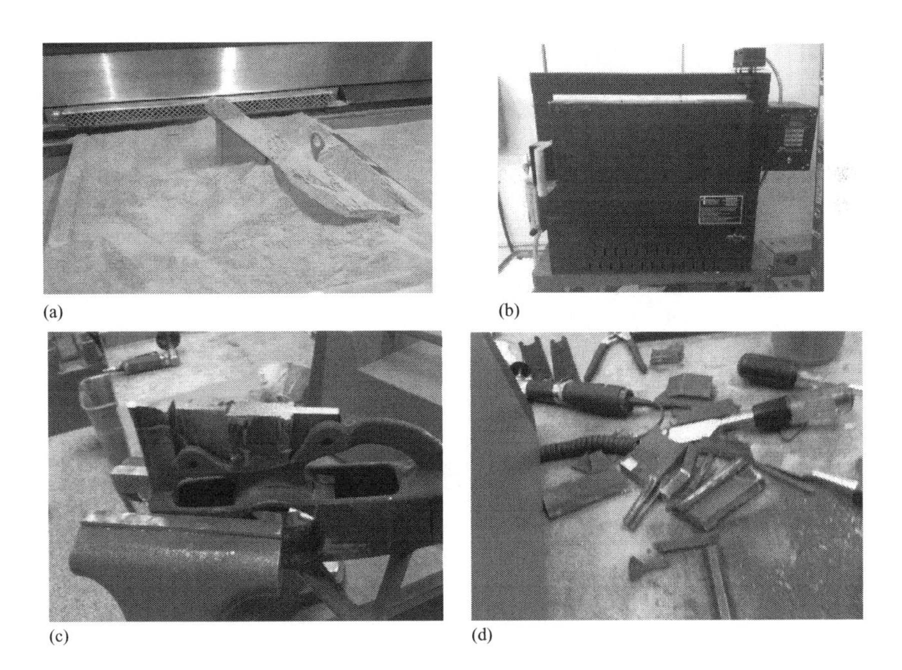

FIGURE 7.20 Images from post-processing of the upright. (a) Excavating part from powder, (b) stress relief on build plate, (c) manual support removal, and (d) support structures after removal.

FIGURE 7.21 Finished upright with supports removed.

as the build plate and upright are excavated from the machine (see Figure 7.20a), and the build plate with the upright still attached to it (see Figure 7.19) was stress relieved in a furnace (see Figure 7.20b). After the upright was removed from the build plate, it was held in a vice so that the support structures can be removed manually (see Figure 7.20c). The support structures end up as scrap after removal from the upright (see Figure 7.20d). The finished upright is shown in Figure 7.21.

7.3.10 Step 10: Inspecting the Part and Compiling Documentation

We had the opportunity to use x-ray CT to scan and inspect the part non-destructively. Granted, x-ray CT scanning may be overkill for a part that does not have any internal cavities/features (e.g., conformal cooling channels) or intricate geometries (e.g., lattice structures); however, it can be useful to look for pores, voids, or other internal defects. No major defects were found; therefore, the CT scan was used to assess the dimensions of the "as-built" part relative to the "as designed" part as shown in Figure 7.22. This is achieved by comparing the CT-scanned volume in Figure 7.22a with the original 3D solid model in Figure 7.22b and highlighting the

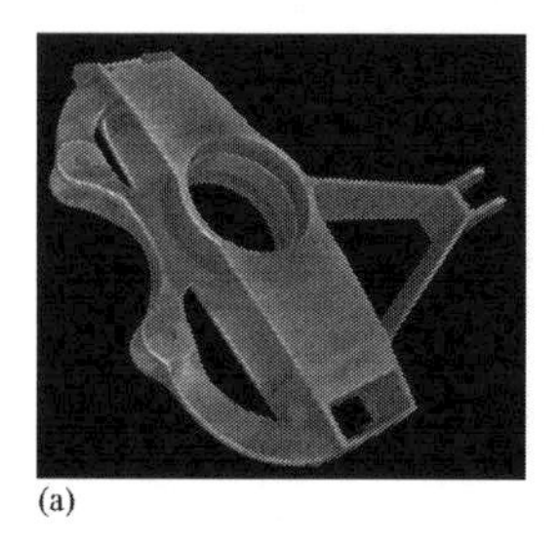

(a)

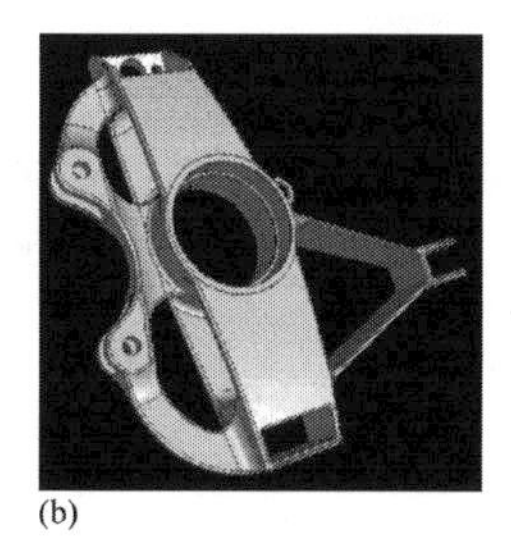

(b)

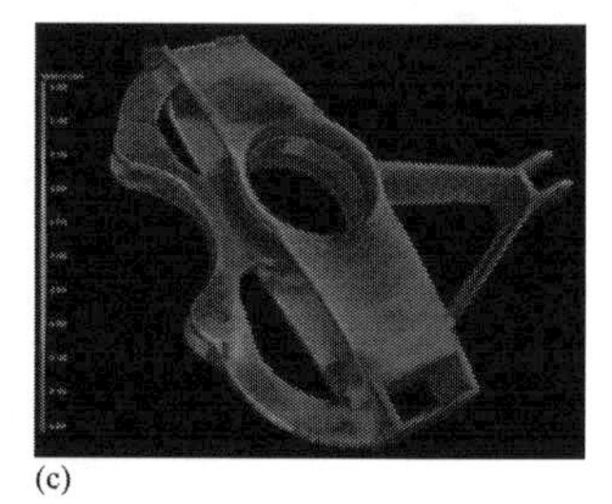

(c)

FIGURE 7.22 Example of dimensional assessment using CT scan and original CAD model. (a) CT scan of upright, (b) original CAD model, and (c) dimensional differences.

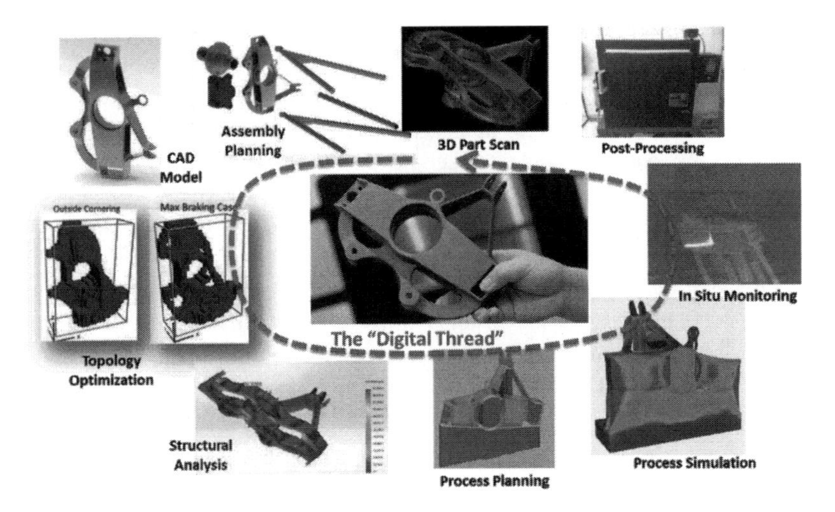

FIGURE 7.23 Example of information contained in the digital thread for an AM part.

dimensions that differ by a certain amount. In Figure 7.22c, the user-specified color-scale ranges from +0.2 mm (over-sized) to −0.2 mm (under-sized).

The CT scan data are then compiled with all of the CAD models, analyses, simulations, results, and other documentation for the upright. This information constitutes the "digital thread" for the upright (see Figure 7.23), an important digital history of the part from conception through production. If sensors were placed on the part to monitor its use, then the digital thread would include usage data that could be used to improve the topology optimization and FEA used to design future uprights.

7.4 PROGRESSION OF DESIGN FOR AM

While many designers and engineers might like to jump right to topology optimization to design a new part for AM, managers in established companies tend to be more risk averse. They will typically want to have an "apples to apples" comparison between an AM-made part and the conventionally manufactured part before they make the leap to more intricate geometries, complex lattice structures, bio-inspired shapes, etc. that can only be made with AM. We have found that companies tend to go through the following three phases when designing for AM.

- *Phase 1: Replicate with AM* - In this first phase, the goal is to reproduce a part exactly with AM and compare it to the conventionally manufactured part in terms of performance, quality, cost, etc. Because the part is being replicated exactly, the geometry is fixed, and the material is the same; so, the AM part will not weigh any less or be any stronger (i.e., the designer has no freedom to lightweight the part or optimize the geometry because it is an exact replica). Worse, the AM part will likely cost more to fabricate given the cost of material feedstock (e.g., wrought titanium costs about \$25–\$50/kg, whereas titanium powder costs \$250–\$500/kg). Fortunately, there will likely be a lead-time advantage when replicating the part with AM, and designers need to manage everyone's expectations carefully during this first phase given the limited benefits that are possible when replicating a part with AM.

- *Phase 2: Adapt for AM* - In this second phase of design for AM, the designer has the freedom to adapt the part for AM. The AM part still has to achieve the same functionality, but non-functional features can be modified to make it less costly and difficult to use AM (e.g., thin walls, overhangs, bulky cross-sections, high aspect ratio features). Designers may also be able to add features that enhance part quality (e.g., lattice structures to reduce weight, conformal cooling channels to improve heat transfer) when adapting for AM. This will help improve the value proposition for AM, namely, performance will start to improve, costs will start to reduce, and the lead-time advantages will increase even more.
- *Phase 3: Optimize for AM* - In this third phase, the designer has the freedom to redesign the part completely for AM or design a new part from scratch. In this phase, designers can take full advantage of AM by using generative design tools, topology optimization, lattice (or cellular) structures, bio-inspired designs, etc. to "design for functionality" rather than "design for manufacturing." During this third phase, part performance can be optimized, cost can be minimized, and lead-time advantages can be maximized and leveraged for flexibility, agility, and rapid responsiveness. Because the geometry is optimized for AM from the start, the designer can easily mitigate the limitations inherent in any AM process.

The "Replicate for AM → Adapt for AM → Optimize for AM" phases often encountered when designing for AM provides companies with a "Crawl → Walk → Run" strategy to navigate AM implementation. As designers proceed from one phase to the next, they learn how to harness the freedoms enabled by AM while avoiding the pitfalls that can restrict AM's potential. In the end, the AM-optimized part may look nothing like the original part as shown in Figure 7.24. This example comes for Jesse Boyer, an AM Fellow at Pratt & Whitney, and it shows how they have evolved from direct substitution with AM (i.e., replicate) to minor modifications to improve producibility (i.e., adapt) to a design optimized for weight, cost, and performance (i.e., optimize). Additional examples follow to highlight some of the benefits that can be obtained in each phase.

7.4.1 Phase 1: Replicate with AM

When replicating with AM, the lead-time advantages associated with **direct substitution** of a traditionally made part by an equivalent AM part can provide many benefits. For instance, AM can be used for **functional prototyping** even if conventional manufacturing techniques will be used for production eventually. This can reduce product development times significantly or enable more design iterations.

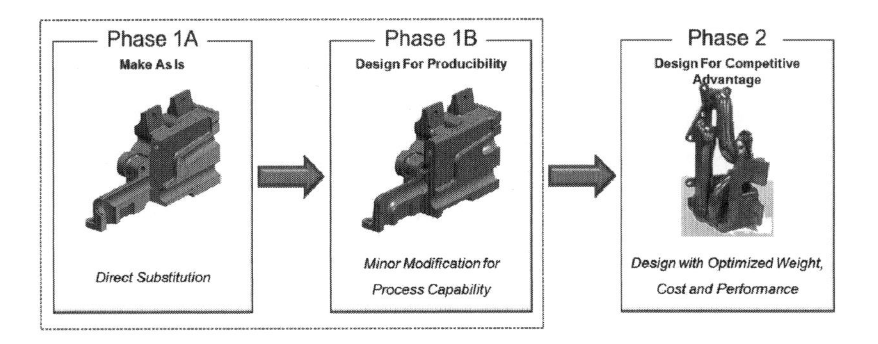

FIGURE 7.24 Example of Replicate-Adapt-Optimize phases of design for AM. (Courtesy of Pratt & Whitney.)

For instance, Siemens boasts a 90% reduction in prototyping lead-time of its gas turbine blades using AM [16]. Other companies are using AM as **supply chain leverage** to reduce the cost of existing components made by traditional means [17].

Spares and replacement parts can also be a benefit for direct part substitution with AM. This is particularly relevant to older, legacy systems that may no longer be supported by the original equipment manufacturer (or worse, the original supplier may no longer exist). Automotive companies such as Porsche,[54] Mercedes-Benz Trucks,[55] and Volvo Construction Equipment[56] have each announced how they will leverage AM for spare parts and replacements. Likewise, Siemens will use AM to make spare parts for its trains,[57] and GE Transportation set the goal of fabricating up to 250 locomotive parts with AM.[58] Meanwhile, Whirlpool has partnered with Spare Parts 3D to offer 3D printed replacement parts,[59] and Etihad Airways will work with EOS and BigRep to improve its maintenance, repair, and overhaul services.[60] Directed energy deposition processes are well suited for part **repair and restoration** and other maintenance, repair, and overhaul services in the aerospace industry, provided that regulatory challenges can be addressed.[61]

Replicating with AM can also be used to **validate the AM workflow**. For example, the Navy recently partnered with Dr. Edward (Ted) Reutzel and his team in

[54] https://newsroom.porsche.com/en/company/porsche-classic-3d-printer-spare-parts-sls-printer-production-cars-innovative-14816.html
[55] https://additivemanufacturingtoday.com/mercedes-benz-trucks-begins-3d-printing-replacement-metal-parts
[56] https://3dprint.com/208226/volvo-ce-3d-printing/
[57] https://www.siemens.com/innovation/en/home/pictures-of-the-future/industry-and-automation/additive-manufacturing-spare-parts-for-the-rail-industry.html
[58] https://www.engineering.com/3DPrinting/3DPrintingArticles/ArticleID/17773/GE-Transportation-Jumps-on-the-3D-Printing-Train.aspx
[59] https://3dprint.com/230707/whirlpool-partners-with-spare-parts-3d/
[60] https://3dprintingindustry.com/news/eos-enters-3d-printing-aircraft-mro-partnership-with-etihad-airways-144290/
[61] https://www.mro-network.com/manufacturing-distribution/future-additive-manufacturing-mro

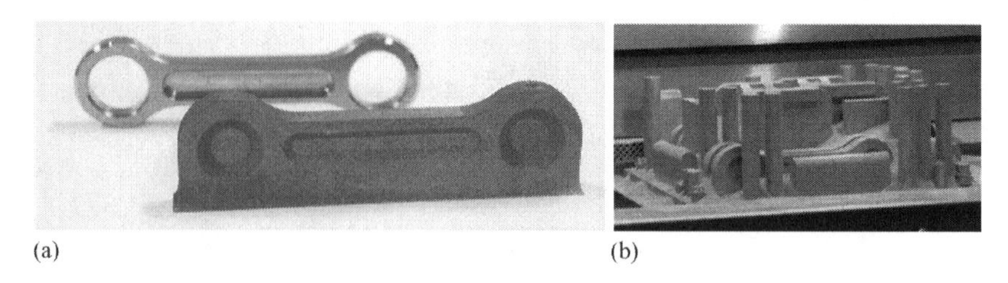

(a) (b)

FIGURE 7.25 The link on the Marine MV-22 Osprey was replaced with an equivalent AM part. (a) Link as printed (front) and after post-processing (rear), and (b) build plate with link and witness coupons.

Penn State's CIMP-3D to develop, test, and fly the first flight critical component on the U.S. Marine MV-22 Osprey helicopter.[62] The geometry and material could not be modified; so, the designers had to determine the best build orientation for the link and fitting (not shown) and then define appropriate machining allowances and specify post-processing requirements. The "as built" link is shown in Figure 7.25a along with the final machined link (background image), and an example production build for the V-22 link and fitting is shown in Figure 7.25b. The final AM build plan included four links, four fittings, and nearly two dozen witness coupons, tensile bars, fatigue specimens, etc. to allow the Navy to conduct testing on the AM parts before installation on the helicopter. Details on the development, testing, and evaluation of the AM link and fitting can be found in [18, 19].

7.4.2 Phase 2: Adapt for AM

Replicating a part with AM quickly reveals the inherent limitations in an AM process, and the designer has to find ways to overcome these limitations. Unfortunately, when replicating a part with AM, the designer often does not have the freedom to make those changes, but that is not the case when adapting for AM. In this phase, the designer can modify some aspects of the part's geometry to balance the restrictive and opportunistic aspects of AM [20]. To do this, designers first target non-functional features and surfaces to eliminate overhangs, thin walls, etc. that can lead to build failures or require excessive post-processing. After these restrictive aspects have been addressed, designers can focus on improving part performance by leveraging the design freedoms associated with AM. For instance, **conformal cooling channels** could be added to **improve heat transfer** in tooling applications [21] or create **novel internal passageways** that lead to innovative designs such as the state-of-the-art faucets from DXV[63] and Kallista.[64]

[62] https://breakingdefense.com/2016/08/osprey-takes-flight-with-3d-printed-part/

[63] https://www.architectmagazine.com/technology/products/these-stunning-faucets-are-3d-printed-in-metal_o

[64] https://www.additivemanufacturing.media/blog/post/direct-metal-printing-rapidly-delivers-innovative-faucet-

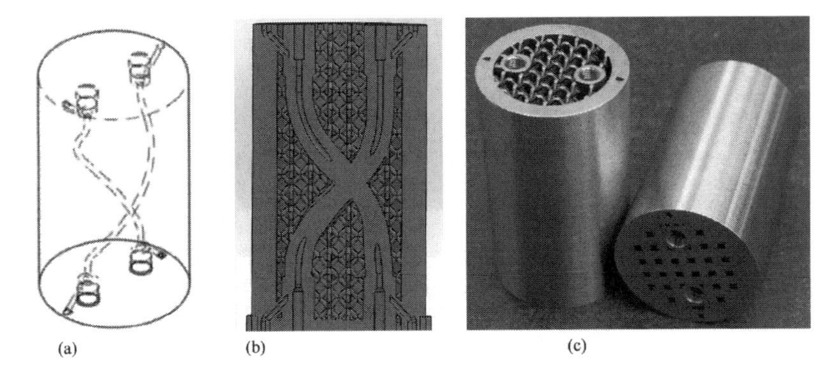

FIGURE 7.26 Example of using a lattice structure to adapt an oil and gas component for AM. (a) Original part, (b) internal lattice structure, and (c) finished AM part.

Lattice structures can come in handy when adapting parts for AM. They are particularly useful when trying to **improve the strength-to-weight ratio** of a part that will be made with AM or to lightweight bulky sections within a part that do not see much loading. Lattices also help **reduce build time** and **improve material utilization**, which can enable material substitutions in some cases. For example, the oil and gas component shown in Figure 7.26 used a lattice structure to lightweight the internal structure of the part, reducing the weight by more than 40% (5.8 lbs vs. 10 lbs) [22]. This component pumps fluid up and down an oil well using centrifugal force, and the internal structure only needs to be strong enough to support the internal passageway, which is difficult to manufacture by traditional means. Because the lattice reduced the amount of material that was needed to make the part, a material substitution was possible. Specifically, a more corrosion resistant alloy was able to be used, increasing down-hole life and enabling new after-market opportunities that were not possible with the original component. This completely changed the lifespan and economics of the part, allowing a more expensive material to be used. The lattice part survived pressure testing of 12,000 psi; for more details on its design, see [22].

Part consolidation is another example of adapting for AM. In this case, the designer is adapting a set of parts for AM, replacing an assembly of components that was too challenging (or costly) to make as a single part with conventional manufacturing methods. While part consolidation strategies for plastic injection modeling have been around for decades, such consolidation was not feasible economically (or technically, in many cases) for metallic components until AM. Now AM enables this for both plastics and metals, and researchers are developing methods to help automate part consolidation with AM [23]. The highly publicized GE LEAP nozzle[65] consolidated ~20 components in the nozzle into 1 unit that weighed 25% less using laser-based powder bed fusion. They have since used AM to reduce 855 parts to 12 AM parts on their Advanced Turboprop[66] engine. While the former example eliminates ~19 components from the nozzle assembly, the latter example potentially eliminates

[65] https://www.ge.com/reports/mind-meld-ge-3d-printing-visionary-joined-forces/
[66] https://www.ge.com/reports/mad-props-3d-printed-airplane-engine-will-run-year/

10–15 suppliers from the supply chain. Thus, AM can help not only **reduce assembly costs**, but also **reduce supply chain complexity** where the compounding effects are even more pronounced. These savings can be significant and are often substantial enough to justify the increased cost of material feedstock needed for AM.

Benefits of part consolidation often include **weight savings** as well as **reduced assembly time**, touch labor, and qualification/testing. For instance, NASA reported that the baffle on Pratt & Whitney's J-2X Heavy Lift Rocket Engine reduced lead-time from over 9 months to 9 days by using laser-based powder bed fusion.[67] The part had **better structural integrity** by eliminating the welds required to make the original baffle. Similarly, NAVAIR and the CIMP-3D team at Penn State [24] used laser-based powder bed fusion to reduce a hydraulic manifold assembly from 17 components to 1 part as shown in Figure 7.27. This reduced the weight by 60%, the height by 53%, and the cost by 20%, and it eliminated five potential leakage points, which pleased the hydraulics engineer. The AM manifold had its own challenges as the twisting internal passageways became clogged with debris due to their intricate curvature. Luckily, this was caught during an x-ray CT scan of the manifold (see Figure 7.27c), and the debris was removed before pressurizing and testing the new manifold; for more details, see [24].

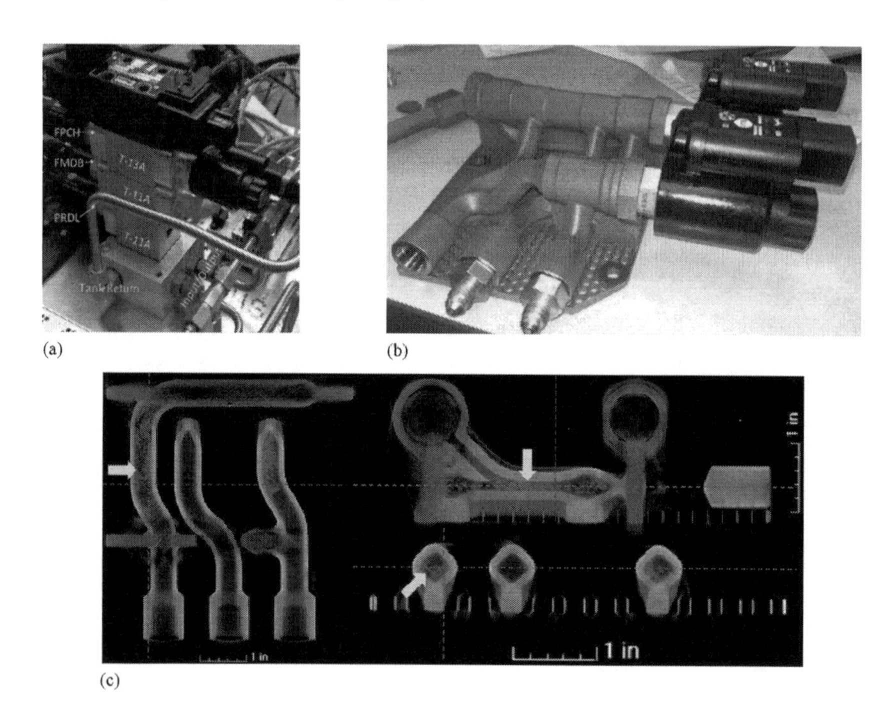

FIGURE 7.27 Hydraulic manifold before and after part consolidation with AM. (a) Original 17-piece assembly, (b) one-piece AM manifold with hydraulic fittings, and (c) CT scan showing debris in internal passageways of one-piece AM manifold. (From Schmelzle, J. et al., *ASME J. Mech. Design*, 137, 111404, 2015.)

[67] https://www.nasa.gov/exploration/systems/sls/j2x/3d_print.html

7.4.3 Phase 3: Optimize for AM

While adapting for AM mainly focuses on improving the manufacturability of an existing part with AM, optimizing for AM enables the designer to start from scratch and thus maximize the benefits of AM. **Topology optimization** tools like those used for the upright in Section 7.3 are readily available for this phase of design for AM, and examples of using topology optimization to **minimize the weight** of structures abound in the literature [14,25]. Meanwhile, GrabCAD has worked with companies like GE,[68] Alcoa,[69] and United Launch Alliance[70] to provide industry-inspired examples to challenge designers to optimize parts for AM.

Designers need to be aware that many commercially available topology optimization tools do not yet take many AM process limitations into account yet (e.g., overhangs, thin walls). As a result, the designer faces new tradeoffs that must be resolved when designing for AM. For instance, the topology optimized design for the upright in Section 7.3 is the lightest weight structure given the loading conditions, yet over 55% of the build time went into fabricating the support structures. If the designer is willing to change the geometry to reduce overhangs and consider a different build orientation, then support structures can be eliminated almost entirely as shown in Figure 7.28. Details on the software tools and process used to redesign the upright can be found in [26].

To overcome these limitations, **generative design tools** seek to help designers realize the full potential of different AM processes. While many software providers are simply rebranding their topology optimization tools as generative design tools, others are taking a different approach. For instance, Autodesk combined high-performance cloud computing with artificial intelligence to rapidly explore design and material options while considering AM's constraints.[71] For instance, General Motors

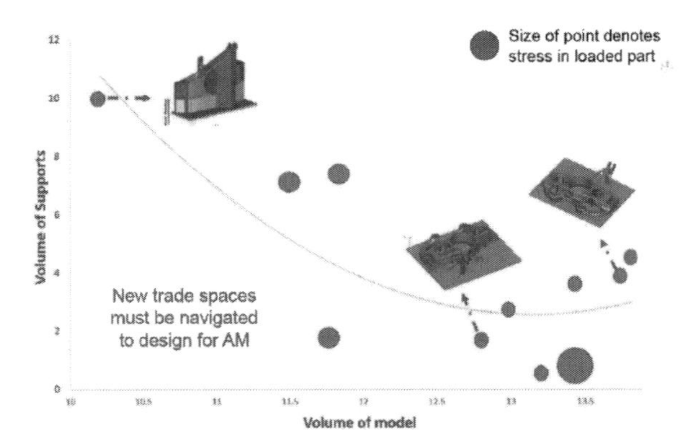

FIGURE 7.28 Tradeoff between light weight design and support structures needed for upright. (From Reddy, S.N. et al., "Application of topology optimization and design for additive manufacturing guidelines on an automotive component," *ASME Design Engineering Technical Conferences: Design Automation Conference*, ASME, Charlotte, NC, 2016.)

[68] http://grabcad.com/challenges/ge-jet-engine-bracket-challenge
[69] https://grabcad.com/challenges/airplane-bearing-bracket-challenge/entries
[70] https://grabcad.com/challenges/3-2-1-liftoff-ula-rocket-hardware-challenge/entries
[71] https://www.autodesk.com/solutions/generative-design

used generative design tools to develop a one-piece organic-looking seat bracket that replaced an assembly of eight parts.[72] Additional generative design examples from Airbus,[73] Stanley Black & Decker,[74] and others can be found on Autodesk's website.

Better software tools are also available for creating **variable lattice structures** for AM parts. While most lattices are uniform, variable lattices change the size, shape, and/or spacing of the unit cell to **maximize the strength-to-weight ratio**. For instance, the variable density lattice structure on the brake pedal shown in Figure 7.29 was optimized by the team at nTopology using their Element Pro[75] software and then additively manufactured by Renishaw on their laser-based powder bed fusion system. The example is available as a case study on nTopology's website.[76]

Perhaps the "killer app" for AM is **mass customization** and **personalization** of parts and products for individual users. For instance, the majority of hearing aids are now customized with AM,[77] and FormLabs even offers 3D printed custom earbuds for your headphones.[78] Align Technology makes more than 220,000 custom orthodontic braces and aligners each day with AM.[79] Gillette recently announced, Razor Maker[80] in partnership with FormLabs to 3D print personalized handles for their razors.[81] Robot Bike Company has worked with Renishaw to produce custom mountain bikes with AM,[82] and Local Motors has been using AM to produce crowd-sourced vehicles for over a decade.[83]

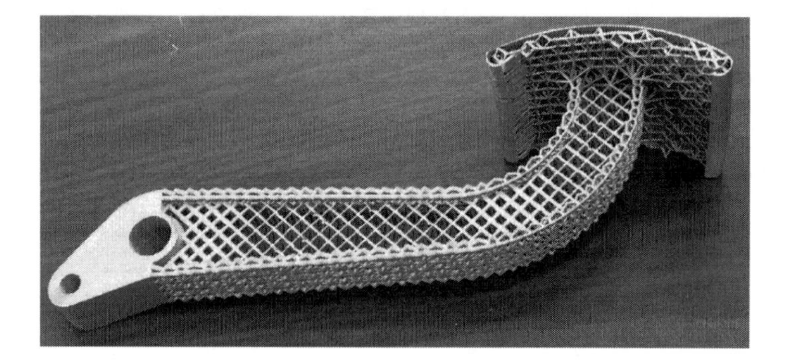

FIGURE 7.29 Brake pedal with optimized lattice structure designed by nTopology and printed by Renishaw.

[72] https://media.gm.com/media/us/en/gm/news.detail.html/content/Pages/news/us/en/2018/may/0503-lightweighting.html
[73] https://www.autodesk.com/customer-stories/airbus
[74] https://enterprisehub.autodesk.com/articles/stanley-black-decker-seized-on-generative-design
[75] https://www.ntopology.com/
[76] https://www.ntopology.com/blog/2017/11/6/element-pro-case-study-f1-brake-pedal
[77] https://www.forbes.com/sites/rakeshsharma/2013/07/08/the-3d-printing-revolution-you-have-not-heard-about/#794fcdf61a6b
[78] https://formlabs.com/industries/audiology/
[79] https://www.forbes.com/sites/tjmccue/2017/09/14/3d-printing-moves-align-technology-toward-1-3-billion-in-sales/#4385cbc75378
[80] https://razor-maker.com/
[81] https://formlabs.com/blog/gillette-uses-3d-printing-to-unlock-consumer-personalization/
[82] http://robotbike.co/technology/
[83] https://digit.hbs.org/submission/local-motors-is-crowd-sourcing-your-next-car/

In fact, they now offer Olli, a "co-created, self-driving electric and cognitive shuttle" that uses a 3D printed chassis.[84] Meanwhile, organizations like Enabling the Future are using 3D printing to make custom protheses for children with missing hands and limbs,[85] and generative design tools from nTopology, Frustrum, and Paramatters are enabling designers to create custom medical devices and implants with AM.[86]

AM makes this possible because it changes the economics of production, enabling **economies of one** [27]. Take Vortic Watch Co. for example, a start-up that uses AM to produce custom cases for pocket watches.[87] By using AM, they can **print-on-demand** and in **small batches** as shown in Figure 7.30. AM enables

FIGURE 7.30 Workflow and steps to produce custom watch cases with AM. (Courtesy of Vortic Watch Co. and Imperial Machine & Tool Co.)

[84] https://localmotors.com/meet-olli/
[85] http://enablingthefuture.org/
[86] https://3dprintingindustry.com/news/defining-generative-design-software-implementation-ntopology-frustum-paramatters-134663/
[87] https://slm-solutions.us/wp-content/uploads/2017/10/SLMSolutionsCaseStudy_VorticWatches Imperial.pdf

manufacturing-as-a-service and requires **little to no tooling**, which benefits entrepreneurs and small companies alike.

7.5 CLOSING REMARKS

In summary, anyone can buy an AM system, fill it with material, and hit "go" to start a build. Understanding the implications of AM on a design and then designing a part for AM to harness the full potential of layer-by-layer manufacturing is what will set one company apart from its competitors. AM's competitive advantage resides in the ability to design for AM and use design to mitigate AM's inherent limitations, which the AM hardware providers do not like to admit exist. Everyone says "complexity is free" when it comes to AM, but if you do not know how to design for AM, then your ability to achieve that "free complexity" will be very expensive.

Regardless of the AM process, the workflow generally follows the ten basic steps discussed in this chapter:

- Step 1: Generating a 3D Solid Model
- Step 2: Converting to an STL File
- Step 3: Specifying the Build Plan
- Step 4: Slicing the Part into Layers
- Step 5: Creating the Toolpath
- Step 6: Finalizing the Process Plan
- Step 7: Readying the AM System
- Step 8: Executing the Build
- Step 9: Post-Processing the Part
- Step 10: Inspecting the Part and Compiling Documentation

The time and effort required for each of these steps will vary by process and material as well as whether AM is being used for prototyping or creating an end-use part. Many of the limitations that we were able to overlook or ignore when prototyping are no longer inconsequential when producing a part that gets installed on an airplane or car—or implanted in someone's hip. The design specifications must be met, and performance requirements must be satisfied just like any other part made by conventional manufacturing. AM does not alter physics or the realities of business; it just gives designers a new means to realize their creativity and be innovative.

The AM part must also be economically viable. While we have not talked specifically about the economics of AM (see, [28,29] for recent summaries of AM cost models), the three phases of design for AM outlined in Section 7.4 provide insight into the benefits that designers can achieve replicating with AM (Phase 1) versus adapting for AM (Phase 2) versus optimizing for AM (Phase 3). As the designer gets more freedom to leverage the benefits of AM and overcome its inherent limitations, the value proposition improves and AM becomes more economical and competitive with conventional manufacturing. As such, companies need to learn how to crawl, walk, and then run with AM as they begin their journey with this new manufacturing technology.

Finally, the parts and examples highlighted in this chapter are but a few of the thousands of cases and components made with AM. The breadth and depth of the AM research is staggering now, and those interested in design for AM can learn more in recent review articles (e.g., [30,31]). Meanwhile, 3D Printing Industry,[88] Additive Manufacturing Magazine,[89] TCT Magazine,[90] and Metal AM,[91] are good sources to keep up on the latest AM software and hardware advances.

ACKNOWLEDGMENTS

The examples in this chapter would not have been possible without the contributions of several of my peers, students, and collaborators at the Pennsylvania State University, the Applied Research Laboratory, and CIMP-3D. I would especially like to thank Dr. Nick Meisel for his feedback on a draft of this chapter as well as Vincent Maranan, S. Nithin Reddy, Prof. Sanjay Joshi, Prof. Todd Palmer, Corey Dickman, Dr. Edward (Ted) Reutzel, and Griffin Jones for sharing information about the examples discussed in this chapter. I am also grateful to Jesse Boyer (Pratt & Whitney), R.T. Custer (Vortic Watch Co.), and Christian Joest, Joe Sinclair, and Wes Hart (Imperial Machine & Tool Co.) for allowing me to use images from their work.

QUESTIONS

1. Why is it important for the designers to know the AM workflow? Identify 2–3 ways that each of the 10 steps might impact the part being designed.
2. The AM workflow consists of 10 steps to get from 3D solid model to a finished part. What software tools are available to support the designer at each of these steps?
3. When converting a 3D solid model into an STL file, what impact does changing the angular deviation have on the dimensional integrity of the 3D printed part?
4. What are the advantages of the AMF and 3MF file formats compared to STL?
5. What are the most important tradeoffs that designers have to make when determining a part's build orientation?
6. Which AM processes need support structures? Which ones do not? Why?
7. If you were allowed to adapt the design of the blower wheel for AM, how would you modify the geometry so that it could be printed horizontally without all of the support structures inside the fins? What effect will this have on print time and material usage?
8. When slicing a part into layers, what is the advantage of thicker layers? What is the drawback? What is the advantage of thinner layers? What is the drawback?

[88] https://3dprintingindustry.com/
[89] https://www.additivemanufacturing.media/
[90] https://www.tctmagazine.com/
[91] https://www.metal-am.com/

9. Why does the toolpath/scan pattern vary from one layer to the next?
10. How does post-processing cost and time vary for each of the AM processes?
11. How can a part be designed to reduce the extent of post-processing required for metal parts made using laser-based powder bed fusion?
12. Topology optimization minimizes the weight of a structure for a given set of loading and boundary conditions. What happens when the part sees loading that was not considered during the optimization?
13. Why do you think the SAE Formula car upright was printed in the orientation shown? How could it be redesigned to reduce support structures in the given build orientation?
14. How does the value proposition for AM change as a company moves from Replicate with AM to Adapt for AM to Optimize with AM? Does this also apply to start-up companies?
15. What is unique about AM that enables mass customization to be cost effective?

REFERENCES

1. ASTM International, 2012, *Standard Terminology for Additive Manufacturing Technologies*, Designation F2792–12a, ASTM International, West Conshohocken, PA, 3 pgs.
2. ASTM International, 2016, *Standard Specification for Additive Manufacturing File Format (AMF) Version 1.2*, ISO/ASTM 52915:2016(E), ASTM International, West Conshohocken, PA, 15 pgs.
3. Calignano, F., 2014, "Design Optimization of Supports for Overhanging Structures in Aluminum and Titanium Alloys by Selective Laser Melting," *Materials and Design*, 64, 203–213.
4. Sames, W. J., List, F. A., Pannala, S., Dehoff, R. R. and Babu, S. S., 2016, "The Metallurgy and Processing Science of Metal Additive Manufacturing," *International Materials Reviews*, 61(5), 315–360.
5. Seifi, M., Gorelik, M., Waller, J., Hrabe, N., Shamsaei, N., Daniewicz, S. and Lewandowski, J. J., 2017, "Progress Towards Metal Additive Manufacturing Standardization to Support Qualification and Certification," *JOM*, 69(3), 439–455.
6. Everton, S. K., Hirscha, M., Stravroulakis, P., Leach, R. K. and Clare, A. T., 2016, "Review of In-Situ Process Monitoring and In-Situ Metrology for Metal Additive Manufacturing," *Materials & Design*, 95, 431–445.
7. Spears, T. G. and Gold, S. A., 2016, "In-Process Sensing in Selective Laser Melting (SLM) Additive Manufacturing," *Integrating Materials and Manufacturing Innovation*, 5(2), 16–40. http://link.springer.com/article/10.1186/s40192-016-0045-4.
8. Gao, H., Kaweesa, D. V., Moore, J. and Meisel, N. A., 2017, "Investigating the Impact of Acetone Vapor Smoothing on the Strength and Elongation of Printed ABS Parts," *JOM*, 69(3), 580–585.
9. Simpson, T. W., 2018, "Why Do the Bills Keep Coming?" *Modern Machine Shop*, https://www.mmsonline.com/blog/post/why-do-the-bills-keep-coming, 12, 38, 40.
10. Todorov, E., Spender, R., Gleeson, S., Jamshidinia, M. and Kelly, S. M., 2014, *Non-Destructive Evaluation (NDE) of Complex Metallic Additive Manufactured (AM) Structures*, Air Force Research Laboratory, Materials and Manufacturing Directorate, AFRL-RX-WP-TR-2014–0162, Wright Patterson Air Force Base, OH.

11. Waller, J. M., Parker, B. H., Hodges, K. L., Burke, E. R. and Walker, J. L., 2014, *Nondestructive Evaluation of Additive Manufacturing: State-of-the-Discipline Report*, NASA Langley Research Center, NASA/TM-2014-218560, Hampton, VA.
12. Maranan, V. J. C., 2013, "Design and Processing of an Additive Manufacturing Component," Undergraduate Honors Thesis, Engineering Science and Mechanics, Penn State University, University Park, PA.
13. Bendsøe, M. and Sigmund, O., 2003, *Topology Optimization, Theory, Methods and Applications*, Springer Verlag, New York.
14. Sigmund, O. and Maute, K., 2013, "Topology Optimization Approaches," *Structural and Multidisciplinary Optimization*, 48(6), 1031–1055.
15. Barclift, M., Joshi, S., Simpson, T. W. and Dickman, C. J., 2016, "Cost Modeling and Depreciation for Reused Powder Feedstock in Powder Bed Fusion Additive Manufacturing," *2016 Solid Freeform Fabrication Conference*, University of Texas-Austin, Austin, TX, 2007–2028.
16. Siebert, M., 2017, "Breakthrough with 3D Printed Gas Turbine Blades," *Pictures of the Future: The Magazine for Research and Innovation*, August 19, 2016, https://www.siemens.com/innovation/en/home/pictures-of-the-future/industry-and-automation/additive-manufacturing-3d-printed-gas-turbine-blades.html, Accessed 7 February 2019.
17. Marchese, K., Crane, J. and Haley, C., 2015, *3D Opportunity for the Supply Chain: Additive Manufacturing Delivers: Driving Supply Chain Transformation*, Deloitte Insights, https://www2.deloitte.com/insights/us/en/focus/3d-opportunity/additive-manufacturing-3d-printing-supply-chain-transformation.html, Accessed 7 February 2019.
18. Kasprzak, J. M., Lass, A. B. and Miller, C. E., 2017, "Development, Test, and Evaluation of Additively Manufactured Flight Critical Aircraft Components," *AHS International 73rd Annual Forum & Technology Display*, Fort Worth, TX, AHS International.
19. Salois, G., 2016, "Operation Warfighter," *FF Journal*, 13(8), 32–49.
20. Laverne, F., Segonds, F., Anwer, N. and Le Coq, M., 2015, "Assembly Based Methods to Support Product Innovation in Design for Additive Manufacturing: An Exploratory Case Study," *ASME Journal of Mechanical Design*, 137(12), 121701 (8 pgs).
21. Brooks, H. and Brigden, K., 2016, "Design of Conformal Cooling Layers with Self-Supporting Lattices for Additively Manufactured Tooling," *Additive Manufacturing*, 11(1), 16–22.
22. Kantanareddy, S. N. R., Roh, B. M., Simpson, T. W., Joshi, S., Dickman, C. J. and Lehtihet, E. A., 2016, "Saving Weight with Metallic Lattice Structures: Design Challenges with a Real-World Example," *2016 Solid Freeform Fabrication Conference*, Austin, TX, University of Texas-Austin, 2139–2154, http://sffsymposium.engr.utexas.edu/sites/default/files/2016/171-Kantareddy.pdf, Accessed 23 August 2019.
23. Yang, S., Tang, Y. and Zhao, Y. F., 2015, "A New Part Consolidation Method to Embrace the Design Freedom of Additive Manufacturing," *Journal of Manufacturing Processes*, 20, 444–449.
24. Schmelzle, J., Kline, E. V., Dickman, C. J., Reutzel, E. W., Jones, G. and Simpson, T. W., 2015, "(Re)Designing for Part Consolidation: Understanding the Challenges of Metal Additive Manufacturing," *ASME Journal of Mechanical Design*, 137(11), 111404 (12 pgs).
25. Orme, M. E., Gshweitl, M., Ferrari, M., Madera, I. and Mouriaux, F., 2017, "Designing for Additive Manufacturing: Lightweighting through Topology Optimization Enables Lunar Spacecraft," *ASME Journal of Mechanical Design*, 139(10), 100905 (6 pgs).
26. Reddy, S. N., Maranan, V., Simpson, T. W., Palmer, T. and Dickman, C. J., 2016, "Application of Topology Optimization and Design for Additive Manufacturing Guidelines on an Automotive Component," *ASME Design Engineering Technical Conferences: Design Automation Conference*, ASME, Charlotte, NC, Paper No. IDETC2016–59719.

27. Petrick, I. J. and Simpson, T. W., 2013, "3D Printing Disrupts Manufacturing: How Economies of One Create New Rules of Competition," *Research-Technology Management*, 56(6), 12–16.
28. Thomas, D. S. and Gilbert, S. W., 2014, *Costs and Cost Effectiveness of Additive Manufacturing*, National Institute of Standards and Technology, NIST Special Publication 1176, Gaithersburg, MD, https://nvlpubs.nist.gov/nistpubs/SpecialPublications/NIST.SP.1176.pdf.
29. Lindemann, C. and Koch, R., 2016, "Cost Efficient Design and Planning for Additive Manufacturing Technologies," *27th Annual International Solid Freeform Fabrication Symposium*, Austin, TX, University of Texas at Austin, 93–112.
30. Thompson, M. K., Moroni, G., Vaneker, T., Fadel, G., Campbell, R. I., Gibson, I., Bernard, A., Schulz, J., Graf, P., Ahuja, B. and Martina, F., 2016, "Design for Additive Manufacturing: Trends, Opportunities, Considerations, and Constraints," *CIRP Annals: Manufacturing Technology*, 65, 737–760.
31. Gao, W., Zhang, Y., Ramanujan, D., Ramani, K., Chen, Y., Williams, C. B., Wang, C. C. L., Shin, Y. C., Zhang, S. and Zavattien, P. D., 2015, "The Status, Challenges, and Future of Additive Manufacturing in Engineering," *Computer-Aided Design*, 69(12), 65–89.

8 Multifunctional Printing
Incorporating Electronics into 3D Parts Made by Additive Manufacturing

Dishit Paresh Parekh, Denis Cormier, and Michael D. Dickey

CONTENTS

8.1 INTRODUCTION

This chapter discusses the use of additive manufacturing (AM) for creating three-dimensional (3D) objects that contain electronic components.

According to the ASTM definition,[1] AM or rapid prototyping (RP) is the process of joining materials to make objects from 3D model data, usually layer upon layer. More colloquially, AM and RP are referred to as *3D printing*. In addition to layer-based AM processes, there are *direct-writing* techniques (DW) in which functional and/or structural materials are precisely deposited onto a substrate in digitally defined locations.[2,3] DW techniques are most often used to print electronic elements onto the surface of an existing part, although there is growing interest in hybrid techniques that allow printed electronics within the bulk of additively manufactured parts. In this chapter, we use the term AM to describe the general field of layer-based manufacturing, and we use DW when describing methods used to deposit functional electronic materials.[4]

Although many AM techniques exist, they all employ the same basic steps. These steps are as follows: (1) create a computer model of the design; (2) slice the model into thin cross-sectional layers; (3) construct the model physically one layer atop another; and (4) clean and finish the part, as shown in Figure 8.1.

Most commercial AM machines are designed to create objects from a specific class of material (typically either polymers,[5] ceramics, or metals). There are exceptions, such as tools that print polymers that contain ceramic or metal particles. Constructing 3D parts with electronic functionality often requires integrating multiple types of materials, which causes fabrication challenges for AM processing. We call these processes *multifunctional 3D printing* because of the need to incorporate multiple materials to achieve multiple functions (typically a synergistic combination of electrical functionality with either form factor or mechanical properties).

The approach for integrating electronics into a 3D object depends entirely on the desired function and the level of sophistication of the electronics involved. In this

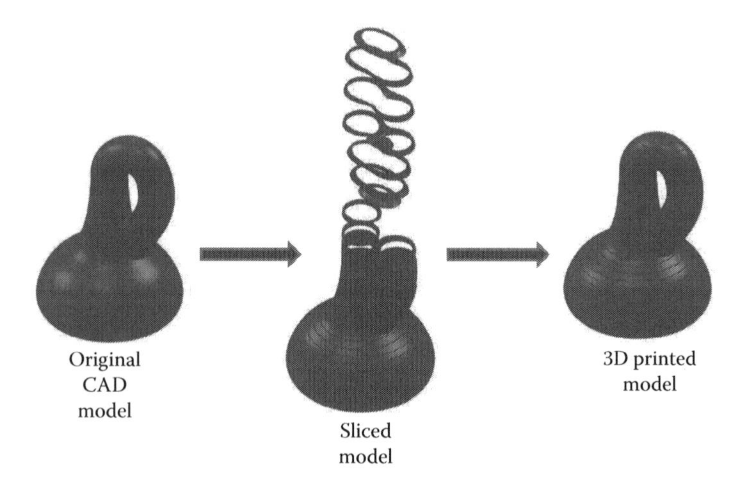

FIGURE 8.1 Example of an additive process. A 3D CAD model is broken up into thin slices, which are then printed in an additive manufacturing process in a layer-by-layer fashion to create a 3D printed part.

chapter, we categorize three broad approaches that utilize AM for integrating electronics with 3D printed parts: (1) hybrid chip-insertion approaches, (2) surface DW approaches, and (3) freeform multi-material 3D printing approaches. Most AM techniques are not able to print electronic materials with satisfactory properties (e.g., purity, crystallinity, doping, and charge mobility) at high enough resolution to compete with conventional electronic devices such as integrated circuits (ICs), light emitting diodes (LEDs), transistors, and solar cells among others. Thus, sophisticated electronics are often fabricated separately and then integrated into a 3D printed part by robot, by hand, or by some other transfer technique. This approach (Category 1) can provide sophisticated functionality, but does not fall within the spirit of AM and requires methods to connect the individual components. DW techniques are capable of 2D printing moderate quality, passive electronic structures (resistors, inductors, capacitors). In principle, these processes (Category 2) could print electronic components as layers in 3D objects, but there are significant challenges with using these 2D techniques in 3D space. It is not the intent of this chapter to review DW techniques since excellent resources exist on this topic.[2,6–12] Instead, we will introduce DW techniques within the context of 3D printing and highlight the challenges. Finally, we review some emerging techniques for printing electronic materials in a manner that is truly 3D (Category 3).

To appreciate the challenges associated with multifunctional printing, we first briefly review what *electronics* means. We discuss the merits of electronics made by AM, the reasons for incorporating electronics into 3D objects, and the new possibilities for electronics made possible by AM. To help motivate the challenges of integrating electronics into 3D parts, we describe briefly the conventional methods for making sophisticated electronics. We then discuss printed electronics, which offer inexpensive approaches for creating low-grade electronics. DW techniques are a subset of printed electronic strategies and the only ones that align conceptually with AM. Thus, we focus most of our discussion on these techniques and discuss challenges of using these techniques within the context of AM. We categorize three general strategies for integrating electronics into 3D objects and spend the remainder of the chapter discussing these strategies. We discuss the capabilities of each technique for depositing materials to achieve the final desired component with appropriate feature sizes and resolution along with the required electronic functionality. Finally, there are many resources on DW and AM, including this book. For this reason, several sections are intentionally concise and we point the reader to appropriate references in these instances. In contrast, there has been very little work in the academic literature on integrating electronics into 3D printed objects. A theme of this chapter is elucidating the challenges and opportunities associated with the various approaches for integrating electronics into 3D objects constructed by AM to inspire and guide future work on this topic.

8.2 WHAT ARE ELECTRONICS AND WHY DO WE NEED THEM IN 3D OBJECTS?

Electronics are devices that provide one or more electrical functions (e.g., communications, sensing, computation, memory) via the use of circuits and electrical components. Examples of electronic components include the fundamental passive circuit elements (resistor, inductor, capacitor, memristor), conductors (wires, interconnects, antennas), insulators, transistors, sensors, and microelectromechanical systems (MEMS) devices (useful for switching, actuating, and sensing).

Electronics are ubiquitous in modern society in the form of cell phones and computers, and are incorporated in many appliances, cars, and toys, to name but a few examples. The electronics industry is driven by products that are smaller, thinner, lighter, faster, and more cost-effective. Because of the importance of electronics in our society and the growing interest in the *Internet of things* (IoT) (i.e., the notion that objects of interest will someday be interconnected wirelessly), it is natural to understand there is—and will continue to be—a great demand for electronics in 3D printed objects. AM is a growing market in every manufacturing sector with a global AM market of $1.8 billion in 2012 that is expected to reach $3.5 billion by 2017[13] and $7 billion by 2025[14]; the integration of electronics will hence only add more value.

AM has several capabilities that are, in principle, well-suited for building electronic components. First, AM generates minimal waste and allows for RP of new designs. Second, it can create complex form factors, spanning structures, out-of-plane geometries, and embedded electronics. Hence, AM enables the creation of intricate and conformal electronics that are structurally integrated into a manufactured part. This attribute may minimize cable interconnects and redundant electronics packaging, resulting of a reduction of mass and assembly complexity in the final electronic component. It may also allow for the creation of entirely new objects containing electronics[15] that simply are not possible to form using 2D approaches.

More than a half-century has gone into the art of building sophisticated components for computers and electronics. It is sensible to consider these mature approaches as a way to integrate electronics into 3D objects. In the next section, we describe these methods briefly to illustrate why it is not straightforward to use these advanced techniques for integrating electronics into 3D objects.

8.3 CONVENTIONAL FABRICATION OF ELECTRONICS

Since the invention of ICs in 1959,[16] the electronics industry has focused on miniaturization, which has faithfully followed Moore's Law.[17,18] Miniaturization lowers the cost per unit (e.g., per number of transistors) and therefore allows the consumer more buying power. Miniaturization also allows devices, such as cell phones, to be possible by fitting large amounts of computing power into a small form factor.

ICs consist of conductors, resistors, insulators, and semiconductors integrated and patterned spatially into complex circuits and devices on a planar substrate. The techniques to build these devices are inherently 2D and involve a number of steps such as deposition (e.g., physical vapor deposition, sputtering), removal (e.g., etching, sputtering), patterning (e.g., photolithography), and modification (e.g., ion implantation for doping) of electronic materials. The quest for miniaturization corresponds with an increase in complexity and sophistication of the fabrication processes used to build ICs. There are several excellent monographs and reviews on fabrication methods utilized to create electronics.[19–26]

Photolithography is the cornerstone process utilized to pattern the components in a computer chip. AM is effectively a patterning technique, and thus, it is prudent to

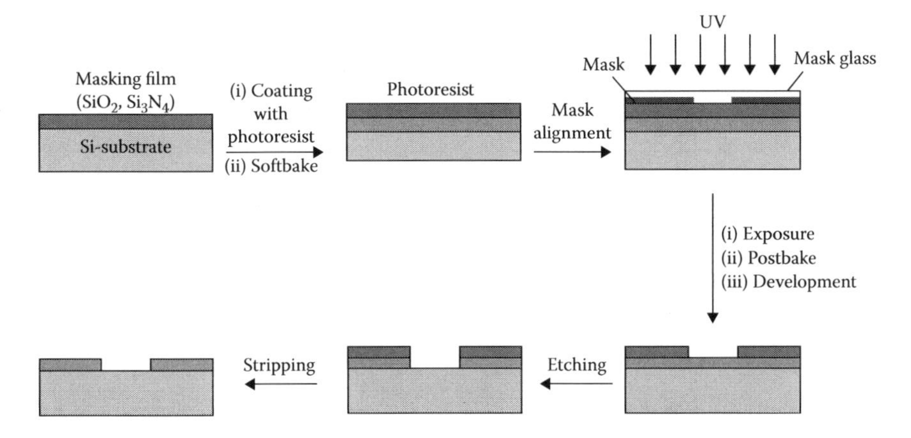

FIGURE 8.2 A schematic of photolithography, which is a conventional subtractive manufacturing process used to make ICs. (Adapted from Optical Issues in Photolithography. *Connexions.* http://cnx.org/content/m25448/latest/.)

briefly discuss photolithography as a basis for comparison. An example of the photolithographic process is shown in Figure 8.2.[27] Photolithography utilizes patterns of light to chemically modify the solubility of thin polymer films coated on a surface. Photolithography is an inherently 2D process because the light used to expose the polymer has a single focal plane. This limitation, along with the need to coat, expose, and remove polymer, makes photolithography essentially incompatible with AM. Although academic research provides many unconventional approaches[8,25,28–35] to pattern electronic materials to overcome some of the limitations of photolithography, it still remains the backbone of the semiconductor industry.

There are several reasons conventional electronic fabrication techniques are not compatible with AM. State-of-the-art electronics (e.g., ICs, hard drives) rely critically on ultra-pure materials, well-defined interfaces, and crystalline semiconductors with precise doping concentration. To date, AM has not been capable of meeting any of these requirements. In addition, these sophisticated planar processes are carried out in a clean room environment on flat substrates (typically precisely engineered silicon wafers), whereas AM is often done under ambient conditions on complex substrates. Also, most AM techniques produce features with resolutions (of ~10–100 μm) that are nearly 3–4 orders of magnitude larger than those produced by photolithography. For these reasons, sophisticated electronic components are typically built separately and then placed onto or into 3D objects; we refer to this approach as *Category 1*. For example, cell phones are currently built in this modular manner in which the case, display, battery, and electronics are all built separately and assembled. There is, however, a need to interconnect these electronic components to other components in the object. Passive components (capacitors, resistors, and inductors) are commonly connected to printed circuit boards (PCBs) by soldering using surface mount technologies (SMTs) or are embedded into multilayered low-temperature co-fireable ceramic packages (LTCC).[36] In the case of a cell phone, these connections can be made using wires, solder, and flexible circuit boards in the void space of the phone, but AM has

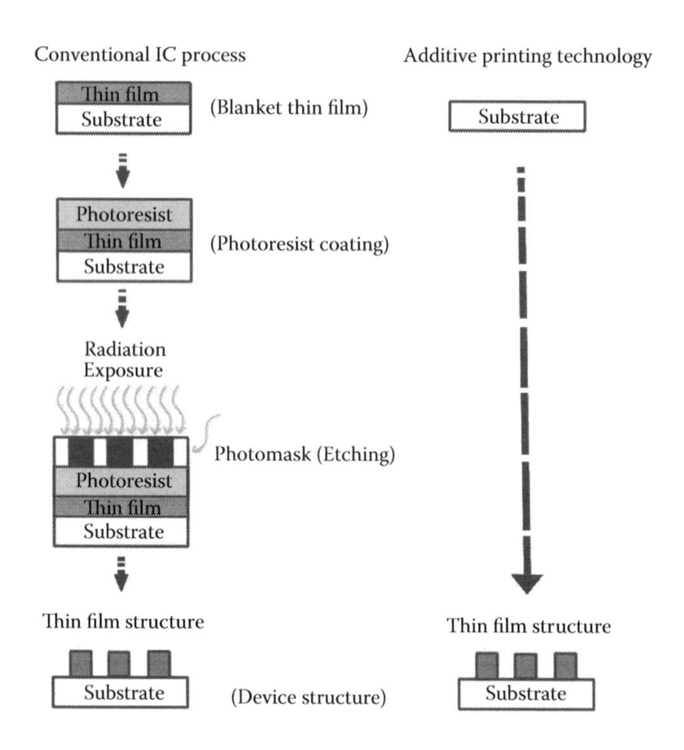

FIGURE 8.3 Process comparison of subtractive IC processing and additive printing approach for thin film device development. (Adapted from Joshi, P.C. et al., Direct digital additive manufacturing technologies: Path toward hybrid integration, In *Future Instrum. Int. Workshop*, 1–4, 2012.)

the potential to directly define these connections in a continuous process to minimize assembly and create more complex, customizable parts.

In summary, conventional electronics processing uses expensive, sophisticated equipment in clean room environments to pattern ultra-pure materials onto planar substrates of a limited geometry using numerous processing steps including those requiring large temperature excursions. There is a need for additional tools to directly and inexpensively print electronic materials by eliminating the masking and etching steps, even if there is a trade-off in quality relative to the state of the art.

The ability to additively print electronics reduces the material waste, energy consumption,[37] and processing time and steps relative to IC processing, as shown in Figure 8.3.[38] The next section describes this class of printed electronics. These inherently 2D methods have features that may be useful for integrating electronics into 3D objects.

8.4 PRINTED ELECTRONICS

Printed electronics are formed by directly dispensing or patterning functional inks to define electronic components onto a wide range of substrates. The main aim of printed electronics is to reduce the manufacturing cost of electronics per unit area using lessexpensive, all-additive printing methods. Low-cost printed electronics[39,40] have gained a great deal of interest over the past 10 years because of their promise to greatly reduce the cost of many electronic applications and the ability to print on larger, unconventional surfaces (e.g., flexible substrates, large displays). The Flexible Electronics Forecast report projects that the market size for printed electronics will grow from $2.2 billion in 2011 to $6.5 billion in 2017 and to $44.2 billion in 2021.[41]

Printed electronics present a trade-off between cost and performance relative to conventional electronics, as described in Table 8.1.[42] There are several challenges associated with printed electronics. First, the materials must start in the form of an ink that can be deposited. Once deposited, most of these inks need some post-processing, such as thermal treatments, to obtain the final desired properties, which limits the substrate onto which the features are printed and creates additional processing steps. The deposition techniques are much lower in resolution than photolithography (typically tens of microns or larger). There is a general trade-off between the feature size and throughput of printed electronic processes (Figure 8.4).[43]

Despite the inferior performance of printed electronics relative to conventional electronics, there are several applications for low-cost printed electronics including radio frequency identification (RFID) tags,[44] chemical and electronic sensors,[45] displays,[46] smart cards,[47] packaging,[48,49] and PCBs/keypads.[50,51] The common element among all these applications is the fact that they do not need either ultra-fast circuitry or ultra-dense circuitry such as that in ICs. The selection of the appropriate printing method is determined by the physical specifications and resolution of the printed materials that comprise the electronic devices, the substrate size and composition, the desired throughput, and economic and technical considerations of the final printed products. Two of the most promising printing technologies for fabricating low-cost printed electronics include inkjet printing and gravure printing. Inkjet printing[52–55] is a well-known technology that makes use of multiple droplet dispensers that deposit individual drops to form patterns on a substrate. Gravure printing[56–58] uses a cylinder featuring an etched pattern, onto which the ink is deposited. Rolling the cylinder over a substrate transfers the pattern from the cylinder to the substrate in a manner that is compatible with high throughput roll-to-roll processing. There are a number of other methods for printing electronics including screen printing,[59,60]

TABLE 8.1
Printed and Conventional Electronics as Complementary Technologies

Printed Electronics	Conventional Electronics
Long switching times	Extremely short switching times
Low integration density	Extremely high integration density
Large printing area	Small printing area
Process compatible with flexible substrates	Process compatible with rigid substrates
Simple and cheap fabrication process	Sophisticated and expensive fabrication process

Source: Wikipedia. File:ComplementaryTechnologies.png. *Wikipedia Free Encycl.* http://en.wikipedia.org/wiki/File:ComplementaryTechnologies.png.

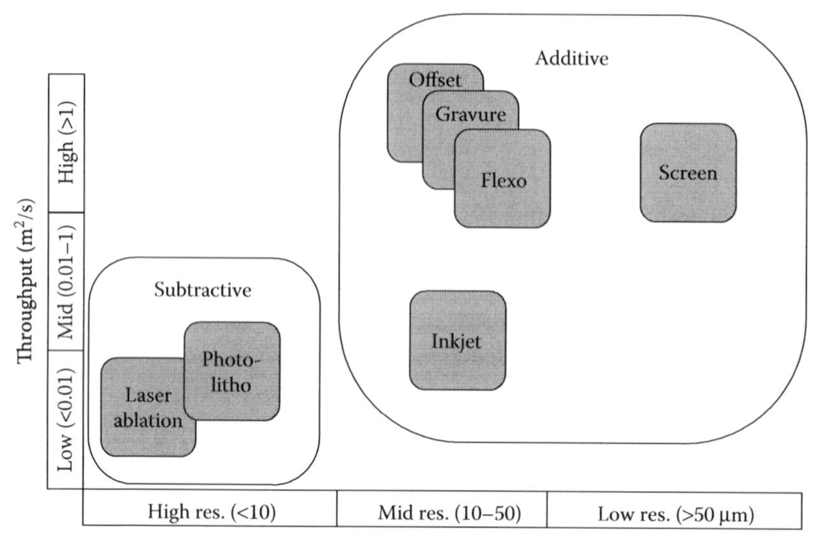

FIGURE 8.4 Comparison of various printing technologies. (Adapted from Chang, J. et al. *2012 IEEE 55th Int. Midwest Symp. Circuits Syst.*, 582–585, 2012.)

flexography,[61,62] and offset lithography[63] that are utilized to fabricate devices such as solar cells, organic light-emitting diodes (OLEDs), and RFIDs.[39]

It is possible to print both organic and inorganic materials as inks to form conductors, semiconductors, dielectrics, or insulators. These ink materials must be available as either a liquid, solution, dispersion, or suspension.[64]

Organic ink materials include conjugated polymers[65] or small molecules that possess conducting, semiconducting, electroluminescent, photovoltaic, and other properties that can be exploited in printed electronics. These organics[66–69] are commercially available in different formulations and have been deposited using inkjet printing,[70] gravure printing,[71] flexography,[72] screen printing,[73] and offset lithography.[74] Organic electronics are used commercially in OLEDs and organic molecules are also used in liquid crystal displays.[75] There are several excellent reviews and books on organic electronics.[76–83] Although organic materials offer appealing properties such as mechanical flexibility and tunability of properties via chemical modifications (e.g., light color in OLEDs),[84] they offer poor charge carrier mobility[65] and hence the electrical performance lags behind that of conventional siliconbased electronics. In addition, organic materials are prone to oxidation,[85] which limits the longevity of devices built using these materials.

Inorganic ink materials such as dispersions of metallic or semiconductor micro- and nano-particles including silver and silicon[86] may also be printed. The silver and gold particles can be deposited using, for example, inkjet printing,[87,88] flexography,[89] and offset lithography.[90] After deposition, these particles need to be sintered to form stable conductive structures using either conventional heating (often at temperatures >200°C[91]), laser sintering,[92,93] flash exposure to light, or microwave sintering.[94] The resulting sintered structures generally cannot match the electrical properties of equivalent bulk materials, but can be deposited relatively easily using these techniques.

There is also a growing interest in printed carbon nanotubes[95–99] and graphene[100–104] because of their remarkable electrical properties. These materials can serve as conductors or semiconductors, but there are challenges with making, purifying, and patterning these materials on a large scale. There are several excellent reviews and books on carbon-based electronic materials.[98,105–112]

Many of the techniques introduced in this section—such as gravure printing—are not directly compatible with the AM of 3D objects because they are not *digital* (i.e., they cannot dispense inks into arbitrary patterns on demand). In addition, most of the methods discussed in this section are intended only for 2D substrates such as plastic films, foils, and paper and cannot be used for AM on complex, non-planar, or 3D substrates. Thus, we limit our focus to only the subset of printed electronics techniques that are digital (i.e., DW techniques, such as inkjet printing) because of their potential to be compatible with 3D printing.

8.5 DIRECT-WRITING OF ELECTRONICS

DW techniques[2,3,12,113] are a subset of printed electronics processes to precisely deposit functional and/or structural materials on to a substrate in *digitally defined* locations to form simple 2D or complex conformal (3D) structures. These locations are determined by a computer-controlled translation stage, which moves a pattern-generating device, for example, ink deposition nozzle or laser writing optics, to create materials with controlled architecture and composition. DW technologies complement the conventional manufacturing techniques such as photolithography for applications that need rapid turnaround and/or pattern iteration, for reducing the environmental impact,[38] for conformal patterning and modeling intricate components, circuits, and sub-assemblies.[114]

DW methods can be used for both conductive and non-conductive materials[2,7,10,12,115–119] including many of the materials introduced in the previous section. DW does not require expensive tooling, dies, or lithographic masks required in conventional electronics fabrication making it a low-cost, consumer-friendly process. In general, DW techniques have the following attributes:

1. The resolution of the features printed by DW ranges from ~250 to 0.1 μm.
2. DW techniques are compatible with various classes of materials including metals, ceramics, polymers, and biological materials such as cells. These materials can be used as powders, slurries, or suspensions depending on the technique employed for printing.
3. DW can print directly passive components such as conductors, resistors, inductors, and dielectrics.
4. DW can increase the reliability of electronic components since no soldering is needed to connect the circuitry.
5. DW can print directly on a variety of substrates including flexible surfaces such as paper and plastics.

6. Since DW deposits materials on-demand, it produces minimal waste and has a small environmental footprint. It can also achieve dramatic weight, cost, space, and inventory savings.
7. DW requires fewer steps compared to conventional subtractive processing and thereby reduces the prototyping time, which is helpful in products with short life-cycle (e.g., electronics).
8. DW allows for front-end inventiveness (due to reduction in the turnaround time) and back-end processing for design revisions thus reducing the time to market of parts from weeks to days.
9. Specialty parts can be built *on the fly* in small volumes without the need for mass production setup and capital investments.

Although DW techniques have many desirable attributes, there exist some key challenges when using it to print electronics into or onto 3D objects.

8.6 WHY DOES DW NOT TRANSLATE READILY TO 3D PRINTING?

AM is based on depositing and stacking materials one layer at a time. It therefore seems intuitive that DW techniques—which have been developed to printing electronics in 2D—could be utilized for depositing layers of electronic materials as part of the AM process of building a 3D device. DW processes have several limitations that make it challenging to translate into 3D printing electronics:

1. Electronics are composed of multiple materials and therefore would necessitate multiple nozzles for dispensing. Each material may have different processing requirements.
2. The minimum feature size of DW is about 4–5 orders of magnitude poorer than photolithography (c.f. Figure 8.4)[43] and therefore cannot produce circuitry as sophisticated as conventional electronics.
3. DW requires fluidic inks that can be dispensed with controlled viscosity, surface tension, and solid content. Dispersions of inorganic or metal inks have the tendency to clog creating non-uniform depositions on the substrate.
4. DW inks often contain solvents and additives that require time to dry, which can slow processing speeds. These solvents can also interact with the underlying substrate in an undesirable manner and the ink needs to be tuned to properly wet the underlying substrate. Additives can improve the printability of inks, but can also affect the electronic functionality of the printed structures.[66]
5. DW inks often require thermal post-processing for attaining the desired electrical and mechanical properties, which limits the variety of substrates onto which the inks may be printed and adds processing steps.
6. Adhesion of DW materials to the AM substrates and external circuitry is decided by factors such as substrate swelling, surface energy and wettability, roughness, and chemical interactions between them.
7. The thickness of layers of *electronic materials* patterned by DW is often much thinner than the layers of polymer deposited by AM, which leads to

incommensurate length scales and the need to do multiple layers of printing by DW to equal one by AM.

8. DW techniques are generally poor at making self-supporting or spanning structures. Features printed by DW usually need an underlying substrate for support.
9. DW techniques are intended primarily for smooth 2D, planar substrates, whereas 3D parts produced by AM are often rough.
10. Multifunctional materials often have different physical properties. For example, materials with different thermal expansion coefficients can lead to stress and ultimately delamination.
11. The feature quality obtained after DW printing multiple layers of a single material in 3D can suffer due to factors such as unstable print head temperatures or possible smearing, which can occur due to unsolidified previous layers.

8.7 CATEGORIES FOR GENERATING ELECTRONICS IN 3D OBJECTS

Up to this point, the chapter has focused on the reasons why existing strategies for fabricating electronics do not easily translate to 3D printing. These challenges motivate the categorization of approaches for integrating electronics in 3D parts into three categories as follows:

1. *Hybrid chip insertion approach:* Section 8.3 established that conventional electronic processing is capable of fabricating complex circuitry, but in a manner that is not compatible with AM. Thus, one approach is to create the electronic components separately using high-throughput, high-resolution processing and then to insert the resulting components (which are usually *off-the-shelf* and commercially available) onto or into a 3D printed object. This hybrid approach is disruptive to the printing process and breaks away from the spirit of 3D printing, but it has the advantage of providing sophisticated circuitry into 3D parts. There are, however, two challenges: (1) How does one transfer the electronics? and (2) How does one connect the electronic components?
2. *Surface DW approach:* This technique uses DW methods to print electronic components directly onto the surface of 3D objects created by AM. Typically, approaches in this category have additional challenges that go beyond those encountered while writing onto 2D substrates. It also requires the ability to print multiple (two or more) materials in parallel or sequence.
3. *Freeform multi-material 3D printing approach:* These techniques can produce true, self-supporting 3D parts (e.g., arches, out-of-plane structures, spanning structures) on their own using DW, but are often used in conjunction with other techniques to make useful parts because of the limited materials pallet that exists today.

We first briefly discuss hybrid chip insertion methods (Category 1) and then organize the remainder of this chapter by relevant DW techniques (Categories 2 and 3).

8.7.1 Hybrid Chip Insertion Approach (Category 1)

8.7.1.1 Pick and Place

Conventional electronics (e.g., ICs) provide sophisticated functionality using fabrication techniques that are not compatible with AM processing. One strategy for integrating electronics is to simply take off-the-shelf parts and place them into a 3D printed part. Although this approach is disruptive to the printing process and the least scalable approach, it is the simplest. There are few examples of academic research that focus on this technique.[15,120–125] The main challenge is finding ways to connect individual components (e.g., batteries, LEDs, CPUs, memory elements) within a 3D printed device beyond the conventional means of wire bonding and soldering. The approaches of Category 2 have the potential to address these challenges using processes that are compatible with AM.

8.7.1.2 Ultrasonic Consolidation for Embedding Electronic Structures

Ultrasonic consolidation (UC)[126–128] deposits metallic foils using ultrasonic energy to make metallic parts that could potentially contain electronic structures. UC[129] bonds each foil layer to the substrate (or previously deposited layers) using an oscillating sonotrode as shown in Figure 8.5[130,131] that applies heat, pressure, and friction to produce solid-state bonds. After bonding, an integrated three-axis CNC milling machine may be used to produce the desired contours for each layer.

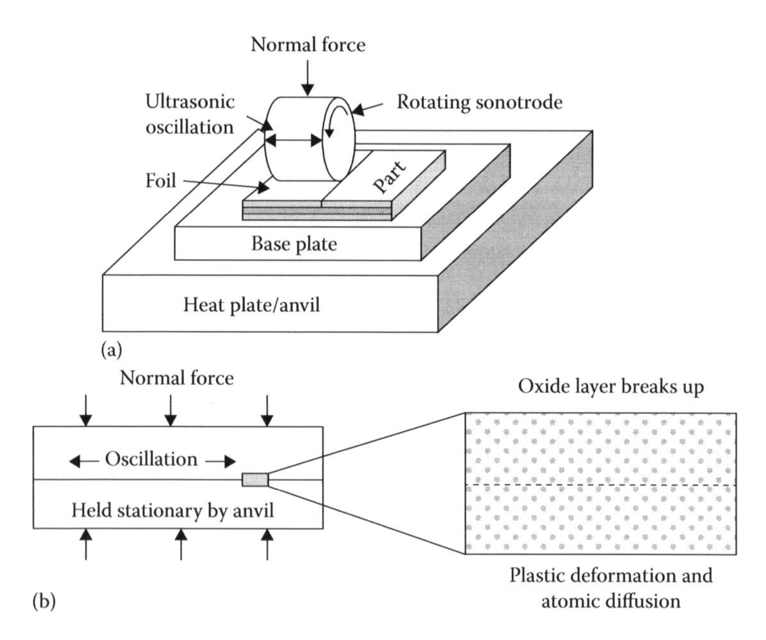

FIGURE 8.5 Schematic showing the UC process. (a, b) UC works by vibrating a foil material against a base substrate with a sonotrode layer by layer. (Adapted from Obielodan, J.O. et al., *J. Mater. Process Technol.*, 211, 988–995, 2011.)

Most metal parts fabricated by conventional AM techniques use powder bed fusion (PBF) processes, which sinter metal powders to form a coherent solid structure.[132] These processes require large processing temperatures that are not compatible with most electronic components. UC can make complex metal parts with high-dimensional accuracy and surface finish at low temperatures without the need for high temperature sintering. The UC process can be performed at various temperatures ranging from room temperature to ~200°C. Generally, a temperature of 150°C is used, which is a relatively low temperature compared to other metal fabrication processes.[132] In addition, PBF processes are often done in inert environments to avoid powder explosions or undesired oxidative reactions[132]; UC does not require an inert enclosure. Since UC does not involve melting/sintering; the dimensional errors due to shrinkage, residual stresses, and distortion that are typically caused by high temperature processing[133] are less pronounced than with some other processes.

Electronic components can be manufactured directly into a solid metal structure (using DW)[126] and subsequently completely enclosed to form an embedded structure (using UC).[134] Although these two processes are usually done separately, it could, in principle, be possible to integrate a DW deposition head onto a UC apparatus, which would eliminate the need to move the part between apparatuses. UC can construct various metallic objects with complex internal passageways, objects made up of multiple materials (engineering materials such as SiC fibers; Fe; Ni; Cu; and dissimilar combinations such as Al/brass, Al/stainless steel, and Al/Ni),[133,135] and objects integrated with wiring, fiber optics, and sensors.

8.7.1.3 Integrating Circuitry Using Transfer Printing

Transfer printing techniques are motivated by the desire to independently fabricate electronics using the most suitable conventional fabrication techniques on the most appropriate substrate (called a *donor* substrate), transfer them temporarily onto a elastomeric stamp (this stamp is sometimes called a *transfer* substrate—TS), and then later transfer them onto a less conventional *receiver* or *device* substrate (DS—i.e., the substrate where the electronics will ultimately reside). This technique transfers only the thin active layer of electronic components, which differs from the pick-and-place (Section 8.7.1.1) transfer of electronics in relatively bulky, finished packaging.

Figure 8.6[136] describes the transfer printing process. The transferred components are called the *printed layer* (PL); the PL is not actually *printed*; rather, it is fabricated using conventional fabrication techniques on the donor substrate. For example, electronic components (PL) may be created on a silicon wafer (a donor substrate) and then later transferred as a thin film onto a plastic sheet (DS) using an elastomeric stamp (TS). To our knowledge, little work has been done using transfer techniques to deposit electronics onto or into 3D objects, but it is possible in principle. Many of the transfer techniques rely on fabricating circuits on donor substrates that allow the features to be released to a temporary polymer TS.[136–140] The transfer printing process is relatively simple and thus compatible with many different materials.

There are a number of methods to release the patterned features that reside as a thin, top layer on the donor substrate including undercutting via etching. The transfer of a PL from a TS to a DS is done via conformal contact, which is

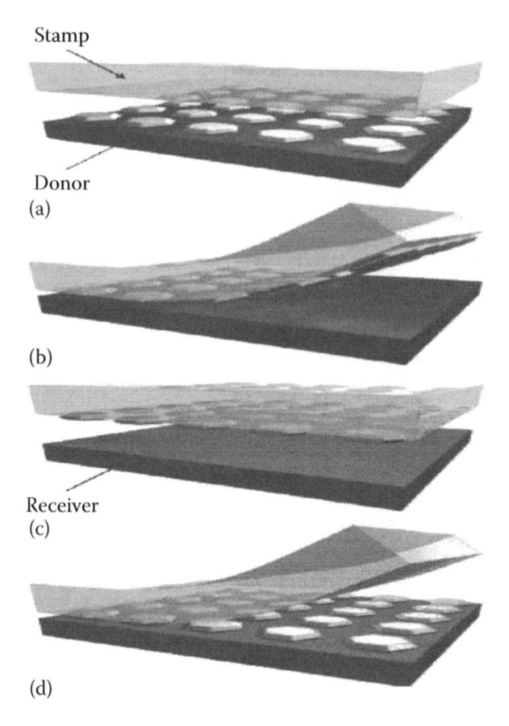

FIGURE 8.6 Schematic illustration of the generic process flow for transfer printing solid objects. (a) Laminating a stamp (TS) against a donor substrate and then quickly peeling it away, (b) pulling the microstructures (PL) from the donor substrate onto the stamp (TS), (c) contacting the stamp (TS) to another substrate (DS), and then (d) slowly peeling it away transfers the microstructures (PL) from the stamp (TS) to the receiver (DS). (Adapted from Meitl, M.A. et al., *Nat. Mater.*, 5, 33–38, 2006.)

driven by generalized adhesion forces that are typically dominated by van der Waals interactions[141–143] but may also be controlled by the viscoelastic properties of the TS.

Transferring a PL from one surface to another requires differential adhesion. If the work of adhesion (W_A) at the DS/PL interface is larger than the TS/PL interface, the printable layer will remain on the device substrate upon removing the TS. However, if the work of cohesion (W_C) of the PL is less than the work of adhesion for the printable layer with both substrates, that is, W_C(PL)<W_A(DS/PL) and W_A(TS/PL), then the printable layer will be only partially transferred. Furthermore, if during the transfer printing process, the transfer substrate makes contact with the device substrate, then, in addition, the work of adhesion between the substrates must be less than the work of cohesion of both the substrates.

A practical implementation of the above requirements can be established using chemical surface treatments. Adhesion between two surfaces can also be controlled kinetically,[136] that is, the peeling rate of the TS from the DS impacts the adhesion of the PL owing to the viscoelastic behavior of the TS. A limitation of such an approach is the possible pattern distortion from using soft stamp materials. This method is similar to the parallel pick-andplace technology[144–147] that is compatible with extremely thin, fragile device components, originally developed for manipulating individual silicon transistors.

8.7.1.4 Laser-Assisted Transfer

Transfer printing of electronic materials from a donor film to the device substrate can also be carried out using laser-assisted transfer. Transfer printing can be assisted by high power lasers,[148–152] which serve as a precision thermal source to locally heat regions of the substrate to high temperatures. The donor film is comprised of several components—a laser transparent substrate (e.g., polystyrene, polyvinyl acetate, or polyethylene), a thin polymeric film, which has a high optical coefficient of absorption to the wavelength of the laser, and a thin layer of metals (e.g., nickel, gold, silver, among others) deposited by physical vapor deposition.

A schematic of this transfer method is shown in Figure 8.7.[148] During exposure to a laser, the absorptive layer is vaporized and propels the metal composite toward the target substrate. When the film reaches the substrate, it forms a cold weld or pressure bond.

This method can transfer a variety of electronic-grade materials for microelectronics manufacturing such as passive components, conductors, and batteries by supplying materials on a ribbon or donor film. The feature sizes are of the order of 25 μm due to the small spot size and high energy possible with commercial laser systems. An appeal of this approach is that it delivers quality electronic materials onto a target substrate without the need to heat the substrate. It does, however, require expensive optics and control systems for maintaining the appropriate gap between the ribbon and the target substrate.

8.7.1.5 Connecting the Transferred Components

The techniques within this section place or transfer electronic components onto a target substrate, but these components often still need to be interconnected.

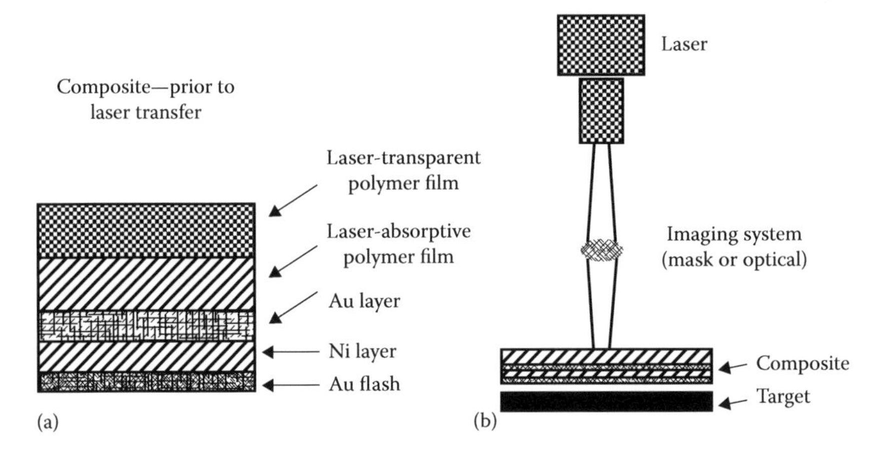

FIGURE 8.7 (a, b) Schematic of laser-assisted transfer technology. (Adapted from Zhang, J. et al., In *Direct-Write Technol. Rapid Prototyp.*, Piqué, A., 33–54, Academic Press, 2002.)

The next section discusses DW techniques that, in principle, could be used to connect discrete components electrically.

8.7.2 Surface DW Approach (Category 2)

One approach for building electronics into a 3D object is to use AM to fabricate parts (typically composed of plastic or ceramic) that have surfaces onto which the electronics circuits may be printed or patterned using DW techniques. This combination of AM approaches to form multifunctional 3D structures of arbitrary and complex form with directly integrated printed electronics offers unique functionality while addressing the cost/performance demands of the manufacturing technology. There are in general two categories of DW techniques: (1) those that can deposit functional materials in a single deposition step, and (2) those that deposit materials that have to be subsequently processed (at low/high temperatures) to induce controlled and reproducible functionality. DW techniques are usually classified based on the deposition mechanism. Figure 8.8 shows the classification of the various DW techniques.[2]

We only discuss DW techniques that have been utilized for electronics. Each technique differs in resolution, writing speed, 3D and multi-material capabilities, operational environment (gas requirement, pressure, and temperature), and what kind of final structures can be built. The common feature to all techniques is their dependence on high-quality starting materials, with specially tailored chemistries and/or physical properties such as viscosity, density, rheology, surface tension/wetting properties, mean particle geometry, size and distribution, coefficient of thermal expansion, solids loading, and porosity. The selection of materials must be based on additional chemical and process factors such as solvent and binder removal, the reactivity, chemical compatibility, stress development, sintering rate, and the direct-write tool used.

The starting materials termed as *pastes* or *inks* may consist of combinations of powders, nanopowders, flakes, surface coatings, organic precursors, binders, vehicles, solvents, dispersants, and surfactants. These materials can serve as conductors[153,154]

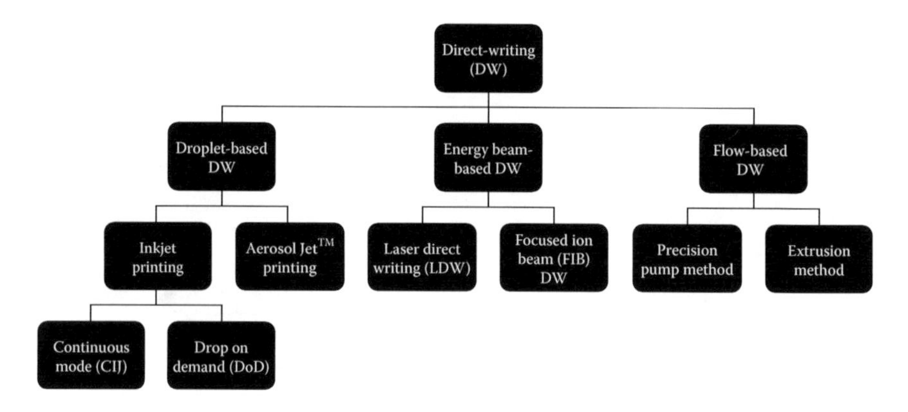

FIGURE 8.8 Classification of surface direct-write (DW) methods. (Adapted from Hon, K.K.B. et al., *CIRP Ann. Manuf. Technol.*, 57, 601–620, 2008.)

(based on silver/gold/copper/palladium/copper or alloys), resistors[155,156] (based on polymer thick film and ruthenium oxide), and dielectrics[157,158] to make various passive electronic components on low-temperature flexible substrates such as plastics, paper, and fabrics. The field is advancing rapidly, and no single source covers all the different technologies. We will briefly discuss the most relevant techniques and point the reader to additional resources as needed.

8.7.2.1 Droplet-Based DW

Droplet-based DW consists mainly of two techniques: (1) Inkjet printing and (2) Aerosol Jet™ printing.

8.7.2.1.1 Inkjet Printing

8.7.2.1.1.1 Process Inkjet-based DW[6,12,54,132,159,160] is a highly advanced technique. Inkjet printing is a droplet-based technology that places materials where they are needed via the ejection of liquid material from a single nozzle or multiple nozzles to precise locations by thermal or piezoelectric actuation. For this process, the jetted material must be a liquid with appropriate physical properties including viscosity, surface tension, and density. Inkjet printing is appealing because it is low cost, high speed and involves non-contact processing. There are two key types of inkjet printing technologies, continuous ink jetting (CIJ) and drop on demand (DoD). In case of CIJ printing, the ink reservoir is pressurized to ensure that a continuous stream of material passes through a nozzle which breaks up into individual uniform droplets as it issues from the orifice according to the waveform generated by the piezoelectric transducer. In the case of DoD printing, droplets may be ejected from the nozzle using a joule heater to heat and volatilize ink or by deforming a piezoelectric material in the print head. Both methods create the required pressure difference for dispensing the droplet on demand. Thermal and piezoelectric techniques are the most common approaches for DoD printing, but although the droplets can also be ejected by other methods such as electrostatic and acoustic actuation.[161] Both CIJ and DoD have four stages of development[162] when a droplet impacts the substrate (kinematic, spreading, relaxation, and wetting), which are time-dependent and controlled by physical forces like inertia, viscosity, and surface tension. In the earlier stages of impact, inertial forces dominate and viscous forces are weak. After impact, capillary (surface tension) forces become more important. Thus, these factors must be considered when designing new inkjet printing applications.

8.7.2.1.1.2 Materials, Writing Speed, and Resolution Inkjet systems usually work best with low viscosity materials (up to 100 mPa-s) that have low interfacial tension[6,163–167] (~20 dynes/cm). Various materials are compatible with inkjet printing including metal particle suspensions (gold/silver/copper/aluminum), ceramic particle suspensions, and electronic/optical materials such as epoxies, solders, and organometallics, among others. Since inkjet printing is a non-contact deposition method, it is compatible with a variety of substrates including metals, ceramics, polymers, and silicon. The volumetric dispense rate[2,113,168] of a single nozzle is typically on the order of 0.3 mm^3/s (DoD) and 60 mm^3/s (CIJ), which can be increased by using an array of nozzles. The resolution of inkjet-based

DW[2,113,132,169] is measured in terms of the droplet size that can range from ~20 μm to 1 mm for CIJ printing and ~15–200 μm for DoD printing.

8.7.2.1.1.3 Applications in Electronics Inkjet-based DW has been used in the field of electronic manufacturing since the 1980s and recently in printed electronics[55,170–176] such as solar cells, transistors, OLEDs, RFIDs, MEMS, and wireless communication, among others. For example, the DoD inkjet approach can micro-deposit organic light-emitting polymers[70] and phosphors, solder bumps,[177,178] spacer balls, electrical interconnects, and adhesive sealant/bond lines in the manufacture of display panels.[179,180] Both CIJ and DoD methods have been used to achieve solder drops from tens to hundreds of micrometers in diameter, with the capability to produce continuous lines and tracks as well as discrete spheres or dots as shown, for example, in Figure 8.9.[177] Figure 8.10[181] shows an electrostatic rotary motor where the electrodes are printed in five layers of silver nanoparticle ink with a 100 μm line trace width and the diameter of the whole device is 25 mm. Using a mixture of metal nanoparticles to form a core-shell dispersion, inkjet printing has also been able to print RFID antennas onto photo paper.[182]

8.7.2.1.1.4 Challenges The major challenge in inkjet printing is to control the four stages of development, which are dynamic in nature, using process parameters like impact velocity and initial drop diameter in order to achieve uniform printing. The drop formation is also affected by inertia, viscosity, and surface tension.[166] For example, undesirable splashing can occur if the drop formed is large with high impact velocity and the fluid has low surface tension and viscosity. Hence, it becomes necessary to study these parameters[162,183] in detail. The development stages decide whether we obtain homogenous CIJ/DoD or a mixture of these processes, which directly affects the quality of the printed structure.

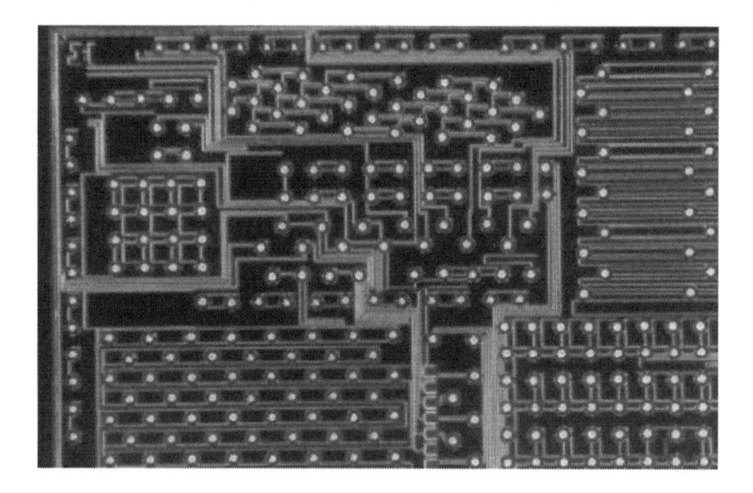

FIGURE 8.9 Image showing solder bumps (70 μm diameter) deposited by DoD inkjet printing onto an IC test substrate. (Adapted from Hayes, D.J. et al., *J. Microelectron. Packag. Soc.*, 1999.)

FIGURE 8.10 Optical micrograph of an inkjet printed electrostatic rotary motor. (Adapted from Fuller, S.B. et al., *J. Microelectromech. Syst.*, 11, 54–60, 2002.)

Electronic features with high conductivity need high concentrations of metal particles; however, high concentrations make the ink more viscous and therefore difficult to jet.[164] Typical formulations contain 30%–40% metal by weight although up to 50% loadings can be found. These higher percent loadings implement very small particle sizes which create the challenge of agglomeration that reduces jetting reliability. Inkjet printing can also be carried out by the direct deposition of bulk metals in liquid form using elevated temperatures to melt the metals. Bulk metals with higher melting points pose significant challenges for print head design since they have large surface tensions, they tend to oxidize, and they can easily ruin the piezoelectric transducer. This problem can be avoided by using other actuation methods such as direct pneumatic ejection[184] via DoD printing. Metallic particles suspended in a suitable fugitive liquid can be printed by inkjet processes and are used for both structural and electrical applications. In these applications, small particles are usually favored as the suspensions formed are more stable (i.e., the particles do not sediment), which lowers the chance of nozzle clogging. In addition, these particles have a high surface area-to-volume ratio which thus requires lower post-processing sintering temperatures.[153,181] Nanoparticle metal ink (Ag/Cu/Al/Ni) suspensions in which the solvent does not evaporate but cures to form a binder usually have lower conductivity than bulk metal inks, which affects the electrical performance in the final device.[37] Other challenges exists as covered in Section 8.6.

8.7.2.1.2 Aerosol Jet™ Printing

8.7.2.1.2.1 Process As the name implies, Aerosol Jet™ is the spraying of inks composed of small droplets dispersed in a liquid as seen in Figure 8.11.[2]

The process contains two main components: the atomizer and the deposition head. The atomizer is either an ultrasonic or pneumatic device that generates a dense vapor

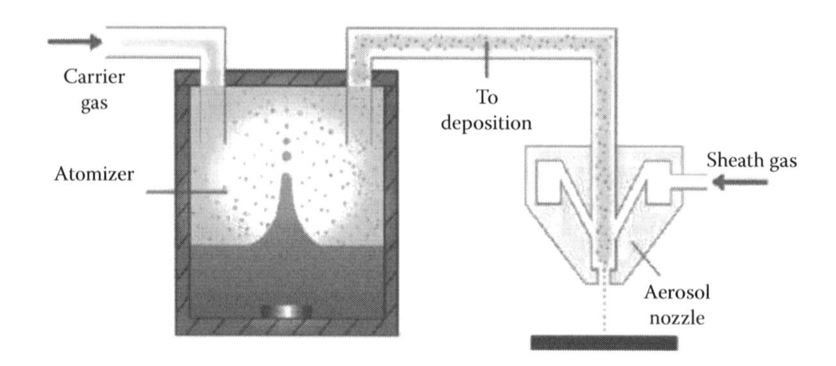

FIGURE 8.11 Schematic of Aerosol Jet™ printing system. (Adapted from Hon, K.K.B. et al., *CIRP Ann. Manuf. Technol.*, 57, 601–620, 2008.)

(mist) of material droplets. A carrier gas such as nitrogen passes through the atomizer to transfer the mist into the deposition head. The resulting annular flow leaves the deposition head through a nozzle onto the substrate. Aerosol Jet™ printing is particularly well suited to 3D applications as its deposition head can be mounted to a five-axis positioning stage to follow the contour of the substrate at a fixed stand-off distance (typically 1–5 mm).[113] In addition, it is possible to obtain fine feature definition as the aerosol consists of a high density of micro-droplets that are aerodynamically focused to produce lines as narrow as ~10 μm.[185] The features can deposit easily and conformally over non-planar surfaces.

8.7.2.1.2.2 Materials, Writing Speed, and Resolution An appeal of Aerosol Jet™ printing is that it can print features with fine resolution using a variety of processing materials and more clog-resistant nozzles, which is challenging in inkjet printing systems. Aerosol Jet™ systems can handle materials in a wide viscosity range, that is, between 0.001 and 2.5 Pa-s, and thus enables a wide range of materials[186] that can be deposited including metals, alloys, polymers, adhesives, and organic electronics, among others. The deposition rate obtainable from a single nozzle is about 0.25 mm^3/s, which can be increased by using an array of nozzles. The resolution of Aerosol Jet™ printing is measured in terms of line width that ranges from 10 to 150 μm with thickness ranging from 10 nm to 5 μm.[2,185,186]

8.7.2.1.2.3 Applications in Electronics Aerosol Jet™ printing is widely used to print conductive traces using gold, silver, or other nanoparticles inks. Conductors can be formed by printing a seed layer, followed by electroless plating.[187] Embedded resistors can be fabricated by printing polymer thick film pastes. Figure 8.12 shows an example of a strain gauge sensor.[188] For 3D surfaces with larger surface profiles, the Aerosol Jet™ system makes use of three-axis printing as shown in Figure 8.13.[186]

8.7.2.1.2.4 Challenges This printing technique requires inks that can form aerosols, which may present a limitation for certain materials. For substrates with high coefficient of thermal expansion, thermal mismatch between the ink and substrate material can lead to cracking or delamination of the printed metallic circuit material, although this is a challenge of all multifunctional printing processes.

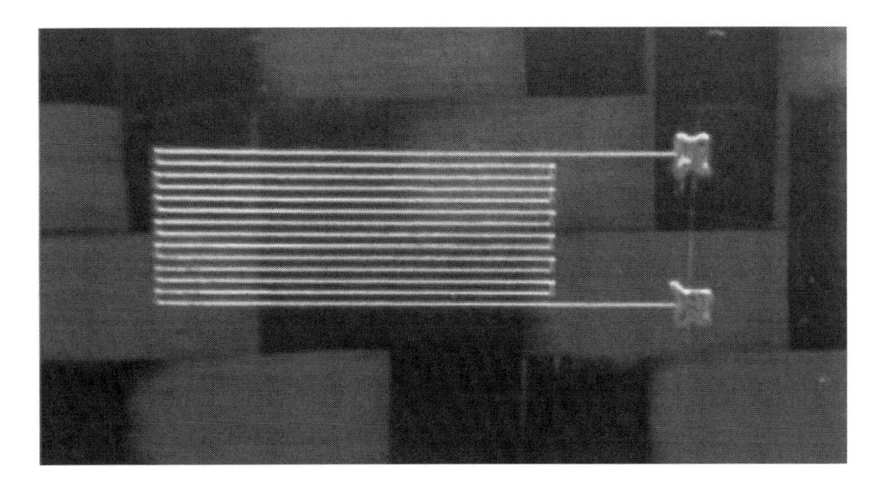

FIGURE 8.12 Image of an aerosol jet printed silver strain gauge on carbon fiber composite. (Courtesy of Optomec, Inc.; Adapted from Hedges, M. & Marin, A.B., 3D aerosol jet® printing-adding electronics functionality to RP/ RM, In *Proc DDMC 2012 Conf.*, Berlin, Germany, 2012. http://aerosoljet.com/downloads/Optomec_NEOTECH_DDMC_3D_Aerosol_Jet_Printing.pdf.)

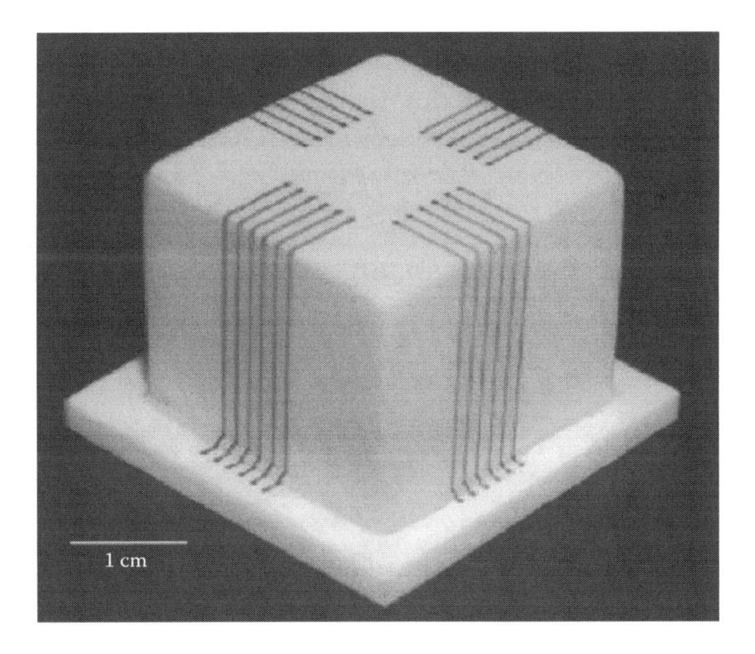

FIGURE 8.13 Image of 3D silver interconnects (150 µm line width) written over an alumina cube. (Courtesy of Optomec, Inc; Adapted from King, B. & Renn, M., Aerosol jet® direct write printing for mil-aero electronic applications. In *Palo Alto Colloq. Lockheed Martin*, 2009. http://www.optomec.com/downloads/Optomec_Aerosol_Jet_Direct_Write_Printing_for_Mil_Aero_Electronic_Apps.pdf.)

Rough and porous surfaces can severely affect the behavior of the deposited ink as such surfaces have relatively high surface energy, which makes it difficult to form a clean and uniform deposit. This effect is aggravated when the surface roughness is much larger than the ink thickness as it affects the quality of the printed lines. This issue can be avoided by pre-machining the rough areas where the electrical circuit is to be printed, although that adds extra processing steps.

8.7.2.2 Energy Beam-Based DW

Energy beam-based DW consists of all those processes that use high-power laser/ion/electron beam sources[148–152,189] as a mode for deposition of a variety of electronic-grade materials on the substrate. These sources precisely heat local regions of the substrate to high temperatures for sintering and curing while taking care to avoid melting of the substrate. The feature sizes are of the order of 25 μm due to the small spot size and high energy made possible by commercial laser systems. These techniques are divided into two broad categories depending on the source of the energy beam used for DW: (1) laser DW (LDW), which uses photons, and (2) focused ion beam (FIB) DW (discussed in Category 3), which uses ions as the energy source. There are many different varieties of these techniques, but almost none of them have been used for electronics in 3D objects. These methods have not been used for integrating electronics due to large processing temperatures involved which can easily affect any surrounding electronic circuitry. This disadvantage also puts a limitation on materials and substrates used in the deposition process.

8.7.2.2.1 Laser DW

8.7.2.2.1.1 Process LDW uses a laser beam to create complex 3D structures with selfsupporting features having fine resolutions without the use of expensive masks or lithographic methods. Laser writing techniques[9,190–202] create patterned materials through gas-phase deposition, ablation, selective sintering, or reactive chemical processes that include several methods such as thin film consolidation,[203,204] laser chemical vapor deposition[199] (LCVD), laser ablation,[193,194] laser-enhanced electroless plating[205,206] (LEEP), laser-induced forward transfer[207,208] (LIFT) and backward transfer[209] (LIBT), laser-guided DW[12] (LGDW), flow-guided DW[12] (FGDW), matrix-assisted pulsed laser direct-write[195,196] (MAPLE), two/multi-photon polymerization (MPP),[191,201] and selective laser sintering[9,197,198] (SLS), among others. All of these techniques utilize lasers to localize energy as a means to modify, deposit, or remove material.

8.7.2.2.1.2 Materials, Writing Speed, and Resolution Each technique places different demands on the laser writing tools and the physicochemical properties of the material being patterned. A wide range of materials can be deposited including metals, ceramics, polymers, semiconductors, and composites, among others which have been reviewed in the methods referenced earlier. The deposition rate or the writing speed is different for each method and the precursor or feed material used at the start of the process. With the exception of ablative approaches, most of the other techniques are capable of generating complex 3D structures with self-supporting features[191,210] at resolutions comparable to those achieved by various ink-based techniques.

8.7.2.2.1.3 Applications in Electronics LDW was introduced in the 1980s to enable the fabrication of micro-electronic circuits with 1D to 2D features and it was then developed in the 1990s to enable the creation of 3D features for applications such as photonic crystals and MEMS. More recently, these techniques have been employed in various devices such as microcapacitors, interconnects, phosphor displays, co-planar transistors, and resistors, among others. An example of a spiral inductor[36] fabricated using contact transfer is shown 500 μm in Figure 8.14. Contact transfer is a derivative of MAPLE in which the mixture material is coated onto a flexible transparent backing film and dried at low temperatures (normally below 100°C) to form a dry, flexible ribbon that is vacuum chucked and placed in direct contact with the receiving substrate in a conformal manner. After the laser irradiation, the ribbon is peeled off and materials remain on the substrate in the laser-defined areas.

8.7.2.2.1.4 Challenges Most of the LDW techniques are limited by the availability of volatile metal-organic or inorganic materials, contamination of the deposited materials, and the need for expensive and specialized reaction chambers, vacuum equipment, and lasers. The processing temperatures (≤400°C) are not suitable for the fabrication of high-quality crystalline materials required for state-of-the-art electronic performance of the final device, even with laser sintering. The presence of porosity in the powders used for DW processes can severely affect the electric performance as it can reduce the effective dielectric constant by almost an order of magnitude.[211] Also, in gas-phase deposition, the morphology and the electrical conductivity of the deposited features are generally inferior when compared with the bulk material and the adhesion of the material to the substrate can be poor and difficult to control.

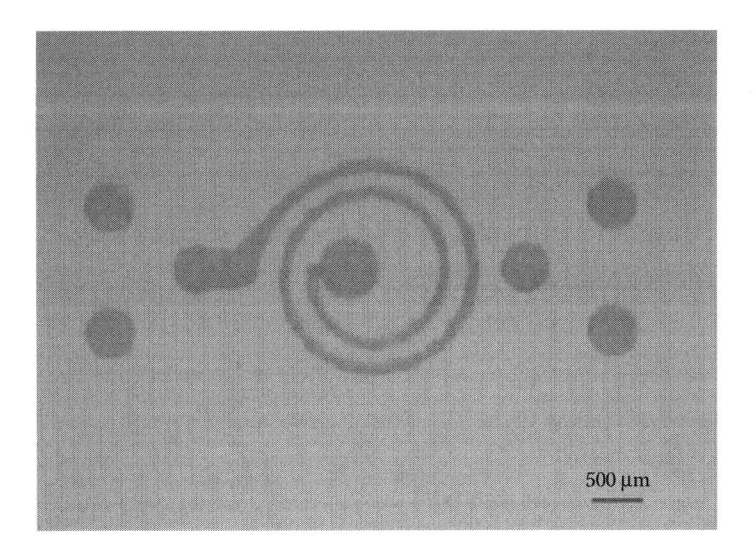

FIGURE 8.14 Photograph showing a spiral inductor deposited by contact transfer technique (derivative of MAPLE DW). (Adapted from Zhang, C. et al., *Microelectron. Eng.*, 70, 41–49, 2003.)

8.7.2.3 Flow-Based DW

Flow-based DW methods consist of processes that require positive mechanical pressure using a pump, air pressure, or extrusion to achieve precise micro-dispensing through a syringe tip leading to a continuous flow of ink, paste, or slurries on to the substrate. Extrusion systems are favorable for integration with AM because the DW tool tip is maneuverable, can dispense in different orientations, and can process high volumes of material. Flow-based DW approaches consist of two main techniques: (1) precision pump method, and (2) extrusion method.

8.7.2.3.1 Precision Pump Method

8.7.2.3.1.1 Process The precision pump method is a direct-dispensing tool integrated with nScrypt's novel pump called smart pump.[212] This pump is based on Sciperio's microdispense direct-write technology developed through the DARPA (US Defence Advanced Research Projects Agency) MICE program (Mesoscopic Integrated Conformal Electronics) in 2002. With this pump, dispensing is initiated via the opening of a valve that allows the material to be dispensed through the dispensing tip onto the substrate. The valve retracts to stop dispensing, and this retraction results in suction of material back into the dispensing nozzle thus resulting in cleaner, more precise printed features.

8.7.2.3.1.2 Materials, Writing Speed, and Resolution The pump is capable of dispensing very small volumes of the materials down to 20 pL and within a wide range of viscosities from 1 to 1,000,000 mPa-s, which makes it highly versatile with respect to the materials that can be used and the patterns that can be drawn with them. The maximum writing speed can be varied from 0.1 to 300 mm/s depending on the material and the application. The resolution measured in terms of the line width varies from ~25 μm to 3 mm depending on the material and the ceramic tip orifice diameter.

8.7.2.3.1.3 Applications in Electronics The process has a wide range of applications including, conductors, resistors, optics, adhesives, sealants, frit, solders, encapsulants, wire bonding, underfilling, flip-chip bumping, and MEMS. This system is capable of writing on highly non-conformal surfaces to make 3D structures. An example of non-conformal writing is shown in Figure 8.15.[212]

8.7.2.3.1.4 Challenges The line width obtained from the pump is directly dependent on the material, which is often a slurry or paste, as well as the nozzle diameter. Generally, the minimum line width is at least 10 times larger than the average particle size in the specific paste.[213] Second, the flow rate is sensitive to the dispensing height (distance between the substrate and the dispensing tip) for a constant applied pressure. Thus, in order to maintain the line width, consistency, and accuracy, the dispensing height must be maintained at a constant value. This is generally accomplished by laser scanning the substrate and then dynamically adjusting the tool path's z-axis height based upon measured variations in surface height.

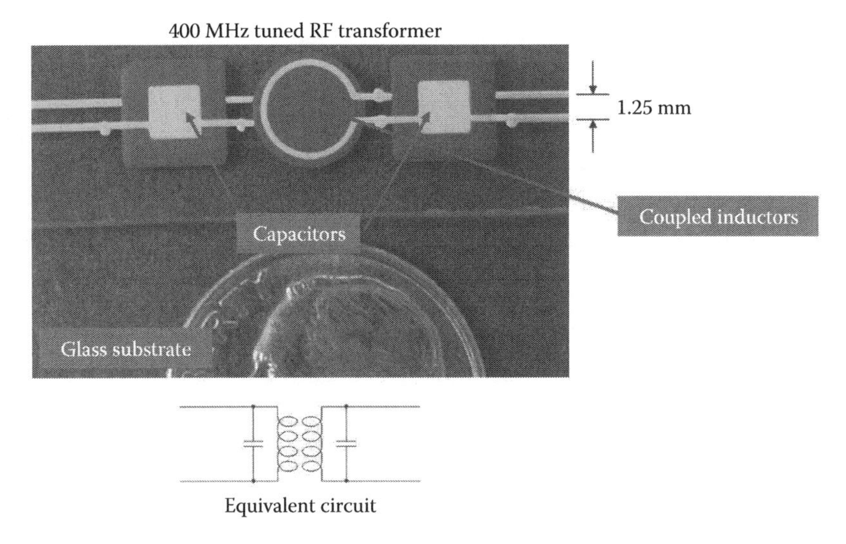

FIGURE 8.15 Image showing capacitors and coupled inductors dispensed on an uneven surface by the precision pump method to fabricate a printed RF device. (Adapted from Li, B. et al., Robust direct-write dispensing tool and solutions for micro/meso-scale manufacturing and packaging, In *ASME 2007 Int. Manuf. Sci. Eng. Conf.*, 715–721, *American Society of Mechanical Engineers*, 2007. http://proceedings.asmedigitalcollection.asme.org/proceeding.aspx?articleid = 1598546.)

8.7.2.3.2 Extrusion Method

8.7.2.3.2.1 Process In this method, the flowable material/ink—typically in the form of liquid, particulate slurry, or molten polymer filament—is loaded into a syringe which is then connected to the writing head that moves with the help of a computer-controlled transition stage to create materials with controlled architecture and composition. Depending on the material used for deposition, several methods are available for patterning materials in 3D, which includes robotic deposition,[214] 3D printing,[215] fused deposition modeling[216,217] (FDM), curved-layer FDM (CLFDM),[218] and micropen writing,[219] among others.

8.7.2.3.2.2 Materials, Writing Speed, and Resolution The extrusion systems can dispense fluidized materials with viscosities up to 5000 Pa-s and are mostly used to dispense metal-based inks consisting of metal particles or flakes dispersed in a volatile solvent that evaporates after being dispensed. The metals used are most commonly gold or silver due to their resistance to oxidation. Less expensive carbon-based inks can also be dispensed with these systems, although these inks are less conductive than metal inks. Typical writing speeds are of the order of 300 mm/s and the resolution measured in terms of the line width varies from 50 μm to 2.5 mm depending on the method employed for DW.

8.7.2.3.2.3 Applications in Electronics Extrusion methods for DW can be used to fabricate functional electronics on AM substrates. These devices

FIGURE 8.16 A photograph of multiple strain gauge devices made up of gallium–indium on glass. These patterns were written using a 119 μm inner diameter nozzle. Inset shows a larger scale directly written strain gauge on glass. Scale bars are 5 mm in length. Device and pads were written using a 379 μm inner diameter nozzle. (Adapted from Boley, J.W. et al., *Adv. Funct. Mater.*, 2014.)

include embedded, wearable and conformal electronics,[220] batteries,[221] and discrete electronics.[125] Electronics can also be printed on flexible substrates such as paper[222] to form functional electronic components,[223] including thermochromic displays,[224] disposable RFID tags,[225,226] and cellulose-based batteries,[227] to name but a few examples.

Figure 8.16[117] shows an application of conductive strain gauge devices formed by extruding liquid metal on a glass. Because the metal is liquid, these strain gauges can also be directly written and encapsulated in elastomeric substrates such as polydimethylsiloxane (PDMS) to obtain stretchable electronics.

8.7.2.3.2.4 Challenges For extrusion DW applications, the metal content of the ink is a critical factor. Higher loading results in higher particle density and thus better conductivity. However, inks with high metal loadings are more viscous and require more pressure to dispense. Their high particle content can lead to clogging issues in the dispensing nozzles and limit the minimum feature size. Thus, using larger nozzle tip diameters represents a trade-off; it reduces clogging but increases the minimum achievable feature size. The printing quality is also affected by several other factors such as the surface energy, ink formulations, roughness, and wettability of the substrate, which have been enlisted in Section 8.6. An additional difficulty is the fabrication of multifunctional complex electronic devices due to non-conformal geometry of these devices which helps to focus the discussion to the next section where we discuss techniques for 3D printing electronics.

8.7.3 Freeform Multi-material 3D Printing Approach (Category 3)

Freeform multi-material 3D printing approaches are capable of producing free-standing, out-of-plane structures with electronic functionality. Most of the techniques are inspired by DW extrusion techniques. The use of inks that can form out-of-plane structures distinguishes the examples in Category 3 from those in Category 2. The resulting structures—typically metals—can be encased in different materials such as polymers, elastomers, or ceramics to obtain functional electronic devices.

8.7.3.1 Omnidirectional Printing

8.7.3.1.1 Process

Colloidal suspensions (e.g., silver nanoparticles in water) are extruded from a syringe.[10,116,221,228–232] Due to the shear yielding rheology of these inks, the material only flows when it is sheared and the printed structures stabilize due to the rheological properties of the inks. As a result, this process can create mechanically stable, out-of-plane structures. As the solvent (e.g., water) evaporates, the structures become even more stable. The structures become conductive as the particles move closer together, but ultimately need to be sintered to improve the conductivity closer to that of bulk metal.

Metal 3D structures can also be electrodeposited[233] to write pure copper and platinum in an ambient air environment to fabricate high density and high quality, complex, microscale and nanoscale metallic structures like interconnects. The electrodeposition relies on an electrolyte-containing micropipette with a microscopic dispensing nozzle as the working toolbit. The electrodeposition is initiated within the substrate surface confined by the meniscus between the dispensing nozzle and the substrate surface using an appropriate electrical potential between the electrolyte contained in the micropipette and the substrate surface.

8.7.3.1.2 Materials, Writing Speed, and Resolution

3D structures consisting of continuous solids, with high aspect ratio (e.g., parallel walls), or self-supporting features that must span gaps in the underlying layers can be fabricated through controlled ink composition, rheological behavior, and printing parameters. This emerging technique has been demonstrated with inks that are typically formulated from shear-thinning concentrated colloidal suspensions or gels,[10,116,234–238] fugitive organic/polymeric[10,238–242] melts, composites,[10,243–245] polyelectrolyte,[10,239,241,246–248] and sol-gel[249,250] building blocks suspended or dissolved in a liquid or heated to create a stable, homogeneous ink with the desired rheological (or flow) behavior. The important rheological parameters for a given ink design include its apparent viscosity, yield stress under shear and compression, and viscoelastic properties (i.e., the shear loss and elastic moduli), which are tailored for the specific direct-write technique of interest. The writing speed is usually decided by the DW method chosen. The DW of 3D periodic architectures with filamentary features ranges from hundreds of micrometers (~250 μm) to sub-micrometer in size (~0.1 μm).

8.7.3.1.3 Applications in Electronics

The omnidirectional structures can be utilized in wire bonding to fragile 3D devices,[116,251] spanning antennas,[228] batteries,[221] and interconnects[116,252] for solar cell

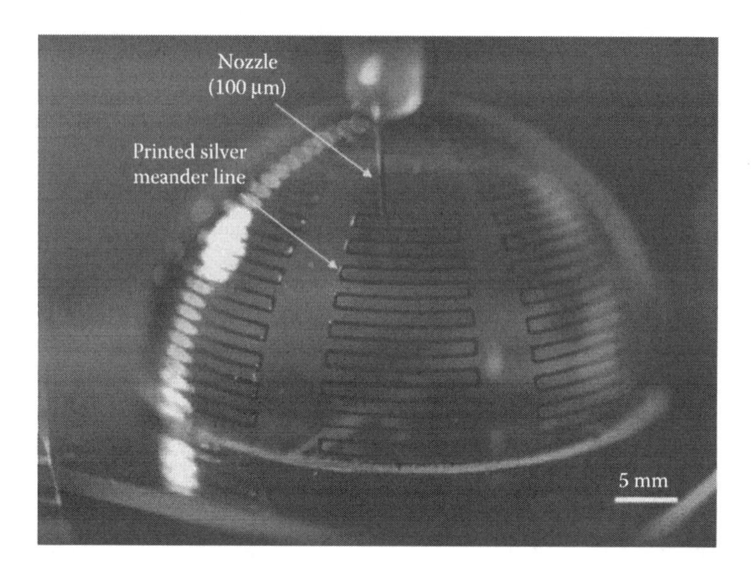

FIGURE 8.17 Optical image captured during conformal printing of electrically small antennas on a hemispherical glass substrate. (Adapted from Adams, J.J. et al., *Adv. Mater.*, 23, 1335–1340, 2011.)

and LED arrays. These structures can also act as functional composites,[253] microfluidic networks,[240] and templates for photonic band-gap materials[254] and inorganic–organic hybrid structures.[239] Figure 8.17 demonstrates the conformal printing of a small 3D antenna.[228] A 100 µm metal nozzle is used to print meander-line patterns on the surface of a glass hemisphere.

Omnidirectional printing has also been exploited to create free-standing interconnects as shown in Figure 8.18[116] for commercially available gallium nitride LED arrays. This ability to print out-of-plane enables the microelectrodes to directly cross pre-existing patterned features through the formation of spanning arches. Such conformal printing of conducting features enables several applications, including flexible,[173,255] implantable[256] and wearable[257] antennas, electronics, and sensors.

It is also possible to direct write 3D microvasculature networks out of polymers as fugitive inks, which can then be backfilled or infiltrated with electronic inks.[240] First, these inks must flow through a fine deposition nozzle under high shear, yet be self-supporting under ambient conditions. Second, the ink scaffold must maintain its shape during resin infiltration and curing. Finally, the ink scaffold must liquefy at modest temperatures to facilitate its removal from the polymer matrix, leaving behind an interconnected network of microchannels.

8.7.3.1.4 Challenges

Most of the challenges faced with omnidirectional printing are the same as those observed in DW extrusion-based techniques discussed in Section 8.7.2.3.2. Most of the shortcomings are unique to the ink. For example, colloidal gel-based inks require significant applied pressures to induce flow during deposition and suffer clogging problems when the nozzle-to-particle diameter is reduced to below ~100.[235,236] Most

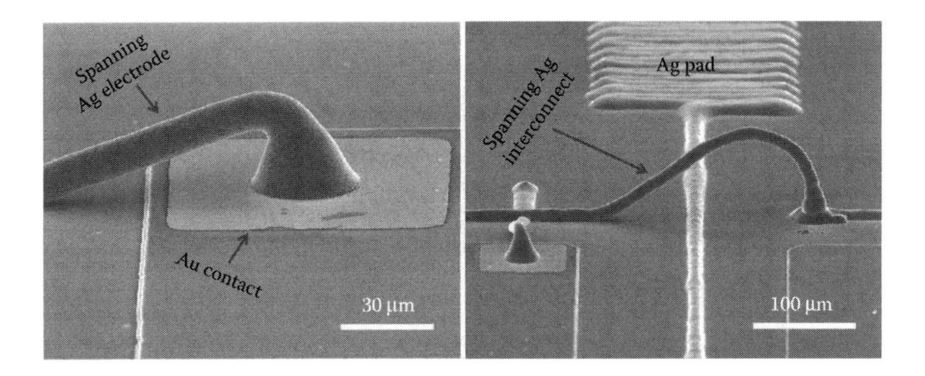

FIGURE 8.18 SEM images of a silver interconnect arch printed on a gold contact pad (80 by 80 μm) (left) and over an electrode junction (right). (Adapted from Ahn, B.Y. et al., *Science*, 323, 1590–1593, 2009.)

of the materials used for omnidirectional printing need sintering at high temperatures for densification and reduction of porosity in the final structure. The resulting structures can be quite conductive, but are not at the same level as bulk metals.

8.7.3.2 Liquid Metal Printing

8.7.3.2.1 Process

Liquid metals offer the electrical and thermal benefits of metals combined with ease of printing to enable the 3D printed fabrication of soft, flexible, and stretchable devices at room temperature. It is possible to direct write a low viscosity liquid metal (e.g., gallium and its alloys, such as eutectic gallium–indium alloy[258]) at room temperature into a variety of stable free-standing 3D microstructures[259,260] (cylinders with aspect ratios significantly beyond the Rayleigh stability limit,[261] 3D arrays of droplets, out-of-plane arches, wires as shown in Figure 8.19). These liquid metals have low viscosities (similar to water) at room temperature and are therefore easy to extrude. They form mechanically stable structures despite being liquids due to the presence of a surface oxide that forms spontaneously and rapidly.

The general approach for printing these liquid metal microstructures is by applying modest pressures to a syringe needle that extrudes the liquid metal wire onto a substrate controlled by a motorized translation stage. In addition to extruding wires, it is possible to form free-standing liquid metal microstructures using three additional methods: (1) rapidly expelling the metal to form a stable liquid metal filament, (2) stacking droplets, and (3) injecting the metal into microchannels and subsequently removing the channels chemically.

8.7.3.2.2 Materials, Writing Speed, and Resolution

The writing is done using a binary eutectic alloy of gallium and indium (EGaIn, 75% Ga 25% In by wt.), but in principle, any alloy of gallium will work. EGaIn is liquid at room temperature (m.p. ~15.7°C) with metallic conductivity one order less than silver and copper.[262] The liquid metal exhibits a negligible vapor pressure and low toxicity. Upon exposure to air, the metal instantaneously forms a thin passivating

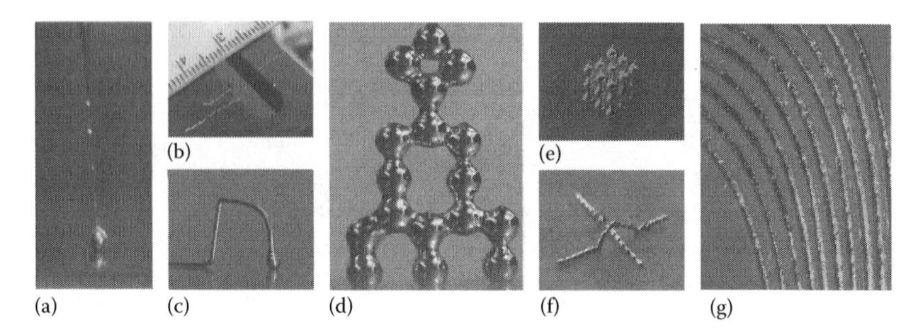

FIGURE 8.19 DW of liquid metal 3D structures. Photographs of the diverse free-standing, liquid metal microstructures that can be direct printed at room temperature. (a) Liquid metal ejected rapidly from a glass capillary forms a thin wire. (b) These fibers are strong enough to suspend over a gap despite being composed of liquid. (c) A free-standing liquid metal arch. (d) A tower of liquid metal droplets. (e) A 3D cubic array of stacked droplets. (f) A metal wire and an arch composed of liquid metal droplets. (g) An array of in-plane lines of free-standing liquid metal fabricated by filling a microchannel with the metal and dissolving away the mold. Scale bars represent 500 μm. (Adapted from Ladd, C. et al., *Adv. Mater.*, 25, 5081–5085, 2013.)

skin composed of gallium oxide,[258] and the electrical resistance remains largely unaffected because the skin is thin.[263] In addition, the liquid metal adheres to most surfaces and alloys with many metals to form ohmic contacts. The writing speed or the draw rate is not yet known. The smallest components fabricated to date are ~10 μm.

8.7.3.2.3 Applications in Electronics

These techniques are used in conjunction with other methods to 3D print conductive devices[117,264–269] that are soft, flexible, and stretchable—properties that may be useful for creating stretchable electronics, soft robotics, and electronic skins. These extrusion methods can also be used for connecting electronic devices by embedding the microstructures in various substrates such as PDMS to form flexible and stretchable electronic devices. An example of such a structure is shown in Figure 8.20.[259]

8.7.3.2.4 Challenges

This approach is relatively new, and the process is not yet well understood. In the context of inkjet printing, the combination of surface tension and the surface oxide makes these alloys non-printable without modifying either the material or the atmosphere. Additionally, the fast-forming oxide layer can potentially clog a nozzle orifice, further increasing the difficulty of printing Ga–In alloys. Finally, the resulting structures are liquid and therefore have to be embedded in a supporting structure for practical applications.

8.7.3.3 Focused Ion Beam DW

8.7.3.3.1 Process

In FIB DW, a low energy ion beam (10–50 keV) generated from a liquid gallium source is used to bombard the precursor gas on the surface of the substrate in a

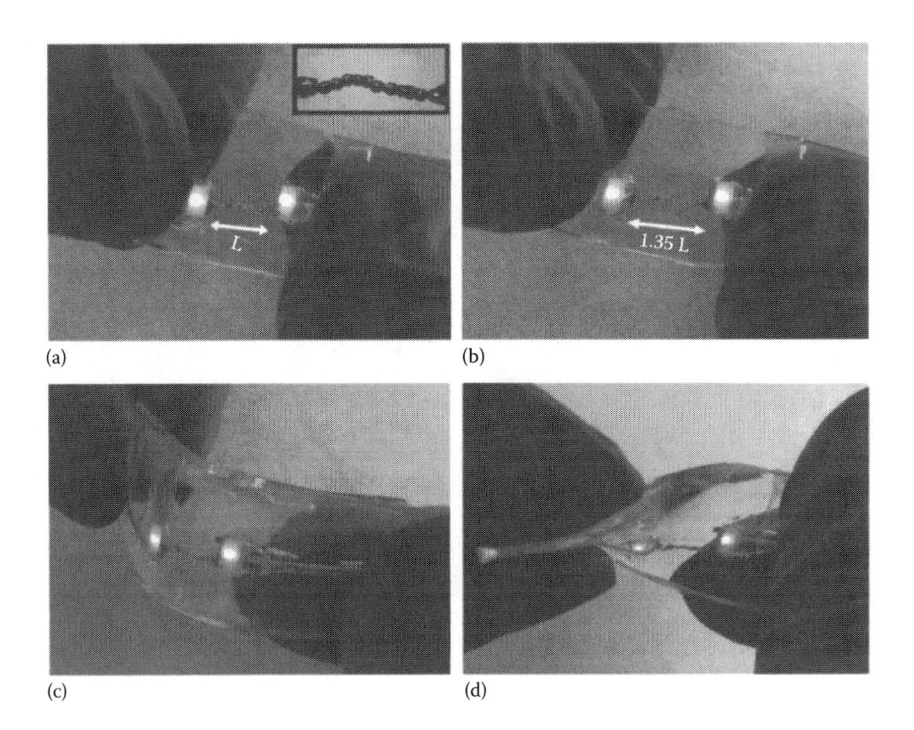

FIGURE 8.20 Stretchable interconnects formed by DW. (a) A prototype device composed of two LEDs connected by a stretchable wire bond and embedded in PDMS (Inset: Microscopy image of the liquid metal wire bonds). (b–d): The fluidic property of the metal wire in the elastomer allows elasticity (b) and flexibility (c, d) of the device and keeps its electrical continuity. (Adapted from Ladd, C. et al., *Adv. Mater.*, 25, 5081–5085, 2013.)

vacuum environment. The precursor gas breaks down, resulting in the deposition of material (typical metals). The deposition method is capable of conformal deposition and forming 3D microstructures.[270–279] At larger energies, the ions can also be utilized to etch the substrate.

8.7.3.3.2 Materials, Writing Speed, and Resolution

FIB DW is usually used for depositing conductors such as gold, aluminum, copper, and platinum along with insulators using organometallic precursor gases. The deposition rate is lower than the LDW methods but it offers higher resolution with the minimum feature that can be produced being of the order of 80 nm. The minimum thickness is about 10 nm and the aspect ratios are between five and ten.[270]

8.7.3.3.3 Applications in Electronics

FIB systems are used for micromachining due to their ability to precisely add or remove the materials. This allows thermographic *in situ* process monitoring and imaging for navigation, alignment, and inspection using electron-beam melting (EBM) technology.[280,281] This feature is extremely important in the semiconductor industry for mask repair, failure analysis, and IC prototype rewiring. Figure 8.21[276]

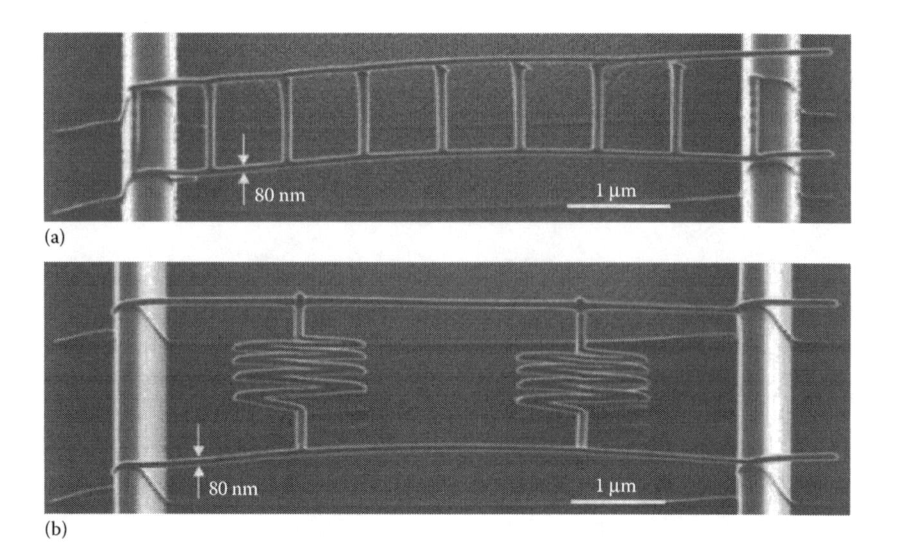

FIGURE 8.21 SIM images of carbon nanostructures prepared by FIBID: (a) free-space wiring with a bridge shape, (b) free-space wiring with parallel resistances. (Adapted from Morita, T. et al., *J. Vac. Sci. Technol. B*, 21, 2737–2741, 2003.)

shows scanning ion microscopy (SIM) images of two nanostructures: a bridge and a parallel-resistance component made of amorphous diamond-like carbon containing a Ga core deposited using FIB-induced deposition (FIBID). This result demonstrates that a 3D electrical circuit can be fabricated in free-space by FIB.

8.7.3.3.4 Challenges

FIB is done in a vacuum environment in a tool that has limited space and can therefore typically only accommodate small substrates. It is therefore not compatible with 3D printing. Features deposited by FIB are not pure because of the organic contaminants that arise primarily from the organometallic precursors used for the deposition of metal. The resistivity of these deposits is about one or two orders of magnitude higher than those of pure metal. Due to the slow writing process, the applications are restricted to low volume production like repair works. Ions from the FIB process can also induce damage of under-lying layers.

8.8 CONCLUSION

This chapter discusses approaches to integrate electronics into 3D printed objects. We characterized these approaches into three categories which have been summarized in Tables 8.2 and 8.3.

The first category builds electronic components separately and places them into the 3D printed object. This hybrid approach goes against the spirit of 3D printing, but provides the sophistication of modern electronics. It remains a challenge to find ways to connect these components using AM principles.

TABLE 8.2

Comparison of All the Approaches Used for Integrating Electronics into 3D Parts Excluding Laser Direct-Write Techniques

Approach	Deposition Method	Mechanism for Integration	Deposition Rate/ Writing Speed	Materials and Viscosity (μ)	Resolution or Minimum Feature Size	3D Periodic Structures
Hybrid chip insertion (Category 1)	Ultrasonic Consolidation (UC)	Deposition of metallic foils using ultrasonic energy	To 50 mm/s	Metals and alloys like Ni, Fe, Cu, brass, and steel	Foil thickness ~0.1 mm	N/A
	Transfer printing	Transfer patterns from TS to DS via differential adhesion	To 10 cm/s	Metallo-organics and conductive polymers	Pattern size ~12 μm	N/A
Surface DW (Category 2)	Inkjet printing (CIJ)	Deposition of liquid droplets by break-up of continuous jet	To 60 mm^3/s with a single nozzle	Liquid with μ ~2–10 mPa-s; can contain small particles	Droplet size ~20 μm–1 mm (typically 150 μm)	No
	Inkjet printing (DoD)	Deposition of individual liquid droplets when required	To 0.3 mm^3/s with a single nozzle	Liquid with μ ~10–100 mPa-s; can contain small particles	Droplet size ~15–200 μm	No
	Aerosol jet printing	Kinetic bombardment of atomized droplets	To 0.25 mm^3/s with a single nozzle	Materials with μ < 2.5 mPa-s that can be atomized	Line width ~10–150 μm, thickness ~10 nm–5 μm	Yes
	Precision pump	Precision micro-dispensing pump with suck-back action	Typically 50 mm/s (up to 300 mm/s)	Liquid, paste and slurry materials with μ up to 1000 Pa-s	Line width ~25 μm–3 mm	Yes
	Extrusion	Deposition of materials via syringe-based flow	Typically 25 mm/s	Liquid, paste and slurry materials with μ up to 500 Pa-s	Line width ~50 μm–2.5 mm	Yes
Freeform multi-material 3D printing (Category 3)	Omnidirectional printing	Extrusion of concentrated inks through fine cylindrical nozzles	Typically 6 mm/s	Liquid, paste, and slurry materials with μ up to 500 Pa-s	0.1–250 μm	Yes
	Liquid metal printing	Deposition of microstructures via extrusion of liquid metals at room temperature	Not yet known	Gallium-based alloys with μ up to 2 mPa-s	~10 μm	Yes
	Focused ion beam (FIB) DW	Ion-induced deposition of precursor gas molecules	Typically 0.05 $\mu m^3/s$	Metals and insulators	Line width ~80 nm–20 μm	Yes

TABLE 8.3
Comparison of All Laser Direct-Write Techniques

Deposition Method	Mechanism	Deposition Rate/ Writing Speed	Materials and Viscosity (μ)	Resolution or Minimum Feature Size	3D Periodic Structures
Thin film consolidation	Melting, fusion onto substrates	10–2000 μm/s	Metals/ceramics on metal/ceramic substrates	10–50 μm	No
LIFT and MAPLE DW	Transfer of material by kinetic energy of vaporizing organic binders	Typically 3–50 mm/s (up to 500 mm/s)	Metals, ceramics, semiconductors, polymers, composites	10–100 μm	No
LEEP	Thermal decomposition of the liquid	0.1–80 μm/s	Metals and ceramics on inorganic substrates	2–12 μm	No
Laser-activated electroplating	Accelerated chemical reaction by local high temperatures	Typically 0.1–10 m/s (up to 2.5 m/s)	Metals on metallic substrates	0.1–300 μm	No
LCVD	Decomposition of gases after vaporization and condensation takes place	Typically 50–100 μm/s (up to 5 mm/s)	Metals, semiconductors and ceramic such as Al, W, Si, Al_2O_3, WC	1–20 μm	Yes
LIBT	Physical vapor/liquid deposition after laser irradiation through transparent medium	10–100 mm/s	Metals and ceramics on transparent substrates	5–200 μm	No
LGDW	Laser-assisted deposition of generated aerosol using optical forces	To 1 m/s	Non-absorbent droplets and solid particulates with $\mu < 2.5$ mPa-s	2 μm	No
FGDW	Gas flow-assisted deposition of generated aerosol using hydrodynamic forces	To 0.25 mm^3/s	Atomizable fluids and colloids with $\mu < 2.5$ mPa-s	25 μm	No
TPP	Photopolymerization of UV-curable resin at laser focus within matrix	To 100 μm/s	Photo-sensitive acrylate polymers	≤100 nm	Yes
SLS	Locally sinters and binds the powder bed	To 35 mm/h	Polymers, metals, alloy mixtures and composites in powder form	100 μm	No

The second category is inspired by the variety of DW techniques developed originally to print electronic materials onto 2D substrates. Because AM is accomplished by building a 3D object one layer at a time, it seems sensible that these techniques could be adapted for 3D printing, yet there are many challenges that arise as discussed in this chapter.

The third category captures emerging techniques that can print freeform 3D electronic components. These techniques have yet to be implemented commercially, but may inspire new approaches to 3D printing electronic materials.

In summary, there has been very little work done on 3D printing of electronics—particularly those with high levels of sophistication—and it is the hope that this chapter will inspire new approaches to enable new electronic components built using AM. An ultimate goal of this field would be to develop a single system capable of rapid (in scale of hours or less) prototyping/manufacturing that can deposit a wide variety of materials (conductors, insulators, semiconductors, ferrites, ruthenates, metals, ferroelectrics, glasses, polymers, etc.) for customized, robust, electronic components at low substrate temperatures in a conformal manner on virtually any substrate (paper, plastic, ceramics, metals, etc.).

REFERENCES

1. F42 Committee. *Terminology for Additive Manufacturing Technologies* (ASTM International, 2012). http://enterprise.astm.org/SUBSCRIPTION/filtrexx40.cgi?+REDLINE_ PAGES/F2792.htm.
2. Hon, K.K.B., Li, L. & Hutchings, I.M. Direct writing technology: Advances and developments. *CIRP Ann. Manuf. Technol.* **57**, 601–620 (2008).
3. Church, K.H., Fore, C. & Feeley, T. Commercial applications and review for direct write technologies. *MRS Online Proc. Libr.* **624**, null–null (2000).
4. Holshouser, C. et al. Out of bounds additive manufacturing. *Adv. Mater. Process.* **171**, 15–17 (2013).
5. Hofmann, M. 3D printing gets a boost and opportunities with polymer materials. *ACS Macro. Lett.* **3**, 382–386 (2014).
6. Calvert, P. Inkjet printing for materials and devices. *Chem. Mater.* **13**, 3299–3305 (2001).
7. Morissette, S.L., Lewis, J.A., Clem, P.G., Cesarano, J. & Dimos, D.B. Direct-write fabrication of Pb(Nb, Zr, Ti)O_3 devices: Influence of paste rheology on print morphology and component properties. *J. Am. Ceram. Soc.* **84**, 2462–2468 (2001).
8. Piner, R.D., Zhu, J., Xu, F., Hong, S.H. & Mirkin, C.A. "Dip-pen" nanolithography. *Science* **283**, 661–663 (1999).
9. Bourell, D., Marcus, H., Barlow, J. & Beaman, J. Selective laser sintering of metals and ceramics. *Int. J. Powder Metall.* **28**, 369–381 (1992).
10. Lewis, J.A. Direct ink writing of 3D functional materials. *Adv. Funct. Mater.* **16**, 2193–2204 (2006).
11. Kim, N.-S. & Han, K.N. Future direction of direct writing. *J. Appl. Phys.* **108**, 102801 (2010).
12. Piqué, A. & Chrisey, D.B. *Direct-write Technologies for Rapid Prototyping Applications: Sensors, Electronics, and Integrated Power Sources*. Academic Press, San Diego, CA, 2002.
13. MarketsandMarkets. Additive Manufacturing Market—Forecasts 2012–17: (MarketsandMarkets) (2013). http://www.marketsandmarkets.com/Market-Reports/additive-manufacturing-medicaldevices-market-843.html.

14. IDTechEx. Applications of 3D Printing 2014–2024 (IDTechEx) (2014). http://www.idtechex. com/research/reports/applications-of-3d-printing-2014-2024-000385.asp.
15. Macdonald, E. et al. 3D Printing for the rapid prototyping of structural electronics. *IEEE Access* **2**, 234–242 (2014).
16. Powers, T.R. *The Integrated Circuit Hobbyist's Handbook*. Newnes, High Text Publications, Inc., Solana Beach, CA, 1995.
17. Moore, G.E. Cramming more components onto integrated circuits. *Proc. IEEE* **86**, 82–85 (1998).
18. Schaller, R.R. Moore's law: Past, present and future. *IEEE Spectr.* **34**, 52–59 (1997).
19. Madou, M.J. *Manufacturing Techniques for Microfabrication and Nanotechnology*. CRC Press, Boca Raton, FL, 2011.
20. Qin, D., Xia, Y. & Whitesides, G.M. Soft lithography for microand nanoscale patterning. *Nat.Protoc.* **5**, 491–502 (2010).
21. Xia, Y. & Whitesides, G.M. Soft lithography. *Annu. Rev. Mater. Sci.* **28**, 153–184 (1998).
22. Colburn, M. et al. Step and flash imprint lithography: A new approach to high-resolution patterning. *SPIE Proc. Emerg. Lithogr. Tech. III.* **3676**, 379–389 (1999).
23. Resnick, D.J. et al. Imprint lithography for integrated circuit fabrication. *J. Vac. Sci. Technol. B* **21**, 2624–2631 (2003).
24. Thompson, L.F. et al. Introduction to microlithography (1983). http://pubs.acs.org/isbn/9780841207752.
25. Gates, B.D. et al. New approaches to nanofabrication: Molding, printing, and other techniques. *Chem. Rev.* **105**, 1171–1196 (2005).
26. Costner, E.A., Lin, M.W., Jen, W.-L. & Willson, C.G. Nanoimprint lithography materials development for semiconductor device fabrication. *Annu. Rev. Mater. Res.* **39**, 155–180 (2009).
27. Barron, A.R. & Ball, Z. Optical Issues in Photolithography. *Connexions*. http://cnx.org/content/ m25448/latest/
28. Xia, Y.N., Rogers, J.A., Paul, K.E. & Whitesides, G.M. Unconventional methods for fabricating and patterning nanostructures. *Chem. Rev.* **99**, 1823–1848 (1999).
29. Rogers, J.A. & Lee, H.H. *Unconventional Nanopatterning Techniques and Applications*. John Wiley & Sons, Hoboken, NJ, 2008.
30. Gates, B.D., Xu, Q.B., Love, J.C., Wolfe, D.B. & Whitesides, G.M. Unconventional nanofabrication. *Annu. Rev. Mater. Res.* **34**, 339–372 (2004).
31. Zhao, X.-M., Xia, Y. & Whitesides, G.M. Soft lithographic methods for nano-fabrication. *J. Mater. Chem.* **7**, 1069–1074 (1997).
32. Odom, T.W., Thalladi, V.R., Love, J.C. & Whitesides, G.M. Generation of 30 – 50 nm structures using easily fabricated, composite PDMS masks. *J. Am. Chem. Soc.* **124**, 12112–12113 (2002).
33. Kim, K.-H., Moldovan, N. & Espinosa, H.D. A nanofountain probe with sub-100nm molecular writing resolution. *Small* **1**, 632–635 (2005).
34. Lewis, A. et al. Fountain pen nanochemistry: Atomic force control of chrome etching. *Appl. Phys. Lett.* **75**, 2689–2691 (1999).
35. Hwang, K., Dinh, V.-D., Lee, S.-H., Kim, Y.-J. & Kim, H.-M. Analysis of line width with nano fountain pen using active membrane pumping. In *2nd IEEE Int. Conf. NanoMicro Eng. Mol. Syst.* 759–763 (2007). http://yadda.icm.edu.pl/yadda/element/bwmeta1.element.ieee-000004160431.
36. Zhang, C. et al. Laser direct-write and its application in low temperature Co-fired ceramic (LTCC) technology. *Microelectron. Eng.* **70**, 41–49 (2003).
37. Williams, E.D., Ayres, R.U. & Heller, M. The 1.7 kilogram microchip: Energy and material use in the production of semiconductor devices. *Environ. Sci. Technol.* **36**, 5504–5510 (2002).

38. Joshi, P.C. et al. Direct digital additive manufacturing technologies: Path towards hybrid integration. In *Future Instrum. Int. Workshop.* 1–4 (2012). doi:10.1109/FIIW.2012.6378353.
39. Cantatore, E. *Applications of Organic and Printed Electronics: A Technology-Enabled Revolution.* Springer, Spring street, NY, 2012.
40. Chang, J., Zhang, X., Ge, T. & Zhou, J. Fully printed electronics on flexible substrates: High gain amplifiers and DAC. *Org. Electron.* 15, 701–710 (2014).
41. Das, R., Harrop, P. & Printed, O. Flexible electronics forecasts. *Play. Oppor.* **2021**, 2011.
42. Wikipedia. File:ComplementaryTechnologies.png. *Wikipedia Free Encycl.* (2008). http://en.wikipedia. org/wiki/File:ComplementaryTechnologies.png.
43. Chang, J., Ge, T. & Sanchez-Sinencio, E. Challenges of printed electronics on flexible substrates. In *IEEE 55th Int. Midwest Symp. Circuits Syst.* Boise, ID, Circuits and Systems, 582–585 (2012).
44. Subramanian, V. et al. All-printed RFID tags: Materials, devices, and circuit implications. *19th Int. Conf. VLSI Des. 2006 Held Jointly 5th Int. Conf. Embed. Syst. Des.* 6 (2006). doi:10.1109/ VLSID.2006.34.
45. Subramanian, V., Lee, J.B., Liu, V.H. & Molesa, S. Printed electronic nose vapor sensors for consumer product monitoring. *Solid-State Circuits Conf. 2006 ISSCC 2006 Dig. Tech. Pap. IEEE Int.* 1052–1059 (2006). doi:10.1109/ISSCC.2006.1696148.
46. Curling, C.J. Prototype line for manufacturing flexible active matrix displays. In *5th Annu. Flex.Disp. Conf.* USDC Flexible Displays & Microelectronics, Plastic Logic Unlimited, Cambridge, 6–9. (February 2006)
47. Finkenzeller, K. In *RFID Handbook.* 347–359. John Wiley & Sons, Chichester, UK, 2010. doi:10.1002/9780470665121.ch12/summary.
48. Berggren, M., Nilsson, D. & Robinson, N.D. Organic materials for printed electronics. *Nat.Mater.* **6**, 3–5 (2007).
49. Blackwell, G.R. *The Electronic Packaging Handbook.* CRC Press, Boca Raton, FL, 1999.
50. Siegel, A.C. et al. Foldable printed circuit boards on paper substrates. *Adv. Funct. Mater.* **20**, 28–35 (2010).
51. Subramanian, V. et al. Printed electronics for low-cost electronic systems: Technology status and application development. *Solid-State Circuits Conf. 2008 ESSCIRC 2008 34th Eur.* 17–24 (2008). doi:10.1109/ESSCIRC.2008.4681785.
52. Yin, Z., Huang, Y., Bu, N., Wang, X. & Xiong, Y. Inkjet printing for flexible electronics: Materials, processes and equipments. *Chin. Sci. Bull.* **55**, 3383–3407 (2010).
53. Molesa, S., Redinger, D.R., Huang, D.C. & Subramanian, V. High-quality inkjet-printed multilevel interconnects and inductive components on plastic for ultra-low-cost RFID applications. *PROC 769 H831* **769**, 2003.
54. Singh, M., Haverinen, H.M., Dhagat, P. & Jabbour, G.E. Inkjet printing: Process and its applications. *Adv. Mater.* **22**, 673–685 (2010).
55. Ko, S.H. et al. All-inkjet-printed flexible electronics fabrication on a polymer substrate by low temperature high-resolution selective laser sintering of metal nanoparticles. *Nanotechnology* **18**, 345202 (2007).
56. Hrehorova, E. et al. Gravure printing of conductive inks on glass substrates for applications in printed electronics. *J. Disp. Technol.* **7**, 318–324 (2011).
57. Gravure Association America; Gravure Education Foundation, *Gravure: Process and Technology.* Gravure Association of America, Gravure Education Foundation, Denver, NC, 1991. http:// catalog.lib.ncsu.edu/record/NCSU2266370.
58. Pudas, M. Gravure-offset printing in the manufacture of ultra-fine-line thick-films for electronics (2004). http://jultika.oulu.fi/Record/isbn951-42-7303-6.

59. Krebs, F.C. et al. A complete process for production of flexible large area polymer solar cells entirely using screen printing: First public demonstration. *Sol. Energy Mater. Sol. Cells* **93**, 422–441 (2009).
60. Pardo, D.A., Jabbour, G.E. & Peyghambarian, N. Application of screen printing in the fabrication of organic light-emitting devices. *Adv. Mater.* **12**, 1249–1252 (2000).
61. White, A. High quality flexography: A literature review. (1998).
62. Blayo, A. & Pineaux, B. Printing processes and their potential for RFID printing. *Proc. 2005 Jt. Conf. Smart Objects Ambient Intell. Innov. Context-Aware Serv. Usages Technol.* 27–30. ACM, New York, 2005. doi:10.1145/1107548.1107559.
63. Ramsey, B.J., Evans, P.S.A. & Harrison, D. A novel circuit fabrication technique using offset lithography. *J. Electron. Manuf.* **7**, 63–67 (1997).
64. Bao, Z. Materials and fabrication needs for low-cost organic transistor circuits. *Adv. Mater.* **12**, 227–230 (2000).
65. Heeger, A.J., MacDiarmid, A.G. & Shirakawa, H. The Nobel Prize in chemistry, 2000: Conductive polymers. *Stockh. Swed. R. Swed. Acad. Sci.* (2000).
66. Krebs, F.C. Fabrication and processing of polymer solar cells: A review of printing and coating techniques. *Sol. Energy Mater. Sol. Cells* **93**, 394–412 (2009).
67. Krebs, F.C. *Polymeric Solar Cells: Materials, Design, Manufacture.* DEStech Publications, Lancaster, PA, 2010.
68. Søndergaard, R., Hösel, M., Angmo, D., Larsen-Olsen, T.T. & Krebs, F.C. Roll-to-roll fabrication of polymer solar cells. *Mater. Today* **15**, 36–49 (2012).
69. Søndergaard, R.R., Hösel, M. & Krebs, F.C. Roll-to-Roll fabrication of large area functional organic materials. *J. Polym. Sci. Part B Polym. Phys.* **51**, 16–34 (2013).
70. Bharathan, J. & Yang, Y. Polymer electroluminescent devices processed by inkjet printing: I. Polymer light-emitting logo. *Appl. Phys. Lett.* **72**, 2660–2662 (1998).
71. Mäkelä, T., Jussila, S., Vilkman, M., Kosonen, H. & Korhonen, R. Roll-to-roll method for producing polyaniline patterns on paper. *Synth. Met.* **135–136**, 41–42 (2003).
72. Mäkelä, T. et al. Utilizing roll-to-roll techniques for manufacturing source-drain electrodes for all-polymer transistors. *Synth. Met.* **153**, 285–288 (2005).
73. Bock, K. Polymer electronics systems: Polytronics. *Proc. IEEE* **93**, 1400–1406 (2005).
74. Zielke, D. et al. Polymer-based organic field-effect transistor using offset printed source/drain structures. *Appl. Phys. Lett.* **87**, 123508 (2005).
75. Sheraw, C.D. et al. Organic thin-film transistor-driven polymer-dispersed liquid crystal displays on flexible polymeric substrates. *Appl. Phys. Lett.* **80**, 1088 (2002).
76. Klauk, H. *Organic Electronics: Materials, Manufacturing, and Applications.* John Wiley & Sons, Weinheim, Germany, 2006.
77. So, F. *Organic Electronics: Materials, Processing, Devices and Applications.* CRC Press, Boca Raton, FL, 2010.
78. Iwamoto, M., Kwon, Y.-S. & Lee, T. *Nanoscale Interface for Organic Electronics.* World Scientific, Singapore, 2011.
79. Klauk, H. *Organic Electronics II: More Materials and Applications.* John Wiley & Sons, Weinheim, Germany, 2012.
80. Forrest, S.R. & Thompson, M.E. Introduction: Organic electronics and optoelectronics. *Chem.Rev.* **107**, 923–925 (2007).
81. Kelley, T.W. et al. Recent progress in organic electronics: Materials, devices, and processes. *Chem. Mater.* **16**, 4413–4422 (2004).
82. Shaw, J.M. & Seidler, P.F. Organic electronics: Introduction. *IBM J. Res. Dev.* **45**, 3–9 (2001).
83. Dimitrakopoulos, C.D. & Malenfant, P.R.L. Organic thin film transistors for large area electronics. *Adv. Mater.* **14**, 99–117 (2002).
84. Moliton, A. & Hiorns, R.C. Review of electronic and optical properties of semiconducting π-conjugated polymers: Applications in optoelectronics. *Polym. Int.* **53**, 1397–1412 (2004).

85. De Leeuw, D.M., Simenon, M.M.J., Brown, A.R. & Einerhand, R.E.F. Stability of n-type doped conducting polymers and consequences for polymeric microelectronic devices. *Synth. Met.* **87**, 53–59 (1997).
86. Britton, D.T. & Härting, M. Printed nanoparticulate composites for silicon thick-film electronics. *Pure Appl. Chem.* **78**, 1723–1739 (2009).
87. Perelaer, J., de Gans, B.-J. & Schubert, U.S. Ink-jet printing and microwave sintering of conductive silver tracks. *Adv. Mater.* **18**, 2101–2104 (2006).
88. Noh, Y.-Y., Zhao, N., Caironi, M. & Sirringhaus, H. Downscaling of self-aligned, all-printed polymer thin-film transistors. *Nat. Nanotechnol.* **2**, 784–789 (2007).
89. Siden, J., Olsson, T., Koptioug, A. & Nilsson, H.-E. Reduced amount of conductive ink with gridded printed antennas. In *5th Int. Conf. Polym. Adhes. Microelectron. Photonics Polytronic,* 86–89 (2005). doi:10.1109/POLYTR.2005.1596493.
90. Harrey, P.M., Evans, P.S.A., Ramsey, B.J. & Harrison, D.J. Interdigitated capacitors by offset lithography. *J. Electron. Manuf.* **10**, 69–77 (2000).
91. Cheng, K. et al. Ink-jet printing, self-assembled polyelectrolytes, and electroless plating: Low cost fabrication of circuits on a flexible substrate at room temperature. *Macromol. Rapid Commun.* **26**, 247–264 (2005).
92. Chung, J., Ko, S., Bieri, N.R., Grigoropoulos, C.P. & Poulikakos, D. Conductor microstructures by laser curing of printed gold nanoparticle ink. *Appl. Phys. Lett.* **84**, 801–803 (2004).
93. Choi, T.Y., Poulikakos, D. & Grigoropoulos, C.P. Fountain-pen-based laser microstructuring with gold nanoparticle inks. *Appl. Phys. Lett.* **85**, 13–15 (2004).
94. Nüchter, M., Ondruschka, B., Bonrath, W. & Gum, A. Microwave assisted synthesis: A critical technology overview. *Green Chem.* **6**, 128–141 (2004).
95. Kordás, K. et al. Inkjet printing of electrically conductive patterns of carbon nanotubes. *Small* **2**, 1021–1025 (2006).
96. Zhou, Y., Hu, L. & Grüner, G. A method of printing carbon nanotube thin films. *Appl. Phys.Lett.* **88**, 123109 (2006).
97. Meitl, M.A. et al. Solution casting and transfer printing single-walled carbon nanotube films. *Nano Lett.* **4**, 1643–1647 (2004).
98. Ishikawa, F.N. et al. Transparent electronics based on transfer printed aligned carbon nano-tubes on rigid and flexible substrates. *ACS Nano* **3**, 73–79 (2009).
99. Rowell, M.W. et al. Organic solar cells with carbon nanotube network electrodes. *Appl. Phys. Lett.* **88**, 233506 (2006).
100. Eda, G., Fanchini, G. & Chhowalla, M. Large-area ultrathin films of reduced graphene oxide as a transparent and flexible electronic material. *Nat. Nanotechnol.* **3**, 270–274 (2008).
101. Liang, X., Fu, Z. & Chou, S.Y. Graphene transistors fabricated via transfer-printing in device active-areas on large wafer. *Nano Lett.* **7**, 3840–3844 (2007).
102. Allen, M.J. et al. Soft transfer printing of chemically converted graphene. *Adv. Mater.* **21**, 2098–2102 (2009).
103. Song, L., Ci, L., Gao, W. & Ajayan, P.M. Transfer printing of graphene using gold film. *ACS Nano* **3**, 1353–1356 (2009).
104. Chen, J.-H. et al. Printed graphene circuits. *Adv. Mater.* **19**, 3623–3627 (2007).
105. Tans, S.J., Verschueren, A.R.M. & Dekker, C. Room-temperature transistor based on a single carbon nanotube. *Nature* **393**, 49–52 (1998).
106. Baughman, R.H., Zakhidov, A.A. & Heer, W.A. de Carbon nanotubes: The route toward applications. *Science* **297**, 787–792 (2002).
107. Werth, J., O'Donnell, S., Lamensdorf, D., Marshall, J. & Teig, L. Carbon nanotube-based electronic devices (2011). http://www.google.com/patents/US8013247.
108. Fecht, H.J. & Brühne, K. *Carbon-based Nanomaterials and Hybrids: Synthesis, Properties, and Commercial Applications.* CRC Press, Boca Raton, FL, 2014.

109. Capano, M. *Special Section: Carbon-based Electronic Materials*. Springer, Pennsylvania, 2009.
110. McEuen, P.L., Fuhrer, M.S. & Park, H. Single-walled carbon nanotube electronics. *IEEE Trans. Nanotechnol.* **1**, 78–85 (2002).
111. Avouris, P., Chen, Z. & Perebeinos, V. Carbon-based electronics. *Nat. Nanotechnol.* **2**, 605–615 (2007).
112. Snow, E.S., Novak, J.P., Campbell, P.M. & Park, D. Random networks of carbon nanotubes as an electronic material. *Appl. Phys. Lett.* **82**, 2145–2147 (2003).
113. Zhang, Y., Liu, C. & Whalley, D. Direct-write techniques for maskless production of microelectronics: A review of current state-of-the-art technologies. in *Int. Conf. Electron. Packag. Technol. High Density Packag. 2009 ICEPT-HDP 09*, 497–503 (2009). doi:10.1109/ICEPT.2009.5270702.
114. Chrisey, D.B. The power of direct writing. *Science* **289**, 879–881 (2000).
115. Aydemir, N. et al. Direct writing of conducting polymers. *Macromol. Rapid Commun.* **34**, 1296–130 (2013). doi:10.1002/marc.201300386.
116. Ahn, B.Y. et al. Omnidirectional printing of flexible, stretchable, and spanning silver microelectrodes. *Science* **323**, 1590–1593 (2009).
117. Boley, J.W., White, E.L., Chiu, G.T.-C. & Kramer, R.K. Direct writing of gallium-indium alloy for stretchable electronics. *Adv. Funct. Mater.* (2014). doi:10.1002/adfm.201303220.
118. Farahani, R.D. et al. Direct-write fabrication of freestanding nanocomposite strain sensors. *Nanotechnology* **23**, 085502 (2012).
119. Gao, Y., Li, H. & Liu, J. Direct writing of flexible electronics through room temperature liquid metal ink. *PLoS One* **7**, e45485 (2012).
120. Lopes, A.J., MacDonald, E. & Wicker, R.B. Integrating stereolithography and direct print technologies for 3D structural electronics fabrication. *Rapid Prototyp. J.* **18**, 129–143 (2012).
121. Navarrete, M. et al. *Integrated Layered Manufacturing of a Novel Wireless Motion Sensor System with GPS, Presented at 18th Symposium on Solid Freeform Fabrication*, Austin, TX. vii–ix, 1-609, U.S. Government or Federal Rights License. August 6–8(2007).
122. Palmer, J.A. et al. Realizing 3-d interconnected direct write electronics within smart stereo-lithography structures. *ASME 2005 International Mechanical Engineering Congress and Exposition on Electronic and Photonic Packaging, Electrical Systems Design and Photonics, and Nanotechnology*, Orlando, FL, November 5–11, 287–293 (2005). doi:10.1115/IMECE2005-79360.
123. Medina, F. et al. Hybrid manufacturing: Integrating direct-write and stereolithography. In *Proc. 2005 Solid Free. Fabr.* (2005). http://edge.rit.edu/edge/P10551/public/SFF/SFF%20 2005. %20Proceedings/Manuscripts%202005/05-Medina.pdf.
124. Espalin, D., Muse, D.W., MacDonald, E. & Wicker, R.B. 3D Printing multifunctionality: Structures with electronics. *Int. J. Adv. Manuf. Technol.* **72**, 963–978 (2014).
125. Medina, F. et al. Integrating multiple rapid manufacturing technologies for developing advanced customized functional devices. In *Rapid Prototyp. Manuf. 2005 Conf. Proc.*, Michigan, Emerald Group Publishing Limited, Bingley, 10–12 (2005).
126. Robinson, C.J., Stucker, B., Lopes, A.J., Wicker, R.B. & Palmer, J.A. Integration of direct-write (DW) and ultrasonic consolidation (UC) technologies to create advanced structures with embedded electrical circuitry. In *Proc. 17th Annu. Solid Free. Fabr. Symp.*, University of Texas, Austin, TX, 60–69. Society of Manufacturing Engineers, Austin, TX, 2006.
127. Siggard, E.J., Madhusoodanan, A.S., Stucker, B. & Eames, B. Structurally embedded electrical systems using ultrasonic consolidation (UC). In *Proc. 17th Solid Free. Fabr. Symp.* Austin, TX, August (American Welding Society, Austin, TX, 2006). http://edge.rit.edu/edge/P10551/public/SFF/SFF%202006%20Proceedings/Manuscripts/07-Siggard.pdf.

128. Siggard, E.J. *Investigative Research Into the Structural Embedding of Electrical and MechanicalSystems Using Ultrasonic Consolidation (UC)* (ProQuest, 2007). http://books.google.com/ books?hl = en&lr = &id = AH1f7mmAGAcC&oi = fnd&pg = PR3&dq = Structurally+embe dded+electrical+systems+using+ultrasonic+consolidat ion+(UC)&ots = BxodKLQaKt&sig = hUxyjagNV7Iikx9GiYhsFjNDkLc.
129. White, D.R. Ultrasonic consolidation of aluminum tooling. *Adv. Mater. Process.* **161**, 64–65 (2003).
130. Ram, G.J., Yang, Y., George, J., Robinson, C. & Stucker, B. Improving linear weld density in ultrasonically consolidated parts. In *Proc. 17th Solid Free. Fabr. Symp.* 692–708. WH Freeman, Austin, TX, 2006. http://utwired.engr.utexas.edu/lff/symposium/proceedingsArchive/pubs/ Manuscripts/2006/2006-60-Ram.pdf.
131. Obielodan, J.O. et al. Optimization of the shear strengths of ultrasonically consolidated Ti/Al 3003. dual-material structures. *J. Mater. Process. Technol.* **211**, 988–995 (2011).
132. Gibson, I., Rosen, D.W. & Stucker, B. *Additive Manufacturing Technologies: Rapid Prototyping to Direct Digital Manufacturing.* Springer, Spring street, NY, 2010.
133. Ram, G.D.J., Robinson, C., Yang, Y. & Stucker, B.E. Use of ultrasonic consolidation for fabrication of multi-material structures. *Rapid Prototyp. J.* **13**, 226–235 (2007).
134. Mosher, T. & Stucker, B. In *Space 2004 Conf. Exhib.* American Institute of Aeronautics and Astronautics, Reston, VA, 2004. doi:10.2514/6.2004-6117.
135. Janaki Ram, G.D., Yang, Y. & Stucker, B.E. Effect of process parameters on bond formation during ultrasonic consolidation of aluminum alloy 3003. *J. Manuf. Syst.* **25**, 221–238 (2006).
136. Meitl, M.A. et al. Transfer printing by kinetic control of adhesion to an elastomeric stamp. *Nat Mater* **5**, 33–38 (2006).
137. Lee, C.H., Kim, D.R. & Zheng, X. Fabricating nanowire devices on diverse substrates by simple transfer-printing methods. *Proc. Natl. Acad. Sci.* **107**, 9950–9955 (2010).
138. Lee, C.H., Kim, D.R. & Zheng, X. Fabrication of nanowire electronics on nonconventional substrates by water-assisted transfer printing method. *Nano Lett.* **11**, 3435–3439 (2011).
139. Weisse, J.M., Lee, C.H., Kim, D.R. & Zheng, X. Fabrication of flexible and vertical silicon nanowire electronics. *Nano Lett.* **12**, 3339–3343 (2012).
140. Keum, H. et al. Silicon micro-masonry using elastomeric stamps for three-dimensional microfabrication. *J. Micromech. Microeng.* **22**, 055018 (2012).
141. Hsia, K.J. et al. Collapse of stamps for soft lithography due to interfacial adhesion. *Appl. Phys. Lett.* **86**, 154106 (2005).
142. Huang, Y.Y. et al. Stamp collapse in soft lithography. *Langmuir* **21**, 8058–8068 (2005).
143. Chen, H., Feng, X., Huang, Y., Huang, Y. & Rogers, J.A. Experiments and viscoelastic analysis of peel test with patterned strips for applications to transfer printing. *J. Mech. Phys. Solids* **61**, 1737–1752 (2013).
144. Huang, Y., Duan, X., Wei, Q. & Lieber, C.M. Directed assembly of one-dimensional nanostructures into functional networks. *Science* **291**, 630–633 (2001).
145. Smith, P.A. et al. Electric-field assisted assembly and alignment of metallic nanowires. *Appl. Phys. Lett.* **77**, 1399–1401 (2000).
146. Collins, P.G., Arnold, M.S. & Avouris, P. Engineering carbon nanotubes and nanotube circuits using electrical breakdown. *Science* **292**, 706–709 (2001).
147. Whang, D., Jin, S., Wu, Y. & Lieber, C.M. Large-scale hierarchical organization of nanowire arrays for integrated nanosystems. *Nano Lett.* **3**, 1255–1259 (2003).
148. Zhang, J., Szczech, J., Skinner, J. & Gamota, D. In *Direct-Write Technol. Rapid Prototyp.* (Piqué, A.) 33–54. Academic Press, San Diego, CA, 2002. http://www.sciencedirect.com/ science/article/pii/B9780121742317500543.
149. Williams, R.T., Wrisley, D.B., Jr & Wu, J.C. Pressure bonding layer of noble metal to base metal substrate. (1991).

150. Drew, R.F., Jr, Schmidt, F.J.S. & Vanduynhoven, T.J. Laser deposition of metal upon transparent materials. (Defensive publication 1979). http://www.google.tl/patents/CA1105093A1?cl=en.
151. Owen, M.D. Laser system and method for plating vias (1997). https://www.google.com/patents/US5614114.
152. Chrisey, D.B. & Hubler, G.K. Pulsed laser deposition of thin films. In *Pulsed Laser Depos. Thin Films* (Chrisey D.B & Hubler, G.K) 648 (Wiley-VCH 2003).
153. Park, B.K., Kim, D., Jeong, S., Moon, J. & Kim, J.S. Direct writing of copper conductive patterns by ink-jet printing. *Thin Solid Films* **515**, 7706–7711 (2007).
154. Woo, K., Kim, D., Kim, J.S., Lim, S. & Moon, J. Ink-jet printing of Cu–Ag-based highly conductive tracks on a transparent substrate. *Langmuir* **25**, 429–433 (2009).
155. Modi, R., Wu, H.D., Auyeung, R.C.Y., Gilmore, C.M. & Chrisey, D.B. Direct writing of polymer thick film resistors using a novel laser transfer technique. *J. Mater. Res.* **16**, 3214–3222 (2001).
156. Pique, A. et al. A novel laser transfer process for direct writing of electronic and sensor materials. *Appl. Phys. Mater. Sci. Process.* **69**, S279–S284 (1999).
157. Kaydanova, T. et al. Direct-write inkjet printing for fabrication of barium strontium titanate-based tunable circuits. *Thin Solid Films* **515**, 3820–3824 (2007).
158. Young, D. et al. Dielectric properties of oxide structures by a laser-based direct-writing method. *J. Mater. Res.* **16**, 1720–1725 (2001).
159. Sirringhaus, H. et al. High-resolution inkjet printing of all-polymer transistor circuits. *Science* **290**, 2123–2126 (2000).
160. De Gans, B.J., Duineveld, P.C. & Schubert, U.S. Inkjet printing of polymers: State of the art and future developments. *Adv. Mater.* **16**, 203–213 (2004).
161. Hudd, A. Inkjet printing technologies. In *Chem. Inkjet Inks*, 3–18. World Scientific Publishers, Singapore, 2010.
162. Rioboo, R., Marengo, M. & Tropea, C. Time evolution of liquid drop impact onto solid, dry surfaces. *Exp. Fluids* **33**, 112–124 (2002).
163. Derby, B. & Reis, N. Inkjet printing of highly loaded particulate suspensions. *MRS Bull.* **28**, 815–818 (2003).
164. Tekin, E., Smith, P.J. & Schubert, U.S. Inkjet printing as a deposition and patterning tool for polymers and inorganic particles. *Soft. Matter.* **4**, 703–713 (2008).
165. Wang, T. & Derby, B. Ink-jet printing and sintering of PZT. *J. Am. Ceram. Soc.* **88**, 2053–2058 (2005).
166. Magdassi, S. Ink requirements and formulations guidelines. In *Chem. Inkjet Inks*. Word Scietific Publishers, Singapore 2010. doi:10.1142/ 9789812818225_0002.
167. Kodas, T.T. et al. Low viscosity precursor compositions and methods for the deposition of conductive electronic features (2010). http://www.google.com/patents/US20070120096.
168. Perez, K.B. & Williams, C.B. *Combining Additive Manufacturing and Direct Write for Integrated Electronics—A Review.* (2013). http://utwired.engr.utexas.edu/lff/symposium/proceedingsArchive/pubs/Manuscripts/2013/2013-77-Perez.pdf.
169. Wallace, D.B., Royall Cox, W. & Hayes, D.J. In *Direct-Write Technol. Rapid Prototyp.* (Piqué, A.) 177–227. Academic Press, San Diego, CA, 2002. http://www.sciencedirect.com/science/article/pii/ B9780121742317500609.
170. Nir, M.M. et al. Electrically conductive inks for inkjet printing. In *Ink Jet Inks* (2010). http:// http://www.worldscientific.com/doi/pdf/10.1142/9789812818225_0012.
171. Korvink, J.G., Smith, P.J. & Shin, D.-Y. *Inkjet-Based Micromanufacturing* (John Wiley & Sons, Weinheim, Germany, 2012).
172. Magdassi, S. *The Chemistry of Inkjet Inks.* **20** (World Scientific Publishers, Singapore, 2010).
173. Forrest, S.R. The path to ubiquitous and low-cost organic electronic appliances on plastic. *Nature* **428**, 911–918 (2004).

174. Van Osch, T.H.J., Perelaer, J., de Laat, A.W.M. & Schubert, U.S. Inkjet printing of narrow conductive tracks on untreated polymeric substrates. *Adv. Mater.* **20**, 343–345 (2008).
175. Perelaer, J., Klokkenburg, M., Hendriks, C.E. & Schubert, U.S. Microwave flash sintering of inkjet-printed silver tracks on polymer substrates. *Adv. Mater.* **21**, 4830–4834 (2009).
176. Magdassi, S., Grouchko, M. & Kamyshny, A. Copper nanoparticles for printed electronics: Routes towards achieving oxidation stability. *Materials* **3**, 4626–4638 (2010).
177. Hayes, D.J., Wallace, D.B. & Cox, W.R. MicroJet printing of solder and polymers for multi-chip modules and chip-scale packages, *J. Microelectron. Packag. Soc.*, (1999) http://microfab.com/images/papers/imaps99.pdf.
178. Liu, Q. & Orme, M. High precision solder droplet printing technology and the state-of-the-art. *J. Mater. Process. Technol.* **115**, 271–283 (2001).
179. Cox, W.R. et al. Microjetted lenslet triplet fibers. *Opt. Commun.* **123**, 492–496 (1996).
180. Hayes, D.J., Grove, M.E., Wallace, D.B., Chen, T. & Cox, W.R. Inkjet printing in the manufacture of electronics, photonics, and displays. in **4809**, 94–99 (2002).
181. Fuller, S.B., Wilhelm, E.J. & Jacobson, J.M. Ink-jet printed nanoparticle microelectromechanical systems. *J. Microelectromechanical Syst.* **11**, 54–60 (2002).
182. Grouchko, M., Kamyshny, A. & Magdassi, S. Formation of air-stable copper–silver core–shell nanoparticles for inkjet printing. *J. Mater. Chem.* **19**, 3057–3062 (2009).
183. Yarin, A.L. Drop impact dynamics: Splashing, spreading, receding, bouncing. *Annu. Rev. Fluid Mech.* **38**, 159–192 (2006).
184. Cheng, S.X., Li, T. & Chandra, S. Producing molten metal droplets with a pneumatic dropleton-demand generator. *J. Mater. Process. Technol.* **159**, 295–302 (2005).
185. Kahn, B.E. The M3D aerosol jet system, an alternative to inkjet printing for printed electronics. *Org. Print. Electron.* **1**, 14–17 (2007).
186. King, B. & Renn, M. Aerosol jet® direct write printing for mil-aero electronic applications. In *PaloAlto Colloq. Lockheed Martin* (2009). http://www.optomec.com/wp-content/uploads/2014/04/ Optomec_Aerosol_Jet_Direct_Write_Printing_for_Mil_Aero_Electronic_Apps.pdf.
187. Vongutfeld, R., Acosta, R. & Romankiw, L. Laser-enhanced plating and etching: Mechanisms and applications. *Ibm J. Res. Dev.* **26**, 136–144 (1982).
188. Hedges, M. & Marin, A.B. 3D aerosol jet® printing-adding electronics functionality to RP/ RM. In *Proc DDMC 2012 Conf.*, Berlin, Germany (2012). http://aerosoljet.com/downloads/ Optomec_NEOTECH_DDMC_3D_Aerosol_Jet_Printing.pdf.
189. Horais, B., Love, L. & Dehoff, R. The use of additive manufacturing for fabrication of multifunction small satellite structures. *AIAAUSU Conf. Small Satell.* (2013). http://digitalcommons. usu.edu/smallsat/2013/all2013/64.
190. Arnold, C.B., Serra, P. & Piqué, A. Laser direct-write techniques for printing of complex materials. *Mrs Bull.* **32**, 23–31 (2007).
191. Cumpston, B.H. et al. Two-photon polymerization initiators for three-dimensional optical data storage and microfabrication. *Nature* **398**, 51–54 (1999).
192. Campbell, M., Sharp, D.N., Harrison, M.T., Denning, R.G. & Turberfield, A.J. Fabrication of photonic crystals for the visible spectrum by holographic lithography. *Nature* **404**, 53–56 (2000).
193. Lim, D., Kamotani, Y., Cho, B., Mazumder, J. & Takayama, S. Fabrication of microfluidic mixers and artificial vasculatures using a high-brightness diode-pumped Nd:YAG laser direct write method. *Lab. Chip* **3**, 318–323 (2003).
194. Takesada, M. et al. Micro-character printing on a diamond plate by femtosecond infrared optical pulses. *Jpn. J. Appl. Phys.* **42**, 4613 (2003).
195. Chrisey, D.B. et al. Direct writing of conformal mesoscopic electronic devices by MAPLE DW. *Appl. Surf. Sci.* **168**, 345–352 (2000).

196. Wu, P.K. et al. Laser transfer of biomaterials: Matrix-assisted pulsed laser evaporation (MAPLE) and MAPLE direct write. *Rev. Sci. Instrum.* **74**, 2546–2557 (2003).
197. Deckard, C.R. Method and apparatus for producing parts by selective sintering (1989). http:// www.google.com/patents/US4863538.
198. Kumar, S. Selective laser sintering: A qualitative and objective approach. *JOM* **55**, 43–47 (2003).
199. Wallenberger, F.T. Rapid prototyping directly from the vapor phase. *Science* **267**, 1274–1275 (1995).
200. Zhang, X., Jiang, X.N. & Sun, C. Micro-stereolithography of polymeric and ceramic microstructures. *Sens. Actuators Phys.* **77**, 149–156 (1999).
201. Kawata, S., Sun, H.B., Tanaka, T. & Takada, K. Finer features for functional microdevices: Micromachines can be created with higher resolution using two-photon absorption. *Nature* **412**, 697–698 (2001).
202. Provin, C., Monneret, S., Le Gall, H. & Corbel, S. Three-dimensional ceramic microcomponents made using microstereolithography. *Adv. Mater.* **15**, 994–997 (2003).
203. Gross, M.E., Appelbaum, A. & Schnoes, K.J. A chemical and mechanistic view of reaction profiles in laser direct-write metallization in metallo-organic films. *J. Appl. Phys.* **60**, 529–533 (1986).
204. Gross, M.E., Appelbaum, A. & Gallagher, P.K. Laser direct-write metallization in thin palladium acetate films. *J. Appl. Phys.* **61**, 1628–1632 (1987).
205. Chen, Q.J., Imen, K. & Allen, S.D. Laser enhanced electroless plating of micron-scale copper wires. *J. Electrochem. Soc.* **147**, 1418–1422 (2000).
206. Shrivastva, P.B., Boose, C.A., Kolster, B.H., Harteveld, C. & Meinders, B. Selective metallization of alumina by laser. *Surf. Coat. Technol.* **46**, 131–138 (1991).
207. Bohandy, J., Kim, B. & Adrian, F. Metal-deposition from a supported metal-film using an excimer laser. *J. Appl. Phys.* **60**, 1538–1539 (1986).
208. Bohandy, J., Kim, B., Adrian, F. & Jette, A. Metal-deposition at 532-nm using a laser transfer technique. *J. Appl. Phys.* **63**, 1158–1162 (1988).
209. Mir-Hosseini, N., Schmidt, M.J.J. & Li, L. Growth of patterned thin metal oxide films on glass substrates from metallic bulk sources using a Q-switched YAG laser. *Appl. Surf. Sci.* **248**, 204–208 (2005).
210. Kuebler, S.M. et al. Optimizing two-photon initiators and exposure conditions for threedimensional lithographic microfabrication. *J. Photopolym. Sci. Technol.* **14**, 657–668 (2001).
211. McNeal, M.P., Jang, S.-J. & Newnham, R.E. Particle size dependent high frequency dielectric properties of barium titanate. *Proc 10th IEEE Int Symp Appl Ferroelectr* **2**, 837–840 (1996).
212. Li, B., Clark, P.A. & Church, K.H. Robust direct-write dispensing tool and solutions for micro/meso-scale manufacturing and packaging. In *ASME Int. Manuf. Sci. Eng. Conf.* 715–721 (American Society of Mechanical Engineers, 2007). http://proceedings. asmedigitalcollection. asme.org/proceeding.aspx?articleid=1598546.
213. Mott, M. & Evans, J.R.G. Zirconia/alumina functionally graded material made by ceramic ink jet printing. *Mater. Sci. Eng. A* **271**, 344–352 (1999).
214. Cesarano III, J.C. & Calvert, P.D. Freeforming objects with low-binder slurry (2000). http:// www.google.com/patents/US6027326.
215. Sachs, E., Cima, M., Williams, P., Brancazio, D. & Cornie, J. Three dimensional printing: Rapid tooling and prototypes directly from a CAD model. *J. Eng. Ind.* **114**, 481–488 (1992).
216. Crump, S.S. Apparatus and method for creating three-dimensional objects (1992). http://www. google.com/patents/US5121329.
217. Mireles, J. et al. Development of a fused deposition modeling system for low melting temperature metal alloys. *J. Electron. Packag.* **135**, 011008–011008 (2013).

218. Diegel, O., Singamneni, S., Huang, B. & Gibson, I. Curved layer fused deposition modeling in conductive polymer additive manufacturing. *Adv. Mater. Res.* **199–200**, 1984–1987 (2011).
219. Cai, Z., Li, X., Hu, Q. & Zeng, X. Study on thick-film PTC thermistor fabricated by micro-pen direct writing. *Microelectron. J.* **39**, 1452–1456 (2008).
220. Castillo, S., Muse, D., Medina, F., MacDonald, E. & Wicker, R. Electronics integration in conformal substrates fabricated with additive layered manufacturing. In *Proc. 20th Annu. Solid Free. Fabr. Symp.*, University of Texas, Austin TX, 730–737 (2009).
221. Sun, K. et al. 3D printing of interdigitated LI-ion microbattery architectures. *Adv. Mater.* **25**, 4539–4543 (2013).
222. Russo, A. et al. Pen-on-paper flexible electronics. *Adv. Mater.* **23**, 3426–3430 (2011).
223. Ahn, B.Y., Lorang, D.J. & Lewis, J.A. Transparent conductive grids via direct writing of silver nanoparticle inks. *Nanoscale* **3**, 2700–2702 (2011).
224. Siegel, A.C., Phillips, S.T., Wiley, B.J. & Whitesides, G.M. Thin, lightweight, foldable thermochromic displays on paper. *Lab. Chip* **9**, 2775–2781 (2009).
225. Dragoman, M., Flahaut, E., Dragoman, D., Al Ahmad, M. & Plana, R. Writing simple RF electronic devices on paper with carbon nanotube ink. *Nanotechnology* **20**, 375203 (2009).
226. Jung, M. et al. All-printed and roll-to-roll-printable 13.56-MHz-operated 1-bit RF tag on plastic foils. *IEEE Trans. Electron Devices* **57**, 571–580 (2010).
227. Nyström, G., Razaq, A., Strømme, M., Nyholm, L. & Mihranyan, A. Ultrafast all-polymer paper-based batteries. *Nano Lett.* **9**, 3635–3639 (2009).
228. Adams, J.J. et al. Conformal printing of electrically small antennas on three-dimensional surfaces. *Adv. Mater.* **23**, 1335–1340 (2011).
229. Ahn, B.Y. et al. Planar and three-dimensional printing of conductive inks. *J. Vis. Exp. Jove* (2011). doi:10.3791/3189.
230. Lewis, J.A. & Gratson, G.M. Direct writing in three dimensions. *Mater. Today* **7**, 32–39 (2004).
231. Walker, S.B. & Lewis, J.A. Reactive silver inks for patterning high-conductivity features at mild temperatures. *J. Am. Chem. Soc.* **134**, 1419–1421 (2012).
232. Lewis, J.A., Smay, J.E., Stuecker, J. & Cesarano, J. Direct ink writing of three-dimensional ceramic structures. *J. Am. Ceram. Soc.* **89**, 3599–3609 (2006).
233. Hu, J. & Yu, M.-F. Meniscus-confined three-dimensional electrodeposition for direct writing of wire bonds. *Science* **329**, 313–316 (2010).
234. Lewis, J.A. Colloidal processing of ceramics. *J. Am. Ceram. Soc.* **83**, 2341–2359 (2000).
235. Smay, J.E., Cesarano, J. & Lewis, J.A. Colloidal inks for directed assembly of 3-D periodic structures. *Langmuir* **18**, 5429–5437 (2002).
236. Smay, J. E., Gratson, G. M., Shepherd, R. F., Cesarano, J. & Lewis, J. A. Directed colloidal assembly of 3D periodic structures. *Adv. Mater.* **14**, 1279–1283 (2002).
237. Guo, J.J. & Lewis, J.A. Aggregation effects on the compressive flow properties and drying behavior of colloidal silica suspensions. *J. Am. Ceram. Soc.* **82**, 2345–2358 (1999).
238. Therriault, D., Shepherd, R.F., White, S.R. & Lewis, J.A. Fugitive inks for direct-write assembly of three-dimensional microvascular networks. *Adv. Mater.* **17**, 395–399 (2005).
239. Xu, M., Gratson, G.M., Duoss, E.B., Shepherd, R.F. & Lewis, J.A. Biomimetic silicification of 3D polyamine-rich scaffolds assembled by direct ink writing. *Soft Matter* **2**, 205–209 (2006).
240. Therriault, D., White, S.R. & Lewis, J.A. Chaotic mixing in three-dimensional microvascular networks fabricated by direct-write assembly. *Nat. Mater.* **2**, 265–271 (2003).
241. Gratson, G.M., Xu, M. & Lewis, J.A. Microperiodic structures: Direct writing of three-dimensional webs. *Nature* **428**, 386–386 (2004).

242. Lebel, L.L., Aissa, B., Khakani, M.A.E. & Therriault, D. Ultraviolet-assisted direct-write fabrication of carbon nanotube/polymer nanocomposite microcoils. *Adv. Mater.* **22**, 592–596 (2010).
243. Michna, S., Wu, W. & Lewis, J.A. Concentrated hydroxyapatite inks for direct-write assembly of 3-D periodic scaffolds. *Biomaterials* **26**, 5632–5639 (2005).
244. Smay, J.E., Cesarano, J., Tuttle, B.A. & Lewis, J.A. Directed colloidal assembly of linear and annular lead zirconate titanate arrays. *J. Am. Ceram. Soc.* **87**, 293–295 (2004).
245. San Marchi, C., Kouzeli, M., Rao, R., Lewis, J.A. & Dunand, D.C. Alumina–aluminum inter-penetrating-phase composites with three-dimensional periodic architecture. *Scr. Mater.* **49**, 861–866 (2003).
246. Cesarano III, J. & Calvert, P.D. US Patent 6,027,326, 2000; Li, Q. & Lewis, J.A. *Adv Mater* **15**, 1639 (2003).
247. Li, Q. & Lewis, J. A. Nanoparticle inks for directed assembly of three-dimensional periodic structures. *Adv. Mater.* **15**, 1639–1643 (2003).
248. Gratson, G.M. & Lewis, J.A. Phase behavior and rheological properties of polyelectrolyte inks for direct-write assembly. *Langmuir* **21**, 457–464 (2005).
249. Ahn, B.Y., Lorang, D.J., Duoss, E.B. & Lewis, J.A. Direct-write assembly of microperiodic planar and spanning ITO microelectrodes. *Chem. Commun.* **46**, 7118–7120 (2010).
250. Duoss, E.B., Twardowski, M. & Lewis, J.A. Sol-gel inks for direct-write assembly of functional oxides. *Adv. Mater.* **19**, 3485–3489 (2007).
251. Guo, X. et al. Two and three-dimensional folding of thin film single-crystalline silicon for photovoltaic power applications. *Proc. Natl. Acad. Sci. U. S. A.* **106**, 20149–20154 (2009).
252. Yoon, J. et al. Ultrathin silicon solar microcells for semitransparent, mechanically flexible and microconcentrator module designs. *Nat. Mater.* **7**, 907–915 (2008).
253. Smay, J.E., Iii, J.C., Tuttle, B.A. & Lewis, J.A. Piezoelectric properties of 3-X periodic Pb(ZrxTi1–x) O3–polymer composites. *J. Appl. Phys.* **92**, 6119–6127 (2002).
254. Gratson, G.M. et al. Direct-write assembly of three-dimensional photonic crystals: Conversion of polymer scaffolds to silicon hollow-woodpile structures. *Adv. Mater.* **18**, 461–465 (2006).
255. Sun, Y. & Rogers, J.A. Inorganic semiconductors for flexible electronics. *Adv. Mater.* **19**, 1897–1916 (2007).
256. Kim, D.-H. et al. Dissolvable films of silk fibroin for ultrathin conformal bio-integrated electronics. *Nat. Mater.* **9**, 511–517 (2010).
257. Hamedi, M., Forchheimer, R. & Inganäs, O. Towards woven logic from organic electronic fibres. *Nat. Mater.* **6**, 357–362 (2007).
258. Dickey, M.D. et al. Eutectic gallium-indium (EGaIn): A liquid metal alloy for the formation of stable structures in microchannels at room temperature. *Adv. Funct. Mater.* **18**, 1097–1104 (2008).
259. Ladd, C., So, J.-H., Muth, J. & Dickey, M.D. 3D printing of free standing liquid metal microstructures. *Adv. Mater.* **25**, 5081–5085 (2013).
260. Trlica, C., Parekh, D. P., Panich, L., Ladd, C. & Dickey, M. D. 3-D printing of liquid metals for stretchable and flexible conductors. **9083**, 90831D–90831D–10 (2014).
261. Rayleigh, Lord. On the capillary phenomena of jets. *Proc. R. Soc. Lond.* **29**, 71–97 (1879).
262. French, S.J., Saunders, D.J. & Ingle, G.W. The system gallium-indium. *J. Phys. Chem.* **42**, 265–274. (1937).
263. Cademartiri, L. et al. Electrical resistance of Ag-TS-S(CH_2)(n-1)CH_3//Ga_2O_3/EGaIn tunneling junctions. *J. Phys. Chem. C* **116**, 10848–10860 (2012).
264. Baldwin, D.F., Deshmukh, R.D. & Hau, C.S. Gallium alloy interconnects for flip-chip assembly applications. *IEEE Trans. Compon. Packag. Technol.* **23**, 360–366 (2000).

265. Chiechi, R.C., Weiss, E.A., Dickey, M.D. & Whitesides, G.M. Eutectic gallium-indium (EGaIn): A moldable liquid metal for electrical characterization of self-assembled monolayers. *Angew. Chem. Int. Ed.* **47**, 142–144 (2008).
266. Cheng, S., Rydberg, A., Hjort, K. & Wu, Z. Liquid metal stretchable unbalanced loop antenna. *Appl. Phys. Lett.* **94**, 144103–144103–3 (2009).
267. Palleau, E., Reece, S., Desai, S.C., Smith, M.E. & Dickey, M.D. Self-healing stretchable wires for reconfigurable circuit wiring and 3d microfluidics. *Adv. Mater.* **25**, 1589–1592 (2013).
268. Tabatabai, A., Fassler, A., Usiak, C. & Majidi, C. Liquid-phase gallium–indium alloy electronics with microcontact printing. *Langmuir* **29**, 6194–6200 (2013).
269. Lu, T., Finkenauer, L., Wissman, J. & Majidi, C. Rapid prototyping for soft-matter electronics. *Adv. Funct. Mater.* (2014). doi:10.1002/adfm.201303732.
270. Nagel, D.J. In *Direct-Write Technol. Rapid Prototyp.* (Piqué, A.) 557–679 (Academic Press, 2002). http://www.sciencedirect.com/science/article/pii/B9780121742317500725.
271. Campbell, A.N. et al. *Electrical and Chemical Characterization of FIB-Deposited Insulators* (Sandia National Labs, Albuquerque, NM, 1997). http://www.osti.gov/bridge/product.biblio.jsp?osti_id = 532558.
272. Edinger, K., Melngailis, J. & Orloff, J. Study of precursor gases for focused ion beam insulator deposition. *J. Vac. Sci. Technol. B* **16**, 3311–3314 (1998).
273. Edinger, K. in *Direct-Write Technol. Rapid Prototyp.* (Piqué, A.) 347–383. Academic Press, 2002. http://www.sciencedirect.com/science/article/pii/B9780121742317500658.
274. Tjerkstra, R.W., Segerink, F.B., Kelly, J.J. & Vos, W.L. Fabrication of three-dimensional nanostructures by focused ion beam milling. *J. Vac. Sci. Technol. B* **26**, 973–977 (2008).
275. Olivero, P. et al. Ion-beam-assisted lift-off technique for three-dimensional micromachining of freestanding single-crystal diamond. *Adv. Mater.* **17**, 2427–2430 (2005).
276. Matsui, S. In *Springer Handb. Nanotechnol.* (Bhushan, P.B.) 211–229 Springer, Berlin, Germany, 2010. doi:10.1007/978-3-642-02525-9-7.
277. Matsui, S. in *Springer Handb. Nanotechnol.* (Bhushan, P.B.) 179–196. Springer, Berlin, Germany, 2007. doi:10.1007/978-3-540-29857-1_6.
278. Matsui, S. Three-dimensional nanostructure fabrication by focused-ion-beam chemical vapor deposition. *SPIE Conf. Optomech. Micro/Nano Devices Comp. II.* Boston, MA, October 1, 6376, 637602–637615 (2006). http://proceedings.spiedigitallibrary.org/proceeding.aspx?articleid=1332448.
279. Morita, T. et al. Free-space-wiring fabrication in nano-space by focused-ion-beam chemical vapor deposition. *J. Vac. Sci. Technol. B* **21**, 2737–2741 (2003).
280. Dinwiddie, R.B., Dehoff, R.R., Lloyd, P.D., Lowe, L.E. & Ulrich, J.B. Thermographic in-situ process monitoring of the electron-beam melting technology used in additive manufacturing. *SPIE Proc. Thermo. Therm. Infra. Appl.* XXXV. **8705**, 87050K–87050K–9 (2013). doi:10.1117/12.2018412.
281. Dehoff, R. et al. Case study: Additive manufacturing of aerospace brackets. *Adv. Mater. Process.* **171**, 19–22 (2013).

9 Industrial Implementation of Additive Manufacturing

Edward D. Herderick and Clark Patterson

CONTENTS

9.1 INTRODUCTION

Over the past 30 years, AM technology has matured beyond rapid prototyping to become a viable route to producing industrial parts in high-performance metals and polymers.

The maturation of AM technologies is accelerating a transition from the traditional *design for manufacturing* model to a new paradigm of *manufacturing by design*. In this new paradigm, industrial designers are fully enabled to design components based on functionality, rather than limits of assembly technologies.

The goal of this chapter is to present the current state of the art, seed ideas for where the technology can be implemented today, and provide thoughts on where the technology will be in the future. The scope is for industrial applications using structural properties of the materials with an emphasis on metal and polymers owing to their high level of technology readiness.

9.2 APPLICATION OF ADDITIVE TECHNOLOGIES FOR INDUSTRIAL PRODUCTS

There are three key tenets for creating an industrial product: design, materials, and manufacturing processes.[1] The shape-making capability of AM technologies has captured the imagination of the design community since its earliest days for rapid creation of form, fit, and function prototypes. The key hurdle toward industrial implementation is the availability of engineering materials and corresponding

TABLE 9.1
ASTM Defined AM Processes, Example Vendors, and Pros/Cons of Technology Class

ASTM Defined Technology Category	Description	Example Vendor Technology	Advantages for Industrial Parts Production	Challenges for Industrial Parts Production
Binder jetting	Liquid bonding agent is selectively deposited to join powder metal	ExOne M-Flex, Voxeljet	Broadest range of materials, cost effective	Composite microstructure (usually), intensive post-processing may be required
Directed energy deposition	Focused thermal energy is fuses materials by melting as material is deposited	Optomec LENS, Lincoln Electric Hybrid Laser-Arc, Sciaky EBDM	In situ alloying, hardfacing, repair	Difficult to manufacture direct parts
Powder bed fusion	Thermal energy selectively fuses regions of powder bed	EOS DMLS, 3D Systems SLS, Arcam EBM	Engineering materials coupled with high resolution, highest density structures (as printed metals with >99.5% density achievable)	Relatively slow, relatively expensive, limited build volume of parts
Material extrusion	Material is selectively dispensed through a nozzle or orifice	Stratasys Fused Deposition Modeling	Engineering polymers, large builds, fast	Surface roughness due to raster, requires support material
Material jetting	Droplets of build material are selectively deposited	Objet Connex, Optomec Aerosol Jet	Fast, high resolution, multiple materials, inexpensive	Requires materials compatible with jetting (stricter for structural than functional)
Sheet lamination	Sheets of material are bonded to form an object	Fabrisonic VHP-UAM, CAM-LEM	Very large builds (6 ft × 6 ft × 6 ft possible), composites of metal alloys and electronic materials	z-Axis strength penalty, maturing technology
Vat photopolymerization	Liquid photopolymer in vat is selectively cured by light-activated polymerization	3D Systems Stereolithography	Fast, very high resolution, inexpensive	Lack of engineering polymers

technical data that designers can use to create products. Manufacturing processes and materials are firmly linked and share a synergistic interaction where innovation in one leads to further innovation in the other.[2] It is this link for innovation that is driving industrial applications for AM.

Table 9.1 is a brief review of the ASTM designated AM technologies including advantages and disadvantages with respect to industrial implementation.[3] A clear demarcation can be made between those technologies that are currently able to manufacture engineering materials. Practically speaking, for engineering classes of polymers those technologies are material extrusion and powder bed fusion. For metals, those processes are binder jetting, powder bed fusion, directed energy deposition, and sheet lamination.

9.3 DIRECT PART FABRICATION IN ENGINEERING THERMOPLASTICS

The largest current application set for industrial implementation of AM for direct parts is for parts manufactured using thermoplastics.[4] Polymeric air moving duct work made of engineering polyamides manufactured using laser powder bed fusion 3D system's selective laser sintering (3DS SLS) has been implemented by Boeing for several years.[5] That example demonstrated that air moving applications are ideal candidates for AM implementation. The parts are non-load bearing, limiting their structural requirements, and include complex shapes that are difficult to machine or injection mold.

Figures 9.1 through 9.3 show examples of generic ducts made of Ultem 9085 and manufactured using material extrusion on a Stratasys Fortus fused deposition modeling (FDM™) industrial grade AM platform. Ultem 9085 made using FDM has a high tensile strength (10,390 psi), has a high heat deflection temperature of 333°F, and is flame-smoke-toxicity certified.[6] Efforts are underway to provide designers a statistically significant data set for Ultem 9085 FDM to support further implementation.[7]

Figure 9.1 shows an image of a thin-walled hot air moving duct. The wall thickness is only a few tool path passes thick, on the order of 0.050″. Parts can be made leak tight using post-process vapor smoothing that joins any incongruities and smooths z-axis roughness. Parts can be made lighter and more complex than those that could be injection molded. Figure 9.2 shows an air-guiding grate. The vanes in the center of this duct have internal curvature that could not be machined or would be die-locked using traditional injection molding. Figure 9.3 is a high temperature duct attachment interfacing between a round and a flat, rectangular shape. Similar to Figure 9.2, this part would be difficult to machine or difficult to injection mold and is enabled by the shape-making capability of AM.

The ability to select and manufacture different materials is key attribute of AM. Using the same digital file, with modified tool paths, the same part design can be made more economically using commodity materials or with a higher performance material in low volumes to serve different markets on an as-needed basis without dramatic impacts on inventory. Figure 9.4 demonstrates this principle for product

FIGURE 9.1 Thin walled hot air moving duct made from Ultem 9085 using Stratasys FDM™. (Copyright Rapid Prototype and Manufacturing, LLC, Avon Lake, Ohio, 2014.)

FIGURE 9.2 Air guiding grate made from Ultem 9085 using Stratasys FDM™. (Copyright Rapid Prototype and Manufacturing, LLC, Avon Lake, Ohio, 2014.)

design, in it are three single to four-port nozzles showing shape-making capability of AM. On the left is a part made using standard white acrylonitrile butadiene styrene (ABS), on the center is Ultem 9085, and on the right is carbon fiber filled poly-ether-imide all manufactured using FDM. In this way, design teams can source and select properties of components specifically for their application need without significant manufacturing process changes.

There are other emerging applications for manufacturing polymers using AM for functional applications. Many of these are in the medical device space, Figures 9.5

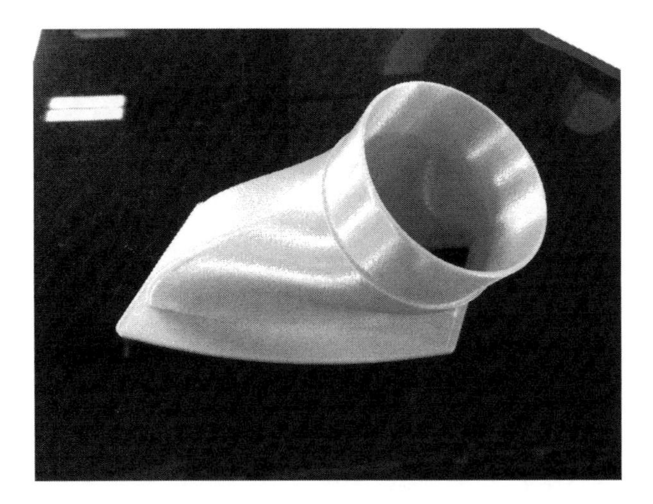

FIGURE 9.3 High temperature duct attachment interfacing between a round and a flat, rectangular shape made from Ultem 9085 using Stratasys FDM™. (Copyright Rapid Prototype and Manufacturing, LLC, Avon Lake, Ohio, 2014.)

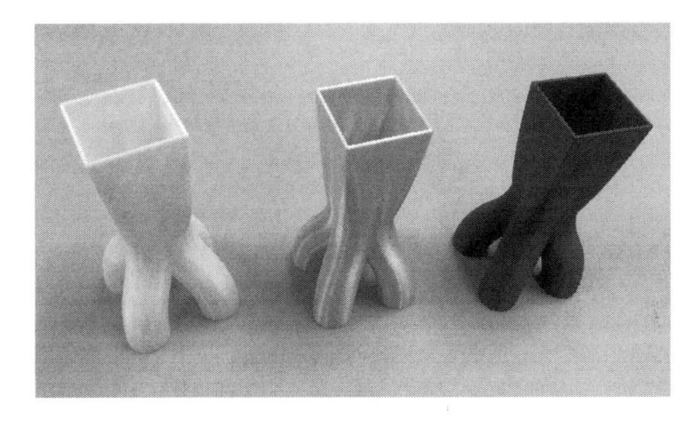

FIGURE 9.4 Three single- to four-port nozzles showing shape-making capability of additive manufacturing. On the left is a part made using standard white ABS, center is Ultem 9085, and on the right is carbon-fiber-filled poly ether imide all manufactured using FDM™. (Copyright Rapid Prototype and Manufacturing, LLC, Avon Lake, Ohio, 2014.)

and 9.6 include examples for medical imaging applications. During normal operation, computed tomography (CT) and positron emission tomography (PET) scanners emit X-rays used to create images that require shielding and filtering. Historically, this has been done using lead or other high z metal shielding made using casting or machining. These applications show a strong value proposition for AM because of their relatively low volumes and complex geometries. The medical imaging components in Figures 9.5 and 9.6 were made from tungsten-loaded poly-carbonate printed using FDM. Figure 9.5 is a mounting bracket and Figure 9.6 is a mounting cup for electronics in a CT scanner. In this case, the tungsten metal in the composite shields

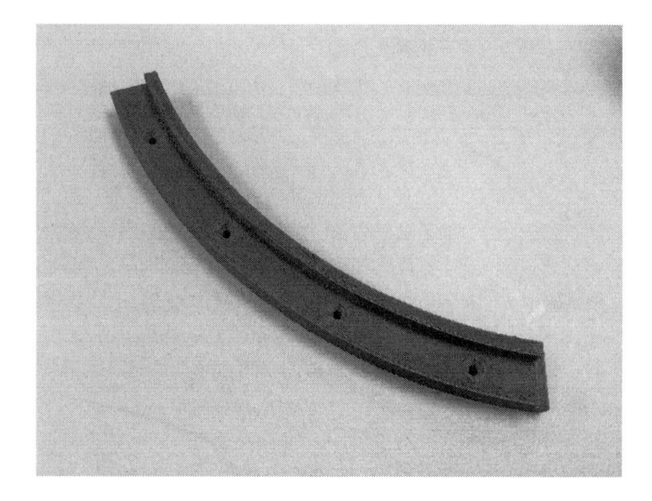

FIGURE 9.5 Mounting bracket made from tungsten loaded polycarbonate printed using FDM™. (Copyright Rapid Prototype and Manufacturing, LLC, Avon Lake, Ohio, 2014.)

FIGURE 9.6 Mounting cup for electronics in a CT scanner made from tungsten loaded polycarbonate printed using FDM™. (Copyright Rapid Prototype and Manufacturing, LLC, Avon Lake, Ohio, 2014.)

the X-rays and the overall material has been tailored to match the shielding characteristics of lead metal castings. AM offers reduced inventory costs, manufacturing without tooling, and higher shape complexity allowing for reduced footprint in the medical imaging devices.

Industrial applications for printed thermoplastics will continue to emerge as more engineering materials are introduced. There are more than 8,000 commercially

FIGURE 9.7 Forming tool set made from chopped carbon fiber loaded poly-ether-imide. (Copyright Rapid Prototype and Manufacturing, LLC, Avon Lake, Ohio, 2014.)

available injection moldable thermoplastics[8]; however, there are less than 20 commercially available AM polymers. This implementation will be particularly valuable for loaded composites and other high-value polymers. An example of an AM composite material component is shown in Figure 9.7, which is a forming tool set made from chopped carbon fiber loaded poly-ether-imide. This material has higher stiffness, a lower CTE (more closely matched to aluminum), and higher thermal conductivity than the related Ultem 9085 and demonstrates the types of materials that are in product development now and will be commercially available over the next 1–3 years.

9.4 APPROACHES TO INDIRECTLY MANUFACTURING PARTS

Applications where AM built parts are indirectly used to manufacture industrial products, such as injection molding and casting, present ripe areas for implementation since the new processes themselves do not need to be recertified.

Injection molding is a mature industry that is highly competitive on performance, cost, and delivery schedule. The core and cavity tool that provide the shape-making capability are themselves expensive to manufacture and require long lead times on the order of months in many cases and therefore require strong business cases to fulfill orders. This means that in general, injection molding is only economical for large volumes of parts on the order of 1,000s or more where the tool cost and schedule can be amortized over many parts. This makes short runs of parts below 1,000 on quick turnaround timelines a high-value niche market that innovative businesses are driven to serve. AM has a strong value proposition for these applications.

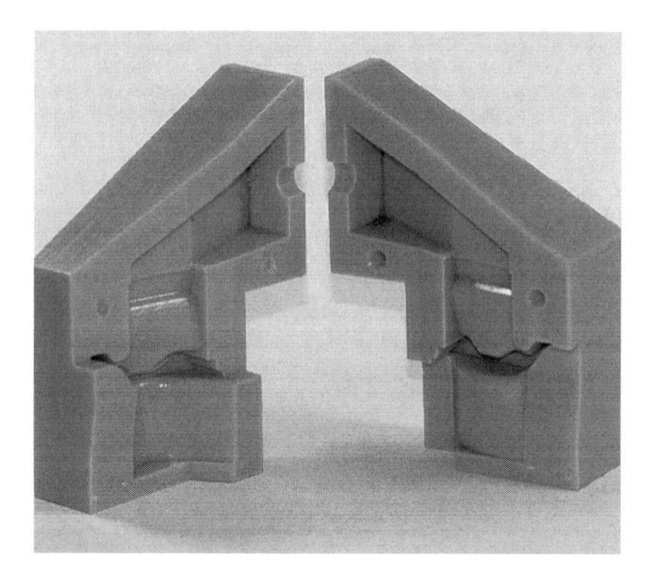

FIGURE 9.8 Example of a short-run injection molding tool made from ABS using Objet material jetting technology. (Copyright Rapid Prototype and Manufacturing, LLC, Avon Lake, Ohio, 2014.)

Figure 9.8 is an example short-run tool made using objet material jetting technology. The two pieces form the cavity into which a polymer is injected thereby forming a component. The material jetting technology uses ABS-like polymers that can be printed in a few hours and can last for a short run of 100–200 parts. The main challenge is that the thermal conductivity of the polymer is lower than standard metal tools increasing cycle times so that the tool is not damaged. For short-run parts, cycle time is typically not a deciding factor when compared to delivery time. When higher temperature or chemical resistance is required, Ultem 9085 manufactured using FDM can be used for higher stress molds and longer part runs. Figure 9.9 is a core and cavity set for thermoplastic elastomer molding. In this example, there are rails on the cavity set that allow for changing the tab features on the top and bottom of the rubber piece.

Although useful, there are many applications where polymer molds will not effectively meet injection molding requirements. For the right value proposition, printed metal injection mold tooling, this includes when tooling is not available quickly enough or with difficult to mold polymers or part geometries. The cost to manufacture may be higher for the printed tool, but printing the tool allows integration of cooling channels that speed cycle times and improve quality leading to holistic life cycle cost reduction. Figure 9.10 shows an example injection molding tool for a fitting made of maraging steel using laser powder bed fusion on an EOS DMLS platform.

Another industrial application of AM technology for indirectly making components is printing of sand molds and cores for metal casting using the binder jetting process.

FIGURE 9.9 Injection mold core and cavity set for thermoplastic elastomer molding made from Ultem 9085 using Stratasys FDM™. (Copyright Rapid Prototype and Manufacturing, LLC, Avon Lake, Ohio, 2014.)

FIGURE 9.10 Example injection molding tool for a fitting made of maraging steel using laser powder bed fusion on an EOS DMLS™ platform. (Copyright Rapid Prototype and Manufacturing, LLC, Avon Lake, Ohio, 2014.)

Figures 9.11 and 9.12 are images of an aluminum sand casting mold made using binder jetting on a Voxeljet vx200 platform. This application is particularly appealing for AM implementation as the same materials, that is, foundry sand and foundry resin, used in the standard process are used in the printing process. Of particular interest are complex core structures used for casting of aluminum fluid moving pump housings.

FIGURE 9.11 Closed aluminum sand casting mold made using binder jetting on a Voxeljet vx200 platform. (Copyright Rapid Prototype and Manufacturing, LLC, Avon Lake, Ohio, 2014.)

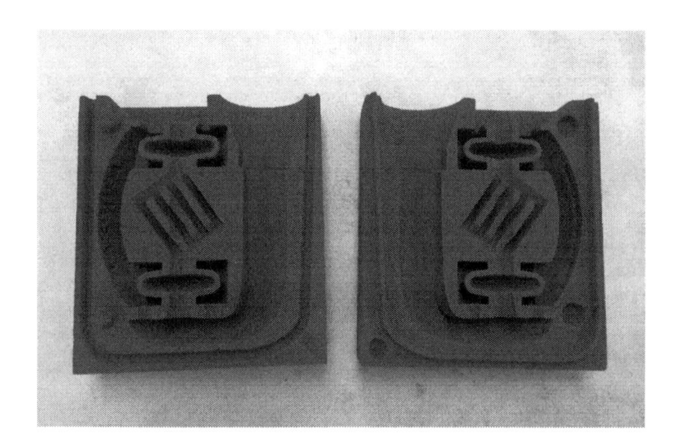

FIGURE 9.12 Open aluminum sand casting mold made using binder jetting on a Voxeljet vx200 platform. (Copyright Rapid Prototype and Manufacturing, LLC, Avon Lake, Ohio, 2014.)

9.5 DIRECT PART FABRICATION IN METALS

The ability to directly produce metal alloy parts using AM is among the fastest growing sectors for the technology and has captured the popular imagination.[9] As the technology has matured, industrial manufacturers have developed a series of use cases where metal parts are difficult or expensive to source with long lead times due to complex castings, machining costs, and diminished supply chain capacity. AM enables turnaround times on the order of weeks without several months long wait for tooling. Furthermore, part consolidation provides a huge benefit as metal assemblies

FIGURE 9.13 Stainless steel PH1 (15-5 Cr-Ni) impeller prototypes bonded to the build plate in an EOS DMLS™ M280 build chamber. (Copyright Rapid Prototype and Manufacturing, LLC, Avon Lake, Ohio, 2014.)

FIGURE 9.14 High-performance automotive rocker arm made from maraging steel MS1 on an EOS DMLS™ platform. (Copyright Rapid Prototype and Manufacturing, LLC, Avon Lake, Ohio, 2014.)

that were originally brazed or joined together can be made in a single piece allowing for greater design flexibility.

In the laser powder bed fusion process for manufacturing metal parts, the use of support materials and structures relative to part orientation is a key determinant of part manufacturability. Figure 9.13 shows several stainless steel PH1 (15-5 Cr-Ni) impeller prototypes bonded to the build plate in the build chamber. The parts are welded to the build plate and support materials are used to promote thermal conduction of heat away from the build layer and prevent warping due to residual stresses. Figure 9.14 is a high-performance automotive rocker arm made from maraging steel

MS1 (a Ni precipitation hardened tool steel alloy). The lighter areas are support material left on the part to demonstrate how supports are required to manufacture open areas in the z direction. The value proposition for this particular application is rapid turnaround time to meet race deadlines and design freedom for new and spare parts.

Figure 9.15 is a functional gear prototype made from PH1 stainless steel. This application use case for laser powder bed fusion is driven by the ability to print a build plate with varying parts for rapid testing of different designs. The conventional method would require machining each design from rod stock. Furthermore, initial production runs could be completed using AM prior to full-scale tooling implementation.

Metal AM parts are also finding strong use cases for fluid moving applications for similar reasons as engineered polymers as discussed in the previous section. Figure 9.16 is a pair of water cooling channels in MS1 using laser powder bed fusion.

FIGURE 9.15 Functional gear prototype made from PH1 stainless steel on an EOS DMLS™ platform. (Copyright Rapid Prototype and Manufacturing, LLC, Avon Lake, Ohio, 2014.)

FIGURE 9.16 Pair of water cooling channels made from MS1 on an EOS DMLS™ platform. (Copyright Rapid Prototype and Manufacturing, LLC, Avon Lake, Ohio, 2014.)

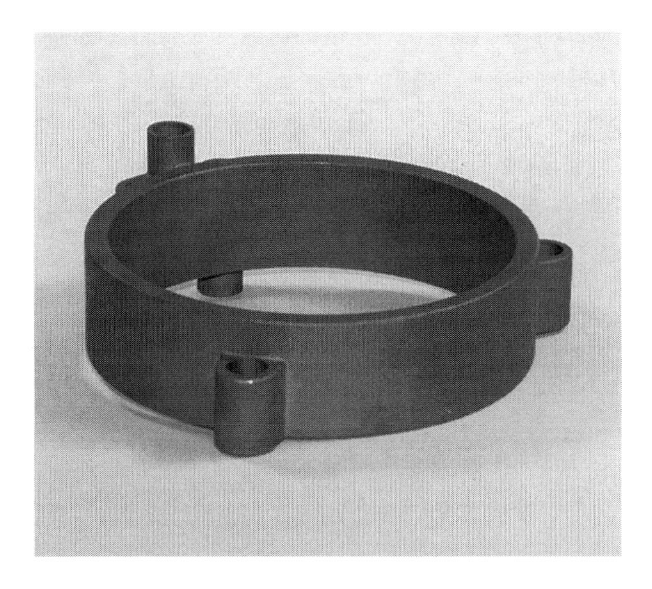

FIGURE 9.17 Complex water cooling assembly made from MS1 on an EOS DMLS™ platform. (Copyright Rapid Prototype and Manufacturing, LLC, Avon Lake, Ohio, 2014.)

The original part design was cast in three separate components and brazed together, where this is printed in a single part. Special considerations like chamfers on edges and part orientation are required to print with open cooling channels inside the part for water flow because of support material considerations as demonstrated in Figures 9.13 and 9.14. Figure 9.17 is another more complex water cooling assembly that was printed in a single part using laser powder bed fusion. The original design called for six cast components that were then welded together, which was cost prohibitive. In this case, using AM to eliminate the joining steps to make a single part actually made the part cost effective in a way that conventional processes could not.

Another fluid moving application for metal AM parts is high-performance impellers in industrial pumping systems. Figure 9.18 shows an example PH1 impeller from Figure 9.13 after removal of support material and finishing steps. AM allows mechanical designers to select the number and shape of individual impeller vanes with more extreme angles that conventional casting or joining techniques do not allow, which enable higher performance, lighter weight pumping systems. In cases of higher volume, lower margin markets like automotive and heavy truck applications, impellers can be built using stainless steel–bronze composite materials using binder jetting. Where these metal–metal composites are appropriate materials, the binder jetting approach is appealing as it can make larger parts than laser powder bed fusion roughly 10x faster and 1/10th the cost. Figure 9.19 shows an example of a hydraulic fluid moving impeller for a heavy truck application manufactured using binder jetting on an ExOne M-Flex platform.

For materials that are difficult to process using fusion techniques, binder jetting AM offers the capability to make parts that could not be easily printed. One example of this capability is for tungsten polymer composites for medical imaging

FIGURE 9.18 Example of a fluid moving impeller made from PH1 on an EOS DMLS™ platform. (Copyright Rapid Prototype and Manufacturing, LLC, Avon Lake, Ohio, 2014.)

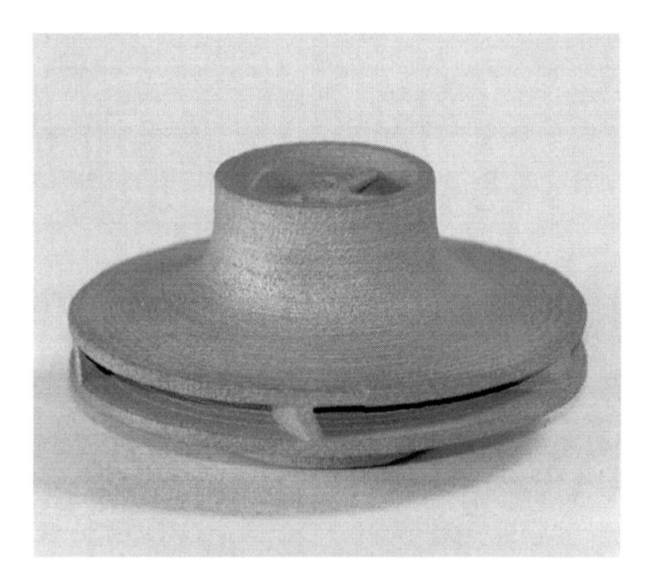

FIGURE 9.19 Example of a hydraulic fluid moving impeller for a heavy truck application made from 410 stainless steel–bronze composite using binder jetting on an ExOne M-Flex platform. (Copyright Rapid Prototype and Manufacturing, LLC, Avon Lake, Ohio, 2014.)

applications. Figure 9.20 shows an X-ray shielding bracket and Figure 9.21 an X-ray collimator, both printed using binder jetting on an ExOne M-Flex platform. These materials are being used to replace lead components that shield and direct radiation to reduce patient dose and improve resolution and clarity of imaging techniques. The AM use case is focused on the reduction of costly materials like tungsten polymer, while managing inventory and producing precision parts on an as-needed basis.

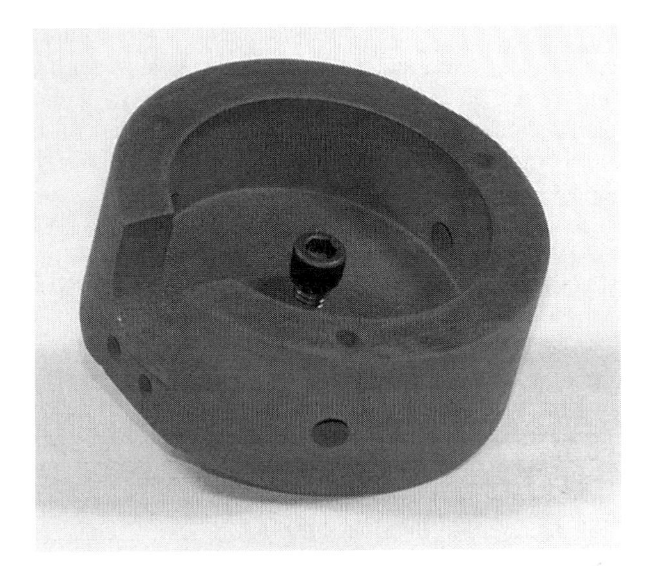

FIGURE 9.20 X-ray shielding bracket made from tungsten—polymer using binder jetting on an ExOne M-Flex platform. (Copyright Rapid Prototype and Manufacturing, LLC, Avon Lake, Ohio, 2014.)

FIGURE 9.21 X-ray collimator made from tungsten—polymer using binder jetting on an ExOne M-Flex platform. (Copyright Rapid Prototype and Manufacturing, LLC, Avon Lake, Ohio, 2014.)

9.6 SUMMARY AND FUTURE TRENDS

AM is perhaps the disruptive manufacturing technology being implemented by industrial manufacturers today. As more engineering materials are introduced, the supply chain will continue to develop use cases to provide greater value to clients. There are still key needs to improve quality and capability of technologies, and corresponding needs for standards development, accreditation, and certification by relevant bodies.[10]

As the suite of AM technologies continues to mature, it is breaking the traditional *design for manufacturing* model and providing the foundation for a new paradigm of *manufacturing by design*. In this new paradigm, industrial designers are fully enabled to design components based on functionality, rather than limits of assembly technologies. In this future paradigm, integrated computational materials engineering will be leveraged to develop new materials for these new processes in parallel to their development and for specific applications. Similarly, the fine line between structural and functional AM will be merged and new devices that support customized medicine and the *Internet of things* will begin to emerge.

REFERENCES

1. National Research Council. *Retooling Manufacturing: Bridging Design, Materials, and Production*. Washington, DC: The National Academies Press, 2004.
2. Schafrik, R. and Sprague, R. Superalloy technology: A perspective on critical innovations for turbine engines. *Key Engineering Materials*, 380, 113, 2008.
3. ASTM Standard F2792, *Standard Terminology for Additive Manufacturing Technologies*. ASTM International, West Conshohocken, PA, 2012, www.astm.org.
4. Wohlers Report. 3D printing and additive manufacturing state of the industry. Annual Worldwide Progress Report, 2014.
5. Lyons, B., Deck, E., and Bartel, A. Commercial aircraft applications for laser sintered poly-amides, SAE Technical Paper ATC-0387, Warrendale, PA: Society of Automotive Engineers International, 2009. doi:10.4271/2009-01-3266.
6. Stratasys. Ultem 9085 FDM™ data sheet, Stratasys, http://www.stratasys.com/~/media/Main/Secure/Material%20Specs%20MS/Fortus-Material-Specs/Fortus-MS-ULTEM9085-01-13-web.ashx, accessed January 7, 2014.
7. Maturation of fused depositing modeling (FDM™) component manufacturing, https://americamakes.us/engage/projects/item/455-maturation-of-fused-depositing-modeling-fdm-component-manufacturing, accessed January 7, 2014.
8. Rosato, D.V. et al. *Injection Molding Handbook,* 3rd edition. Norwell, MA: Kluwer Academic Partners, 2000.
9. The Economist. Print me a jet engine. *The Economist*, November 22, 2012. http://www.economist.com/node/21567145, accessed January 7, 2014.
10. Energetics Incorporated. Measurement Science Roadmap for Metal-Based Additive Manufacturing *Workshop Summary Report*. Columbia, MA: National Institute of Standards and Technology, U.S. Department of Commerce, 2013.

10 Additive Manufacturing for the Space Industry

Christian Carpenter

CONTENTS

10.1 INTRODUCTION

Space exploration is a captivating endeavor, not only for discovering the unknown, but also for overcoming difficult technical challenges. We must take a moment to understand the culture, objectives, and technical challenges of the space industry in order to understand how additive manufacturing can be infused for the benefit of space exploration.

One of the main reasons that we are currently experiencing an infrequent increase in space exploration capabilities is that the culture has become afraid to fail. A culture with a strong aversion to risk accepts long development and production schedules and high life-cycle cost because they are incorrectly perceived to be associated with heightened safety and mission assurance. Improved approaches are forfeited for continuing the status quo or *heritage* processes because new approaches are perceived to increase risk, even though new approaches often reduce risk when infused properly. In order to reinvigorate space exploration and develop robust space exploration programs, this culture must be changed. Technology is unlikely to drive change in the cultural acceptance of risk; however, additive manufacturing is a revolutionary technology that, when infused properly, can significantly reduce development and production schedules and life-cycle cost, thereby causing an opportunity to change what is viewed as normal and acceptable. In his book *Leading Change*, John Kotter recommends eight steps for developing lasting and meaningful change. In addition to describing the potential technical impacts of additive manufacturing on space exploration, this chapter will discuss three of Kotter's steps as applied to the adoption of additive manufacturing including development of a change vision, generation of short-term wins, and incorporation of changes into culture.

The vision for space exploration is in constant flux; however, a few constant themes can be identified which are summarized in the so-called von Braun paradigm. Simply put, the vision for space exploration is the realization of four main objectives:

1. Establishment of a capability to reach low Earth orbit (LEO) routinely and affordably
2. Establishment of a near-Earth, space-based research station
3. Colonization of the Moon
4. Colonization of Mars

To accomplish the vision, we must develop a safe and affordable space transportation architecture. This architecture must encourage rapid development and infusion of new technologies that maximize our potential for exploration and ultimately colonization. The transportation architecture must support a range of mission classes including robotic surveying and science missions, crewed missions, and cargo logistics missions that support crewed missions and pre-deployment of crewed systems. A modular transportation architecture that separates mission phases is required to enable demonstration and adoption of new vehicles (versus the all-up missions of the past) and new business models (e.g., government and commercial space launch providers). Exploration of small bodies such as asteroids and the moons of Mars may well be included in future endeavors; however, a transportation architecture that supports the four vision objectives also efficiently supports exploration of these small bodies. Within the mission architecture, a variety of vehicles are required including launch vehicles, space vehicles, landers, and Moon- and Mars-based infrastructure elements such as habitats and laboratories. Vehicle service life will range from one to five years for small satellite demonstrations to over 20 years of service life for missions such as colonization of Mars. We need to develop serviceable vehicles and subsystems, replacement components, and a logistics infrastructure capable of supporting a long duration supply chain. Modular vehicles that support demonstration and infusion of new vehicle subsystems as well as in situ servicing and replacement will be required to support the architecture and the vision objectives in an affordable and timely manner. The International Space Station (ISS) project has shown how a modular vehicle architecture supports widespread collaboration and incremental building, which enables sustainable long-term exploration and capability development.

There is probably no more significant technical challenge or contributor to cost in space exploration than the simple logistics of moving mass from origin to destination. The reality of gravity and rocket propulsion physics is that significant mass is required to move a payload from Earth to a location in space. The total mass of a mission at the beginning of life is generally treated as mass on the launch pad

(pad mass) that includes all launched mass or initial mass in LEO (IMLEO), which includes all mass released from the launch vehicle into LEO. Due to a number of cultural factors ranging from certification paperwork to complexity of systems, to low volume production, one can generally estimate the cost of a space mission based solely on the pad mass or IMLEO. Barring physics breakthroughs such as wormholes and warp drives, the matter of launched mass must be addressed in order to reduce the cost of space exploration.

The dry mass of a space system includes everything except propellant. The rocket equation shows that dry mass, mission ΔV (change in velocity), acceleration due to gravity, and specific impulse (or I_{sp}, which is a measure of propulsive efficiency similar to gas mileage) determine how much propellant is required to perform a mission. The result of these physics is that a typical mission results in propellant accounting for over 98% of pad mass with the remaining 2% being dry mass. An average satellite IMLEO consists of 50% propellant and 50% dry mass. There are four fundamental technology development areas that one can explore to minimize the mass and thus the cost of space missions. These are listed below in order of increasing impact to space mission cost:

1. *Low-cost systems*: This approach focuses on simply reducing the production cost of space systems.
2. *Low-mass (lightweight) systems*: This approach focuses on reducing dry mass, usually through lightweight materials or designs, but as noted previously reductions in dry mass have a significant effect on required propellant mass.
3. *Advanced propulsion*: This approach focuses on reducing propellant mass through increased specific impulse, but can also result in reduced dry mass through smaller propellant tanks and structural elements, which are a significant component of dry mass.
4. *In Situ Resource Utilization (ISRU)*: This approach reduces propellant mass and dry mass by developing technologies that allow missions to live off the land so that less material needs to be shipped to the destination.

Now that we have a basic understanding of space exploration culture and vision and have defined characteristics for the vehicles and technologies required, we can begin to evaluate additive manufacturing as a solution that enables affordable and sustainable space exploration.

10.1.1 Low-Cost Systems

Space systems are highly complex and manufactured in low volumes, and these characteristics align well with scenarios where additive manufacturing offers great cost benefit. Demonstrated metrics of 50% reduction in cost and schedule for complex, low volume components simply cannot be ignored and even the most risk-averse space product manufacturers have to take notice. Many manufacturers are likely to take advantage of the cost savings offered by switching from subtractive to additive manufacturing, but the real innovators that will help ensure a robust future of space exploration are those that adopt a design-for-additive-manufacturing philosophy. This design philosophy surpasses the basic concept of replacing machined parts with printed parts and transitions to designing systems that fully utilize the capabilities of additive manufacturing including the following:

- Reduced machining cost enabled through designs that minimize the need for build supports and post-machining
- Reduced tooling costs enabled through designs that integrate tooling required for post-machining and assembly
- Reduced labor costs enabled through designs that combine parts to reduce total parts count, joining operations, and assembly hours

Development phase systems are likely the best infusion opportunity for these additive manufacturing philosophies, because the aforementioned culture presents a significant barrier to changing existing fielded systems, even for the sake of cost improvements. Let us now consider the process of transforming a preexisting space product designed for subtractive manufacturing into an additively manufactured product and examine the associated cost benefits and technical challenges. Let us consider a rocket engine subassembly that includes an injector and a thrust chamber, and for the sake of simplicity, let us assume that these components are both made of the same metallic alloy and that the parts fit into the build volume of several existing additive manufacturing machines.

The preexisting fabrication process might begin by procuring a metal plate for the injector and a casting for the thrust chamber. As we transition to the additive manufacturing approach, we have an opportunity to cut the number of procurements in half by procuring a common powder batch. Additionally, the powder can be purchased in volume to accommodate subsequent production of these parts or others that might use the same powder. In addition to reducing the cost of labor to place the orders, we have also reduced the cost of tracking material certification and inventory by aligning the parts on a common material. Rather than inventorying and tracking several sizes of metal, we can track a single powder part number with certifications by lot. This feature of additive manufacturing will likely reduce standing inventories at manufacturers, reducing the costs of holding this inventory. Another key aspect of cost savings in the procurement cycle is the reduced impact of the government specialty metals clause. Because additive processes use powder, it is much easier to locate compliant materials, greatly reducing sensitivity to material sourcing, availability, and price volatility.

An additive manufacturing approach clearly offers us significant cost and schedule in the procurement phase for this scenario. However, several barriers to adopting this process will surface that drive non-recurring costs including, but not limited to the following:

- A need to qualify a new supplier.
- A need to qualify a new material.

- The company may not own the 3D printing machine and therefore must qualify a 3D printing vendor and their process.
- Changing materials and fabrication processes could drive a requirement to re-qualify the product.

These non-recurring cost drivers must be included in the consideration in order to determine if there is actual cost benefit for our scenario. It is possible that these challenges will present significant non-recurring costs and these must be evaluated with several other factors in the broader context of cost savings. Taken alone, the procurement phase may not provide cost savings for a single feature of a product and it may make sense to consider application to several products or product lines to amortize the non-recurring cost impact. However, in a risk-averse culture the need to drive significant change through many products will likely be seen as an insurmountable task and could damage the potential of future infusion of additive manufacturing if a cultural perception develops that suggests additive manufacturing as something that requires significant non-recurring cost or sweeping application. It is therefore highly important to consider all factors and ensure that additive manufacturing is applied to the right product at the right time to realize the significant benefits and gain cultural acceptance. It is clear that the procurement phase presents both opportunity for benefit and risk of implementation for additive manufacturing.

Next, let us assume that the procurement phase analysis presents no show-stoppers and we are now ready to dig into the design improvements. First, we might simply compare the cost of manufacturing the components between the two processes. It would not be surprising to find a 50% cost and schedule reduction and this has been demonstrated previously in real scenarios. While we could stop here and declare success, our intent is to affect systemic cost reductions. In our next step of design considerations, we might consider joining the parts together in CAD and then printing them as a single piece. Eliminating joining processes, such as welding and brazing, can carry significant cost schedule savings while also reducing quality risks and improving mechanical performance. It is important to note here that for new product designs, the elimination of joining processes can also reduce non-recurring costs.

At this point, our design improvements have significantly improved cost and schedule, but we can go further still by designing to reduce tooling and post-machining. Space products often require at least some features with tight tolerances, such as joints, seals, or flow passages, that are not achievable with additive manufacturing. As such post-machining of the parts may be necessary. With our legacy subtractive machining approach, we may find that a significant amount of cost is spent designing, producing, verifying, and tracking tooling. As we transition to additive manufacturing, we have the opportunity to reduce non-recurring and recurring costs of tooling in the following ways:

- Identify critical features that must be machined, but allow loose tolerances and rougher surfaces for non-critical features
- Identify build direction and minimize overhangs and the need for build supports that will require subsequent removal
- Integrate post-machining tooling into the design so that this tooling can be scrap material that does not require inventorying or tracking

At this point, we have developed a design that significantly streamlines production and results in significant life-cycle cost and schedule savings. However, as in the procurement phase we are likely to encounter several barriers to design acceptance that must be addressed including the following:

- How will we verify the new design for thermomechanical performance?
- Do we have sufficient material data to believe our analyses?

These types of questions arise with any new material or manufacturing process, and determining how to qualify the analysis for a specific part will be dependent on the maturity of the analysis methods and material databases. These issues will be discussed later in the chapter, but for now it is appropriate to note that there is significant ongoing investment to answer these questions for additive manufacturing processes. Let us now assume that we have completed our design and are ready to proceed with the manufacturing process development.

With our legacy subtractive manufacturing processes, a significant amount of time was spent creating g-code for machining operations and concerns of non-recurring costs for new code may arise. However, with additive manufacturing the cost of generating machine code is significantly reduced through the use of slicing programs that auto-generate this code making the non-recurring costs quite low. With subtractive processes, there was a significant amount of setup due to required tooling; however, with additive manufacturing, this setup cost is greatly reduced due to the reduced or eliminated tooling accomplished in the design phase. With subtractive manufacturing, there was a cost of waste rejection associated with cutting fluids and scrap material; however, with the additive manufacturing, the cutting fluid is eliminated and scrap material is minimized. In addition, scrap material associated with build supports can be made crushable enabling efficient packaging for lower cost waste disposal. Finally, the cost to rework our subtractively manufactured parts was significant and included difficulties in scheduling queues across several machines to work the parts. With the new additive manufacturing process, we find the cost impact of rework is significantly reduced and may times it is possible to place a rework part into empty space in already-planned builds. It is clear that the fabrication phase offers many benefits, but like other phases there are barriers to implementation including the following:

- How will we develop and control the build configuration? For instance, if build supports are required do they need to be configuration controlled?
- How will we inspect the additively manufactured parts? For instance, the injector holes are critical features and may be difficult or impossible to inspect if inseparable from the thrust chamber.
- How will we qualify, handle, use, and reuse feedstock (powder)? For instance, are we allowed to reuse unconsumed powder from a build and if so, under what conditions?

Resolution of these issues will be discussed later in the chapter, and like the design and analysis phase, there is significant ongoing investment to answer these questions.

It is critical to note, however, that these considerations must be taken into account during the design phase even though they are not encountered until the manufacturing phase.

Let us now assume that we have successfully completed development phase and are ready to transition into production of several units. For space products in transition from development to production, there is typically a significant cost to develop controlled drawings, work instructions, and material traceability. At this point, additive manufacturing provides significant benefits as the major quality parameters are captured in the build files providing an intrinsic quality control set that significantly reduces the cost of transitioning to production.

The single-material printing processes assumed for the presented scenario are currently the most mature and prolific with capabilities ranging from plastics to aerospace metals. Development and production of either replacement parts for legacy systems or new components for development phase systems can be accomplished with these processes. There are few plastics in space systems, but aerospace metals such as steels, aluminum (primarily 6061 T6), and titanium are of high interest. For high-performance components, Inconel and exotics such as moly/rhenium are required. Because components are traditionally machined from a single billet of material, we can reasonably expect, and in fact it is already being realized, that the first infusion of additive manufacturing will be single-material manufacturing. Selective laser melting, electron beam melting, and other powder bed single-material processes are sufficiently accurate that they are already being infused. Freeform fabrication processes such as laser freeform fabrication (LF3) and electron beam free form fabrication (EBF^3) provide capabilities for larger parts, but have some progress to make in feature size before wide infusion is possible for intricate space systems.

Infusion of single material additive manufacturing will certainly have a significant impact on the cost of space products, but we can go farther. In the previously presented scenario, we made the assumption that the injector and thrust chamber were constructed of the same metallic alloy, whereas a more realistic scenario would be that the two are composed of differing alloys. Let us now explore for a moment the next tier of cost savings that can only be achieved with multi-material additive manufacturing. Akin to color printing versus black and white, multi-material additive manufacturing is likely to replace most current additive manufacturing approaches because when qualified, it should reduce or eliminate joining and assembly of components that are some of the largest areas of risk in the manufacturing life cycle. A few additive manufacturing processes, such as LF3 and EBF^3, currently offer this capability in metals and some studies have begun to demonstrate significant successes blending and transitioning between dissimilar materials. Multi-material printing complicates the design and analysis phase because development of models capable of analyzing transitions between materials is still in its infancy and there is a wide range of potential combinations of materials that must be characterized and tested to ensure a path to qualification. Due to the significant opportunity for cost and schedule reductions enabled by multi-material additive manufacturing, as technology progresses we are likely to see significant effort put into establishing tools and validation processes that support this complex capability.

Valves are one of the most expensive and long lead elements of space systems, and as such, a 3D printed valve is currently considered a sort of Holy Grail for space additive manufacturing. Achieving this feat requires a system with characteristics of tight tolerances and the ability to deposit metallic and soft good material systems. Even with the advent of multi-material systems, it is likely that tolerances will require post-machining of printed components. Therefore, the next tier of cost savings may be achieved with multi-material additive and subtractive manufacturing, which we call additive–subtractive manufacturing, or ASM. Here we find a very small base of capability where the limitations of additive manufacturing machines are addressed by integrating some subtractive machining that can be done in-process. The ASM approach enables most of the benefits of additive manufacturing while maintaining high tolerances and smooth surface finish of machined parts.

In summary, additive manufacturing presents significant opportunities to reduce cost and schedule for space systems through the implementation of a design-for-additive-manufacturing philosophy. Many technical challenges also exist to capture these savings and many organizations are working to overcome these challenges. In the next section, we will explore how to take these benefits even further by reducing the mass of space systems, which translates into significant mission level cost savings.

10.1.2 Low-Mass Systems

Let us again consider the scenario of transforming an existing product into an additive manufactured product, but this time we will focus on how additive manufacturing can be used to reduce the mass of the product. It is important to remember that launch costs are approximately $10,000 per pound of mass launched into space, so saving dry mass can significantly reduce cost. For this example, we will consider a notional rocket propulsion system consisting of the following elements:

- Structure
- Propellant tank
- Gimbal
- Rocket engine

Let us first focus on using additive manufacturing to reduce the mass of the structure. The first approach taken with design for additive manufacturing would be to replace solid structure with less dense infill layers such as a honeycomb pattern encapsulated with a thin outer wall. From the outside, the part might look identical to a machined part, but could be over 80% lighter. Modern freeform fabrication machines can implement this approach directly and powder bed machines simply require incorporation of a powder removal method. Algorithms exist that automate the process of creating low density infill layers, and it is likely that in the future, more complex infill algorithms will be developed to optimize the infill for structural strength, stiffness, and mass. This approach to mass reduction can have a significant and immediate impact on spacecraft structure and mechanisms, which typically comprise 10%–20% of dry mass. The same philosophy can be applied to every aspect of the vehicle to enable significant mass reductions throughout a space system.

Next let us focus on the propellant tank, which is a thin-walled pressure vessel, typically made of a metallic alloy. Here we cannot implement the infill approach used on the structure, so the next step in reducing mass is to highly engineer the shape of the tank for optimal mass. Highly coupled CAD, structural, and thermal analysis tools must be employed to determine how to shape the part for maximum strength, stiffness, and mass. The addition of goal-seeking tools, such as genetic algorithms, capable of developing non-intuitive designs could enable highly engineered parts with reduced non-recurring costs. The same tools could be used to engineer subtractively manufactured parts; however, the resulting designs would likely be too expensive to produce using subtractive manufacturing alone making the additive manufacturing community the likely driver of development of these advanced tools. The result of this design philosophy might include areas of thicker or thinner solid material coupled to areas with low infill or complex open-cell shapes. Alternatively, we might find that it is best to implement a thin-walled vessel with an exoskeleton. In this case, merging the structure with the propellant tank may reduce parts count and eliminate areas that would typically carry extra material for attachment between the two system elements. All of the above processes can be applied using single material processes, but as we move into consideration of multi-material additive manufacturing we can expect to see tools that enable the blending or transitioning of materials along our tank wall to optimize mass. Complex algorithms that take into account variables of shape, strength, mass, temperature, alloy, and cost would need to be employed to optimize material systems throughout the part. New alloys are likely to be designed during this process that enable continuous transitions from one metal to another enabling even more highly engineered designs.

This philosophy, like the ones mentioned previously, can be applied throughout the vehicle and is likely to result in substantially increased performance not only for mass, but especially for thermal and cost as well. In the particular case of cryogenic propellant tanks, the industry has the significant challenge of balancing propellant boil-off with tank mass and use of cryocoolers. The ability to design the propellant tank to minimize heat flow into the propellant could have savings not only on the tank mass itself, but on the amount of cryocoolers support needed and the amount of propellant lost due to boil-off. In this case, additive manufacturing may be able to offer new solutions to a difficult and significant challenge.

Next we will consider the engine gimbal and we can certainly apply the aforementioned design philosophies to reduce mass of the component, but with additive manufacturing the most significant mass reductions occur when we consider the system as a whole and it follows that multifunctional system designs will begin to emerge quickly as additive manufacturing design philosophies are applied at the system level. The engine gimbal has a unique requirement for propellant lines and wiring to cross movable joints, which is a problem that we have yet to consider. If one were to imagine a highly engineered beam structure with a significant amount of free space inside, such as we would likely find in our gimbal, the designer is likely to make us of this space by incorporating flow passages, wiring, and so on into the free space in order to leverage the existing structural capabilities of the parent part to reduce the overall mass of the multifunctional system. One example would be embedding the propellant tubes in the wall of the gimbal structure. Another example would be elimination of circular wire sheaths through a sandwich of center conductor, insulator, and then gimbal structure. While elimination of a wire sheath might seem a small mass savings, the amount of wire harnessing in spacecraft is not insignificant and one can expect both mass and cost to significantly decrease. Finally, development of additively manufactured slip rings and propellant swivel joints that allow electricity or fluids to pass through the rotating joints would be both a significant challenge and benefit. If we consider an entire space vehicle, we can find many systems and components that could implement these design philosophies especially solar power systems and environmental control and life support systems.

Finally, let us briefly consider the last component in the system, the rocket engine. Rocket engines operate at high temperature and pressure over many cycles. They are highly complex components that can include valves, turbopumps, injectors, combustion chambers, and nozzles. Optimizing rocket engine mass will require implementing all of the aforementioned processes and will require multi-material design and analysis tools covering metals, ceramics, softgoods, composites, and coatings. The rocket engine optimization will bring the added complexity of incorporation of performance models required to analyze engine performance over a range of conditions to ensure that the additive manufacturing design philosophies are applied in a way that considers all extremes of operation. It is likely that such design and analysis tools are years away, but we can begin by applying our philosophies to rocket engine components with a knowledge of our vision for the future we can make significant near-term progress.

Significant challenges are likely to occur as additive manufacturing design philosophies are applied to space systems. The evolution and validation of complex design and analysis tools must occur in order to enable progress. First, we must improve tools to allow consideration of infilled, open-celled, and exoskeletal designs. Next we must add capabilities for consideration of material transitions such as joint and blended materials. The addition of blended materials brings a complication of requiring some way to qualify the new alloys used in these parts as well as the processes used to fabricate the parts. Typically, qualification of new material systems is a very costly activity. Additive manufacturing can reduce the cost of qualification through low-cost and rapid fabrication of samples; however, in this case we are discussing highly engineered material systems that will have varying lengths, repeatability, and mixture from design to design. Qualification of materials used in this way will present a significant challenge to the culturally accepted norms. It is possible that the cost of space missions will be so significantly reduced that it may become acceptable to fly new materials with validation by protoflight testing only, but alternate methods to qualification are likely required. As components and systems are combined to save mass and cost, the perceived risk of failure is likely to increase. It will be critically important for innovators to consider failure modes and effects when determining what components and systems to merge. All of these considerations point to a common theme that has been propagated for decades, which is that we are moving into a time where systems engineering disciplines and tools will need drastic and significant improvements to enable success.

It should now be clear that additive manufacturing has much to offer in the way of reducing space system dry mass and that the benefits are as significant as the challenges. Efforts made to develop the significant design and analysis tools required to enable new paradigms in space systems will also have great impact on other terrestrial markets, and as such, we should expect that many markets will drive development and implementation of these types of tools enabling space-focused manufacturers a significant opportunity to collaborate and focus on unique problems to the space industry. Next we will consider how additive manufacturing can affect significant propellant mass reductions through the development of more efficient propulsion systems.

10.1.3 Advanced Propulsion

Excluding exotic and theoretical transportation systems such as wormholes and warp drives, the primary types of propulsion that are available to support space exploration are either thermal or electric. In the previous sections, we considered how reducing the cost and mass of propulsion systems through additive manufacturing can provide revolutionary improvements. However, aspects of additive manufacturing can also be brought to bear to improve propulsive efficiency that reduces propellant mass. Because propellant mass can account for up to 98% of pad mass, any improvements in propulsive efficiency can have drastic impacts on the affordability and sustainability of space exploration.

Thermal propulsion systems (sometimes referred to as high thrust propulsion) where specific impulse (gas mileage) is dependent on combustion temperature are limited by the material systems employed within the rocket engine. Storable monopropellants produce a specific impulse typically from 220s to 250s and operate with combustion temperatures ranging from 800°C to 1800°C. At the high end of this range, exotic materials such as rhenium and iridium are required to contain combustion. For these systems, highly engineered additive manufactured components with new alloys can not only improve cost and lead time, but also improve temperature capability and structural strength of rocket engine components, thereby enabling higher temperature propellants and thus higher specific impulse. Storable and cryogenic bipropellants produce specific impulse in the range of 300s to 452s and combustion temperatures rise to over 2700°C where ablative or regeneratively cooled components must be implemented. Here, additive manufacturing can be used to embed coolant passages that would be unaffordable or impossible to implement in subtractively manufactured components. This capability can enable improved specific impulse and there is potential to increase cooling efficiency to the point that lower cost materials could be implemented. At the highest range of thermal propulsion is nuclear thermal propulsion where nuclear fuel heats a working fluid, typically hydrogen, to create thrust and produce specific impulse in the 900s range. Here additive manufacturing of regeneratively cooled passages as well as incorporation of engineered radiation shielding could significantly improve not only the performance of these engines, but also safety of the system.

Solar electric propulsion systems employ electrostatic or electromagnetic forces to produce very high specific impulse ranging from 400s to >10,000s. These systems are sometimes referred to as low thrust propulsion because they are dependent on solar power to create thrust, and modern solar power systems only provide kilowatts of power, which results in thrust levels of millinewtons. As a result of power limited thrust, electric propulsion systems require long duration firings (non-Hohmann), which result in less efficient thrusting and increased total ΔV (typically 2×) compared with high thrust thermal systems. As we discussed previously, increased ΔV requires more propellant; however, electric propulsion systems typically offer 4× increased specific impulse resulting in a net 50% reduction in propellant mass for a typical mission. A 50% propellant mass reduction has an enormous impact on space mission affordability, and as a result, we are seeing a significant increase in the use of electric propulsion. Additive manufacturing can improve electric propulsion system efficiency through 3D printing of improved magnetic structures, electrical insulators, and ion optics increasing engine efficiency. However, electric propulsion systems are already highly efficient typically in the >50% range. The most significant impact to the performance of these systems is improvement in the spacecraft level power to mass ratio that increases the total thrust potential (reducing ΔV and trip time) of the solar electric propulsion system. 3D printed lightweight structures for solar arrays, and high reliability slip rings and SADA drives would all provide significant improvements in solar electric propulsion systems, but the most significant impact to solar electric propulsion system efficiency is the solar cells themselves. 3D printed, high-efficiency, radiation-hardened, low-mass solar cells are the key to significant propellant savings. Additionally, solar electric propulsion systems operate best at high voltage levels, which presents a problem for space solar arrays as voltage increases result in significant risk of arcing and plume interactions. Additive manufacturing processes can provide for fully encapsulated solar cells and solar arrays that eliminate this risk and enable high voltage solar electric propulsion systems to operate at higher efficiency and reduce mass. These benefits could easily drive propellant savings of 10%–20% resulting in significant reductions in space mission mass and cost.

Additive manufacturing has many potential benefits to thermal and electric propulsion systems that we are just beginning to explore, but the net result is the potential for substantial propellant mass reductions. Up to this point, we have focused on how terrestrial additive manufacturing can reduce the mass and cost of space logistics, but in the next section we will consider how space-based additive manufacturing can fundamentally address the root of the problem by reducing the overall need for these logistics operations.

10.1.4 In Situ Resource Utilization

ISRU is a space architecture design philosophy wherein raw materials located at the destination are leveraged to support a mission or campaign rather than solely depending on supplies shipped from the mission origin. This approach transports only what is needed to perform manufacturing at the destination, thereby drastically reducing cargo mass, pad mass, and space mission cost. An extension of this philosophy is in situ manufacturing wherein raw materials and manufacturing equipment are shipped to the destination rather than transporting all components that might be needed. This approach provides lesser, but still significant, mass and cost reductions for destinations where needed materials may not exist. Space-based additive manufacturing processes could play a significant role in both approaches; however, these processes will exist in a different environment from ground-based process including low gravity, vacuum pressure levels, wide temperature ranges, and electrostatic charging, just to name a few. In this section, we will explore what is likely the most significant value proposition for additive manufacturing to long-term space exploration.

Manned space missions are currently confined to LEO where the ISS serves as the primary laboratory for research. The internal environment of the ISS requires consideration of low gravity, but eliminates vacuum pressure and temperature variables. As such, it is not surprising that the first use of space-based in situ additive manufacturing will be demonstrated onboard the ISS. Made In Space, a company located at the NASA Ames Research Park, is currently flying a plastic extrusion 3D printer on the ISS to validate and demonstrate low gravity 3D printing. If successful, the demonstration could have a significant impact on ISS logistics by enabling researchers to ship material to the ISS for in situ manufacturing versus shipping and maintaining an inventory of spare parts. This capability would reduce the total mass shipped to the ISS. Additionally, the ability to print in situ would enable researchers to change their experiments after launch and allow them to be more responsive to research results enabling more value for a given experiment. In the far term, this capability has a significant impact on long duration crewed missions to destinations such as Mars where replacement parts will need to be fabricated en route destinations and during the exploration mission operations.

Powder bed processes are likely to have an issue with the low gravity environment. Spinning the device to create an artificial gravity might be a solution, but loading material and removing parts from the spinning platform could complicate the situation. Electrostatic approaches that provide force onto the powder might be another approach, but could encounter issues in electron beam devices. Powder spray approaches such as LENS™ could also be used; however, the physics of ensuring the powder is delivered to the intended target with no effect of residual spray would be critical. Freeform fabrication techniques such as EBF^3 and LF3 use a wire feed approach and appear well suited to handling the low gravity environment. As low gravity additive manufacturing matures, it is likely that external applications will become a desire. The ISS is equipped with external research points and robotics that could be used for space-based additive manufacturing demonstrations.

The external environment of the ISS adds variables such as changing thermal environments, vacuum pressures (though the ISS does outgas compared to a true deep-space mission and there is a higher level of atomic oxygen in the ISS orbit), and electrostatic charging. As the ISS circles the Earth every 90 minutes, it encounters approximately 60 minutes in Sun and 30 minutes in eclipse. These alternating thermal conditions could cause significant thermal stresses to build up as additive manufactured parts are produced. Initially, it will be possible to perform demonstrations in Sun or eclipse to limit these variations, but eventually they will have to be overcome. Layer-by-layer thermal image recording is being developed for quality purposes in terrestrial applications, but space-based applications my actually need to react to these data in order to produce good parts. The machine may have to rotate or have a rotating build plate in order to more uniformly distribute solar and Earth-reflected heat influx to the part. Forms of composition analysis such as residual gas analysis may be required in order to ensure that contaminants are not introduced during a build due to atomic oxygen levels or outgassing. As the part is fabricated, differential charging could occur on the part or in the machine, and as such, special equipment grounding or charge mitigation devices may be required. Automatic part removal is also likely required in order to simplify the logistics of moving the part to its intended point of use. There are many new variables that will need to be taken into account, but the ISS research platform provides an excellent test bed to develop space-based additive manufacturing. As space-based additive manufacturing technologies are demonstrated, there will be a desire to use them in support of spacecraft servicing.

Spacecraft servicing generally means the on-orbit repair or upgrade of a satellite or spacecraft. This technology will likely be proven in LEO and then later applied to medium Earth orbit and geosynchronous orbit satellites. Modular spacecraft systems are being developed to enable in situ repairs or upgrades to a satellite by a servicing vehicle. Some repair or upgrade operations will be accomplished through replacement or addition of modules; however, in situ manufacturing can provide a capability to create components that the servicing mission did not anticipate, repair non-modular components, or repair damaged components. In situ manufacturing for spacecraft servicing and life extension can significantly reduce pad mass and mission cost by enabling lower cost servicing missions and by reducing the number and mass of satellites put into orbit. Examples might include patching holes created by micrometeorite damage or repair of faulty wiring. In one form of satellite servicing, a life extension spacecraft attaches itself to an existing host satellite in order to provide propulsion and power that enable significantly longer missions for the host satellite. Many existing satellites are not designed for grappling, and the process of a life extension vehicle grappling with a host satellite is highly complex. Additive manufacturing can play a significant role for this mission by enabling in situ fabrication of custom grappling points that enable life extension and/or servicing missions.

Large-scale missions require grand structures that we cannot affordably fabricate on Earth, test, and then deploy into space. One example is the James Webb Space Telescope that has large structures, but must be packaged into the launch vehicle fairing requiring complicated spacecraft design and intricate deployment mechanisms. As space-based additive manufacturing matures, it is probable that low gravity production of optimized mass elements can significantly contribute to reduced pad mass and cost. Space-based additive manufacturing could enable the in situ manufacturing of such spacecraft, and this capability would enable an entirely new approach to space architecture from both a design and a logistics philosophy. In this scenario, raw materials for the spacecraft structure would be shipped to LEO where the spacecraft would be fabricated. Modular system components such as solar array panels, electronics boxes, propulsion systems, and so on would be installed on orbit, possibly prior to close out of the structure similar to how buildings are fabricated on Earth. When complete, a skin could be applied, or even printed into place to close out the spacecraft. In addition to structure, this approach to spacecraft manufacturing would have a significant impact on the specific mass of solar arrays, radiators, and antennas. Additionally, launch loads, which typically drive a significant portion of spacecraft component cost and mass, could be handled much more easily as parts could be optimally packaged, oriented, and protected during launch. Upon arrival in orbit, the components could be unpacked and installed into the satellite. This approach also poses significant benefits for manned missions where large volume habitats are ideal for the crew. Presently, inflatable structures are being explored to create the large volumes, but an in situ manufacturing approach may provide an optimal solution in the future.

It stands to reason that the continued push by mankind to explore space will eventually lead to small space-based research stations, outposts, and settlements beyond LEO. The enormous cost and schedule of the logistics systems required to transport materials from Earth to likely targets such as the Moon, asteroids, and Mars offers a strong business case for ISRU. The benefits highly depend on the materials found at the destination, but the net result is reduced pad mass and mission cost. Assessments of the impacts of ISRU propellant generation have shown that ISRU has the highest impact to mission mass of any other technology. Few studies have explored ISRU for generation of structures, but it is reasonable to think that the impact of building exploration outposts using ISRU would enable maximum reductions to the mass and cost of space exploration. Additive manufacturing processes, especially freeform processes, are likely to play a major role in ISRU structure generation. To date, most space architecture studies have focused on structures shipped to a destination and then assembled on-site. A more affordable approach would be to send robotic scouting missions (called prospectors) to analyze the materials that can be found at a destination as well as the terrain. Following identification of the material types and locations for facilities, Earth-based analogs can be performed to validate additive manufacturing processes required. Once validated, robotic missions can be sent to the destination that would excavate materials that would be fed into freeform additive manufacturing machines. These machines could then print (probably in thick layers) the structures required for the exploration mission. Both the Moon and Mars are covered in fine dust, which may prove useful for powder-based freeform additive manufacturing processes, or for processing into feedstock for extruder processes. The regolith of Earth's Moon and Mars contains silicon dioxide, titanium dioxide, aluminum oxide, iron oxide, magnesium oxide, and calcium oxide, all of which could be processed and used for construction of outposts. Separation and storage of oxygen from the base material would be beneficial for crew air supplies or oxidizer for propulsion.

Destinations such as the Moon and Mars add new variables to the space-based additive manufacturing processes. The new variables are too numerous to mention in this section, but we will briefly explore a few examples. As most destinations of interest are farther away from the sun, it is important to note that larger thermal gradients than typical will be experienced during additive manufacturing. This will have to be taken into account and will likely complicate Earth-based experiments to prove out additive manufacturing processes for these locations. Additionally, destinations such as Earth's Moon present issues of dust accumulation as machines move about on the surface. This effect will require additive manufacturing systems that are tolerant to the dust and will also require some consideration of dust accumulation onto the structure being fabricated. Systems to remove dust during the build may be required. In the case of Mars, which has both dust and significant atmosphere, weather effects will have to be taken into account if processes are to be conducted outside enclosures, which is probable if we are manufacturing structure for an outpost. Finally, the signal delay at these destinations will require that all of these processes be conducted autonomously. This presents an additional complication as on Earth we can pause a crashed build, perform a fix, and then continue or restart; however, this type of capability will have to be included in additive manufacturing machines designed for ISRU applications. On a positive note, solving many of these issues also translates into more robust machines that can be implemented to benefit Earth-based additive manufacturing.

10.2 DEVELOPING CULTURAL ACCEPTANCE

Space products are typically required to operate in extreme environments driving a need for exotic materials and complex designs. The market for space products is characterized by low volumes and long life cycles. The severe cost and reputation impacts of space product failures require thorough qualification of product designs and processes before manufacturers can adopt them for flight use. These characteristics of the space product market have resulted in long development schedules and high costs. In fact, it is not uncommon for a space product to take 10 years to progress from concept to first flight. As a result, new technologies are traditionally infused into space products at a low rate. This has resulted in a large gap between the demand for low cost, fast delivery products and the supply chain's characteristically high cost, long lead products. Manufacturers of space products are experiencing ever increasing pressure from government and commercial customers to conform to the commercial electronics paradigm of delivering smaller, faster, and cheaper products on increasingly shorter timescales. When customer expectations significantly contrast with the capability of a supply base, such as is the case with space products, a revolutionary change must be made to resolve the gap between supply and demand. The development of additive manufacturing technology enables a new paradigm of significantly reduced cost and lead time for space products. It is expected that additive manufacturing for many complex space products will result in greater than 50% reduction in both cost and schedule. Additionally, the ability to fabricate parts in an additive fashion and/or with new materials could enable new designs with improved performance capabilities. As a result, it is not surprising to see space product manufacturers rushing to adopt and infuse additive manufacturing and in some cases emerging as leaders in additive manufacturing processes. This behavior is a testament not only to the technology of additive manufacturing, but to the willingness of space product manufacturers to return to the innovative culture demonstrated in the Apollo era. Additive manufacturing has a unique opportunity to change the paradigm of space products, but cultural acceptance of the design, analysis, manufacturing, and test processes must be developed.

In a risk-averse culture, safety issues are the most costly impact to a program. Resolution of safety issues drives lengthy and costly investigations and ultimately results in more lengthy and costly production processes. Therefore, the most important step that additive manufacturing advocates can take to develop cultural acceptance is to demonstrate and communicate that the processes are safe. The resolution of quality issues has a similar impact in that they require lengthy and costly investigations and generally result in increased production cost and schedules. Therefore, the second most important step in developing cultural acceptance is to demonstrate that additive manufacturing processes reliably produce high-quality components. Long schedules are typically accepted based on a combination of technical challenges associated with development and the lead time to produce a system. However, the length of a program schedule drives a certain standing army cost and therefore shorter schedules can intrinsically reduce mission cost and this standing army can have a more significant than the production cost of the hardware. Additive manufacturing advocates will know that 50% schedule savings are possible, but many used to long schedules will likely see these claims as unrealistic, naive, or impossible. It is therefore highly important for advocates to develop case studies that clearly demonstrate the greatly compressed development and production schedules enabled by additive manufacturing to overcome the preconceived notions and prior experiences of the culture. Last is the actual system development and production cost where cheaper materials and manufacturing approaches have much to offer. Like the schedule scenario, it will be difficult for many that are used to high-cost products to believe that it is possible to achieve 50% cost savings. Again, advocates should develop and communicate case studies to demonstrate the cost-saving metrics to the community. It is important for additive manufacturing advocates to consider safety, quality, schedule, and cost aspects as well as cultural norms in order to determine what change will drive improvements in space mission schedules and life-cycle costs.

10.2.1 Ensuring Safety and Quality: Qualification of Additive Manufacturing Processes

The qualification process is the way in which a product or process demonstrates that it is safe and of high quality. The qualification process must be comprehensive while also ensuring minimum impact to the product schedule and cost. In an ideal scenario, each product to be used in a space mission would be able to demonstrate through test that it is able to successfully complete the mission with significant margin. However in reality, we must also find ways to qualify though inspection and analysis in order to minimize schedule and cost. Space product qualification is a rigorous process and the requirements are highly dependent on the type of product, the launch and operating environments, and, in many cases, the specific end user. When compared to the commercial electronics paradigm that is being requested of the space industry, one begins to discover that a lack of standards exists in the space product market that drives enormous scope, schedule, and cost into the products. For example, each satellite manufacturer may have their own operating voltage and electrical interfaces and one can imagine the high cost that household appliances might attain if not for a standard operating voltage and physical interface. The lack of standards will require additive manufacturing process advocates to consider how to qualify a general process, yet be able to accommodate particular requirements of specific end users. We will now explore some of the requirements for qualification of additive manufacturing processes.

Material and manufacturing processes must also be qualified. Material vendors for additive manufacturing materials must develop and document standards for powder or feedstock composition and physical properties. Lot traceability of materials must be maintained. Processes for proper handling and reclamation of the materials must be established. Data from extensive testing will be required before designers can consider incorporation of a material system into the overall product development process.

Design and analysis processes must be qualified. Initially, design and analysis software used for subtractive manufacturing will suffice for single material additive manufacturing processes with subtractive style designs, but as designs begin to take full advantage of additive manufacturing it is probable that structural and thermal analysis codes will not be able to accurately model the complex geometries, especially in an efficient manner. New codes will likely need to be developed and validated through testing to ensure that produced parts match analysis results. Companies and government organizations are already beginning to understand these challenges and are starting to make progress toward resolutions. A few multi-material printers have already been brought to market; however, the complexity of the design and qualification process along with the cost of these machines is likely to delay their infusion into the aerospace market. Additionally, the usually risk-averse aerospace companies investing in qualification of single-material processes over the next decade are unlikely to make a subsequent large investment to make the technology leap to multi-material machines. Despite the likely delayed infusion, multi-material additive manufacturing has the potential for far greater impacts to aerospace products than single-material printers over the next 10–20 years. As multi-material processes evolve an added dimension of material property, transitions will need to be incorporated and similarly qualified through validation testing. The range of possible geometries and material compositions is significant and acceptance of qualification data would and should be scrutinized for accuracy and range applicability.

Additive manufacturing process qualifications should be conducted in parallel with material, design, and analysis qualifications. Processes need to define and document all build parameters such as speed and power settings. During the build, data on part temperature, composition, build chamber gas levels, and so on need to be captured and saved for later analysis. Completed parts need to be tested for porosity and mechanical strength. Initially, simple geometries may be tested similar to castings, but as geometries become more complex, lot testing may be required. Build supports and infill geometry are currently variables in many processes, but space manufacturers will likely require that these become fixed process parameters as variability in these could significantly affect resulting part mechanical properties.

Qualification of materials, design, analysis, and manufacturing processes is likely to be costly and tedious tasks, but must be completed to ensure high-quality space products. Involvement of all levels of the supply chain including insurance groups, spacecraft operators, spacecraft manufacturers, and component suppliers should be strongly encouraged to ensure buy-in for qualification activities. On the long road to process qualification, near-term flight demonstrations can be accomplished to help improve cultural acceptance.

10.2.2 Demonstrating Short-Term Wins: Enabling Shorter Development Schedules

It is an exciting time in the field of space exploration because CubeSat and SmallSat standards have enabled routine, low–cost access to space. In 2010, over 100 CubeSats were launched, most from Universities. In 2014, the company Planet-Labs alone expected to launch a constellation of 100 units, the largest satellite constellation in history. If this growth trend continues, we can expect >200 deployments per year in the very near future. Because CubeSat and SmallSat systems are low cost and packaged inside a dispenser, more risk can be accepted and lengthy and costly material process analyses can be deferred in favor of test and flight demonstrations. These small platforms enable opportunities for rapid demonstration of additive manufacturing materials and designs on real missions that will generate data and experience required for cultural acceptance by larger spacecraft manufacturers.

At the simplest level, one can use CubeSats to begin exposing additive manufactured materials to the environment of space. The ISS acts as a possible platform wherein a CubeSat can be exposed to space and then brought back into the ISS for subsequent analysis or return to Earth. Follow-on missions might include demonstrating through flight that additive manufactured materials can be used for CubeSat and SmallSat structure. The next step in this evolution might be inclusion of additive manufactured functional components such as circuit boards and valves. High pressure and liquid propellant systems will likely be one of the pinnacles of CubeSat and SmallSat demonstrations due to severe consequences of a failure scenario. Further still, swarms of CubeSats or SmallSats equipped with additive manufacturing heads and proximity operations capabilities could be used to demonstrate freeform space-based additive manufacturing.

CubeSats and SmallSats provide additive manufacturing advocates with an excellent opportunity to develop cultural acceptance with real flight missions at low–cost and on fast–paced schedules. While skeptics may still present scaling challenges, data collected from these small-scale missions will ultimately prove invaluable in the push toward adoption of additive manufacturing for space exploration.

10.2.3 Instilling the Culture: Training the Workforce in Additive Manufacturing

The final step of any change effort is to engrain the change into the culture. This is a difficult task as there are many ways in which this is accomplished. For space missions, there are two ways to accomplish this task. The first method is to fly additive manufactured components on a major mission. Once a component is successfully flown on a major mission, the risk-averse culture kicks in and making any change to the incumbent system is almost impossible and certainly expensive. This approach is good, but highly dependent on a small number of long timescale and high-cost missions. However, a second and highly impactful method is to train the emerging workforce to adopt the new way of doing things. For additive manufacturing, several functional areas will need to have training in order to engrain the process into the culture.

Design and analysis engineers need to be trained on how to design systems and components for additive manufacturing. It can be expected that the resistance to change will be driven by some of the previously mentioned qualification issues. One can expect that designers will be concerned that parts will be too complex and that design costs will increase as specialized build support tooling is added to the work scope. Additionally, designers may become frustrated as processes emerge and they find that redesigns are required to account for process development or incorporation of alternate processes to existing designs. Additive manufacturing advocates can facilitate this change by working early to train designers on the philosophies of design for additive manufacturing, available processes, and case studies showing long-term cost and schedule savings. Additionally, managers can facilitate change by incorporating training on new CAD tools that are being designed with additive manufacturing in mind.

Thermal and structural analysts will share many of the same concerns as design engineers; however, it is likely that analysis tools will lag design tools. Additionally, the design and analysis process is likely to merge in the future and analysts will need to learn more about the design process to develop, capture, and communicate suggestions for improvement as well as risks to quality. Analysts will require training similar to that of designers and it may be necessary to develop common training for both design and analysis functions.

Manufacturing and industrial engineers are likely champions for adoption of additive manufacturing, but incorporating any new processes will be met with at least some level of skepticism. For many additive manufactured parts, there is a reduced ability to inspect parts including porosity, features located inside a part, and surface properties driving difficulty in implementing traditional NDT methods. Additionally, it is probable that many parts will require some level of post-machining driving some to conclude there are minimal cost savings. Advocates will need a clear manufacturing plan that incorporates printing, post-machining, inspection, and assembly in order to convince others that the ideas are sound. Cost assessments will also be needed and as aforementioned may require consideration of more than one part before cost savings are evident. Early training on available processes, pitfalls, new inspection methods, and so on will be critical for gaining acceptance and advocates should be patient while new training is conducted.

Standards and qualification programs will be key to development of cultural acceptance as will early flight demonstrations on low-cost platforms. Despite qualification and flight demonstrations, space product manufacturers will likely take a long time before cultural acceptance of additive manufacturing is attained and pervasive, but early adopters and advocates can take steps to expedite the infusion of additive manufacturing technology through immersive cross-cutting training plans that include coverage of new tools, risks and mitigation methods, detailed manufacturing, inspection and test plans, and through case studies. Working together to develop robust training programs, space industry advocates can drive accelerated cultural acceptance of additive manufacturing processes.

10.3 SUMMARY

Space products are currently expensive and require a decade or more to infuse new technologies. Additive manufacturing enables >50% cost and schedule reduction for most parts driving affordable space exploration. Additive manufacturing enables improved system designs and space mission architectures that drive further toward sustainable and affordable space exploration. Over the next five to ten years, we can expect to see single material processes gain acceptance. In the next 10–20 years we can expect to see multi-material processes infused into space products. In the far term, we can expect to see space-based additive manufacturing in wide use. With robust industry training, qualification programs, and low-cost flight demonstrations, advocates can greatly accelerate cultural acceptance and develop a new paradigm of affordable and sustainable space exploration.

11 Bioprinting—Application of Additive Manufacturing in Medicine

Forough Hafezi, Can Kucukgul, S. Burce Ozler, and Bahattin Koc

CONTENTS

11.1 INTRODUCTION

Over the last few decades, tissue engineering has grown as a new inter- and multidisciplinary scientific field to reinstate damaged or diseased tissue with a combination of functional cells and/or biodegradable scaffolds with engineered biomaterials (Langer and Vacanti 1993). The basic concept of traditional tissue engineering methods is to seed living cells and/or biologically active molecules into a highly porous scaffold to fix and/or regenerate damaged tissues (Bonassar and Vacanti 1998). A highly porous scaffold is required to make both mass transfer and cell incorporation possible. Traditionally, particulate leaching, solution casting, gas foaming, phase separation, freeze drying, and melt molding processes have been used for scaffold fabrication (Leong et al. 2003, Pham et al. 2006, Hafezi et al. 2012). However, these methods generally have no or little control over pore size, geometry, and interconnectivity. Recently, various additive manufacturing techniques have been used to produce scaffolds with controlled micro-architecture and geometry (Yeong et al. 2004, Lee et al. 2008, Peltola et al. 2008, Khoda et al. 2011, Ozbolat et al. 2011, Khoda et al. 2013 and Koc and Khoda 2013). Regardless of the successful scientific outcomes of scaffold-based approaches, these methods still face some challenges. First, the cell growth and proliferation in a scaffold might take a long time because of weak cell adhesion to the scaffold. Second, organs and tissues are generally complex structures containing different cell types. Placement of cells within scaffolds is essentially a random process. It is very challenging to seed different cell types spatially into a scaffold and provide necessary cell-to-cell interaction to form the desired tissue. Moreover, most of scaffold-based approaches suffer from the absence of built-in vascularization. Vascularization and new blood-vessel formation is compulsory to provide nutrients and oxygen to the cells for a successful tissue or organ engineering. Hence, living cells alone or in a combination of biomolecules and biomaterials need to be assembled in three-dimension for a successful tissue or organ engineering.

This chapter focuses on bioprinting, a special additive manufacturing technique, for tissue/organ engineering. Bioprinting or biofabrication is defined as the production of complex living and non-living biological products from living cells, biomolecules, and biomaterials (Derby 2008, Mironov et al. 2009). Bioprinting combines both fundamental science and technologies of additive manufacturing, materials, and biological sciences (Figure 11.1).

In spite of the extensive variety of additive manufacturing (AM) techniques, just few of them are suitable for bioprinting. Similar to other additive manufacturing processes, three-dimensional living structures are built layer-by-layer using bioprinting methods. Most of the bioprinting methods employ inkjet (Nakamura et al. 2005, Xu et al. 2005, Calvert 2007), photolithography/two photon polymerization (Cooke et al. 2003, Arcaute et al. 2006, Schade et al. 2010), direct laser printing (Barron et al. 2004, Yuan et al. 2008, Guillotin et al. 2010), extrusion/deposition (Smith et al. 2007, Skardal et al. 2010), or direct cell printing, self-assembly (Mironov et al. 2003,

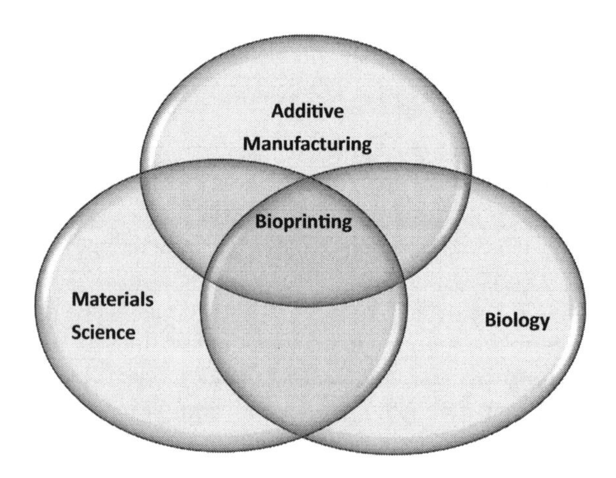

FIGURE 11.1 The main disciplines contributing to the emergence of bioprinting.

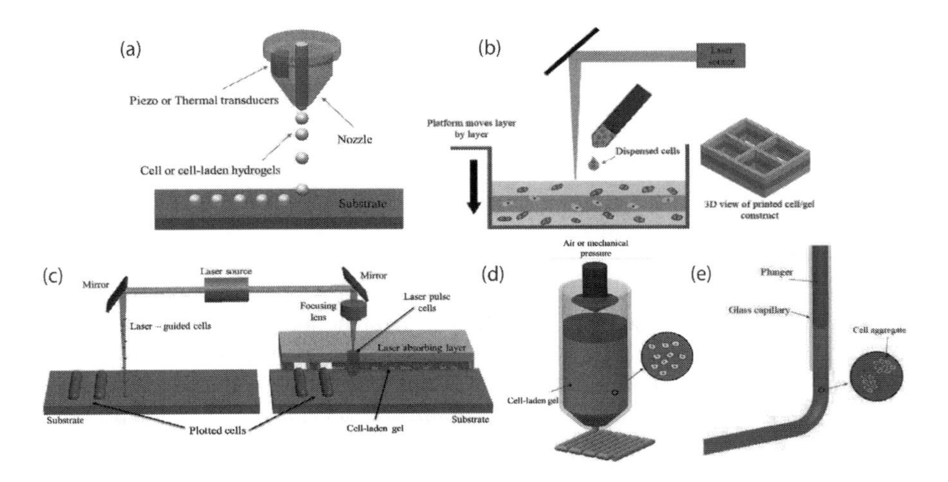

FIGURE 11.2 (a) Thermal and piezoelectric inkjet bioprinting, (b) stereolithography-based bioprinting, (c) laser-based bioprinting setup. (Left) Laser-guided direct cell printing. (Right) The cell-hydrogel compound is propelled forward as a jet by the pressure of a laser-induced vapor bubble. (d) Extrusion/deposition-based bioprinting and (e) direct cell-aggregate deposition.

Mironov et al. 2008, Norotte et al. 2009) techniques. Figure 11.2 shows the schematic representation of various bioprinting methods. The details of each of these methods are discussed below.

11.2 INKJET PRINTING

Inkjet bioprinting is likely one of the most commonly used biofabrication methods because of its distinctive characteristics of high-throughput efficiency, cost effectiveness, and full automation (Boland et al. 2006). Traditionally, inkjet printing is used to reproduce digital image data on a substrate through applying picoliter ink droplets. The technique can be classified into: continuous inkjet, where a stable flow of small droplets made by fluid instability on a passage through a nozzle is either deviated by an electrostatic field onto a substrate or not deviated and gathered for reutilization, and drop-on-demand inkjet, where ink droplets are only produced when needed. Since continuous inkjet needs electrically conducting ink formulations, there is a contamination risk on ink re-cycling so that makes the technique not useful cell printing. Therefore, only drop-on-demand inkjet as shown in Figure 11.2a has been used in bioprinting up to now.

Boland and his colleagues used thermo-sensitive gels to make consecutive layers on top of each other for 3D cellular assemblies with bovine aortal endothelial cells (Boland et al. 2003). They observed that during consecutive culture of the tissue-engineered construct, cells fused to each other within the hydrogel. Cooper and colleagues demonstrated that spatial control of osteoblast differentiation *in vitro* and bone formation *in vivo* is possible via applying inkjet bioprinting technology. They showed that it is possible to create three-dimensional persistent bio-ink patterns of bone morphogenetic protein-2 and its modifiers inactivated within microporous scaffolds via using inkjet bioprinting techniques. Semi-circular patterns of bone morphogenetic protein-2 were printed within human allograft scaffold constructs. They demonstrated that patterns of bone formation *in vivo* were comparable with patterned responses of osteoblastic differentiation *in vitro* (Cooper et al. 2010). Inkjet printers can also be used to deliver drugs or other active biomolecules along with cells. However, shear stress during the process could cause degradation of enzyme activity corresponding to the voltage applied for printing (Cook et al. 2010). Ola Hermanson and colleagues used an inkjet printer to print biologically active macromolecules on poly-acrylamide-based hydrogels, which were subsequently seeded with primary fetal neural stem cells. They observed that inkjet printing can successfully be combined with gene delivery to achieve effective control of stem cell differentiation (Ilkhanizadeh et al. 2007). Although inkjet bioprinting has been one of the most commonly used methods in printing living cells and biomaterials, the technology still faces many challenges. Considering the fact that inkjet printers need cells to be suspended in liquid, cell aggregation and sedimentation are intrinsic weaknesses of the system (Parsa et al. 2010). Because of the high shear stress during printing, cell survivability and degradation of biomolecule activity is another drawback related to inkjet-based bioprinting.

11.3 PHOTOLITHOGRAPHY AND TWO PHOTON POLYMERIZATION

Photolithography or stereolithography is another additive manufacturing technique used for bioprinting (Melchels et al. 2010). Most of the stereolithography-based bioprinting setups used are similar to the ones first developed (Cooke et al. 2003, Dhariwala et al. 2004). Applying a computer-controlled laser beam to sketch a pattern, cell-laden structures are built bottom up from a support platform that lies beneath the resin surface as shown in Figure 11.2b.

Arcaute and her research group showed that it is possible to change the resolution of the cure depth by changing the laser energy, the concentration of poly(ethylene glycol) (PEG) dimethyacrylate as photocrosslinkable material, and the type and concentration of the photoinitiator (Arcaute et al. 2006). In Arcaute et al. (2010), the same group later fabricated a bioactive scaffold via using stereolithography techniques. Human dermal fibroblast cells were seeded on top of the fabricated scaffolds. They applied fluorescent microscopy so that they could observe specific localization of cells in the regions patterned with bioactive PEG (Arcaute et al. 2010).

Two photon-polymerization (2PP) is another emerging photolithography-based technique. In this process, light is applied to trigger a chemical reaction leading to polymerization of a photosensitive material. 2PP initiates the polymerization through irradiation with near-infrared femtosecond laser pulses of 800 nm (Liska et al. 2007).

Recently, Ovsianikov and the colleagues (2011) fabricated 3D PEG diacrylate scaffolds for tissue engineering by means of 2PP. The fabricated scaffold was reproducible, and it was suitable for investigation of cellular processes in three dimensions and for better understanding of *in vitro* tissue formation. The results of their work suggest that 2PP may be used to polymerize poly (ethylene glycol)-based materials into three-dimensional structures with well-defined geometries that mimic the physical and biological properties of the native cell environment (Ovsianikov et al. 2011). The researchers also fabricated methacrylamide modified gelatin scaffolds for tissue engineering applications through using 2PP. The results demonstrated that the fabricated scaffolds are suitable to support porcine mesenchymal stem cell adhesion and subsequent proliferation (Ovsianikov et al. 2011).

11.4 DIRECT LASER PRINTING

Patterning cells with the help of laser light has been one of the first methods for 2D cell patterning. The various different methods by use of laser have been used to move cells for patterning (Figure 11.2c). The most commonly used laser-based direct-write techniques for cellular applications are laser-induced forward transfer (LIFT), absorbing film-assisted laser-induced forward transfer (AFALIFT), laser-guided direct writing (LGDW), matrix-assisted pulsed laser evaporation direct writing and biological laser processing (BioLP). LIFT, AFA-LIFT, BioLP, and matrix-assisted pulsed laser evaporation direct writing have some distinct similarities in methodology for the direct writing of cells. These direct-write techniques utilize laser transparent print ribbons on which one side is coated with cells that are either adhered to a biological polymer through initial cellular attachment or uniformly suspended in a thin layer of liquid (usually cell culture medium mixed with glycerol) or a hydrogel. A receiving substrate is coated with a biopolymer or cell culture medium to maintain cellular adhesion and sustained growth, mounted on motorized stages, and positioned facing the cell-coated side of the ribbon. A pulsed laser beam is transmitted through the ribbon and is used to propel cells from the ribbon to the receiving substrate. The rapid volatilization of the cellular support layer on the ribbon creates the force necessary to allow the cells to cross the small (700–2000 μm) gap between the ribbon and receiving substrate.

Ovsianikov and colleagues applied LIFT for cell printing purposes (Ovsianikov et al. 2010). They demonstrated that in order to control cell migration and cellular interactions within the scaffold, novel technologies capable of producing 3D structures in accordance with predefined design are required. Hence, they first used the 2PP technique for the fabrication of scaffolds. Cells were then seeded into the scaffold by means of LIFT. They showed that with this technique printing of multiple cell types into 3D scaffolds is possible. The combination of LIFT and 2PP provides a route for the realization of 3D multi-cellular tissue constructs and artificial extracellular matrix (ECM) engineered at a micro scale. Guillemot and colleagues applied AFALIFT for cell printing (Guillemot et al. 2011). They demonstrated that applying the AFALIFT technique avoids the printing of undesired debris produced by a metallic interlayer because of potential harm to cell biology in the long term.

LGDW of living cells was developed in 1999 (Odde 1999, Odde et al. 2000). Here, cells drifting by natural convection in the fluid medium were directly deposited onto an untreated glass surface by the laser. The laser beam continuously captures cells as they drifted into the light path, pulls the cells into the center of the beam where the intensity is maximal, and pushes them through the fluid medium along the beam axis onto the target surface. When the desired amount of cells, either a single cell or a number of cells, had been deposited in one spot with a spot size of 10 μm, the focusing lens was translated to move the focal point to a new spot. The terminology "direct writing" indicates that no mask or similar film is used in this process. In 2005, Yaakov Nahmias and colleagues demonstrated that the LGDW can pattern multiple cells types with micrometer resolution on arbitrary surfaces including biological gels. They applied LGDW in order to seed human umbilical vein endothelial cells in two- and three-dimensions with micrometer accuracy. Via patterning human umbilical vein endothelial cells on Matrigel, they could direct their self-assembly into vascular structures along the desired pattern. Finally, co-culturing the vascular structures with hepatocytes resulted in an aggregated tubular structure similar in organization to a hepatic sinusoid (Nahmias et al. 2005). Jason A. Barron and colleagues applied another laser-based writing, matrix-assisted pulsed laser evaporation direct writing to rapidly and accurately deposit mammalian cells (Barron et al. 2004). Recently, Bruce and colleagues also used a similar method, cell deposition microscope based on the laser-guidance technique in which they can micropattern individual cells to specific points on a substrate with high spatial resolution. Their deposition microscope was capable of patterning different cell types onto and within standard cell research devices and providing on-stage incubation for long-term cell culturing (Ma et al. 2011). Similar to other laser-based writing methods, Wu and colleagues applied BioLP in order to fabricate branch/stem structures of human umbilical vein endothelial cells and human umbilical vein smooth muscle cells. They mimicked vascular networks in natural tissue, but also allow cells to develop new and finer structures away from the stem and branches (Wu et al. 2010). Guillotin and his research group also printed human endothelial cells from an alginate ink as well as deposition of nano-particulate hydroxyapatite by BioLP (Guillotin and Guillemot 2011). Pirlo et al. designed a laser cell deposition system where they apply laser to locate single cells at particular points in a multiplicity of in vitro environments. Finally, from the results they concluded, laser cell deposition system can achieve time-specific placement of an individual cell in a cell culture for the systematic investigation of cell-cell and cell-extracellular matrix interactions (Pirlo et al. 2006).

Laser-based cell writing methods are very versatile and can precisely pattern the cells. However, most of these methods are limited to two-dimensional patterning, and it is difficult to fabricate three-dimensional tissue constructs. Another drawback of the laser-based cell writing methods is cell injury due to the process-induced mechanical stress during the cell droplet formation and landing processes. The thermal stress and ultraviolet radiation caused by laser printing could also affect the cell viability.

11.5 EXTRUSION PRINTING

11.5.1 Cell-Laden Printing

The main principle of extrusion-based bioprinting techniques is to force continuous filaments of a material through a nozzle in a controlled manner to construct a 3D structure (Figure 11.2d). For the purpose of extrusion printing of cells, the material normally includes an extremely viscous cell-laden hydrogel (Fedorovich et al. 2007) which has the ability to flow from the nozzle without applying high temperatures. After the material was deposited and became a solid via thermal, physical, or chemical methods, it provides adequate mechanical integrity to fabricate 3D structures. The bioprinter design here is normally simple, including a three-axis robot that controls the motions of either pneumatically or volumetrically driven displacement pens or syringes with a typical nozzle diameter of 150–500 μm. Landers and Mülhaupt at the Freiburg Research Center (Landers et al. 2000) developed an extrusion-based bioprinting process called bioplotting. Bioplotting is a biofabrication technology based on the extrusion of continuous filaments. The 3D fabrication happens in layer-by-layer fashion through the computer-controlled deposition of material on a surface. The dispensing head moves in three dimensions, whereas the fabrication platform is immobile. It is feasible to carry out either discontinuous dispensing of microdots or a continuous dispensing of microstrands. Landers and his colleagues used the developed bioplotter for printing hydrogel scaffolds with encapsulated cells. Fedorovich and his research group also used the similar bioplotter for bone tissue engineering (Fedorovich et al. 2008). They showed maintaining spatially organized, osteo and endothelial progenitor cells in printed grafts after in vivo implantation for the first time (Fedorovich et al. 2011). This 3D bioplotter technology is commercialized by Envisontec (envisiontec.com).

Smith and his colleagues also used a commercial extrusion-based printer developed by nScrypt (nscrypt.com), and they deposited cold (2°C–10°C) solutions of either human fibroblasts encapsulated in Pluronic-F127 or bovine aortic endothelial cells (BAECs) in type I collagen onto heated substrates where solidification of the printed structures was induced by thermal gelation of the biopolymers. Applying low temperatures has resulted in low cell viabilities (Smith et al. 2004). Significantly, Smith's work indicated that computer-aided design and manufacturing (CAD/CAM) technology could be used to deposit cell-laden structures mimicking a vascular structure.

Butcher and colleagues applied 3D bioprinting to fabricate living alginate/gelatin hydrogel valve conduits with anatomically correct architecture and direct incorporation of dual cell types. The researchers used a modified fab@home printer for printing hydrogels encapsulated with cells. After bioprinting, encapsulated aortic root sinus smooth muscle cells and aortic valve leaflet interstitial cells were viable within alginate/gelatin hydrogel discs over 7 days in culture (Duan et al. 2013) (Figure 11.3).

Extrusion-based bioprinters have also been used for printing stem cell-laden hydrogels. Xu et al. demonstrated that adipose-derived stem cells printed in gelatin/alginate/fibrinogen gels have the ability to differentiate into endothelial cells at the walls of printed channels (Xu et al. 2009). They used a bioprinter with two nozzles controlled by a computer called cell-assembler for printing a 3D structure, which consisted of square grids and orderly channels. The same group also used adipose-derived stromal cells combined within a gelatin/alginate/fibrinogen hydrogel to form a vascular-like network and hepatocytes combined gelatin/alginate/chitosan were placed around it for fabrication of complex 3D structures mimicking a liver (Shengjie et al. 2009) (Figure 11.4).

Khademhosseini and his group used a cell-laden printing technique for fabricating microfluidic channels from cell-laden hydrogels (Ling et al. 2007). They demonstrated that the encapsulation of mammalian cells within the bulk material of microfluidic channels may be beneficial for applications ranging from tissue engineering to cell-based diagnostic assays. They presented a technique for fabricating microfluidic channels from cell-laden agarose hydrogels. The channels of different

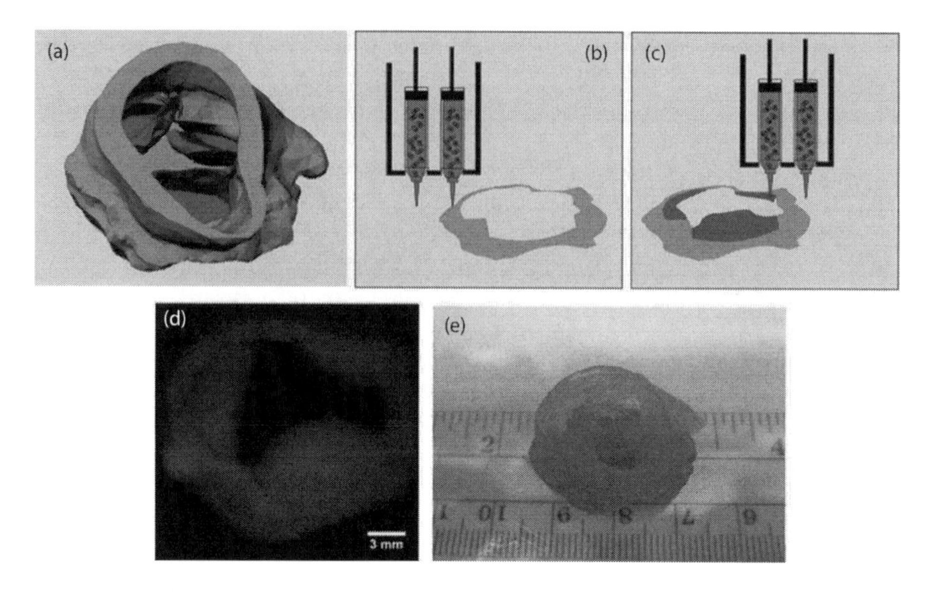

FIGURE 11.3 Bioprinting of aortic valve conduit. (a) Aortic valve model reconstructed from micro-CT images. The root and leaflet regions were identified with intensity thresholds and rendered separately into 3D geometries into STL format (green color indicates valve root and red color indicates valve leaflets); (b, c) schematic illustration of the bioprinting process with dual cell types and dual syringes; (b) root region of first layer generated by hydrogel with SMC; (c) leaflet region of first layer generated by hydrogel with VIC; (d) fluorescent image of first printed two layers of aortic valve conduit; SMC for valve root were labeled by cell tracker green and VIC for valve leaflet were labeled by cell tracker red. (e) as-printed aortic valve conduit. (Reproduced from Duan, B. et al., *J. Biomed. Mater. Res. A*, 101, 1255–1264, 2013. With permission.)

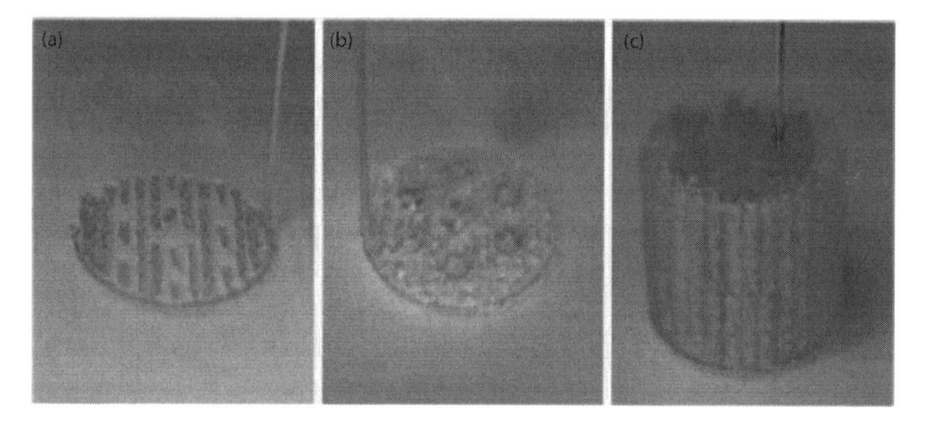

FIGURE 11.4 Sequential deposition of hepatocytes in gelatin/alginate/collagen (a, clear) and ASCs in gelatin/alginate/fibrinogen (b, red) to produce a 3D liver construct (c). (Adapted from Shengjie, L. et al. *J Bioact Compat Polym* 24, 249–265, 2009.)

dimensions were generated, and it was shown that agarose, though highly porous, is a suitable material for performing microfluidics. Cells embedded within the microfluidic molds were well distributed, and media pumped through the channels allowed the exchange of nutrients and waste products. While most cells were found to be viable upon initial device fabrication, only those cells near the microfluidic channels remained viable after 3 days, demonstrating the importance of a perfused network of microchannels for delivering nutrients and oxygen to maintain cell viability in large hydrogels.

There are several challenges related to extrusion-based printing of cell-laden hydrogels. First, hydrogel material used for cell encapsulation could limit cell-cell interaction and ECM formation. Because of weak mechanical properties of hydrogel, the printed structures lack of mechanical integrity. This could limit scale of the constructs printed and make the implantation as well as the maintenance of the printed structure difficult. Hydrogels used for printing tissue constructs could also cause immunogenic reactions after implantation.

11.5.2 Direct Cell Printing/Self-assembly

The concepts of tissue fusion and tissue fluidity are fundamental for the modern organ printing technology based on self-assembly (Steinberg 1963, Foty et al. 1994, Whitesides and Boncheva 2002). Direct cell printing also called self-assembly or scaffold-free bioprinting employs layer-based printing of multi-cellular aggregates or self-assembling tissue spheroids as building blocks (Mironov 2003, Mironov et al. 2008) (Figure 11.2e). The main idea of this technique is formation of tissue fusion of closely placed multi-cellular aggregates or tissue spheroids without using any scaffolds. Tissue fusion is an omnipresence process during embryonic development (Perez-Pomares et al. 2006), and hence this biofabrication technology is considered biomimetic. Cell aggregates or microtissues can be fabricated in pre-designed shapes by seeding and culturing in micro-molded well plates and serve as building blocks to assembly of multi-cellular tissues at a higher level of organization. With the help of computer-controlled automated bioprinting methods, tissue or organ printing technology can be completely automated. Fully developed organ bioprinting could make 3D vascularized functional human organs or living functional organ constructs possible for surgical implantation (Rivron et al. 2009).

Forgacs and colleagues (Norotte et al. 2009) developed a fully biological self-assembly approach through an extrusion-based bioprinting method for scaffold-free small diameter vascular reconstruction. Various vascular cell types, including smooth muscle, endothelial, and fibroblasts cells, were aggregated into discrete units, either multi-cellular spheroids or cylinders of controllable diameter (300–500 mm). These were printed layer-by-layer concurrently with agarose rods, used here as a molding template. The post-printing fusion of the discrete units resulted in single- and double-layered small diameter vascular tubes (outer diameters ranging from 0.9 to 2.5 mm). A unique aspect of the method was the ability to engineer vessels by directly printing multi-cellular aggregates scaffold-free as shown in Figure 11.5. The developed bioprinting technology was commercialized by Organovo (organovo.com).

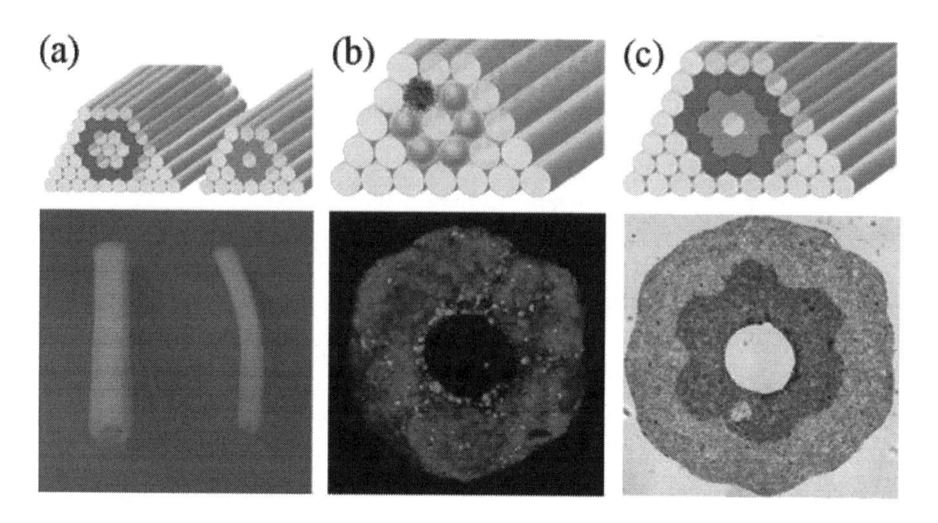

FIGURE 11.5 (a) Design templates (top) and fused constructs (bottom) of different vessel diameters built with cylindrical bioink. (b) The top image shows a template to build a construct with spheroids composed of SMC (red) and ECs (green). A transversal section after fusion (bottom) shows that the lumen is composed predominantly of endothelial cells. (c) Template to construct a double-layered vascular tube (top). The inner layer is constructed of SMC building blocks (green), the second of fibroblast building blocks (red). The transversal section (bottom) shows fusion and the segregation of the two cell types mimicking the media and adventitia of blood vessels. (Reproduced with permission from Jakab, K. et al., *Biofabrication*, 2, 022001, 2010.).

Chang and colleagues (2011) developed a hybrid bioprinting process to create a 3D micro-organ which biomimics the cell's natural microenvironment with enhanced functionality. The researchers used an automated syringe-based direct cell writing hybrid bioprinting process. They first dispensed gel lines and then used the dispensed gels as mold to bioprint microvascular cells and isolated micro-vessel fragments into composite 3D structures. The printed cells and vessel fragments remained viable after incorporation into biohybrid structures.

In direct cell printing, complex tissue constructs with multiple cells can be printed spatially. Since live cells are directly printed without mixing with hydrogels or seeding into scaffolds, there will not be any immunogenic rejection or inflammation problems. Cells can be patterned easily with automated bioprinting processes. However, the printed tissue constructs need vascularization to survive and to form tissue or organ substitutes in vitro. Another challenge is keeping the form of mechanically weak cells or cell aggregates in 3D during printing process.

11.5.3 Direct Cell Printing of Macrovascular Constructs

Scaffold-free tissue engineering of small-diameter tubular grafts has been studied in literature (Norotte et al. 2009). However, the previous research is limited to fabrication of simple small-diameter tubular vascular conduits. We developed novel

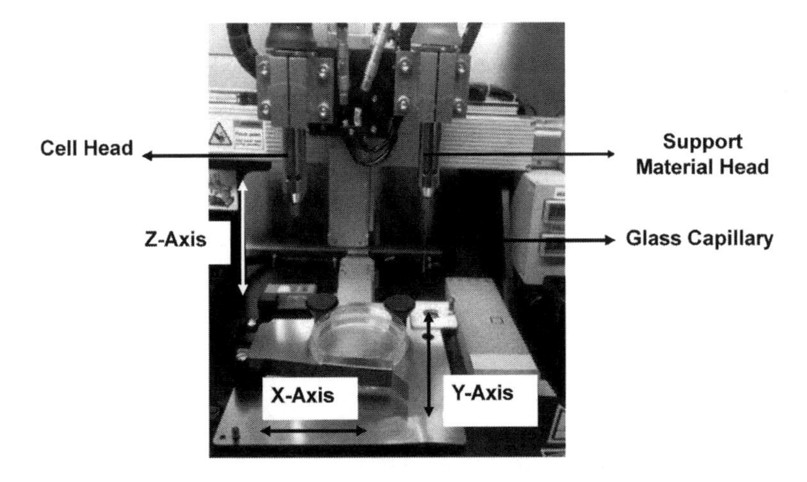

FIGURE 11.6 First commercial bioprinter: Organovo Novogen MMX.

computer-aided algorithms and strategies to model and 3D bioprint a human aortic tissue construct biomimetically (Kucukgul et al. 2015). Cylindrical mouse embryonic fibroblast cell aggregates and agarose-based support structures (hydrogels) were 3D bioprinted using a Novogen MMX Bioprinter as shown in Figure 11.6. The 3D bioprinter has two deposition heads equipped with glass capillaries for cell aggregates and hydrogel biomaterials respectively. Mouse embryonic fibroblast cells were cultured based on general cell culturing protocols. In order to obtain cell pellets for printing, cells were centrifuged at relatively high speed following the detachment of them from the cell culture flask. Once the dense cell pellet was obtained, it was transferred into capillaries for continuous bioprinting.

Since the primary objective of this work is to 3D bioprint an anatomically correct blood vessel, the original geometry of the vessel needs to be obtained and converted into a computer model. Hence, the correct form of the vascular structures can be exactly mimicked in the biofabrication process. A part of the human abdominal aorta model was chosen to illustrate the proposed methods for 3D bioprinting of macrovascular structures. The geometry of the aorta were captured from a sample medical image using a imaging and segmentation process and then transformed into 3D computer model as a stereolithography model. Stereolithography files are constructed with numerous triangles that tessellate the outside surface of the object, here, a part of abdominal aorta. To be able to optimize 3D bioprinting and generate tool path planning, the obtained stereolithography model of the abdominal aorta needs to be represented with parametric surfaces. A novel biomodeling method was developed to create smooth parametric surfaces using the geometric information of these mesh (triangular facet) structures. The developed method generates section curves as well as the center points from the mesh model, then approximates a centerline curve from those center points. Smooth parametric surfaces are then generated along with this centerline curve (Figure 11.7).

After the biomodeling phase, the aorta model was then 3D bioprinted layer-by-layer vertically, with micro capillaries of specific diameter (450 μm), using hydrogel

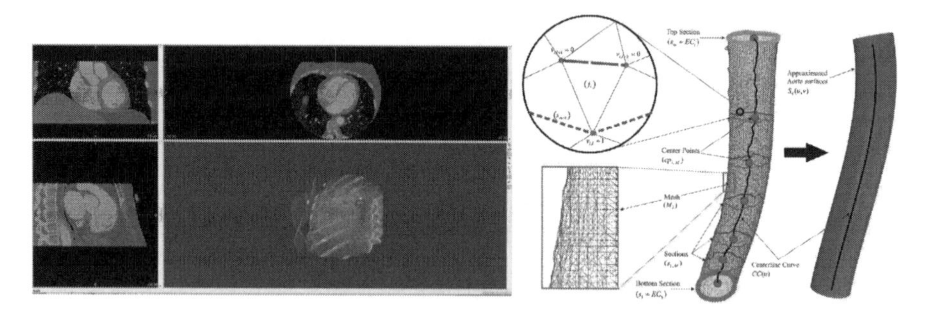

FIGURE 11.7 Biomimetic modeling of aorta directly from medical images.

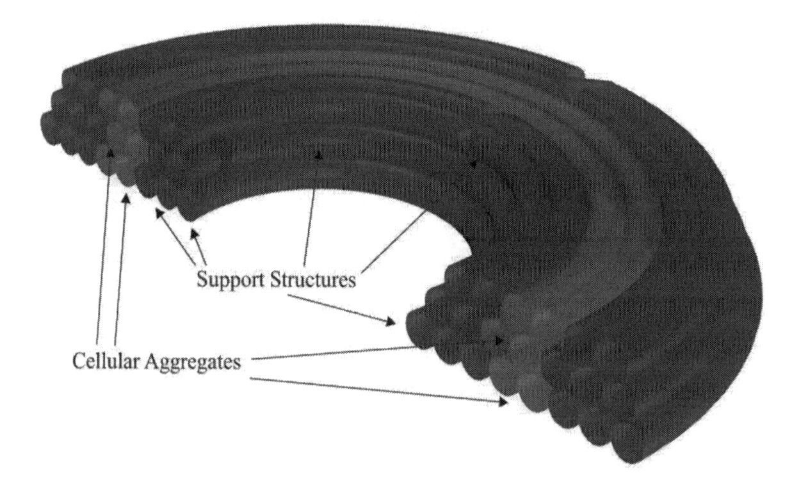

FIGURE 11.8 3D Bioprinting of three consecutive example layers showing support structures (darker color) and cellular aggregates (lighter color).

as supportive material and mouse embryonic fibroblast cells as bioink. Tool path planning for the biofabrication step was generated by developed computer algorithms so the entire printing process could be carried out without any human intervention.

To bioprint an anatomically correct aorta model, mechanically weak cellular aggregates need to be supported by hydrogels in order to preserve the desired/exactly mimicked shape. As shown in Figure 11.8, a novel self-supporting method was developed so that both cellular aggregates (red cylinders) and hydrogels (blue cylinders) are placed on the valleys of the preceding layer to form 3D construct.

After path planning and optimization, cylindrical cellular aggregates and their support structures were printed layer-by-layer using the 3D bioprinter (Figure 11.6). As shown in Figure 11.9, mouse embryonic fibroblast cell aggregates were successfully printed at the valleys formed by the support material (hydrogel). The cells were printed directly by controlling the bioprinter's cell and support heads using the generated scripts. After bioprinting, the printed tissue constructs are kept in an incubator until the cell aggregates are fused and form the tissue construct.

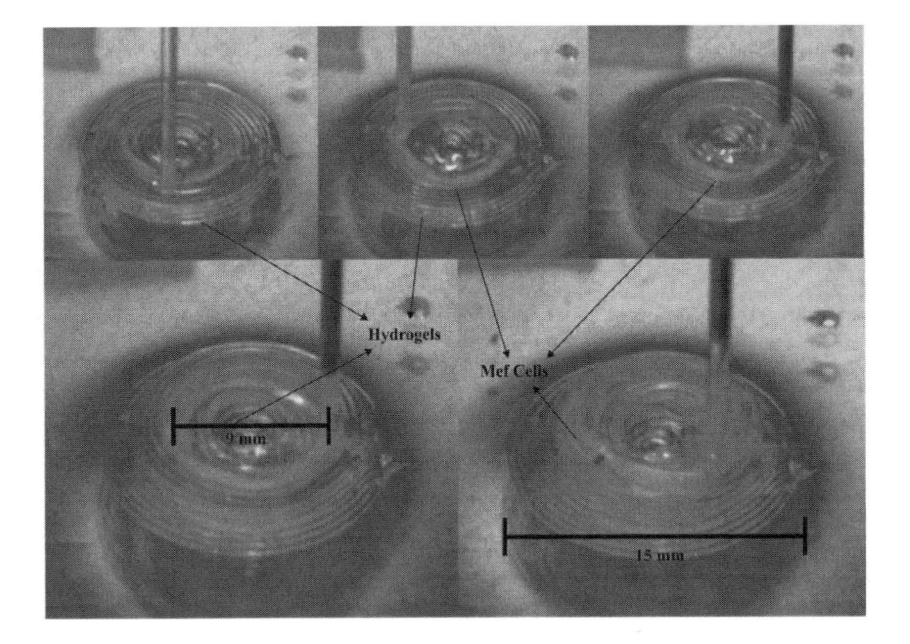

FIGURE 11.9 Direct cell bioprinting for aortic constructs.

11.6 BIOINKS

In previous sections, several most important bioprinting techniques were explained. In this section, material systems so called bioinks used for bioprinting methods will be discussed. Bioinks are the material systems which include at least cells alone or a carrying material encapsulating the cells. Moreover, bioinks could also include biologically active molecules such as growth factors, cytokines, etc., and other stabilizing additives in carrying material. After bioprinting, the cells alone or the encapsulated cells in bioinks forms or regenerates the targeted tissue. Bioinks with hydrogels should be in the solution form and be cross-linked or stabilized during or just after the bioprinting process to keep the final desired shape for the targeted tissue. Therefore, an ideal bioink should possess not only biological properties to form or regenerate the targeted tissue, but also should possess the mechanical and rheological properties suitable for printability of the bioprinting process used. The printability of the bioink depends not only on rheological properties of bioink such as viscosity, shear thinning or thickening properties, and surface tension of the bioink, but also the ability to keep the final desired shape by crosslink after the bioprinting process.

11.6.1 BIOINKS WITH CELL ENCAPSULATION

As mentioned above, the cells can be encapsulated in a carrying biomaterial used as a bioink for the bioprinting process. Therefore, the carrying biomaterial, usually a hydrogel, must be biocompatible, cytocompatible, and also should provide a suitable environment for cell growth and proliferation during and after the bioprinting process. The hydrogels could be made from natural, synthetic biomaterials, or a combination of the both. The most commonly used natural hydrogels for bioinks include alginate, agarose, gelatin, chitosan, collagen, fibrin/fibrinogen, and hyaluronic acid. Matrigel™ is a commercially available hydrogel also used for the bioprinting process. Most of the natural hydrogels are based on polysaccharides, extracellular proteins, or directly from decellularized extracellular matrix. Synthetic hydrogels can be polymerized different chains and structures suitable for bioprinting. Poly (ethylene glycol) and Pluronic are some examples for synthetic hydrogels. More recently, decellularized extracellular matrix (dECM)-based hydrogels were successfully used as bioinks. There are also composite hydrogels that combine at least two of those hydrogels; agarose/chitosan and gelatin/alginate are some examples for those bioinks.

The cross-linking method of hydrogels mainly determines their suitability of the bioprinting process. Table 11.1 lists natural, synthetic, and dECM hydrogels used as a bioink in different types of bioprinting processes and their underlying cross-linking mechanisms.

The chemical, biological, and physical properties of hydrogels could be modified by adding additives during bioink preparation (Nadernezhad et al. 2016, Topuz et al. 2018). The additives could not only improve printability of materials, but also functionality.

TABLE 11.1
Hydrogels for Bioinks

Hydrogel	Bioprinting Method	Cross-linking Method	References
Alginate	Extrusion-based, Laser-assisted, Inkjet-based	Ionic	Chang et al. (2010), Gao et al. (2015), Zhang et al. (2013), Guillotin et al. (2010), Xu et al. (2012), Christensen et al. (2015)
Agarose	Extrusion-based, Laser-assisted, Inkjet-based	Thermal	Moon et al. (2010), Neufurth et al. (2014), Blaeser et al. (2014)
Gelatin/Gelatin Methacrylated (GelMA)	Extrusion-based, Laser-assisted, Inkjet-based	Thermal Photopolymerization (GelMA)	Jia et al. (2014), Zhang et al. (2013), Duan et al. (2014), Bertassoni et al. (2014), Kolesky et al. (2014), McGuigan et al. (2007)
Chitosan	Extrusion-based	Ionic	Ye et al. (2014), Geng et al. (2005)
Collagen	Extrusion-based, Laser-assisted, Inkjet-based	Thermal	Koch et al. (2012), Michael et al. (2013), Lee et al. (2014), Moon et al. (2010), Campos et al. (2015)

(Continued)

TABLE 11.1 (*Continued*)
Hydrogels for Bioinks

Hydrogel	Bioprinting Method	Cross-linking Method	References
Fibrin/fibrinogen	Inkjet-based	Enzymatic	Nakamura et al. (2010), Xu et al. (2013), Gruene et al. (2011)
Hyaluronic acid	Extrusion-based	Photopolymerization	Duan et al. (2014), Rajangam et al. (2013), Park et al. (2014), Skardal et al. (2010), Kesti et al. (2015)
PEG	Extrusion-based Inkjet-based	Thermal Photopolymerization	Schuurman et al. (2011), Rutz et al. (2015), Skardal et al. (2010), Gao et al. (2014)
Pluronic acid	Extrusion-based	Thermal	Smith et al. (2004), Fedorovich et al. (2008), Hockaday et al. (2012)
Decellularized extracellular matrix	Extrusion-based	Self-assembly, thermal	Pati et al. (2014, 2015), Skardal et al. (2015), Toprakhisar et al. (2018)

11.6.2 Bioinks with Cell Aggregates/Pellets, Spheroids

The hydrogels used for encapsulating cells could provide structural stability during and after the printing process. However, the cell carrying material (hydrogel) could block cell-cell interactions or could cause immunogenic reactions because of cross-linking agents. Therefore, cells in the form of aggregates/pellets (Norotte et al. 2009, Kucukgul et al. 2015) or spheroids process (Mironov 2003, Mironov et al. 2003, Mironov et al. 2008) are also directly used for scaffold or encapsulation-free bioprinting. To be able to bioprint, cells should be specially prepared (Ozler et al. 2017, Bakirci et al. 2017) and also supported temporarily by hydrogels (Kucukgul et al. 2015, Ozler et al. 2017, Bakirci et al. 2017) until fusion and maturation of cells.

11.7 CONCLUSION AND FUTURE PROSPECTS

In this chapter, a novel use of additive manufacturing for bioprinting tissue or organs is presented. Various additive manufacturing techniques have been successfully used for developing tissue scaffolds with controlled geometry and microarchitecture. However, possible immunogenic reactions of scaffold materials, difficulty cell seeding, and vascularization problems lead researchers to develop scaffold-free bioprinting methods with live cells for tissue or organ printing. We have discussed several bioprinting processes for engineering replacements for diseased or damaged tissues. For any scaffold-based or scaffold-free bioprinting processes, vascularization of printed tissue constructs to allow oxygen and nutrition delivery and waste removal must be considered. Tissues or organs are complex structures, new methods should also biomimick the geometry and multi-cellular architecture including an extracellular matrix for a successful formation of tissues and organs. Newly developed bioprinting processes should allow not only printing regular cells, but also stem cells as well as active biomolecules to stimulate cells for possible organogenesis. Finally, the developed tissue engineering technologies will have to consider the feasibility of printed 3D tissue constructs for clinical applications to move from bench to bedside.

SUGGESTED FURTHER READING

Ozbolat, I. T., 3D Bioprinting, *Fundamentals, Principles and Applications*, 1st ed., by Ibrahim Ozbolat, Academic Press, Oxford, UK, 2016.

Khademhosseini, A., and Gulden, C-U., 3D Bioprinting in Regenerative Engineering. Principles and applications CRC Press Series. In. *Regenerative Engineering*, 1st ed. by Ali Khademhosseini, Gulden Camci-Unal, CRC Press, Boca Raton, FL, 2018.

Murphy, S. V., and Atala, A., "3D bioprinting of tissues and organs." *Nat Biotechnol* 32 (2014): 773–785.

Selcan, G. O. P., Zhang, Y. S., Khademhosseini, A., and Dokmeci, M. R., "Bioinks for 3D bioprinting: An overview." *Biomater Sci* 6 (2018): 915.

Hospodiuk, M., Madhuri, D., Sosnoki, D., and Ozbolat, I. T., "The bioink: A comprehensive review on bioprintable materials." *Biotechnol Adv* 35 (2017): 217–239.

QUESTIONS

1. Research and describe multi-material bioprinting processes.
2. How would you change the material properties during the bioprinting process.
3. Explain the cross-linking mechanisms for inkjet-based bioprinting process.
4. What are the photo-initiators suitable for extrusion-based bioprinting.
5. Compare and list the advantages and disadvantages of the extrusion-based and inkjet-based bioprinting processes.
6. How would you capture the external geometry of biological objects? Develop a method to capture and design a human ear model from computer-tomography images.
7. Tissue or organs require more than one type of cells and organize them. How would you organize cells during or after bioprinting?
8. Derive the equation for determining the pressure required extruding hydrogels (assume material is Newtonian initially) for extrusion-based bioprinting. How would you change your model if the biomaterial is non-Newtonian?
9. Research and explain submerged bioprinting processes. What are the advantages and disadvantages of submerged bioprinting.

REFERENCES

Arcaute, K., Mann, B. K., Wicker, R. B. "Stereolithography of three-dimensional bioactive poly(ethylene glycol) constructs with encapsulated cells." *Ann Biomed Eng* 34(9) (2006): 1429–1441.

Arcaute, K., Mann, B. Wicker, R. "Stereolithography of spatially controlled multi-material bioactive poly(ethylene glycol) scaffolds." *Acta Biomater* 6(3)(2010): 1047–1054.

Bakirci, E., Toprakhisar, B., Zeybek, M., Ince, G. O., Koc, B. "Cell sheet based bioink for 3D bioprinting applications." *Biofabrication* 9(2)(2017): 024105.

Barron, J. A., Ringeisen, B. R., Kim, H., Spargo, B. J., Chrisey, D. B. "Application of laser printing to mammalian cells." *Thin Solid Films* 453–454(2004): 383–387.

Barron, J. A., Wu, P., Ladouceur, H. D., Ringeisen, B. R. "Biological laser printing: A novel technique for creating heterogeneous 3-dimensional cell patterns." *Biomed Microdevices* 6(2)(2004): 139–147.

Bertassoni, L. E., Cardoso, J. C., Manoharan, V. et al. "Direct-write bioprinting of cell-laden methacrylated gelatin hydrogels." *Biofabrication* 6(2)(2014): 024105.

Blaeser, A., Duarte Campos, D. F., Weber, M., Neuss, S., Theek, B., Fischer, H., Jahnen-Dechent, W. "Biofabrication under fluorocarbon: A novel freeform fabrication technique to generate high aspect ratio tissue-engineered constructs." *Biores Open Access* 2(2014): 374–384.

Boland, T., Mironov, V., Gutowska, A., Roth, E. A., Markwald, R. R. "Cell and organ printing 2: Fusion of cell aggregates in three-dimensional gels." *Anat Rec A Discov Mol Cell Evol Biol* 272(2)(2003): 497–502.

Boland, T., Xu, T., Damon, B., Cui, X. "Application of inkjet printing to tissue engineering." *Biotechnol J* 1(9)(2006): 910–917.

Bonassar, L. J., Vacanti, C. A. "Tissue engineering: The first decade and beyond." *J Cell Biochem Suppl* 30–31(1998): 297–303.

Calvert, P. "Materials science. Printing cells." *Science* 318(5848)(2007): 208–209.

Campos, D. F., Blaeser, A., Korsten, A., Neuss, S., Jäkel, J., Vogt, M., Fischer, H. "The stiffness and structure of three-dimensional printed hydrogels direct the differentiation of mesenchymal stromal cells toward adipogenic and osteogenic lineages." *Tissue Eng Part A* 21(3–4)(2015): 740–756.

Chang, C. C., Boland, E. D., Williams, S. K., Hoying, J. B. "Direct-write bioprinting three-dimensional biohybrid systems for future regenerative therapies." *J Biomed Mater Res B Appl Biomater* 98(1)(2011): 160–170.

Chang, R., Emami, K., Wu, H., Sun, W. "Biofabrication of a three-dimensional liver micro-organ as an in vitro drug metabolism model." *Biofabrication* 2(4)(2010): 045004.

Christensen, K., Xu, C., Chai, W., Zhang, Z., Fu, J., Huang, Y. "Freeform inkjet printing of cellular structures with bifurcations." *Biotechnol Bioeng* 112(5)(2015): 1047–1055.

Cook, C. C., Wang, T., Derby, B. "Inkjet delivery of glucose oxidase." *Chem Commun (Camb)* 46(30)(2010): 5452–5454.

Cooke, M. N., Fisher, J. P., Dean, D., Rimnac, C., Mikos, A. G. "Use of stereolithography to manufacture critical-sized 3D biodegradable scaffolds for bone ingrowth." *J Biomed Mater Res B Appl Biomater* 64(2)(2003): 65–69.

Cooper, G. M., Miller, E. D., Decesare, G. E., et al. "Inkjet-based biopatterning of bone morphogenetic protein-2 to spatially control calvarial bone formation." *Tissue Eng Part A* 16(5)(2010): 1749–1759.

Derby, B. "Bioprinting: Inkjet printing proteins and hybrid cell-containing materials and structures." *J Mater Chem* 18(47)(2008): 5717.

Dhariwala, B., Hunt, E., Boland, T. "Rapid prototyping of tissue-engineering constructs, using photopolymerizable hydrogels and stereolithography." *Tissue Eng* 10(9–10)(2004): 1316–1322.

Duan, B., Kapetanovic, E., Hockaday, L. A., Butcher, J. T. "Three-dimensional printed trileaflet valve conduits using biological hydrogels and human valve interstitial cells." *Acta Biomater* 10(5)(2014): 1836–1846.

Duan, B., Hockaday, L. A., Kang, K. H., Butcher, J. T. "3D bioprinting of heterogeneous aortic valve conduits with alginate/gelatin hydrogels." *J Biomed Mater Res A* 101(5)(2013): 1255–1264.

Fedorovich, N. E., Alblas, J., de Wijn, J. R., et al. "Hydrogels as extracellular matrices for skeletal tissue engineering: State-of-the-art and novel application in organ printing." *Tissue Eng* 13(8)(2007): 1905–1925.

Fedorovich, N. E., De Wijn, J. R., Verbout, A. J., Alblas, J., Dhert, W. J. "Three-dimensional fiber deposition of cell-laden, viable, patterned constructs for bone tissue printing." *Tissue Eng Part A* 14(1)(2008): 127–133.

Fedorovich, N. E., Wijnberg, H. M., Dhert, W. J., Alblas, J. "Distinct tissue formation by heterogeneous printing of osteo and endothelial progenitor cells." *Tissue Eng Part A* 17(15–16)(2011): 2113–2121.

Foty, R. A., Forgacs, G., Pfleger, C. M., Steinberg, M. S. "Liquid properties of embryonic tissues: Measurement of interfacial tensions." *Phys Rev Lett* 72(14)(1994): 2298–2301.

Gao, G., Schilling, A. F., Yonezawa, T., Wang, J., Dai, G., Cui, X. "Bioactive nanoparticles stimulate bone tissue formation in bioprinted three-dimensional scaffold and human mesenchymal stem cells." *Biotechnol J* 9(10)(2014): 1304–1311.

Gao, Q., He, Y., Fu, J., Liu, A., Ma, L. "Coaxial nozzle-assisted 3D bioprinting with built-in microchannels for nutrients delivery." *Biomaterials* 61(2015): 203–215.

Geng, L., Feng, W., Hutmacher, D., San Wong, Y., Tong Loh, H., Fuh, J. "Direct writing of chitosan scaffolds using a robotic system." *Rapid Prototyp J* 11(2)(2005): 90–97.

Gruene, M., Pflaum, M., Deiwick, A., Koch, L., Schlie, S., Unger, C., Wilhelmi, M., Haverich, A., Chichkov, B. N. "Adipogenic differentiation of laser-printed 3D tissue grafts consisting of human adipose-derived stem cells." *Biofabrication* 3(1)(2011): 015005.

Guillemot, F., Guillotin, B., Fontaine, A., et al. "Laser-assisted bioprinting to deal with tissue complexity in regenerative medicine." *MRS Bulletin* 36(12)(2011): 1015–1019.

Guillotin, B., Guillemot, F. "Cell patterning technologies for organotypic tissue fabrication." *Trends Biotechnol* 29(4)(2011): 183–190.

Guillotin, B., Souquet, A., Catros, S., et al. "Laser assisted bioprinting of engineered tissue with high cell density and microscale organization." *Biomaterials* 31(28)(2010): 7250–7256.

Hafezi, F., Hosseinnejad, F., Fooladi, A. A., et al. "Transplantation of nano-bioglass/gelatin scaffold in a non-autogenous setting for bone regeneration in a rabbit ulna." *J Mater Sci Mater Med* 23(11)(2012): 2783–2792.

Hockaday, L. A., Kang, K. H., Colangelo, N. W. et al. "Rapid 3D printing of anatomically accurate and mechanically heterogeneous aortic valve hydrogel scaffolds." *Biofabrication* 4(3)(2012): 035005.

Ilkhanizadeh, S., Teixeira, A. I., Hermanson, O. "Inkjet printing of macromolecules on hydrogels to steer neural stem cell differentiation." *Biomaterials* 28(27)(2007): 3936–3943.

Jakab, K., Norotte, C., Marga, F., Murphy, K., Vunjak-Novakovic, G., Forgacs, G. "Tissue engineering by self-assembly and bio-printing of living cells," *Biofabrication* 2(2)(2010):022001. doi:10.1088/1758-5082/2/2/022001.

Jia, J., Richards, D. J., Pollard, S. et al. "Engineering alginate as bioink for bioprinting." *Acta Biomater* 10(106)(2014): 4323–4331.

Kesti, M., Müller, M., Becher, J., Schnabelrauch, M., D'Este, M., Eglin, D., Zenobi-Wong, M. "A versatile bioink for three-dimensional printing of cellular scaffolds based on thermally and photo-triggered tandem gelation." *Acta Biomater* 11(2015): 162–172.

Khoda, A. K. M., Koc, B. "Functionally heterogeneous porous scaffold design for tissue engineering." *Computer Aided Design* 45(11)(2013): 1276–1293.

Khoda, A. K. M., Ozbolat, I. T., Koc, B. "Designing heterogeneous porous tissue scaffolds for additive manufacturing processes." *Computer Aided Design* 45(12)(2013): 1507–1523.

Khoda, A. K., Ozbolat, I. T., Koc, B. "Engineered tissue scaffolds with variational porous architecture." *J Biomech Eng* 133(1)(2011): 011001.

Koch, L., Deiwick, A., Schlie, S. et al. "Skin tissue generation by laser cell printing." *Biotechnol Bioeng* 109(7)(2012): 1855–1863.

Kolesky, D. B., Truby, R. L., Gladman, A. S., Busbee, T. A., Homan, K. A., Lewis, J. A. "3D bioprinting of vascularized, heterogeneous cell-laden tissue constructs." *Adv Mater* 26(12)(2014): 3124–3130.

Kucukgul, C., Ozler, S. B., Inci Ilyas, Karakas E., Irmak S., Gozuacik D., Taralp, A., Koc B. "3D bioprinting of biomimetic aortic vascular constructs with self-supporting cells." *Biotechnol Bioeng* 112(4)(2015): 811–821.

Landers, R., Mülhaupt, R. "Desktop manufacturing of complex objects, prototypes and biomedical scaffolds by means of computer-assisted design combined with computer-guided 3D plotting of polymers and reactive oligomers." *Macromol Mater Eng.* 282(1)(2000): 17–21.

Langer, R., Vacanti, J. P. "Tissue engineering." *Science* 260(5110)(1993): 920–926.

Lee, J., Cuddihy, M. J., Kotov, N. A. "Three-dimensional cell culture matrices: State of the art." *Tissue Eng Part B Rev* 14(1)(2008): 61–86.

Lee, V., Singh, G., Trasatti, J. P., Bjornsson, C., Xu, X., Tran, T. N., Yoo, S.-S., Dai, G., Karande, P. "Design and fabrication of human skin by three-dimensional bioprinting." *Tissue Eng Part C Methods* 20(6)(2014): 473–484. doi:10.1089/ten. TEC.2013.0335.

Leong, K. F., Cheah, C. M., Chua, C. K. "Solid freeform fabrication of three-dimensional scaffolds for engineering replacement tissues and organs." *Biomaterials* 24(13)(2003): 2363–2378.

Ling, Y., Rubin, J., Deng, Y., et al. "A cell-laden microfluidic hydrogel." *Lab Chip* 7(6)(2007): 756–762.

Liska, R., Schuster, M., Inführ, R., et al. "Photopolymers for rapid prototyping." *J Coat Technol Res* 4(4)(2007): 505–510.

Ma, Z., Pirlo, R. K., Wan, Q., et al. "Laser-guidance-based cell deposition microscope for heterotypic single-cell micropatterning." *Biofabrication* 3(3)(2011): 034107.

McGuigan, A. P., Sefton, M. V. "Modular tissue engineering: Fabrication of a gelatin-based construct." *J Tissue Eng Regen Med* 1(2)(2007): 136–145.

Melchels, F. P., Feijen, J., Grijpma, D. W. "A review on stereolithography and its applications in biomedical engineering." *Biomaterials* 31(24)(2010): 6121–6130.

Michael, S., Sorg, H., Peck, C.-T., Koch, L., Deiwick, A., Chichkov, B., Vogt, P. M., Reimers, K. "Tissue engineered skin substitutes created by laser-assisted bioprinting form skin-like structures in the dorsal skin fold chamber in mice." *PLoS One* 8(3)(2013): e57741.

Mironov, V. "Printing technology to produce living tissue." *Expert Opin Biol Ther* 3(5)(2003): 701–704.

Mironov, V., Boland, T., Trusk, T., Forgacs, G., Markwald, R. R. "Organ printing: Computer-aided jet-based 3D tissue engineering." *Trends Biotechnol* 21(4)(2003): 157–161.

Mironov, V., Kasyanov, V., Drake, C., Markwald, R. R. "Organ printing: Promises and challenges." *Regen Med* 3(1)(2008): 93–103.

Mironov, V., Trusk, T., Kasyanov, V., et al. "Biofabrication: A 21st century manufacturing paradigm." *Biofabrication* 1(2)(2009): 022001.

Moon, S., Hasan, S. K., Song, Y. S. et al. "Layer by layer three-dimensional tissue epitaxy by cell-laden hydrogel droplets." *Tissue Eng Part C* 16(2010): 157–166.

Nadernezhad, A., Khani, N., Skvortsov, G.A., Toprakhisar, B., Bakirci, E., Menceloglu, Y., Unal, S., Koc, B. "Multifunctional 3D printing of heterogeneous hydrogel structures." *Sci Rep* 6(33178)(2016).

Nahmias, Y., Schwartz, R. E., Verfaillie, C. M., Odde, D. J. "Laser-guided direct writing for three-dimensional tissue engineering." *Biotechnol Bioeng* 92(2)(2005): 129–136.

Nakamura, M., Iwanaga, S., Henmi, C., Arai, K., Nishiyama, Y. "Biomatrices and biomaterials for future developments of bioprinting and biofabrication." *Biofabrication* 2(1)(2010): 014110.

Nakamura, M., Kobayashi, A., Takagi, F., et al. "Biocompatible inkjet printing technique for designed seeding of individual living cells." *Tissue Eng* 11(11–12)(2005): 1658–1666.

Neufurth, M., Wang, X., Schröder, H. C., Feng, Q., Diehl-Selfert, B., Ziebart, T., Steffen, R., Wang, S., Müller, W. E. G. "Engineering a morphogenetically active hydrogel for bioprinting of bioartificial tissue derived from human osteoblast-like SaOS-2 cells." *Biomaterials* 35(31)(2014): 8810–8819.

Norotte, C., Marga, F. S., Niklason, L. E., Forgacs, G. "Scaffold-free vascular tissue engineering using bioprinting." *Biomaterials* 30(30)(2009): 5910–5917.

Odde, D. "Laser-guided direct writing for applications in biotechnology." *Trends Biotechnol* 17(10)(1999): 385–389.

Odde, D. J., Renn, M. J. "Laser-guided direct writing of living cells." *Biotechnol Bioeng* 67(3)(2000): 312–318.

Ovsianikov, A., Deiwick, A., Van Vlierberghe, S., et al. "Laser fabrication of three-dimensional CAD scaffolds from photosensitive gelatin for applications in tissue engineering." *Biomacromolecules* 12(4)(2011): 851–858.

Ovsianikov, A., Gruene, M., Pflaum, M., et al. "Laser printing of cells into 3D scaffolds." *Biofabrication* 2(1)(2010): 014104.

Ovsianikov, A., Malinauskas, M., Schlie, S., et al. "Three-dimensional laser micro and nano-structuring of acrylated poly(ethylene glycol) materials and evaluation of their cytoxicity for tissue engineering applications." *Acta Biomaterialia* 7(3)(2011): 967–974.

Ozbolat, I. T., Koc, B. "A continuous multi-material toolpath planning for tissue scaffolds with hollowed features." *Comput Aided Des Appl* 8(2)(2011): 237–247.

Ozler, S. B., Bakirci, E., Kucukgul, C., Koc, B. "Three-dimensional direct cell bioprinting for tissue engineering." *J Biomed Mater Res Part B* 105(8)(2017): 2530–2544.

Park, J. Y., Choi, J.-C., Shim, J.-H., Lee, J.-S., Park, H., Kim, S. W., Doh, J., Cho, D.-W. "A comparative study on collagen type I and hyaluronic acid dependent cell behavior for osteochondral tissue bioprinting." *Biofabrication* 6(3)(2014): 035004.

Parsa, S., Gupta, M., Loizeau, F., Cheung, K. C. "Effects of surfactant and gentle agitation on inkjet dispensing of living cells." *Biofabrication* 2(2)(2010): 025003.

Pati, F., Ha, D.-H., Jang, J., Han, H. H., Rhie, J.-W., Cho, D.-W. "Biomimetic 3D tissue printing for soft tissue regeneration." *Biomaterials* 62(2015): 164–175.

Pati, F., Jang, J., Ha, D.-H., Won Kim, S., Rhie, J.-W., Shim, J.-H., Kim, D.-H., Cho, D.-W. "Printing three-dimensional tissue analogues with decellularized extracellular matrix bioink." *Nat Commun* 5(2014): 3935.

Peltola, S. M., Melchels, F. P., Grijpma, D. W., Kellomaki, M. "A review of rapid prototyping techniques for tissue engineering purposes." *Ann Med* 40(4)(2008): 268–280.

Perez-Pomares, J. M., Foty, R. A. "Tissue fusion and cell sorting in embryonic development and disease: Biomedical implications." *Bioessays* 28(8)(2006): 809–821.

Pham, Q. P., Sharma, U. Mikos, A. G. "Electrospinning of polymeric nanofibers for tissue engineering applications: A review." *Tissue Eng* 12(5)(2006): 1197–1211.

Pirlo, R. K., Dean, D. M., Knapp, D. R., Gao, B. Z. "Cell deposition system based on laser guidance." *Biotechnol J* 1(9)(2006): 1007–1013.

Rajangam. T., An, S. S. A. "Fibrinogen and fibrin based micro and nano scaffolds incorporated with drugs, proteins, cells and genes for therapeutic biomedical applications." *Int J Nanomed* (8)(2013): 3641–3662.

Rivron, N. C., Rouwkema, J., Truckenmuller, R., et al. "Tissue assembly and organization: Developmental mechanisms in microfabricated tissues." *Biomaterials* 30(28)(2009): 4851–4858.

Rutz, A. L., Hyland, K. E., Jakus, A. E., Burghardt, W. R., Shah, R. N. "A multimaterial bioink method for 3D printing tunable, cell-compatible hydrogels." *Adv Mater* 27(9)(2015): 1607–1614.

Schade, R., Weiss, T., Berg, A., Schnabelrauch, M., Liefeith, K. "Two-photon techniques in tissue engineering." *Int J Artif Organs* 33(4)(2010): 219–227.

Schuurman, W., Khristov, V., Pot, M. W., van Weeren, P. R., Dhert, W. J. A., Malda, J. "Bioprinting of hybrid tissue constructs with tailorable mechanical properties." *Biofabrication* 2(3)(2011): 021001.

Shengjie, L., Zhuo, X., Xiaohong, W., et al. "Direct Fabrication of a Hybrid Cell/Hydrogel Construct by a Double-nozzle Assembling Technology." *J Bioact Compat Polym* 24(3) (2009): 249–265.

Skardal, A., Devarasetty, M., Kang, H.-W. et al. "A hydrogel bioink toolkit for mimicking native tissue biochemical and mechanical properties in bioprinted tissue constructs." *Acta Biomater* 25(2015): 24–34.

Skardal, A., Zhang, J., McCoard, L., et al. "Photocrosslinkable hyaluronan-gelatin hydrogels for two-step bioprinting." *Tissue Eng Part A* 16(8)(2010): 2675–2685.

Smith, C. M., Christian, J. J., Warren, W. L., Williams, S. K. "Characterizing environmental factors that impact the viability of tissue-engineered constructs fabricated by a direct-write bioassembly tool." *Tissue Eng* 13(2)(2007): 373–383.

Smith, C. M., Stone, A. L., Parkhill, R. L., et al. "Three-dimensional bioassembly tool for generating viable tissue-engineered constructs." *Tissue Eng* 10(9–10)(2004): 1566–1576.

Steinberg, M. S. "Reconstruction of tissues by dissociated cells." *Science* 141(3579)(1963): 401–408.

Toprakhisar, B., Nadernezhad, A., Bakirci, E., Khani, N., Skvortsov, G. A., Koc, B. "Development of bioink from decellularized tendon extracellular matrix for 3D bioprinting." *Macromol Biosci* 18(10)(2018): e1800024.

Topuz, M., Dikici, B., Gavgali, M., Yilmazer, H. "A review on the hydrogels used in 3D bioprinting." *International Journal of 3D Printing Technologies and Digital Industry* 2(2) (2018): 68–75.

Whitesides, G. M., Boncheva, M. "Beyond molecules: Self-assembly of mesoscopic and macroscopic components." *Proc Natl Acad Sci U S A* 99(8)(2002): 4769–4774.

Wu, P. K., Ringeisen, B. R. "Development of human umbilical vein endothelial cell (HUVEC) and human umbilical vein smooth muscle cell (HUVSMC) branch/stem structures on hydrogel layers via biological laser printing (BioLP)." *Biofabrication* 2(1)(2010): 014111.

Xu, C., Chai, W., Huang, Y., Markwald, R. R. "Scaffold-free inkjet printing of three-dimensional zigzag cellular tubes." *Biotechnol Bioeng* 109(12)(2012): 3152–3160.

Xu, T., Binder, K. W., Albanna, M. Z., Dice, D., Zhao, W., Yoo, J. J., Atala, A. "Hybrid printing of mechanically and biologically improved constructs for cartilage tissue engineering applications." *Biofabrication* 5(1)(2013): 015001.

Xu, M., Yan, Y., Liu, H., Yao, R., Wang, X. "Controlled adipose-derived stromal cells differentiation into adipose and endothelial cells in a 3D structure established by cell-assembly technique." *J Bioact Compat Polym* 24(1 Suppl)(2009): 31–47.

Xu, T., Jin, J., Gregory, C., Hickman, J. J., Boland, T. "Inkjet printing of viable mammalian cells." *Biomaterials* 26(1)(2005): 93–99.

Ye, K., Felimban, R., Traianedes, K., Moulton, S. E., Wallace, G. G., Chung, J., Quigley, A., Choong, P. F., Myers, D. E. "Chondrogenesis of infrapatellar fat pad derived adipose stem cells in 3D printed chitosan scaffold." *PLoS One* 9(6)(2014): e99410.

Yeong, W. Y., Chua, C. K., Leong, K. F., Chandrasekaran, M. "Rapid prototyping in tissue engineering: Challenges and potential." *Trends Biotechnol* 22(12)(2004): 643–652.

Yuan, D., Lasagni, A., Shao, P., Das, S. "Rapid prototyping of microstructured hydrogels via laser direct-write and laser interference photopolymerisation." *Virtual Phys Prototyp* 3(4)(2008): 221–229.

Zhang, T., Yan, K. C., Ouyang, L., Sun, W. "Mechanical characterization of bioprinted in vitro soft tissue models." *Biofabrication* 5(4)(2013): 1758–5082.

Zhang, Y., Yu, Y., Chen, H., Ozbolat I. T. "Characterization of printable cellular micro-fluidic channels for tissue engineering." *Biofabrication* 5(2013): 24004.

12 Additive Manufacturing in Materials Innovation

Mitun Das and Vamsi Krishna Balla

CONTENTS

12.1 INTRODUCTION

During the last two decades, significant progress has been made in the development of complex, net shape structures using novel materials such as composites and functionally graded materials. Some of the products have also been commercialized for various applications in automotive, aerospace, and consumer products. More than 20 additive manufacturing (AM) techniques have been developed, however, the most widely used are stereolithography (SLA), three-dimensional printing (3DP), fused deposition modeling (FDM), selective laser sintering (SLS), selective laser melting (SLM), and laser metal deposition (LMD) (Guo and Leu 2013). As shown in Table 12.1, materials including polymers, ceramics, metals, and their composites have been successfully used in these AM technologies. For example, photo-curable

TABLE 12.1
Important AM Processes and Materials Used for Fabricating Complex Components

AM Process	Feedstock Form	Materials Type	Materials
SLA	Liquid	Polymers	Photo-curable resins.
		Ceramics	Suspensions of SiO_2, ZrO_2, Al_2O_3, and other ceramics.
3DP	Powder	Polymers	Acrylic plastics, wax.
		Ceramics	ZrO_2, SiO_2, Al_2O_3, CaP ceramics, sands, and mullite.
		Composites	Polymer matrix, ceramic matrix, and short fiber-reinforced composites.
SLS	Powder	Polymers	Polymide 12, GF polymide, polystyrene, etc.
		Ceramics	SiO_2, ZrO_2, Al_2O_3, ZrB_2, CaP ceramics, graphite, bioglass, mullite, and sand.
		Composites	Metal matrix, ceramic matrix, polymer matrix, and fiber-reinforced composites.
FDM	Filament or paste	Polymers	PC, ABS, PC-ABS, ULTEM™ (polyetherimide)
		Ceramics	PZT, Si_3N_4, Al_2O_3, SrO_2, SiO_2, CaP ceramics, and mullite.
		Composites	Polymer matrix and short fiber-reinforced composites.
LMD/SLM	Powder	Metals	Stainless steels, CoCrMo, Ti and Ti alloys, Ni-based alloys, Al alloys, and Cu alloys.
		Composites	Metal matrix, ceramic matrix, and particulate-reinforced composites.
		Functionally graded materials	Metal-metal, metal-ceramic, and ceramic-ceramic FGMs.

Source: Modified Data from Guo, N. and Leu, M.C. *Front. Mech. Eng.* 8, 215–243, 2013.

resins in liquid form are used as feedstock in SLA, and FDM uses thermoplastics, thermo-setting plastics, and waxes in filament form. Feedstock materials in powder forms are suitable for SLS, 3DP, and LMD.

The process of fabricating net shape components with commercially available single materials (polymers, metals, and ceramics) using AM techniques has been mostly optimized by respective equipment manufacturers. Therefore, the users simply use process parameters prescribed by the manufacturer while fabricating the components. Further, the use of custom materials in some of these AM machines appears to be difficult, and as a result, developing structures with novel materials such as composites and functionally graded materials becomes difficult as well. For example, for high performance and properties, composites should be fabricated with a high volume fraction of reinforcements, having controlled alignment/orientation (with respect to the load) and appropriate connectivity, and exhibit micro-structural hierarchy if desired. Further, these composites should be manufactured to near/net shape to eliminate problems associated with conventional machining. Such new composite materials are difficult to fabricate using conventional manufacturing routes.

Apart from providing geometrical design flexibility, recent developments in AM technologies have enabled tailored materials design (such as composites, meso-scale compositional variations, functional variation in composition in three dimensions and multi-functional materials) and their effective incorporation into the complex geometrical designs. These capabilities of AM technologies can be effectively utilized to reduce the materials innovation gestation period. Therefore, several research groups across the globe are focusing on creating new and designed materials for a variety of applications for increased functionality, performance, and reduced environmental impact. Considering the above, the first part of the chapter is focused on the use of AM techniques in creating novel materials, such as composites, nanocomposites, and multi-functional materials structures. In the second part, we discuss the novel designs enabled via AM technologies.

12.2 COMPOSITES BY ADDITIVE MANUFACTURING

The use of additive manufacturing capabilities for composite materials fabrication has started in the last decade, but its full potential has yet to be explored. Combining AM with composite fabrication has opened up new possibilities to explore unique tailored materials (Wohlers 2010). Among many available AM technologies, only a few have shown their potential in composites fabrication, which include selective laser sintering/melting, laser engineered net shaping, laminated object manufacturing, stereolithography, fused deposition modeling, three-dimensional printing, and ultrasonic consolidation (Kumar and Kruth 2010). However, a majority of the efforts are focused on the polymer composites, while studies on the metal matrix composites are limited. There has not been much progress reported in the area of manufacturing ceramic matrix composites (CMCs) using AM due to the numerous materials and processing challenges. Therefore, herein, we emphasize more on the metal and ceramic matrix composites fabrication using various AM technologies, but briefly discuss the polymer matrix composites as well.

12.2.1 Metal Matrix Composites

Metal matrix composites (MMCs) with ceramic reinforcements exhibit properties of both the ceramic and metal (Tjong and Ma 2000). The addition of ceramic reinforcements to a metallic matrix is known to improve the specific strength, fracture toughness, stiffness, fatigue, and wear resistance, compared to their metallic counterparts (Mortensen and Llorca 2010). Among continuous or discontinuous reinforcements, discontinuously reinforced MMCs, including both particulates and whiskers or short fibers, have drawn considerable attention due to their relatively lower costs and isotropic properties of composites (Tjong and Ma 2000).

In recent years, laser-based additive manufacturing has been attempted extensively in order to fabricate MMC 3D structures or coatings to improve materials properties (Kumar and Kruth 2010, Hu and Cong 2018). These techniques include direct laser fabrication (DLF) (Li et al. 2009), laser engineered net shaping

(LENS™) (Banerjee et al. 2003a, Balla et al. 2012), direct metal deposition (Hong et al. 2013), and powder bed-based techniques such as electron beam melting (EBM) and SLS/SLM (Lu et al. 2000). The properties of the MMCs depend on the size and size fraction of the reinforcements, as well as the nature of the matrix-reinforcement interfaces (Tjong and Ma 2000). One unique feature of these techniques is their ability to incorporate reinforcements, either ex-situ (wherein reinforcing phases are prepared separately and added to the matrix during composite fabrication) or in-situ (where reinforcing phases are synthesized/prepared during composite fabrication) (Tjong and Ma 2000). In ex-situ MMCs, the size of the reinforcing phase depends on the starting powder, whereas *in situ* synthesized MMCs contain homogeneously distributed finer ceramic phases, which provide enhanced mechanical properties. Further, AM techniques enable the creation of functionally graded composites in complex-shaped bulk MMC components with finer micro-structure that is due to the very fast cooling associated with laser processing.

The selective laser sintering/melting technique has been used to study different ex situ reinforcement MMCs, such as a WC-Co composite (Laoui et al. 2000, Maeda and Childs 2004, Kumar 2009), a WC–10% Co particulate-reinforced Cu matrix composite (Gu and Shen 2006, 2007), (Fe, Ni)-TiC composites (Gåård et al. 2006), Al-7Si-0.3Mg/silicon carbide (SiC) composites (Simchi and Godlinski 2008), SiC particulates-reinforced Al-MMCs (Ghosh et al. 2010, Ghosh and Saha 2011), etc. Gu and Shen (2007) prepared submicron WC–10% Co particulate-reinforced Cu matrix composites using direct metal laser sintering. With increasing reinforcement content, densification of a Cu matrix composite deteriorated (at 20 wt.%), and a heterogeneous micro-structure with significant particulate aggregation was found at 40 wt.% reinforcement. Laser sintering has been used to fabricate ultra high temperature ceramic composites. High density (>92%) multi-layer Ti–ZrB_2 mixtures with hardness up to 11.4 GPa were fabricated by use of laser sintering (Sun and Gupta 2011). Furthermore, *in situ* synthesis of reinforcement via the reaction between the constituent elements of composites during the fabrication process is regarded as a more promising method in order to obtain more homogeneous micro-structures (Tjong 2007). Dadbakhsh and Hao (2012) and Dadbakhsh et al. (2012) studied the selective laser sintering of an Al-matrix composite part from an Al/Fe_2O_3 powder mixture. It was observed that the incorporation of Fe_2O_3 significantly influences the SLM processability of a particulate-reinforced Al-matrix composite. At present, multi-material reinforcements are being proposed for the betterment of Al-MMCs. Ghosh et al. (2014) used the direct metal laser sintering of a premixed powder containing Al, TiO_2, and B_4C, which reacted *in situ* to form Al_2O_3, TiC, and TiB_2 reinforcements in an Al-matrix. Gu et al. (2009a and 2009b) prepared *in situ*-formed TiC-reinforced ($TiAl_3$+Ti_3AlC_2) matrix composites from high-energy ball-milled Ti-Al-C composite powder. An *in situ* TiC-reinforced Cu matrix composite synthesized from Cu–Ti–C and Cu–Ni–Ti–C powder mixtures showed that the addition of Ni-improved wettability between Cu and TiC particulates thus resulted in an improved micro-structure and surface quality of the laser-fabricated parts (Lu et al. 2000). While the above studies show the SLM capabilities in fabricating net shape MMCs, one major drawback of the SLM process is high porosity, which is detrimental to the mechanical properties. In order to achieve full density, trace amounts of rear-earth elements have been added in order to reduce the surface tension of melt (Gu et al. 2007) or the porous product is infiltrated (Kumar 2009). Another processing difficulty associated with the selective laser melting of MMCs is the balling effect, where the fresh molten material do not wet the underlying substrate, thus leading to the formation of broken liquid cylinders and a rough surface after consolidation (Dadbakhsh et al. 2012). The tendency of the balling effect can be reduced by increasing the energy input by increasing the laser power, lowering the scan speed, or decreasing the powder layer thickness (Gu and Shen 2009).

In the case of direct metal deposition techniques, a high power laser locally melts the top surface of the work piece, and then simultaneously a powder (ceramic or metal) is injected into the melt pool. Depending on the interaction with the laser beam and the melt pool temperature, the injected particles may react *in situ* with the molten matrix or be entrapped in the matrix with a minimal chemical reaction. The processing parameters, such as laser power, scan speed, and powder flow, have an influence on the resulting composite micro-structure and, hence, the properties. Several ex situ-reinforced MMC coatings have been deposited using LENS™. Table 12.2 summarizes various materials systems used to create an ex situ-reinforced MMC via laser-based direct metal deposition techniques.

The direct metal deposition techniques have also been used to deposit *in situ* synthesized MMCs. *In situ* synthesized ceramic-reinforced titanium matrix composites have been studied extensively. Among the different types of reinforcements, TiB has been considered the best material for a titanium matrix due to its high elastic modulus, a similar thermal expansion coefficient that minimizes residual stress, and excellent interfacial bonding with titanium matrixes (Banerjee et al. 2003a). Earlier studies on TiB-reinforced Ti MMC coating that was fabricated via the direct metal deposition method, using Ti, B, or TiB_2 as feedstock powders, showed a considerable improvement in tribological properties (Banerjee et al. 2003a, Wang et al. 2008,

TABLE 12.2
Materials Used for the Direct Fabrication of Ex-situ-Reinforced MMC in LENS™

S. No.	Materials	Composites	References
1	WC-Co	MMC	Xiong et al. (2008)
2	WC-12%Co	MMC	Balla et al. (2010)
3	Ti, SiC	TMC	Das et al. (2010)
4	Ti, SiC	TMC	Das et al. (2011)
5	Ti, TiN	TMC	Balla et al. (2012)
6	Ni, TiC	MMC	Zheng et al. (2010)
7	Ni, SiC	MMC	Cooper et al. (2013)
	Ni, Al_2O_3		
	Ni, TiC		
8	Inconel 718, TiC	MMC	Hong et al. (2013)
	Inconel 625, TIC		Hong et al. (2015)
	Inver, TiC		Li et al. (2000)

Attar et al. 2014, Hu et al. 2018). Samuel et al. (2008) demonstrated *in situ*-a formed boride-reinforced Ti-Nb-Zr-Ta alloy for orthopedic application using the LENS™ technique. They found substantial improvement in the wear resistance due to the homogeneous distribution of very fine boride reinforcement in Ti-Nb-Zr-Ta alloys.

Hu et al. fabricated *in situ* a TiB-Ti composite and studied the effects of laser power on the formation of a three-dimensional quasicontinuous network micro-structure. The three-dimensional quasicontinuous network micro-structure enhanced the toughness of the composite. In recent years, SLM has also been used for the processing of MMCs, such as AlSi10Mg + TiC (Gu et al. 2015), Fe + SiC (Song et al. 2014), and Ti + TiB (Attar et al. 2014, 2015, Hu and Li 2017). Attar et al. produced a Ti-TiB composites by an in-situ chemical reaction between the Ti and TiB_2 powders during SLM. Fine needle-like TiB particles that were distributed across the α-Ti matrix were found to enhance the hardness and stiffness of the composite.

In situ synthesized TiB + TiN-reinforced Ti6Al4V alloy composite coatings were successfully deposited on Ti using a premixed Ti6Al4V and BN powder using LENS™ (Das et al. 2012). An *in situ* reaction of BN with Ti formed a novel micro-structure with homogeneously distributed fine reinforcements of TiB and TiN. The micro-structures shown in Figure 12.1 consist of TiB nano rods locally concentrated in the matrix, resulted quasi-continuous network architecture (Das et al. 2014). These LENS™-processed *in situ* synthesized TiB-TiN-reinforced titanium matrix composites exhibited high hardness, high modulus, and wear resistance (Das et al. 2014). Zhang et al. (2011) studied *in situ* (TiB + TiC)/TC4 composites by laser direct deposition of coaxially fed TC4 and B_4C mixed powders. The micro-structure showed needle-like and prismatic TiB and granular TiC with small amounts of the unreacted B_4C. Authors found that the unreacted B_4C weakened its interface bonding with the titanium matrix, thus leading to the deterioration of the mechanical properties. A thin wall structure of the composite is shown in Figure 12.2. Over the last few years, interest in additive manufacturing-based surface modification of biomaterials has increased significantly (Bose et al. 2018a, 2018b, Sahasrabudhe et al. 2018, Sahoo et al. 2018). Different bioactive ceramics, such as calcium phosphate and strontium titanate, were incorporated into a metal matrix to improve bioactivity. The incorporation of hydroxyapatite 1% and 3% wt. into a CoCrMo alloy leads to the formation of a tribolayer, which

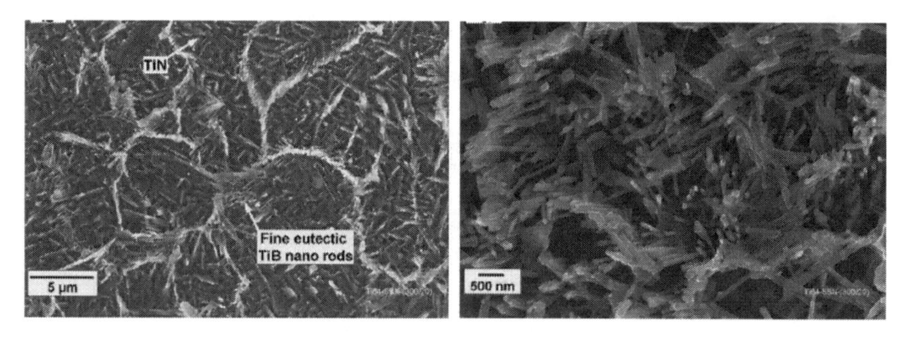

FIGURE 12.1 A FESEM micro-structure of laser-processed TiB-TiN-reinforced titanium matrix composite coatings containing TiB nano rods. (Reprinted with permission from Das, M. et al., *J. Mech. Behav. Biomed. Mater.*, 29:259–271, 2014.)

FIGURE 12.2 A laser direct deposited (TiB + TiC)/TC4 composite thin wall. (Reprinted with permission from Zhang, Y.Z. et al., *J. Mater. Process. Technol.* 211:597–601, 2011.)

protects the leaching of Co and Cr ions during wear (Bose et al. 2018a). The sustained release of Sr^{2+} ions forms a Ti-based implant surface has huge potential for orthopedic and dental applications. Sahoo et al. (2018) developed a laser deposition of $SrTiO_3$-reinforced Ti coatings on a Ti surface, which showed the sustained Sr^{2+} ions release and a significantly high bioactivity as compared to the commercial titanium.

In situ-formed TiC-reinforced Ni-based MMCs were studied by Li et al. (2009) and Gopagoni et al. (2011). *In situ* reaction between elemental titanium and carbon (graphite) within the molten nickel pool allowed for the Ni-TiC composite formation with refined, homogeneously distributed carbide precipitates. This novel *in situ* composite showed enhanced micro-hardness and tribological properties, with comparisons to laser deposited pure Ni. *In situ* synthesized MMCs showed homogeneously distributed hard ceramic phases, which provide hardness, and a metal matrix gives the toughness. The LENS™ technique has also been used for creating functionally graded composite coatings, as well as coatings on complex-shaped parts in a single step, which are not possible in conventional processing.

12.2.2 Ceramic Matrix Composites

CMCs have numerous applications as high temperature and high performance materials in defense, aerospace, and energy conservation sectors (Sommers et al. 2010). However, additive manufacturing of CMCs is not well investigated and very limited data have been reported in the literature. The most popular AM methods for fabricating dense and porous ceramic parts are 3DP, laminated object manufacturing (LOM), SLS, and selective laser gelation (SLG). Among the different ceramics, AM of SiC-based CMCs has been reported widely (Griffin et al. 1996). Laminated object manufacturing is a promising AM technique that is used for fiber-reinforced ceramic matrix composites. For ceramic fabrication using LOM, ceramic tapes usually produced by tape casting are bonded with an adhesive in a layer-by-layer fashion. The bonding between the layers is activated by a heated plate or roller during the LOM process. Klosterman et al. (1999) produced monolithic SiC ceramics as well as SiC/SiC composites using LOM. They reported using ceramic-grade Nicalon fiber-based phenolic prepregs with alternating layers of monolithic ceramic tapes. However, the flexural strength of these CMCs was found to be low due to weak interlayer bonding. Weisensel et al. (2004) studied the fabrication of biomorphous Si/SiC composites using LOM. Porous carbon preforms were made from pyrolyzed paper sheets and additionally, phenolic resin was used as an adhesive, which was subsequently pyrolyzed. The net-shaped structure was later infiltrated with liquid silicon in order to form a Si/SiC composite. The composite exhibited a bending strength between 123 ± 8 MPa and 130 ± 10 MPa.

SLS is the most popular AM approach used since the early 1990s to fabricate ceramic parts. However, ceramic parts fabrication using SLS is more challenging due to its high melting temperature, low or no plasticity, and the low thermal shock resistance of ceramics. The fabrication of ceramic parts using SLS can be broadly divided into two categories of techniques: direct and indirect. In the direct SLS technique, a laser beam heats up the ceramic powder, creating a solid state sintering or melting of the loose powder bed. In indirect SLS processing, the bonding is achieved via a polymer binder. The green ceramic parts are subsequently debinded and sintered in order to enhance both the density and strength. Stevinson et al. (2008) studied an indirect SLS technique to fabricate silicon/silicon carbide composite structures. The resulting net shape Si/SiC composites were observed to be thermally stable. Liu et al. (2011) demonstrated an SLG technique which combines SLS and a sol–gel technique to fabricate ceramic-matrix composite green parts. In this process, stainless steel (316L) powders are mixed with the silica sol, and, subsequently, the silica sol is evaporated using a laser beam, which yields a gel that links particles together to form a composite green part. Figure 12.3 shows the metal-ceramic composite green parts, that were fabricated using SLG, with a surface finish of 32 μm (R_z) and a dimensional variation of 10%.

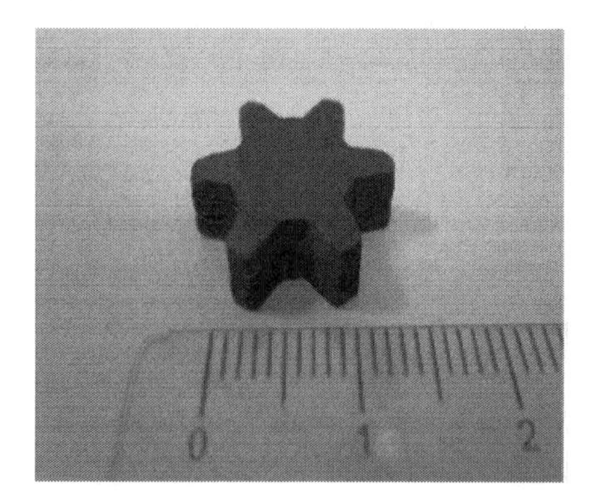

FIGURE 12.3 Ceramic-matrix composite parts that were obtained by using SLG. (Reprinted with permission from Liu, F.H. et al., *Composites: Part B,* 42, 57–61, 2011.)

3DP is another approach for ceramic composites fabrication. Different ceramic-based composites, such as Al_2O_3/Cu-O (Travitzky and Shlayen 1998, Melcher et al. 2006), Si/SiC (Moon et al. 2001), $TiAl_3/Al_2O_3$ (Yin et al. 2006), $Ti_3AlC_2/TiAl_3/Al_2O_3$ (Yin et al. 2007), $NbAl_3/Al_2O_3$ (Zhang et al. 2008), and Al_2O_3/glass (Zhang et al. 2009) have been successfully fabricated using 3DP and subsequent post-processing. In addition, designed porous ceramic preforms that were fabricated using 3DP were processed further using reactive metal melt infiltration in order to create complex ceramic matrix composites with tailored distribution of a reinforcement/matrix. In the reactive melt infiltration process, the micro-structure of the ceramic preforms, e.g., pore size, shape, and interconnectivity, strongly affects the wettability of the infiltrating metal melt on the ceramic preform (Zhang et al. 2008). Fu et al. (2013) demonstrated a gradient macrocellular lattice truss structure (Figure 12.4) from silicon/silicon carbide ceramic composites. The composite structure was fabricated by the 3DP from Si/SiC/dextrin powder blends.

12.2.3 Polymer Matrix Composite

Polymer matrix composites using AM is receiving attention in order to simultaneously build hierarchical materials and net shape structures. In general, to improve mechanical properties, as well as functionality in the polymer, various fillers (fibers, whiskers, platelets, or particles) are incorporated to form polymer matrix composites. A wide range of polymeric materials, such as photosensitive resin, Nylon, elastomer, acrylonitrile butadiene styrene(ABS), and wax have been successfully used in stereolithography (SL), FDM, SLS/SLM, LOM, and 3DP to create complex structures. Among these techniques, SL, FDM, and LOM enable the processing of fiber-reinforced polymer matrix composites.

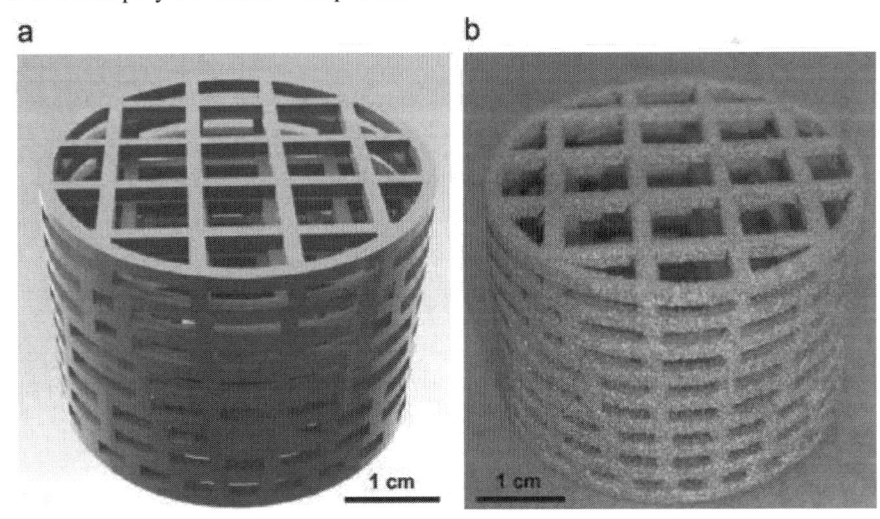

FIGURE 12.4 The 3D printing of SiSiC ceramics with a starting composition of 49.2 vol.% Si, 32.8 vol.% SiC, and 18 vol.% dextrin: (a) CAD design of the lattice structure and (b) reactive-infiltrated SiSiC part. (Reprinted with permission from Fu, Z. et al., *Mater. Sci. Eng. A*, 560, 851–856, 2013.)

In the FDM process, several attempts have been made to incorporate metal and ceramic filler into the feedstock filament for composite fabrication (Nikzad et al. 2011, Masood and Song 2004). Zhong et al. (2001) studied a method of forming a short glass fiber (GF)-reinforced ABS polymer feedstock filament with GF contents that were up to 18 wt.% for use in FDM. It was found that the addition of glass fibers resulted in a higher tensile strength along the longitudinal direction, but the interlayer adhesive strength was low as compared to ABS without reinforcement. Singh and Singh (2014) have introduced varying concentrations of aluminum oxide (Al_2O_3) in nylon fibers in order to fabricate feedstock for FDM. Ryder et al. (2018) fabricated an ABS-420 stainless steel composite using FDM. The metal particle concentration that was up to 15 wt.% showed no reduction of the mechanical property, but it added functionality, such as magnetic properties to the 3D part. The incorporation of common metal powders (aluminum and iron) in polymers [ABS, polypropylene, and polyamide (PA)] enhance the viscosity, which can be improved by using surfactants and plasticizers (Postiglione et al. 2015). The low temperature polymers, such as polylactic acid, polycaprolactone (PCL), and polypropylene with ceramic reinforcement have been fabricated using FDM for tissue engineering (Kalita et al. 2003, Mohan et al. 2017). The bioactive ceramic particles such as hydroxyapatite (HAp), tricalcium phosphate (TCP), and bio-glass added into the polymer matrix could increase the bioactivity as well as mechanical properties of the composite scaffold (Sharma et al. 208). In the last few years, an effort has been made to use high temperature engineering polymer [poly ether ketone and poly ether ketone (PEEK)]-based composites in FDM (Berretta et al. 2017).

SLS/SLM is another highly employed AM technique for polymer matrix composites. Polycarbonate (PC) (Ho et al. 2003), polystyrene (PS) (Yang et al. 2009), and PA (Caulfield et al. 2007) have been widely used as SLS materials. A polyamide-based polymer has been found to be the most frequently studied material for composites. Different micron-scale particles, such as glass beads (Chung and Das 2006), silicon carbide (Hon and Gill 2003), aluminum powders (Mazzoli et al. 2007), and hydroxyapatite (HAp) (Zhang et al. 2008), have been developed for the SLS process. Hon and Gill (2003) demonstrated SiC/PA composites for SLS and found a reduced tensile strength, but an improved stiffness for the composite parts when compared with pure PA parts. Mazzoli et al. (2007) developed a mechanically mixed aluminum/PA composite powder for SLS. These composite SLS parts showed a metallic appearance with a higher dimensional accuracy and stiffness, smoother surface, and better finishing properties, with respect to pure PA SLS parts. A commercially available carbon fiber/polyamide composite powder called CarbonMide® was developed by EOS (Munich, Germany) from the mechanically mixing of pure polyamide powder and carbon fibers (Yan et al. 2011). However, SLS of these composite powders is expected to form an agglomeration of carbon fibers, which causes poor mechanical properties. Yan et al. (2011) developed a new route for carbon fiber (CF)/PA-12 composite powders preparation and showed the manufacturing of high performance components by SLS. Surface-modified carbon fibers with a layer of PA-12 on the surface have been prepared by the dissolution–precipitation process and were found to provide uniform dispersion and good interfacial bonding with the matrix (Figure 12.5). The results indicated that the addition of carbon fibers with good interfacial bonding

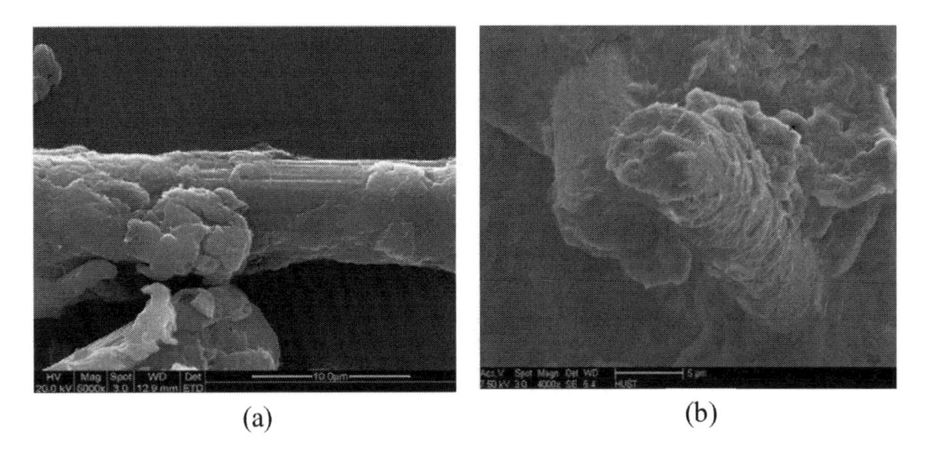

FIGURE 12.5 SEM micro-graphs of (a) 30% CF/PA composite powder and (b) the fractured surfaces of the 30% CF/PA SLS parts. (Reprinted with permission from Yan, C. et al., *Compos. Sci. Technol.*, 71, 1834–1841, 2011.)

greatly enhanced the flexural strength and flexural modulus of sintered components. Kenzari et al. (2012) developed and commercialized Nylon-based composites that were reinforced by Al-based quasicrystals particles for the SLS process. This composite showed promising applications in the rapid manufacturing of complex functional parts with high dimensional accuracy, wear resistance, and reduced friction coefficients. Figure 12.6 shows quasicrystals-reinforced polyamide-based composite parts processed by SLS.

FIGURE 12.6 Some examples of freeform SLS composite parts of a polymer matrix reinforced by quasicrystalline AlCuFeB particles. This SLS part has a volume fraction of porosity lower than 2% and is directly leak-tight without post-impregnation of resin. Courtesy of Ateliers CINI SA and MV2T. (Reprinted with permission from Kenzari, S. et al., *Mater. Des.*, 35, 691–695, 2012.)

SLS, one of the promising AM techniques, is capable of fabricating complex shapes from different biopolymers, PCL, PEEK, and PA. PCL is widely used for laser sintering of materials because of its high thermal stability (Bose et al. 2018b). Xia et al. (2013) fabricated a PCL/HAp composite scaffold. They observed around a 130% improvement on the compressive strength of the PCL scaffold after the incorporation of nano HAp. The fabrication of PEEK-based composites with complex geometries became popular for the aerospace and biomedical fields. Yan et al. (2018) fabricated CF/PEEK composites using SLS. They observed that tensile strength composites reached 109 ± 1 MPa with an elasticity modulus of 7365 ± 468 MPa, which is 85% higher than injection-molded pure PEEK. Several researchers studied bioactive particle-incorporated PEEK mostly using injection molding, compression molding, and cold press sintering (Vaezi et al. 2016). There has not been a large amount of information reported on the fabrication of PEEK composites using SLS (Tan et al. 2003, VonWilmowsky et al. 2008, Vaezi et al. 2016). VonWilmowsky et al. developed bioactive glass and β-TCP-incorporated PEEK using laser sintering. They reported the highest osteoblast cell viability observed in bioactive glass-incorporated PEEK. Additive-manufactured PEEK with osteoconductive and bioactive materials could be a potential material for a load bearing bone substitute.

12.3 NANOCOMPOSITE STRUCTURES BY ADDITIVE MANUFACTURING

The field of nanocomposites brought to it the attention of the scientific world because of its improved material properties. The mechanical, electrical, thermal, optical, electrochemical, and catalytic properties of nanocomposites differ significantly from that of the component materials. The concept of multi-functionality and improved properties in nanocomposites is attributed to homogeneous distribution of nanoscale phases in the matrix. However, key challenges remain for nanocomposite production, including processing, cost, consistency in volume production, high lead time, and oxidative and thermal instability of nanomaterials. The application of AM manufacturing in the production of nanocomposite parts is expected to be a promising approach to alleviate some of these limitations (Campbell and Ivanova 2013). The development of nanocomposites using AM has started very recently and the reported literature is very limited. Moreover, most of the work concentrated on polymer-based nanocomposites, where AM techniques were used to add nanomaterials such as carbon nanotubes, nanowires, quantum dots, metal, or ceramic nano particles into the host polymer matrix. This early development in polymer matrix nanocomposites is due to the ease of incorporating nanomaterials and the enormous progress in polymer-based AM techniques. However, limited work has been reported in the area of metal matrix nanocomposites due to its inherent difficulties for incorporating nanoparticles in metal powder-based AM techniques such as SLS, LENS™, and 3DP.

Despite few successes of nanocomposite development using AM, there are several issues unanswered such as the interaction of nanoparticles with printing material, optimization of process parameters, and synthesis methods for different nanomaterials. Further, thermal stability and the tendency of the agglomeration of nanomaterials in the printing media and the subsequent modification of AM processing conditions are huge challenges to integrate nanomaterials and AM (Ivanova et al. 2013). The limited choice of materials has restricted the wider adoption of these technologies, which can be overcome by developing innovative materials and processes.

12.3.1 Metal Matrix Nanocomposites

Among different nanomaterials, metal nanoparticles possess attractive optical, thermal, and electrochemical properties. Initially, metal nanoparticles were used in the solid freeform fabrication process to densify porous metallic parts that were prepared using 3DP and SLS (Crane et al. 2006). The porous test specimens were printed using 410 SS powders with the particle size ranging from 63 to 90 μm. Nanoparticle suspension was used to infiltrate the porous part, which showed strengthened bonds between particles and also reduced creep and sintering shrinkage up to 60%.

The particulate reinforcement in the titanium matrix composites (TMCs) greatly improved the performance of titanium alloys. The size of the ceramic particle strongly affects the strength, ductility, and failure mode of TMCs. Generally, decreasing the size of ceramic particles to the nanometer level can lead to a substantial improvement in the mechanical properties of TMCs (Mortensen and Llorca 2010). The nanoscale dispersion of the ceramic phase in TMCs tends to introduce novel behaviors owing to high surface-to-volume ratios, which are absent in the conventional composites. The homogeneous distribution and restricting grain coarsening of the nanoscale ceramic reinforcements during processing are the main criteria for the enhanced performance of the nanocomposites. The nanocrystalline TiC-reinforced Ti matrix composites parts were fabricated using SLM by Gu et al. (2011). Figure 12.7 shows

FIGURE 12.7 TiC-reinforced titanium matrix nanocomposite parts were prepared using SLM. (Reprinted with permission from Gu, D. et al., *Compos. Sci. Technol.*, 71, 1612–1620, 2011.)

cubic specimens of nano TiC (50 nm)-reinforced composites that were fabricated using SLM. The coarsening of the TiC phase was found to depend on laser energy density, and nanoscale TiC reinforcements were observed when the laser energy density was below 120 J/mm^3. The SLM-processed TiC/Ti nanocomposite part showed significantly high nanohardness (90.9 GPa), elastic modulus (256 GPa), and wear resistance.

LENS™, a direct metal deposition-based AM technique, has the flexibility to deposit different TMCs with additional flexibility in terms of functional gradation in composition. Das et al. (2014) have synthesized a TiB-TiN-reinforced Ti6Al4V alloy composite using LENS™, where premixed Ti6Al4V and BN powder were used. They found the formation of TiB nanorods in the matrix when composites were deposited at a lower laser energy density (38 J/mm^2). Similar nanometer-scale TiB precipitates were also observed in the case of Ti6Al4V-TiB composites, which had been laser-deposited from a premixed powder consisting of Ti-6Al-4V and elemental boron powders (Banerjee et al. 2005). The fabrication of multi-walled carbon nanotubes to be dispersed in nickel matrix composites has been attempted using LENS™ (Hwang et al. 2008). Bhat et al. (2011) fabricated multi-wall carbon nanotube-reinforced Cu-10Sn alloy composites using LENS™. Composites containing 12 vol.% carbon nanotubes showed more than an 80% increase in the Young's modulus and a 40% increase in the thermal conductivity of a Cu-10Sn alloy.

Furthermore, the addition of metal nanoparticles were found to reduce shrinkage and distortion of 3DP processed parts. Bai et al. (2007) showed that incorporating silver nanoparticles, through a water-based binder system, onto a 3DP of micro-silver powder, significantly reduced the sintering temperature and improved sintering characteristic parts. Direct inkjet printing is an attractive technique of making the desired conductive patterns or electrodes in printed electronic and optoelectronic devices. Further, the addition of nanoparticles in the printing medium could change the rheology of the material, which is very important for AM processing. Ahn et al. (2009) showed the printing of Ag micro-electrodes using the omnidirectional printing of concentrated silver nanoparticles inks. The ink with high solids loading (≥70 wt. %) was achieved by optimizing the silver nanoparticles concentration, size, and distribution. Such ink was found suitable for the printing of a self-supporting micro-electrode with complex shapes.

12.3.2 Polymer Matrix Nanocomposites

SLS is quite popular in fabricating polymer matrix nanocomposites with high accuracy. LS polymer parts have a huge demand in the aerospace, automotive, defense, and medical industries. However, anisotropic mechanical behavior and the limited choice of materials have restricted its progress (Kruth et al. 2007). Several attempts have been made to improve the mechanical or physical properties of the LS polymer using nano fillers, such as clay (Kim and Creasy 2004, Jain et al. 2009, Wahab et al. 2009), nanosilica (Chung and Das 2008), nano-Al_2O_3 (Zheng et al. 2006), and carbon nanofiber (Goodridge et al. 2011). Incorporating nano fillers not only significantly enhance the mechanical properties, but also improve other properties, such as optical properties, thermal conductivity, heat resistance, and flame retardancy or accelerate biodegradability or increase bioactivity of the polymer composite. However, the success of a polymer nanocomposite for SLS is highly dependent on the uniform dispersion of the nanoparticles and good interfacial adhesion between the filler and polymer matrix (Goodridge et al. 2011). Therefore, the preparation of the starting powder for SLS is most important. Coating the filler materials with the base polymer was found to be successful. Zheng et al. (2006) used PS coating on the nano Al_2O_3 particles by an emulsion polymerization technique and used the polymer-coated fillers to reinforce the PS-based composites using SLS. The nanoparticles were found dispersed homogeneously in the matrix and the tensile strength of the nanocomposites improved more significantly (300%) than in the unfilled PS.

It was observed by Chung and Das (2008) that the mechanical mixing of nylon-11 and nanosilica powders used for SLS causes the agglomeration of nanosilica particles in the sintered part. Yan et al. (2009) used a dissolution-precipitation technique to prepare a nanosilica/nylon-12 composite powder for SLS. The nanosilica particles (3 wt.%) were found homogeneously dispersed in the sintered part. The tensile strength, tensile modulus, and impact strength of the nanocomposites increased by 20.9%, 39.4%, and 9.54%, respectively, compared to the neat nylon-12 prepared using an identical dissolution-precipitation technique. Athreya et al. (2010) used 4 wt.% nanosized carbon black powder reinforcement in nylon-12 to process an electrically conductive polymer nanocomposite using SLS. The flexural modulus of nylon-12/carbon black composites was observed to be lower than pure nylon-12 due to the segregation of carbon black in the composite and a weak polymer-filler interface. The electrical conductivity of the nanocomposite was five orders of magnitude higher than that of pure nylon-12 that was processed by SLS.

Further, owing to exceptional mechanical, thermal, and electrical properties, carbon nanotubes and nanofibers have inspired their use as a filler in polymers. A study by Goodridge et al. (2011) demonstrated that the reinforcement of laser-sintered polyamides with carbon nanofibers can increase the strength of a base polyamide. The composite powder containing 3 wt.% carbon nanofibers was prepared using melt mixing and cryogenic milling, which facilitates homogeneous dispersion of the nanofibers within the polymer matrix after sintering. The mechanical behavior of the nanocomposites showed a 22% increase in the storage modulus compared to the base material.

AM techniques have achieved more and more attention for scaffold fabrication in recent decades due to the unique advantages, such as a possibility of defined external and internal architectures of scaffolds, computer controlled fabrication processes, higher accuracy, and reproducibility. Duan et al. (2010) successfully fabricated three-dimensional nanocomposite scaffolds using selective laser sintering. Calcium phosphate (Ca-P) and carbonated hydroxyapatite (CHAp) nanoparticles were incorporated into poly (hydroxybutyrateco-hydroxyvalerate) (PHBV) and poly (L-lactic acid) (PLLA), respectively, to prepare nanocomposite micro-spheres. These nanocomposites micro-spheres (Ca-P/PHBV and CHAp/PLLA) were used to fabricate 3D scaffolds. Figure 12.8 shows four kinds of scaffolds, namely, PHBV, Ca-P/PHBV, PLLA, and CHAp/PLLA, were fabricated using micro-sphere powders. The compressive strength and modulus of the nanocomposites scaffolds

(a) (b)

FIGURE 12.8 (a) 3D scaffolds with different composition (A) PHBV; (B) Ca-P/PHBV; (C) PLLA; and (D) CHAp/PLLA produced by SLS (b) Micro-computed tomography image of a Ca-P/PHBV scaffold. (Reprinted with permission from Duan, B. et al., *Acta Biomater.*, 6, 4495–4505, 2010.)

were found to be higher than their polymer counterparts in dry conditions. The cell proliferation and Alkaline phosphatase (ALP) expression by SaOS-2 cells were enhanced in the case of Ca-P/PHBV scaffolds compared to pure PHBV, while no difference was observed in the cell proliferation between CHAp/PLLA and pure PLLA scaffolds.

SL can be used to build 3D nanocomposite parts using a light-curable photopolymer. In the SL process, a thin layer of photosensitive resin with nanoparticles is cured on the surface of a resin bath using localized UV exposure, which allows the nanoparticles to reinforce in the polymer. The viscosity of the resin can significantly increase by the addition of nanoparticles. Micro-stereolithography was used to fabricate porous scaffolds by mixing the nanosized hydroxyapatite powder in a photo-cross-linkable PDLLA-diacrylate resin (Ronca et al. 2013). With increasing concentration of the nanoparticles, viscosity of the resin increases and stiffness increases in the cured composites. In a study, Gurr et al. (2010) showed *in situ* synthesized calcium phosphate/layered silicate hybrid nanoparticles dispersed in acrylic resin. The nanocomposite was prepared using the rapid prototype process based on the photopolymerization of acrylic resin to improve the property of resins. The nanocomposite materials showed significantly increased stiffness with increasing filler contents both in the green and post-cured state. Duan et al. (2011) used stereolithography to improve the mechanical and thermal properties of photosensitive resin by incorporating nano-TiO_2. The mechanical and thermal stability of the resin was found significantly high when the nano TiO_2 content was at 0.25%. It was observed that the tensile strength was increased by 89% from 25 to 48 MPa, and the tensile modulus increased by 18% from 2,001 to 2,362 MPa, and the flexural strength and the hardness increased by 6% and 5%, respectively. Moreover, the presence of nanomaterials in photopolymer suspension has changed the absorption or refraction ability of UV light, which subsequently alters cure depth and cure shape profile (Ivanova et al. 2013). Therefore, process optimization is required for different nanomaterials when they are incorporated in a stereolithography system.

12.4 FUNCTIONAL MATERIALS

12.4.1 Functionally Graded Materials

FGMs are materials in which the composition and micro-structure changes gradually (gradient) or step-wise (graded) from one side to the other, resulting in a corresponding variation in the properties (Liu and DuPont 2003). Various fabrication techniques, such as chemical vapour deposition/physical vapour deposition (CVD/PVD), plasma spraying, powder metallurgy, and self-propagating high-temperature synthesis are generally used for producing functionally graded materials (Kieback et al. 2003). Most of these techniques are not able to manufacture complex shape parts in a single step, which is highly desirable. In recent years, the AM of dissimilar and graded structures have been receiving more attention due to its huge demand for different applications. Additive manufacturing processes, more specifically, the processes which are capable of delivering multiple materials at a time, are potentially suitable to manufacture complex-shaped FGM parts. Directed metal deposition processes have demonstrated their potential for manufacturing FGMs due to their flexibility in a powder feeding mechanism that can change or mix materials when fabricating multi-material structures (Yakovlev et al. 2005). Several attempts have been made in FGM deposition of different metals and alloys, such as stainless, nickel base superalloys, Co-Cr-Mo alloy, and titanium using the LENS™ technique.

Collins et al. (2003) and Banerjee et al. (2003) deposited a graded binary Ti-V and Ti-Mo alloy using the LENS™ process from a blend of elemental Ti and V (or Mo) powders. Several researchers have studied functionally graded TiC-reinforced metal matrix composites by adjusting processing parameters and real-time variation of the feeding ratio of metal powder to TiC during a laser direct deposition process (Liu and DuPont 2003, Zhang et al. 2008, Wilson and Shin 2012). Crack-free functionally graded TiC/Ti composites, from pure Ti to approximately 95 vol.% TiC, were fabricated using LENS™ by Liu and DuPont (2003). A functionally graded TiC/Inconel 690 composite coating having TiC particles, varied from 0 to 49 vol.%, showed a significant improvement in hardness and wear resistance (Wilson and Shin 2012). A compositionally graded alumina coating (Bandyopadhyay et al. 2007) and a yttria-stabilized zirconia coating (Balla et al. 2007) on a steel and a Ti-TiO_2 structure (Balla et al. 2009) with a composition and structural gradation was fabricated using the LENS™ technique. These graded coatings were found to be superior to conventional homogeneous coatings, as they provide better bonding strength between the coating and the substrate and relatively less residual stress due to gradual compositional variation. As a result, there are no sharp interfaces and the micro-structure and hardness change smoothly, leading to lower stress intensity at the interface. The laser deposition of a full

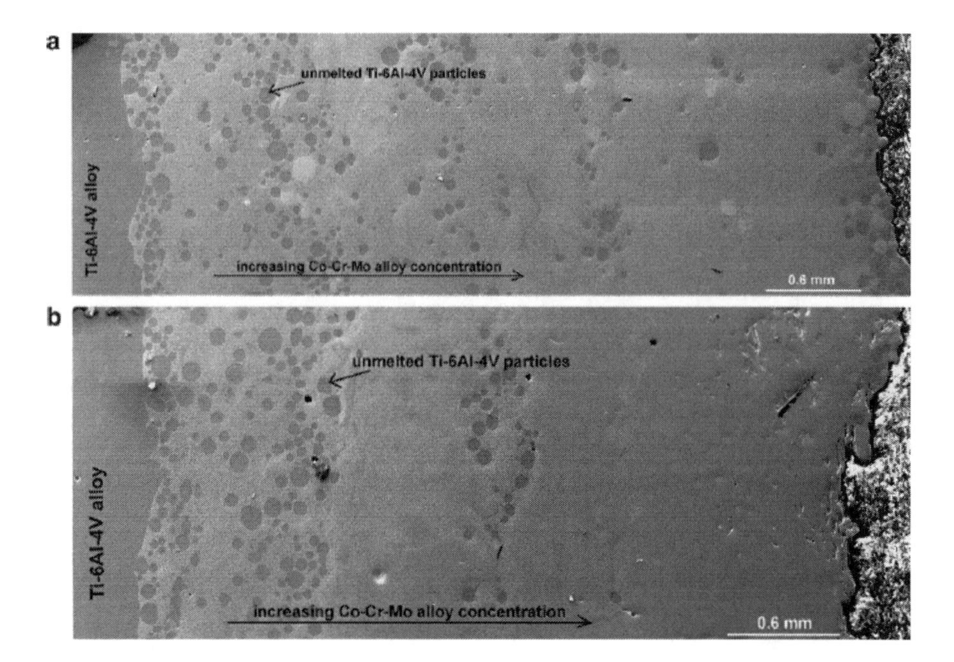

FIGURE 12.9 SEM micro-graph of graded Co-Cr-Mo coatings on Ti6Al4V alloy: (a) 50% Co-Cr-Mo alloy at the surface and (b) 86% Co-Cr-Mo alloy at the surface. (Reprinted with permission from Krishna, B.V. et al., *Acta Biomater.*, 4, 697–706, 2008b.)

CoCrMo coating on titanium was difficult to produce crack free, possibly because of a brittle intermetallic formation and residual stresses generation. Krishna et al. (2008b) demonstrated the crack- and defect-free deposition of functionally graded, hard, and wear-resistant Co-Cr-Mo alloy coating on Ti-6Al-4V using LENS™. The cross-sectional micro-structure of the functionally graded coating is shown in Figure 12.9. In laser direct deposition, multiple powder hoppers were used to deliver different elemental powders individually, and their feed rates were controlled separately to adjust the composition in FGMs. Wang et al. (2007) and Farayibi et al. (2013) prepared titanium-based FGM components by combining powder (TiC or WC) and wire (Ti6Al4V) for direct laser fabrication. During the process, by gradually increasing the powder feed rate and/or decreasing the wire feed rate, the graded composition was fabricated. This technique has the advantage of less wastage due to no mixing of the feedstock materials.

FGM by SLS has been reported by Chung and Das (2006). They created glass bead particulate-filled Nylon-11 composites where filler volume fraction varies from 0% to 30%. In this work, a macroscopic 3D polymer composite part with a one-dimensional material gradient in the build direction was demonstrated in a single uninterrupted SLS run. Schematic of the FGM part with a change in the filler concentration and the corresponding cross-sectional SEM micro-graphs are shown in Figure 12.10. A smooth interfacial region indicates the successful fabrication of Nylon-11 and glass beads FGM.

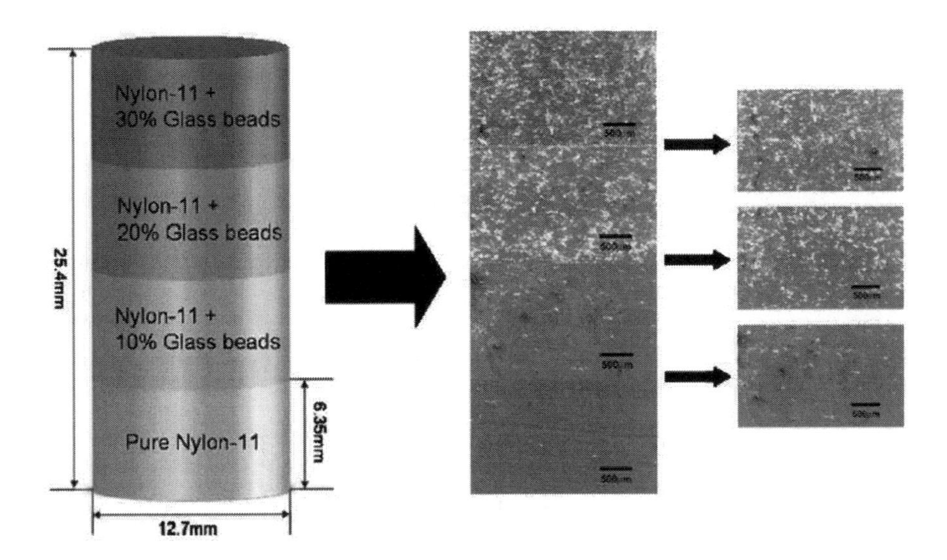

FIGURE 12.10 A schematic description of graded compositions, SEM micro-graph of each composition in the fabricated FGM specimen, and SEM micro-graph of interface of two different compositions. (Reprinted with permission from Chung, H., and Das, S., *Mat. Sci. Eng., A*, 437, 226–234, 2006.)

12.4.2 Materials for Hydrogen Storage

Laser-based additive manufacturing was found to be an attractive technique for deposition of multi-component high entropy alloys. These laser processing techniques have the advantage of a rapid cooling velocity, leading to a significant non-equilibrium solute-trapping effect that avoids component segregation and overcomes solubility limitations. Kunce et al. (2013) successfully synthesized a high entropy alloy (ZrTiVCrFeNi) from elemental powders in a near equimolar ratio using LENS™ technology. The synthesized alloy exhibited good chemical composition after laser deposition, compared to the nominal composition. The as deposited alloy showed a maximum hydrogen capacity of 1.81 wt.% and reduced to 1.56 wt.% after the additional heat treatment.

12.5 DESIGN FREEDOM/AM-ENABLED DESIGNS

Design optimization is very essential for manufacturing optimized products. This is because many of the novel design benefits/aims are compromised due to conventional manufacturing constraints. For example, design and manufacturing of lightweight structures are some of the most important requirements of components for use in aerospace. To achieve this optimization of geometrical structure of components is required and is mostly done by mathematical means. A typical optimized design thus obtained is shown in Figure 12.11, which is not possible to manufacture using traditional manufacturing techniques.

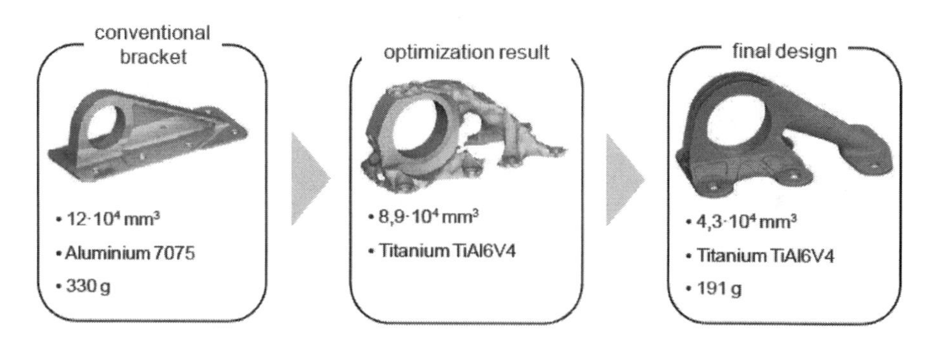

FIGURE 12.11 Mathematic design optimization process. (Open Access from Emmelmann, C. et al., *Phys. Procedia*, 12, 364–368, 2011.)

Emergence of AM technologies enable a the manufacturing of optimized products with improved functionality, in some cases multi-functionality, reduced weight and wastage, and associated energy savings. These product designs are normally of complex shapes, material combinations, and hierarchy (in composition, internal architecture, and micro-structure), some of which could possibly be achieved using current AM technologies. Some of the complex designs enabled by AM technologies are shown in Figure 12.12.

12.5.1 Design and Development of Lattice Structures

For more effective mechanical performance and weight reduction, the periodic arrangement of load bearing cross-sections (struts) has been developed. Such structures/materials are known as lattice structures and exhibit more predictable mechanical behavior (Parthasarathy et al. 2011). These lattice structures exhibit several useful properties (Gibson and Ashby 1997), such as acoustic and thermal insulation, energy absorption, and these properties can be easily tailored by changing the size and shape of struts and total porosity (Gibson and Ashby 1982). Further, these designed materials with lattice structure-type internal macrostructure/architecture are placed in different areas too far from conventional materials in terms of their elastic modulus as a function of density (Figure 12.13). Therefore, lattice structures enable more design options to achieve the desired mechanical or functional properties.

The internal unit cell structure has been tailored to achieve negative Poisson's ratio (Rehme and Emmelmann 2009). The honeycomb structures analyzed for their Poisson's ratio are shown in Figure 12.14. Both the designed exhibited negative Poisson's ratio and cubic chiral architecture have the highest negative Poisson's ratio of −0.2835. The results are primarily dictated by the diameter of the nodes and should be balanced with the length of the struts. Yang et al. (2012a,b) successfully fabricated auxetic mesh structures, shown in Figure 12.15, using electron beam melting (EBM). They found that desired strength or stiffness can be achieved

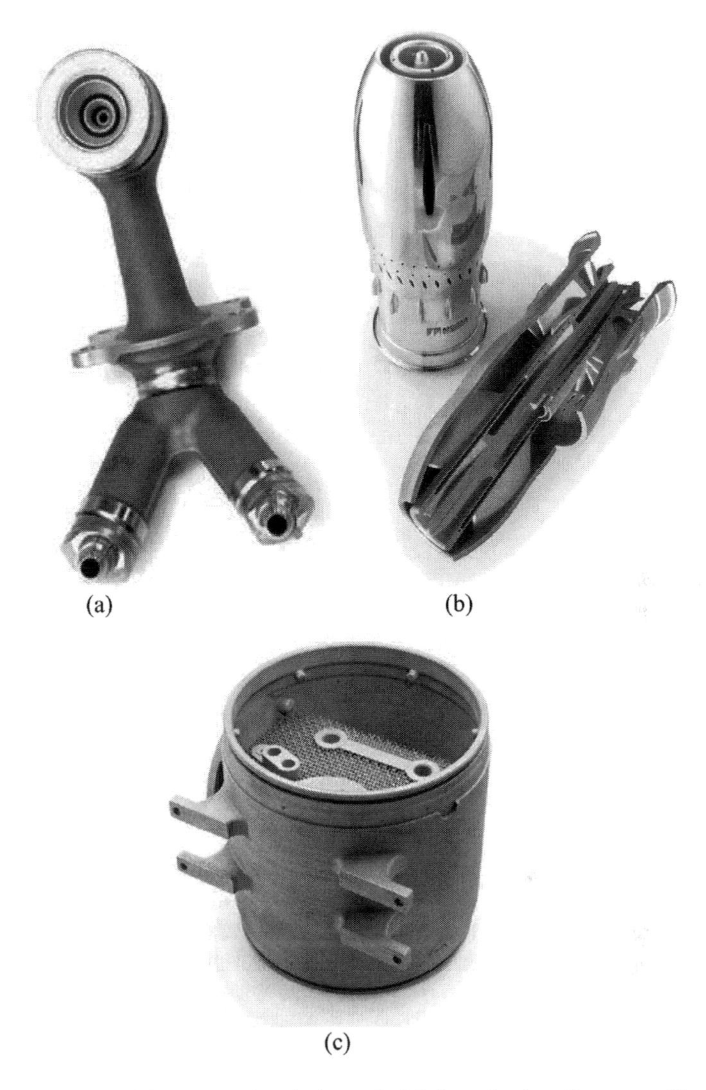

(a) (b) (c)

FIGURE 12.12 (a) GE Leap engine fuel nozzle fabricated using direct metal laser sintering (DMLS). (Courtesy of www.industrial-lasers.com.); (b) component with complex internal cooling channels manufactured by SLS (Courtesy of www.mmsonline.com.); and (c) metal housing with integrated solid and lattice structures (Courtesy of www.arcam.com.)

by changing the strut length, and these properties are high for the structures with high negative Poisson's ratio. The Poisson's ratio has been tailored, changing the re-entrant strut angle and/or ratio of vertical-to-re-entrant strut length (Yang et al. 2012a). The compression and bending tests performed on these structures demonstrated that the bending strength without solid skin is significantly higher than conventional sandwich structures (Yang et al. 2012b). The EBM process-induced

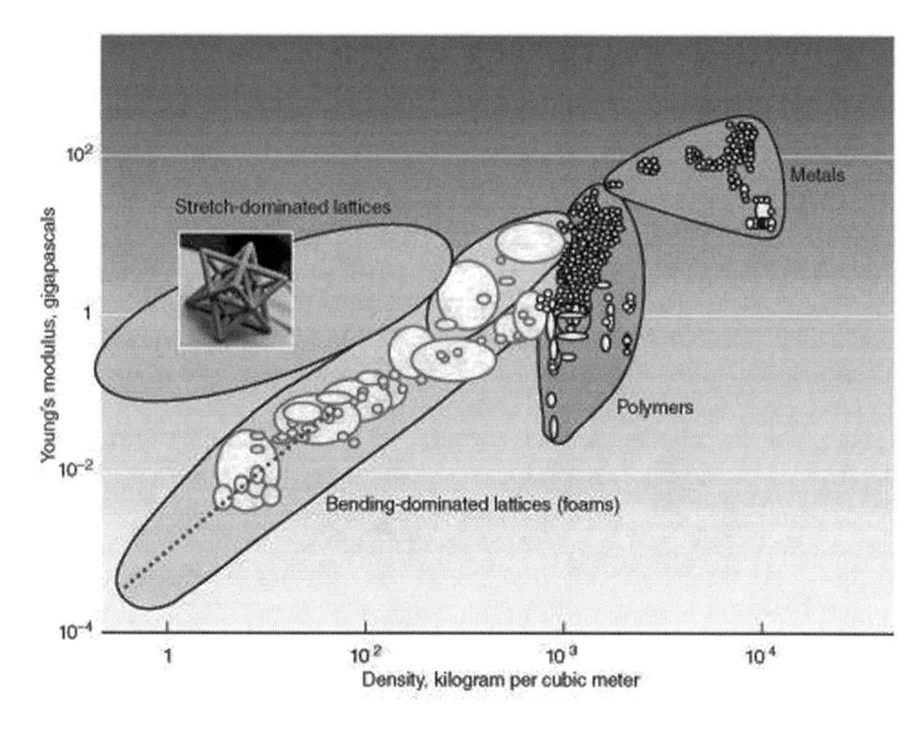

FIGURE 12.13 The Lattice structures position as compared to conventional materials. (Courtesy of https://manufacturing.llnl.gov/additive-manufacturing/designer-engineered-materials.)

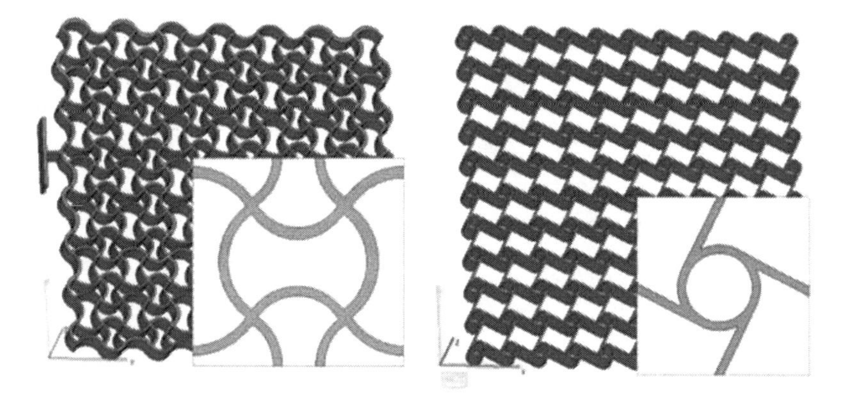

FIGURE 12.14 Tailored honeycomb structures: (Left) cubic sinus wave design and (Right) cubic chiral design. (Open Access from Rehme, O. and Emmelmann, C., *J. Laser. Micro. Nanoen.*, 4, 128–134, 2009.)

defects reduced the compressive strength and energy absorption capacity, but the structures with negative Poisson's ratio may compensate this (Yang et al. 2012b). These experimental results suggest that lattice structures with negative Poisson's ratio show strong potential for use in applications, such as shock/impact absorbers

FIGURE 12.15 Ti6Al4V alloy auxetic structures that were fabricated using EBM. (Reprinted with permission from Yang, L. et al., *Acta Mater.*, 60:3370–3379, 2012a.)

and artificial intervertebral discs, where high shear strength and low resistance to compression are primary requirements (Rehme and Emmelmann 2009).

The effect of a unit cell strut edge design on the impact absorption of lattice structures has also been reported (Brennan-Craddock et al. 2012). Structures with two types of unit cell struts were fabricated using FDM and were compression tested to delineate the differences. The compression behavior of these structures is presented in Figure 12.16. It can be clearly seen that the energy absorption capacity (area under the curve) of a helix strut structure is significantly higher than other structures. This is primarily due to the ability of a helical design to increase strut overall length allowing more deformation before collapse (Brennan-Craddock et al. 2012). These results demonstrate the potential of complex tailored lattice structures that are fabricated using AM as efficient energy absorbing structures.

Another unique design that uses this lattice architecture is conformal lattice structures or spatially variant structures. These structures offer valuable properties, such as high strength-to-weight ratio, predictable load and stress distributions, better mechanical performance, and noise and vibration dampening. The effective use of such conformal lattice structures has been in the reduction of losses in directional-dependent self-collimation (Rumpf et al. 2013). In this report, a spatially variant device was fabricated using FDM to control electromagnetic waves. Figure 12.17a shows such a device which can direct an unguided beam without a significant loss

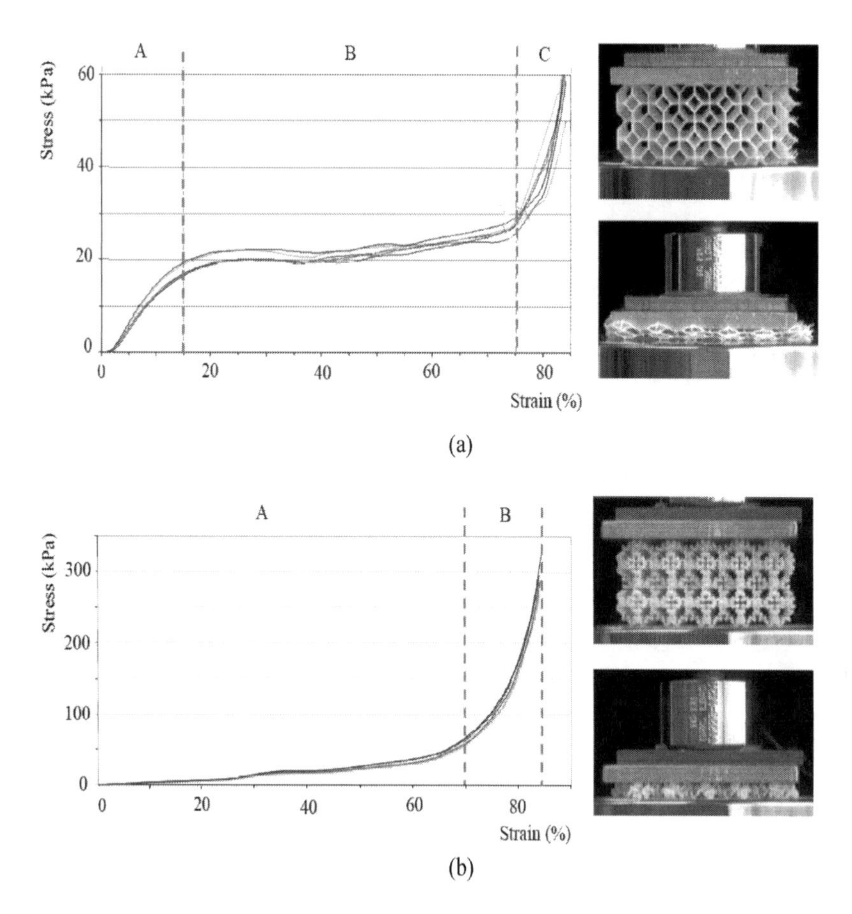

FIGURE 12.16 The compression deformation behavior of lattice structures: (a) structure with straight struts and (b) structure with helical strut—please note significantly high compression stress. (Open Access from Brennan-Craddock, J. et al., *J. Phys: Conference Series,* 382, 012042, 2012.)

due to the conformal positioning and orientation of unit cells without variations in their size and shape. The device has been experimentally tested between 14.8 GHz and 15.8 GHz and was found to exhibit 6.5% fractional bandwidth (Rumpf et al. 2013). The structures with conformal lattice architecture possibly enable the effective control of electromagnetic waves in various applications/devices.

12.5.2 Design Innovations for Medical Applications

It is well known that the natural bone is a highly complex composite comprising different materials and functional gradation in micro-structure, macro-structure, and composition. Therefore, the mechanical properties of bone also change accordingly—elastic modulus of 20 GPa for a dense cortical bone and 0.5 GPa for a highly porous cancellous bone. However, the current artificial implants that replace this natural bone are fully dense and also have elastic modulus that is significantly higher than

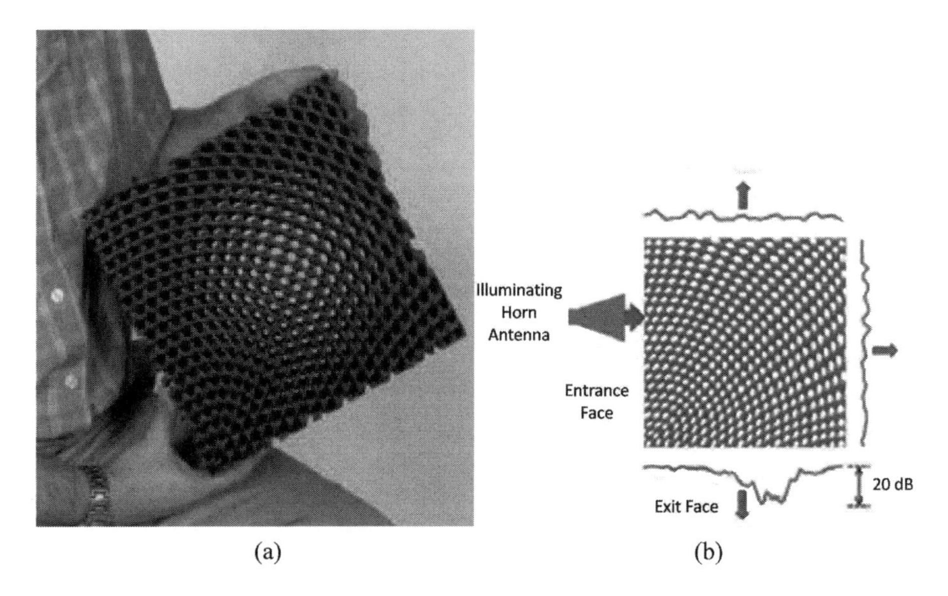

FIGURE 12.17 (a) Spatially variant structure fabricated using FDM and (b) measured profile of the waves around the device. (Open Access from Rumpf, R.C. et al., *Prog. Electromagn. Res.*, 139:1–14, 2013.)

natural bone. This modulus mismatch results in stress shielding leading to bone loss. Therefore, the ideal bone replacement material must have identical geometry and site-specific mechanical and functional properties to that of natural bone. For example, monoblock acetabular shell, shown in Figure 12.18a, with a porous surface on one side (to improve osseointegration) and a wear resistance surface on the other side (articulating against femoral head) can potentially improve the overall in vivo performance and life of artificial implants (España et al. 2010). The functional gradation in porosity, similar to natural bone, is also beneficial for implant long-term stability (Figure 12.18b). Artificial implants with complex internal and external architectures can also be designed to obtain site-specific functions at different locations of the same implant, as shown in Figure 12.18c. However, manufacturing these implants is not possible with traditional manufacturing routes, but with AM processes, such unique implants can be easily fabricated (Figure 12.19).

Porous metals have been proposed to address these stress shielding problems (Krishnaet al. 2007, Xue et al. 2007, Krishna et al. 2008a, Krishna et al. 2009, Bandyopadhyay et al. 2009, 2010, Balla et al. 2010a, DeVasConCellos et al. 2012), but with a drop in mechanical properties as a consequence. However, the regular arrangement of pores in these porous structures was found to greatly improve mechanical properties, while maintaining the desired elastic modulus close to the natural bone (Balla et al. 2010b).

Fabrication, deformation behavior, and mechanical properties of cellular materials with tailored internal micro-architecture for different applications, fabricated using AM technologies, have been reported by several authors (Yan et al. 2012, Ahmadi et al. 2014, Yan et al. 2014a,b). These internal architectures have also been

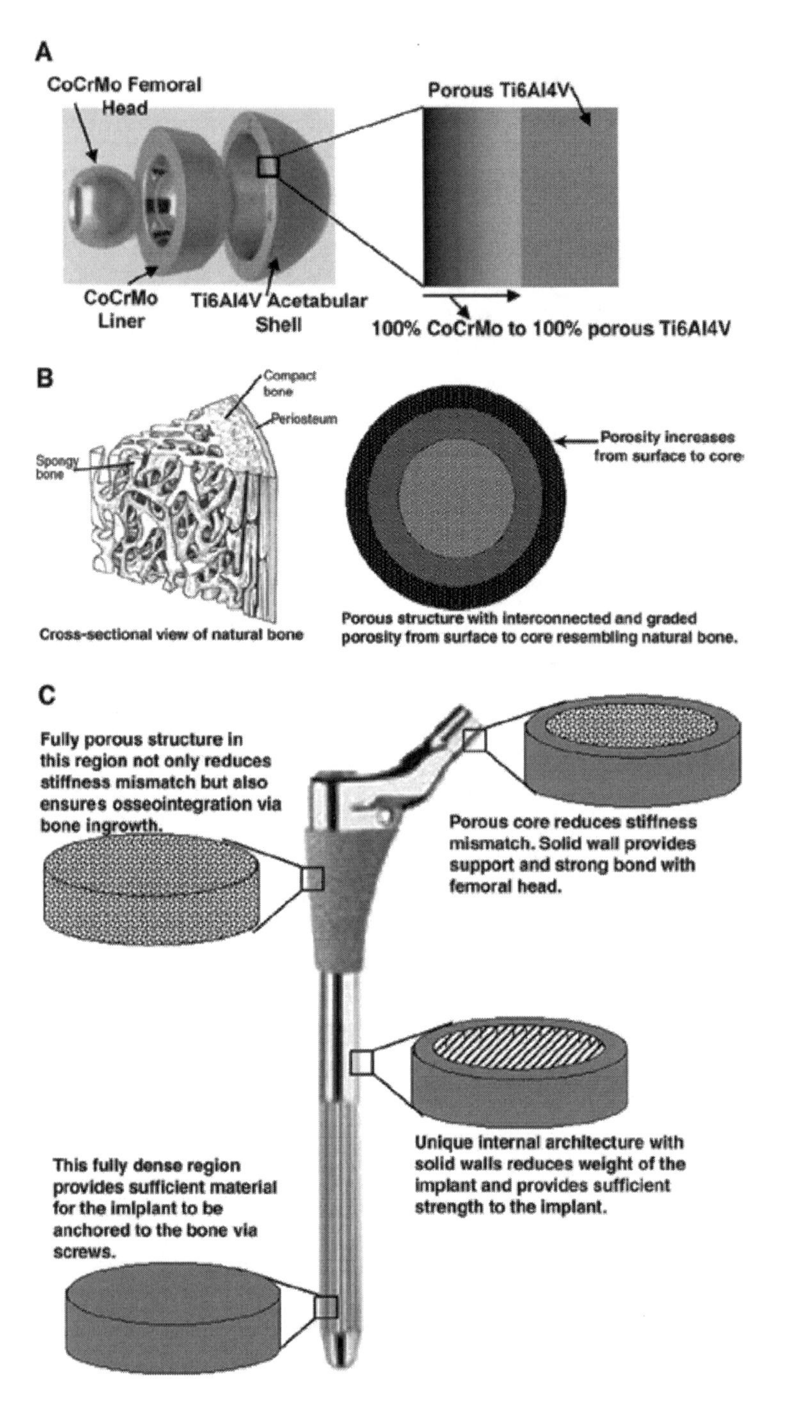

FIGURE 12.18 A schematic showing the monoblock acetabular shell, graded porous structures, and site-specific design of artificial implants. (Reprinted with permission from España, F.A. et al., *Mater. Sci. Eng. C*, 30, 50–57, 2010.)

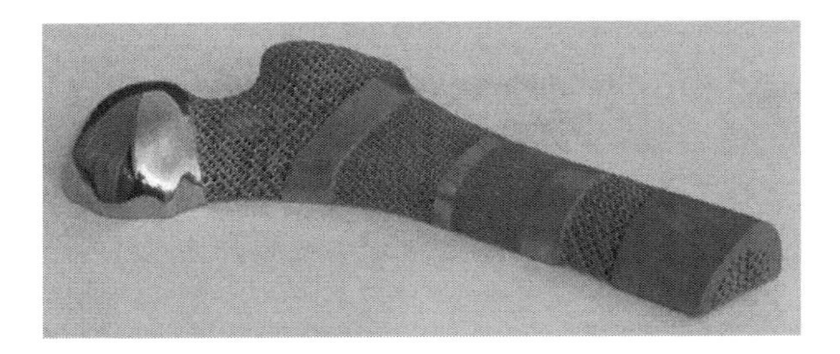

FIGURE 12.19 A typical femoral prototype with different internal architecture manufactured using AM. (Courtesy of www.tctmagazine.com.)

incorporated in artificial load bearing implants with an aim to reduce weight, match stiffness with natural bone, improve osseointegration, and overall long-term stability (Ovidiu et al. 2010, Heinl et al. 2008, Kusakabe et al. 2004, Ghiba et al. 2010, Stoica 2009). Earlier simulations on implants with internal tailored lattice structures showed a clear reduction of stress shielding (Ovidiu et al. 2010), in addition to favorable bone ingrowth into open pores. Another novel approach proposed by Mueller et al. (2012) involves the incorporation of functional cavities and channels into current load bearing implants, such as hip and knee. Such design features provide a local supply of desired materials such as drugs, filler materials, and post-operative inspection of the implants (Mueller et al. 2012). EBM fabricated functional implants with complex lattice structures have been reviewed in Murr et al. (2012a). These implants have been demonstrated to have excellent biocompatibility and to provide paths for osseointegration and antibacterial material loading. Further, the inclusion of lattice architecture in the implants eliminates stress shielding, thereby improving the long-term implant survivability. Tibial components of the knee with incorporated lattice structures are shown in Figure 12.20. The capabilities of AM technologies also enable the fabrication of functionally graded lattice structures mimicking the natural bone. The mechanical properties of cellular implants have been reported by Murr et al. (2012b). These research investigations have shown the strong application potential of these novel implants, but they require significant efforts in terms of in vitro and in vivo trials and testing before these can be used for clinical use.

Being computer-aided design (CAD)-based manufacturing technologies, the AM processes enable the design and development of custom devices and implants directly from computed tomography (CT) or a magnetic resonance imaging (MRI) scan data of a patient. Clinical and experimental results clearly demonstrate that custom fit implants ensure mechanical stability and long-term in vivo success (Fitzpatrick et al. 2011, McCarthy et al. 1997). AM technology has been successfully used to fabricate customized amputation prosthesis with a functional gradation in porosity (DeVasConCellos et al. 2012). Figliuzzi et al. (2012) employed a direct laser metal forming technique to a fabricate root-analogue implant design using 3D projections of the maxilla and residual root (Figure 12.21). The Ti6Al4V alloy implant was manufactured using direct laser metal forming and has been implanted. Perfect matching between the implant and the root was observed, thus improving the stability of the implant after 1 year of follow-up. The implant did not show any indication of pain or

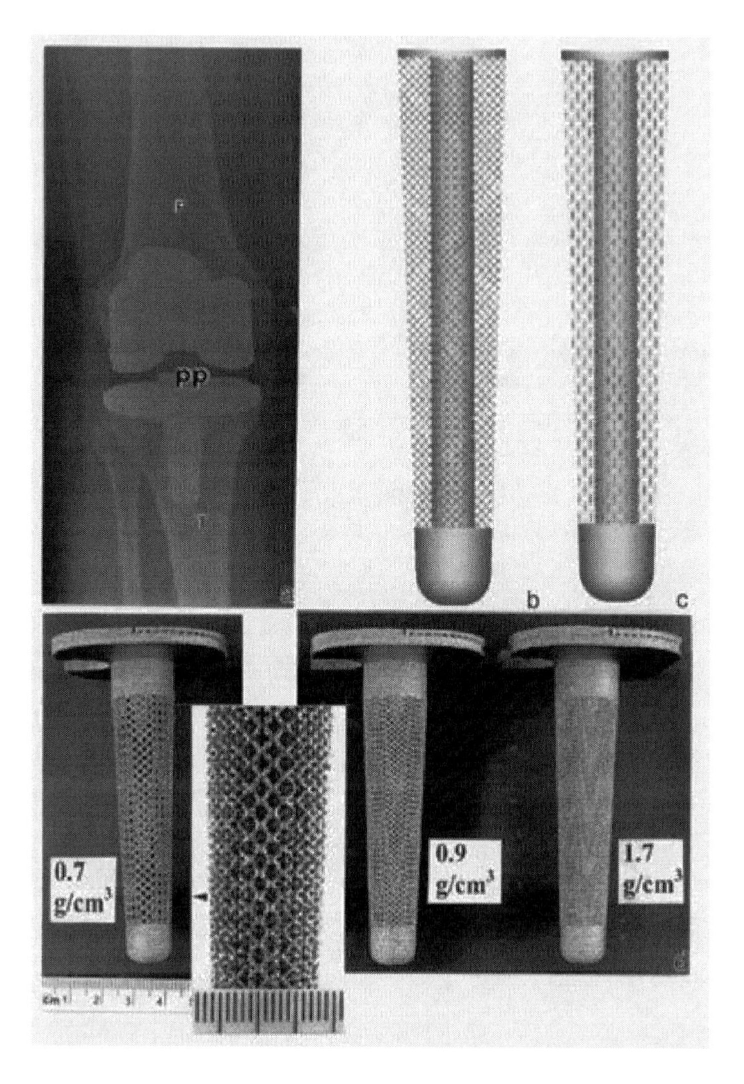

FIGURE 12.20 CAD models of a tibial tray showing a solid core for mechanical support and a cellular shell for bone ingrowth (above), and EBM fabricated components with complex architectures (below). (Open Access from Murr, L.E. et al., *J. Biotechnol. Biomaterial.*, 2, 1000131, 2012a.)

infection. Figure 12.21c shows a good bone-implant integration and stability of the natural bone. The AM capabilities can be effectively exploited to include functional gradation in cellular structures, which can potentially mimic the natural bone structure and, hence, enable a favorable vascularization and early bone formation (Murr et al. 2012a). A typical implant with a functional variation in porosity from the core to the shell fabricated using EBM is presented in Figure 12.22. More efficient mathematical designs for effective mechanical and flow ability properties can also be incorporated in the desired components (Khoda et al. 2013). Future development in

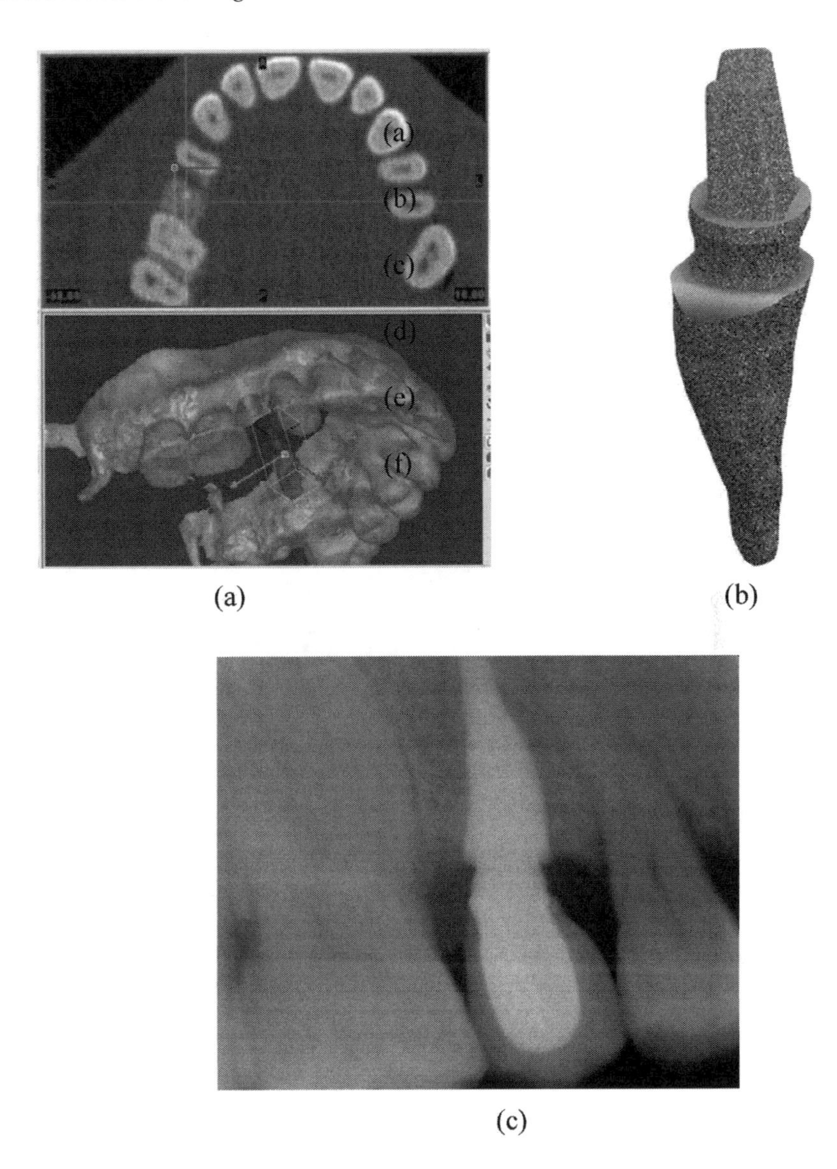

FIGURE 12.21 (a) Designings dental implant using the image data of maxilla and the residual root; (b) custom designed dental root-analogue implant model; and (c) 1 year post-surgery radiograph of a custom dental implant with crown. (Reprinted with permission from Figliuzzi, M. et al., *Int. J. Oral Maxillofac. Surg.*, 41, 858–862, 2012.)

AM technologies are expected to enable the fabrication of complex artificial organs as well (Wang et al. 2013). AM has also been extensively exploited to fabricate tailored and designed tissue constructs (Melchels et al. 2012). Complex structures with desired cell seeding enable fundamental studies to understand the cellular behavior during tissue formation.

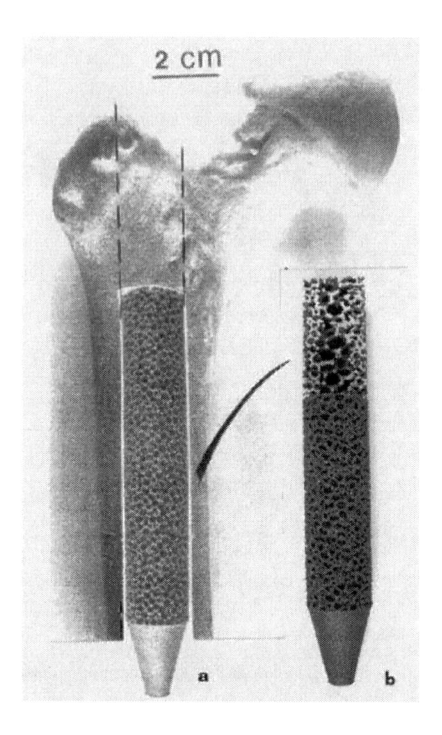

FIGURE 12.22 Intramedullary implant fabricated using EBM (a) and cross-section showing the variation in porosity from core to surface and (b). (Open Access from Murr, L.E. et al., *J. Biotechnol. Biomaterial.*, 2, 1000131, 2012a.)

12.5.3 Multi-functional Devices

Recently, advances in AM enabled fabrication of integrated systems such as embedded electronics, electrical circuits, and sensors in mechanical structure/parts. Further, these new systems may also consist of complex/conformal shapes made with a variety of materials (multi-materials) with/without functional gradation in the composition (Vaezi et al. 2013). Such multi-functional devices can be designed and manufactured on-demand using a combination of current AM technologies. For example, Lopes et al. (2012) attempted to integrate stereolithography and direct printing (DP) processes in creating 3D polymeric structures with up to 555 embedded timer circuits. The overall process consists of multiple starts and stops between SL, direct printing and intermediate processes, where SL has been used to create a main supporting structure and direct printing for conductive circuits. Several processing steps are currently manual and therefore require further developments in order to enable the automatic fabrication of complex 3D structure with embedded circuits and inter connects (Lopes et al. 2012).

Very recently, the problems associated with the materials used in SL (for producing structures with electronic circuits, such as long-term durability, functionality, and conductive inks with low curing temperatures) have been addressed by a novel technology that uses FDM and direct printing or thermal embedding technology (Espalin et al. 2014). The use of FDM replaces polymers with thermoplastics having high strength, and conductive copper wires embedded into the substrates, using thermal technology to enable fabrication of devices with superior performance and robustness as compared to SL-based processes (Lopes et al. 2012). However, it was found that FDM-based processes require other techniques, such as direct wire technology for printing electronic circuits in addition to subtractive processes, such as micromachining to achieve the desired feature resolution (Espalin et al. 2014). The capability of such multi-3D systems has been effectively used to fabricate a CubeSat module (Figure 12.23), and the systems were found to provide significant improvements in overall performance (Espalin et al. 2014).

Inkjet printing was used to fabricate split-ring resonator arrays on a flexible polyimide substrate (Walther et al. 2009). This study demonstrated that inkjet printing is an agile processing route to deposit metamaterial structures on a variety of substrates for gigahertz to terahertz frequencies (Figure 12.24). The circuits were printed using 20 wt.% silver nanoparticle suspension, and the printed polyimide substrates were then heated at 220°C before testing. The performance of these split-ring resonator arrays was comparable to that of conventionally processed arrays, but the variation was relatively high.

In the medical field, there is a growing interest and demand for minimally invasive and even non-invasive surgeries. In fact, for ideal minimally invasive surgeries,

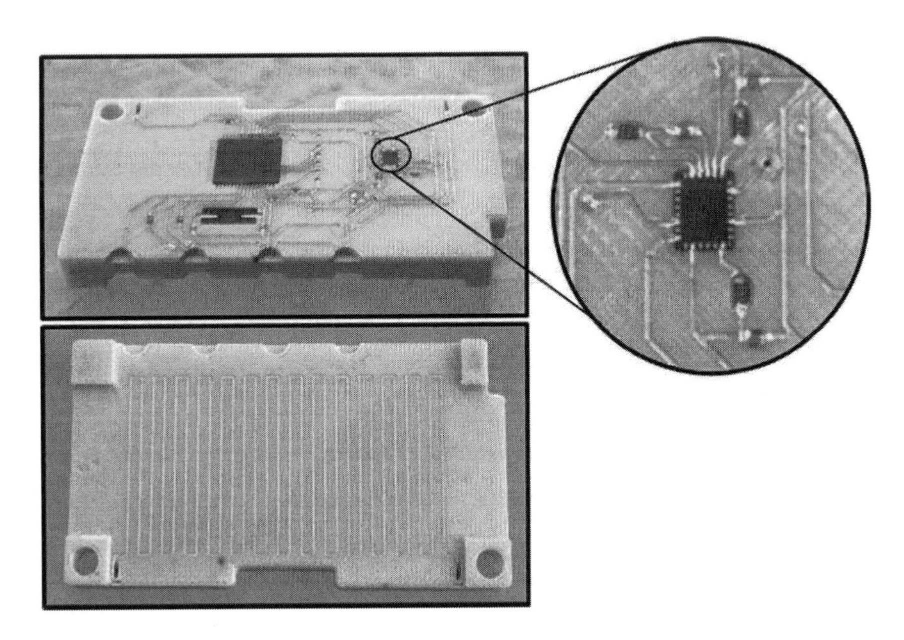

FIGURE 12.23 A typical CubeSat module fabricated using hybrid technique (FDM, direct printing, and micro-machining). (Reprinted with permission from Espalin, D. et al., *Int. J. Adv. Manuf. Technol.*, 72, 963–978, 2014.)

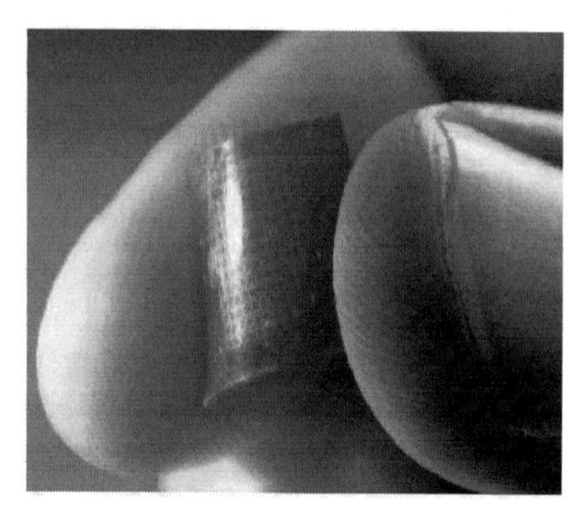

FIGURE 12.24 Conductive circuits on a flexible substrate printed using the inkjet printing process. (Reprinted with permission from Walther, M. et al., *Appl. Phys. Lett.*, 95, 251107, 2009.)

the surgical tools must be as small as possible, sometimes the dimensions could be in micro-meters. Conventionally manufacturing technologies are not suitable for fabricating miniature surgical devices or tools. Recently, electrochemical FABrication technology has been identified as one suitable manufacturing technology for miniature surgical tools. This technology was found to have extremely high geometrical resolution and is capable of producing micro-devices with several individual moving and assembled parts (Cohen et al. 2010). It was demonstrated that devices with small features up to 4 μm can be easily fabricated using electrochemical FABrication, and it is the only technology that can produce miniature metal devices with micron level features and moving mechanisms (Cohen et al. 2010). The technology can also produce miniature sensors for military applications, micro-fluidic devices, etc.

Micro-stereolithography (MSL) is another AM technique that has the capability to procure micro-devices. With an ability to use multiple materials and extremely fine feature resolution, the technology has been effectively used in the development of micro-electromechanical systems. In addition to structural support, the resins have been added with the desired filler material to achieve the desired functionality. A functional composite material consisting of magnetic nanoparticles added to the resin has been reportedly used to build a micro-flow sensor device (Leigh et al. 2011). Such micro-devices find applications where space restrictions are very high (Figure 12.25). The details and capabilities of other micro-AM processes are discussed by Vaezi et al. (2013) in greater detail.

12.6 INDUSTRIAL PERSPECTIVE

Although the above discussion demonstrates a strong application potential for AM technologies to produce materials and components with multi-functionality, the adaptation of these technologies into mainstream manufacturing requires careful consideration of several aspects. Albeit due to extremely complex nature of

FIGURE 12.25 Miniature flow sensor with functional composite rotor made using micro-stereolithography. (Reprinted with permission from Leigh, S.J. et al., *Sensors and Actuat. A*, 168:66–71, 2011.)

integration of AM with current component designs, materials, and production, there still exists strong industrial opportunity. Generally, current AM technologies offer "complexity for free," and therefore it must be given priority in addition to production volume (Fera et al. 2018). The increased complexity, designed based on the functionality, can certainly improve the overall product performance (Merkt et al. 2012). As discussed above, the complexity of parts that can be manufactured using AM include (Rosen et al. 2007): (i) shape complexity and (ii) material complexity. The incorporation of these complexities in the product provides a significant energy savings throughout the product cycle. In the production phase, the benefits arise due to redesigning the system and eliminating expensive tools. The enhanced performance, energy savings, and reduced delivery times are key considerations in the service phase of the product. Finally, at the end of life, the savings can be realized in terms of overall less energy input, due to the reduced amount of material used per part. It appears that significant research efforts are being made in the area of shape complexity, while the research on materials complexity to exploit the capabilities of AM is still at infancy. For example, the shape complexity can be defined using several parameters, such as volume ratio, thickness ratio, depth ratio, etc. (Joshi and Ravi 2010). Recently, such geometrical complexity correlated with the energy consumption in EBM showed that the complexity is free in AM (Martin et al. 2016). However, similar approaches to define the materials complexity and their relation with part production with AM are still lacking. Recently, a review by Bandyopadhyay and Traxel (2018), proposed a simple process workflow integrating process optimization and simulation to manufacture novel components consisting of a significant variation in geometry, composition, and functionality. More studies along these lines are required to realize the full benefits of design and materials complexity offered by AM from an industrial perspective.

12.7 SUMMARY

The existing AM techniques have demonstrated their capabilities to create novel structures with unique geometrical design, as well as the incorporation of tailored/designed materials into the structures. In particular, cellular structures and components with embedded lattice structures show strong application potential in various sectors. For example, large structures with designed internal lattice structures provide significant weight savings, while maintaining strength and other service requirements of aerospace industries. These structures are also good candidate materials for energy absorption systems in automobiles. Such structures provide site-specific functional requirements for orthopedic implants, thus improving their in vivo life. The ideal combination of multiple AM technologies appears to be an effective approach in creating multi-functional devices and structures with embedded sensors.

In spite of significant developments and improvements in AM technologies, materials innovation using AM appears to be still in its embryonic stage. Comprehensive processing, micro-structure, and property correlations are required. A major hurdle in designed materials development using AM is non-availability of the desired materials. Although promising results have been reported in developing metal, ceramic, and polymer matrix composites via AM techniques utilizing existing and new feedstock materials, several materials-related challenges remain to be addressed. Even existing materials are not optimized or designed for AM technologies. An important requirement in the development of new materials for AM is the preparation of the feedstock materials with the desired characteristics suitable for a specific AM technique. Finally, as with challenges, the rewards are also extremely high for the integration of nanomaterials and AM.

SUGGESTED FURTHER READING

Bandyopadhyay, A., Traxel, K.D. 2018. Invited review article: Metal-additive manufacturing—Modeling strategies for application-optimized designs. *Addit. Manuf.* 22: 758–774.

Bose, S., Robertson, S.F., Bandyopadhyay, A. 2018a. Surface modification of biomaterials and biomedical devices using additive manufacturing. *Acta Biomaterialia*. 66:6–22.

Campbell, T.A., Ivanova, O.S. 2013. 3D printing of multifunctional nanocomposites. *Nano Today* 8:119–120.

Fera, M., Macchiaroli, R., Fruggiero, F., Lambiase, A. 2018. A new perspective for production process analysis using additive manufacturing: Complexity vs production volume. *Int J AdvManuf Technol*. 95:673–685.

Hu,Y., Cong, W. 2018. A review on laser deposition-additive manufacturing of ceramics and ceramic reinforced metal matrix composites. Ceramics International. https://doi.org/10.1016/j.ceramint.2018.08.083.

Kieback, B., Neubrand, A., Riedel, H. 2003. Processing techniques for functionally graded materials. *Mater. Sci. Eng. A* 362:81–106.

Mohan, N., Senthil, P., Vinodh, S., Jayanth, N. 2017. A review on composite materials and process parameters optimisation for the fused deposition modelling process, *Virtual and Phys. Prototyping*. doi:10.1080/17452759.2016.1274490.

Mortensen, A., Llorca, J. 2010. Metal matrix composites. *Annu. Rev. Mater. Res.* 40:243–270.

Sommers, A., Wang, Q., Han, X., T'Joen, C., Park, Y., Jacobi, A. 2010. Ceramics and ceramic matrix composites for heat exchangers in advanced thermal systems: A review. *Appl. Therm. Eng.* 30(11–12):1277–1291.

Vaezi, M., Srisit, C., Brian, M., Shoufeng, Y. 2013. Multiple material additive manufacturing: Part 1: A review. *Virtual and Phys. Prototyping* 8(1):19–50.

QUESTIONS

What is the difference between ex situ and *in situ* composites?
What are the advantages of *in situ* composites?
What are the advantages of laser-based surface modification?
Why surface modification is required for biomedical applications?
Name a few biopolymers which can be processed using additive manufacturing.
Name a suitable AM technique used for printed electronic and optoelectronic devices.
What is the mechanism for preparing porous structure using deposition-based and powder-bed-based AM technologies?
Why is a porous structure required for biomedical implants?

REFERENCES

Ahmadi, S.M., Campoli, G., Yavari, G.A., Sajadi, B., Wauthle, R., Schrooten, J., Weinans, H., Zadpoor, A.A. 2014. Mechanical behavior of regular open-cell porous biomaterials made of diamond lattice unit cells. *J. Mech. Behav. Biomed. Mater.* 34:106–115.

Ahn, B.Y., Duoss, E.B., Motala, M.J. et al. 2009. Omnidirectional printing of flexible, stretchable, and spanning silver microelectrodes. *Science* 323:1590–1593.

Athreya, S.R., Kalaitzidou, K., Das, S. 2010. Processing and characterization of a carbon black-filled electrically conductive nylon-12 nanocomposites produced by selective laser sintering. *Mater. Sci. Eng. A* 527:2637–2642.

Attar, H., Bönisch, M., Calin, M., Zhang, L.C., Scudino, S., Eckert, J. 2014. Selective laser melting of *in situ* titanium-titanium boride composites: Processing, microstructure and mechanical properties. *Acta Mater.* 76:13–22.

Attar, H., Löber, L., Funk, A., Calin, M., Zhang, L.C., Prashanth, K.G., Scudino, S., Zhang, J., Eckert, Y.S. 2015. Mechanical behavior of porous commercially pure Ti and Ti–TiB composite materials manufactured by selective laser melting. *Mater. Sci. Eng. A* 625:350–356.

Bai, J.G., Creehan, K.D., Kuhn, H.A. 2007. Inkjet printable nanosilver suspensions for enhanced sintering quality in rapid manufacturing. *Nanotechnology*. 18:185701–185705.

Balla, V.K., Bhat, A., Bose, S., Bandyopadhyay, A. 2012. Laser processed TiN reinforced Ti6Al4V composite coatings. *J. Mech. Behav. Biomed. Mater.* 6:9–20.

Balla, V.K., Bose, S., Bandyopadhyay, A. 2010. Microstructure and wear properties of laser deposited WC–12%Co composites, *Mater Sci. Eng. A* 527(24–25):6677–6682.

Balla, V.K., DeVasConCellos, P.D., Xue, W., Bose, S., Bandyopadhyay, A. 2009. Fabrication of compositionally and structurally graded Ti–TiO_2 structures using laser engineered net shaping (LENS), *Acta Biomaterialia*. 5(5):1831–1837.

Balla, V.K., Bandyopadhyay, P.P., Bose, S., Bandyopadhyay, A. 2007. Compositionally graded yttria-stabilized zirconia coating on stainless steel using laser engineered net shaping (LENS™). *Scripta Mater.* 57:861–864.

Balla, V.K., Bodhak, S., Bose, S., Bandyopadhyay, A. 2010a. Porous tantalum structures for bone implants: Fabrication, mechanical and in vitro biological properties. *Acta Biomaterialia* 6(8):3349–3359.

Balla, V.K., Bose, S., Bandyopadhyay, A. 2010b. Understanding compressive deformation in porous titanium. *Philos. Mag.* 90(22):3081–3094.

Bandyopadhyay, A., España, F.A., Balla, V.K., Bose, S., Ohgami, Y., Davies, N.M. 2010. Influence of porosity on mechanical properties and in vivo response of Ti6Al4V implants. *Acta Biomaterialia* 6(4):1640–1648.

Bandyopadhyay, A., Krishna, B.V., Xue, W., Bose, S. 2009. Application of laser engineered net shaping (LENS) to manufacture porous and functionally graded structures for load bearing implants. *J. Mater. Sci.: Mater. Med.* 20(S1):S29–S34.

Bandyopadhyay, A., Traxel, K.D. 2018. Invited review article: Metal-additive manufacturing: Modeling strategies for application-optimized designs. *Addit. Manuf.* 22:758–774.

Bandyopadhyay, P.P., Balla, V.K., Bose, S., Bandyopadhyay, A. 2007. Compositionally graded aluminum oxide coatings on stainless steel using laser processing. *J. Am. Ceram. Soc.* 90(7):1989–1991.

Banerjee, R., Collins, P.C., Bhattacharyya, D., Banerjee, S., Fraser, H.L. 2003. Microstructural evolution in laser deposited compositionally graded α/β titanium-vanadium alloys. *Acta Mater.* 51:3277.

Banerjee, R., Collins, P.C., Genc¸, A., Fraser, H.L. 2003a. Direct laser deposition of *in situ* Ti-6Al-4V-TiB composites. *Mater. Sci. Eng. A* 358:343–349.

Banerjee, R., Genc, A., Collins, P.C., Fraser, H.L. 2005. Nanoscale TiB precipitates in laser deposited Ti-Matrix composites. *Scripta Mater.* 53:1433–1437.

Berretta, S., Davies, R., Shyng, Y.T., Wang, Y., Ghita, O. 2017. Fused deposition modelling of high temperature polymers: Exploring CNT PEEK composites. *Polym. Test.* 63:251–262.

Bhat, A., Balla, V.K., Bysakh, S., Basu, D., Bose, S., Bandyopadhyay, A. 2011. Carbon nanotube reinforced Cu–10Sn alloy composites: Mechanical and thermal properties. *Mater. Sci. Eng. A* 528(22–23):6727–6732.

Bose, S., Banerjee, D., Shivaram, A., Tarafder, S., Bandyopadhyay, A. 2018b. Calcium phosphate coated 3D printed porous titanium with nanoscale surface modification for orthopedic and dental applications, *Mater. Des.* 151:102–112.

Bose, S., Robertson, S.F., Bandyopadhyay, A. 2018a. Surface modification of biomaterials and biomedical devices using additive manufacturing. *Acta Biomaterialia.* 66:6–22.

Brennan-Craddock, J., Brackett, D., Wildman, R., Hague, R. 2012. The design of impact absorbing structures for additive manufacture. *J. Phys.: Conference Series* 382:012042.

Campbell, T.A., Ivanova, O.S. 2013. 3D printing of multifunctional nanocomposites. *Nano Today* 8:119–120.

Caulfield, B., McHugh, P.E., Lohfeld, S. 2007. Dependence of mechanical properties of polyamide components on build parameters in the SLS process. *J. Mater. Process. Technol.* 182:477–488.

Chung, H., Das, S. 2006. Processing and properties of glass bead particulate-filled functionally graded nylon-11 composites produced by selective laser sintering. *Mat. Sci. Eng. A* 437:226–234.

Chung, H., Das, S. 2008. Functionally graded nylon-11/Silica nanocomposites produced by selective laser sintering. *Mat. Sci. Eng. A* 487:251–257.

Cohen, A., Chen, R., Frodis, U., Wu, M.-T., Folk, C. 2010. Microscale metal additive manufacturing of multi-component medical devices. *Rapid Prototyping J.* 16(3):209–215.

Collins, P.C., Banerjee, R., Banerjee, S., Fraser, H.L. 2003. Laser deposition of compositionally graded titanium–vanadium and titanium–molybdenum alloys. *Mater. Sci. Eng. A* 352(1):118–128.

Cooper, D.E., Blundell, N., Maggs, S., Gibbons, G.J. 2013. Additive layer manufacture of Inconel 625 metal matrix composites, reinforcement material evaluation. *J. Mater. Process. Technol.* 213(12):2191–2200.

Crane, N.B., Wilkes, J., Sachs, E., Allen, S.M. 2006. Improving accuracy of powder-based SFF processes by metal deposition from a nanoparticle dispersion. *Rapid Prototyping J.* 12:266–274.

Dadbakhsh, S., Hao, L. 2012. *In situ* formation of particle reinforced Al matrix composite by selective laser melting of Al/Fe_2O_3 powder mixture. *Adv. Eng. Mater.* 14: 45–48.

Dadbakhsh, S., Hao, L., Jerrard, P.G.E., Zhang, D.Z. 2012. Experimental investigation on selective laser melting behaviour and processing windows of *in situ* reacted Al/Fe_2O_3 powder mixture. *Powder Technol.* 231:112–121.

Das, M., Balla, V.K., Basu, D., Bose, S., Bandyopadhyay, A. 2010. Laser processing of SiC-particle-reinforced coating on titanium. *Scripta Mater.* 63(4):438–441.

Das, M., Balla, V.K., Basu, D., Manna, I., Kumar, T.S.S., Bandyopadhyay, A. 2012. Laser processing of *in situ* synthesized TiB–TiN-reinforced Ti6Al4V alloy coatings. *Scripta Mater.* 66(8):578–581.

Das, M., Bhattacharya, K., Dittrick, S.A. et al. 2014. *In situ* synthesized TiB–TiN reinforced Ti6Al4V alloy composite coatings: Microstructure, tribological and in-vitro biocompatibility, *J. Mech. Behav. Biomed.Mater.* 29:259–271.

Das, M., Bysakh, S., Basu, D., Kumar, T.S.S., Balla, V.K., Bose, S., Bandyopadhyay, A. 2011. Microstructure, mechanical and wear properties of laser processed SiC particle reinforced coatings on titanium, *Surf. Coat. Tech.* 205(19):4366–4373.

DeVasConCellos, P., Balla, V.K., Bose, S., Bandyopadhyay, A., Fugazzi, R., Dernell, W.S. 2012. Patient specific implants for amputation prostheses: Design, manufacture and analysis. *Vet. Comp. Orthop. Traumatol.* 25(4):286–296.

Duan, B., Wang, M., Zhou, W.Y., Cheung, W.L., Li, Z.Y., Lu, W.W. 2010. Three-dimensional nanocomposite scaffolds fabricated via selective laser sintering for bone tissue engineering. *Acta Biomater.* 6:4495–4505.

Duan, Y., Zhou, Y., Tang, Y., Li, D. 2011. Nano-TiO_2-modified photosensitive resin for RP, *Rapid Prototyping J.* 17(4):247–252.

Emmelmann, C., Sander, P., Kranz, J., Wycisk, E. 2011. Laser additive manufacturing and bionics: Redefining lightweight design. *Physics Procedia* 12:364–368.

Espalin, D., Muse, D.W., MacDonald, E., Wicker, R.B. 2014. 3D Printing multifunctionality: Structures with electronics. *Int. J. Adv. Manuf. Technol.* 72:5–8:963–978. doi:10.1007/s00170-014-5717-7.

España, F.A., Balla, V.K., Bose, S., Bandyopadhyay, A. 2010. Design and fabrication of CoCrMo based novel structures for load bearing implants using laser engineered net shaping. *Mater. Sci. Eng. C* 30(1):50–57.

Farayibi, P.K., Folkes, J.A., Clare, A.T. 2013. Laser deposition of Ti-6Al-4V wire with WC powder for functionally graded components. *Mater. Manuf. Process.* 28:514–518.

Fera, M., Macchiaroli, R., Fruggiero, F., Lambiase, A. 2018. A new perspective for production process analysis using additive manufacturing—complexity vs production volume. *Int J AdvManuf Technol.* 95:673–685.

Figliuzzi, M., Mangano, F., Mangano, C. 2012. A novel root analogue dental implant using CT scan and CAD/CAM: Selective laser melting technology. *Int. J. Oral Maxillofac. Surg.* 41:858–862.

Fitzpatrick, N., Smith, T.J., Pendegrass, C.J., Yeadon, R., Ring, M., Goodship, A.E., Blunn, G.W. 2011. Intraosseous transcutaneous amputation prosthesis (ITAP) for limb salvage in 4 dogs. *Vet. Surg.* 40(8):909–925.

Fu, Z., Schlier, L., Travitzky, N., Greil, P. 2013. Three-dimensional printing of SiSiC lattice truss structures. *Mater. Sci. Eng. A* 560:851–856.

Gåård, A., Krakhmalev, P., Bergström, J. 2006. Microstructural characterization and wear behavior of (Fe, Ni)–TiC MMC prepared by DMLS. *J. Alloys Compd.* 21:166–171.

Ghiba, M.O., Prejbeanu, R., Vermesan, D. 2010. The mechanical behavior of a mini hip endoprosthesis with a lattice structure tail. *Revista de ortopedie si traumatologie a Asociatiei de ortopedie Romano-Italiano-Spaniole* 2(18):101–104.

Ghosh, S.K., Bandyopadhyay, K., Saha, P. 2014. Development of an in-situ multi-component reinforced Al-based metal matrix composite by direct metal laser sintering technique: Optimization of process parameters. *Mater. Charact.* 93:68–78.

Ghosh, S.K., Saha, P. 2011. Crack and wear behavior of SiC particulate reinforced aluminium based metal matrix composite fabricated by direct metal laser sintering process. *Mater Des.* 32:139–145.

Ghosh, S.K., Saha, P., Kishore, S. 2010. Influence of size and volume fraction of SiC particulates on properties of ex situ reinforced Al–4.5Cu–3Mg metal matrix composite prepared by direct metal laser sintering process. *Mater. Sci. Eng. A* 527:4694–701.

Gibson, L.J., Ashby, M.F. 1982. The mechanics of three-dimensional cellular materials. *Proc. R. Soc. Lond., Series A (Mathematical and Physical Sciences)* 382:43–59.

Gibson, L.J., Ashby, M.F. 1997. *Cellular Solids: Structure and Properties.* Cambridge, UK: Cambridge University Press.

Goodridge, R.D., Shofner, M.L., Hague, R.J.M., McClelland, M., Schlea, M.R., Johnson, R.B., Tuck, C.J. 2011. Processing of a polyamide-12/carbon nanofibre composite by laser sintering. *Polym. Test.* 30:94–100.

Gopagoni, S., Hwang, J.Y., Singh, A.R.P. et al. 2011. Microstructural evolution in laser deposited nickel–titanium–carbon *in situ* metal matrix composites. *J. Alloys. Compd.* 509:1255–1260.

Griffin, E.A., Mumm, D.R., Marshall, D.B. 1996. Rapid prototyping of functional ceramic composites. *The Amer.Ceram. Soc. Bull.* 75(7):65–68.

Gu, D., Hagedorn, Y.-C., Meiners, W., Wissenbach, K., Poprawe, R. 2011. Nanocrystalline TiC reinforced Ti matrix bulk-form nanocomposites by selective laser melting (SLM): Densification, growth mechanism and wear behavior, *Compos. Sci. Technol.* 71:1612–1620.

Gu, D., Shen, Y. 2006. WC–Co particulate reinforcing Cu matrix composites produced by direct laser sintering. *Mater. Lett.* 60(29–30):3664–3668.

Gu, D., Shen, Y. 2007. Influence of reinforcement weight fraction on microstructure and properties of submicron WC–Co_p/Cu bulk MMCs prepared by direct laser sintering. *J. Alloys. Compd.* 431(1–2):112–120.

Gu, D., Shen, Y. 2009. Balling phenomena in direct laser sintering of stainless steel powder: Metallurgical mechanisms and control methods. *Mater. Des.* 30(8):2903–2910.

Gu, D., Shen, Y., Meng, G. 2009a. Growth morphologies and mechanisms of TiC grains during Selective Laser Melting of Ti–Al–C composite powder. *Mater. Lett.* 63(29): 2536–2538.

Gu, D., Shen, Y., Zhao, L., Xiao, J., Wu, P., Zhu, Y. 2007. Effect of rare earth oxide addition on microstructures of ultra-fine WC–Co particulate reinforced Cu matrix composites prepared by direct laser sintering. *Mater. Sci. Eng. A* 445–446:316–322.

Gu, D., Wang, Z., Shen, Y., Li, Q., Li, Y. 2009b. In-situ TiC particle reinforced Ti–Al matrix composites: Powder preparation by mechanical alloying and Selective Laser Melting behavior. *Appl. Surf. Sci.* 255(22): 9230–9240.

Gu, D.D., Wang, H.Q., Dai, D.H., Yuan, P.P., Meiners, W., Poprawe, R. 2015. Rapid fabrication of Al-based bulk-form nanocomposites with novel reinforcement and enhanced performance by selective laser melting. *Scripta Mater.* 96:25–28.

Guo, N., Leu, M.C. 2013. Additive manufacturing: Technology, applications and research needs, *Front. Mech. Eng.* 8(3):215–243.

Gurr, M., Thomann, Y., Nedelcu, M., Kübler, R., Könczöl, L., Mülhaupt, R. 2010. Novel acrylic nanocomposites containing in-situ formed calcium phosphate/layered silicate hybrid nanoparticles for photochemical rapid prototyping, rapid tooling and rapid manufacturing processes. *Polymer* 51:5058–5070.

Heinl, P., Müller, L., Körner, C., Singer, R.F., Müller, F.A. 2008. Cellular Ti–6Al–4V structures with interconnected macro porosity for bone implants fabricated by selective electron beam melting. *Acta Biomaterialia.* 4(5):1536–1544.

Ho, H.C.H., Cheung, W.L., Gibson, I. 2003. Morphology and properties of selective laser sintered Bisphenol A polycarbonate. *Ind. Eng. Chem. Res.* 42:1850–1862.

Hon, K.K.B., Gill, T.J. 2003. Selective laser sintering of SiC/polyamide composites. *CIRP Ann Manuf. Technol.* 52:173–176.

Hong, C., Gu, D., Dai, D. et al. 2013. Laser metal deposition of TiC/Inconel 718 composites with tailored interfacial microstructures. *Opt. Laser Technol.* 54:98–109.

Hong, C., Gu, D.D., Dai, D.H. et al. 2015. Laser additive manufacturing of ultrafine TiC particle reinforced Inconel 625 based composite parts: Tailored microstructures and enhanced performance. *Mater Sci. Eng. A* 635:118–128.

Hu, Y., Cong, W., Wang, X., Li, Y., Ning, F., Wang, H. 2018. Laser deposition-additive manufacturing of TiB-Ti composites with novel three-dimensional quasi-continuous network microstructure: Effects on strengthening and toughening. *Composites Part B* 133:91–100.

Hu, Y., Li, J. 2017. Selective laser alloying of elemental titanium and boron powder: Thermal models and experiment verification, *J. Mater. Process. Tech.* 249:426–432.

Hu,Y., Cong, W. 2018. A review on laser deposition-additive manufacturing of ceramics and ceramic reinforced metal matrix composites, Ceramics International, https://doi.org/10.1016/j.ceramint.2018.08.083

Hwang, J.Y., Neira, A., Scharf, T.W., Tiley, J., Banerjee, R. 2008. Laser-deposited carbon nanotube reinforced nickel matrix composites. *Scripta Mater.* 59(5):487–490.

Ivanova, O., Williams, C., Campbell, T. 2013. Additive manufacturing (AM) and nanotechnology: Promises and challenges. *Rapid Prototyping J.* 19(5):353–364.

Jain, P.K., Pandey, P.M., Rao, P.V.M. 2009. Selective laser sintering of clay reinforced polyamide, *Polym. Compos.* 31(4):732–743.

Joshi, D., Ravi, B. 2010. Quantifying the shape complexity of cast parts. *Comput-Aided Des Appl.* 7(5):685–700.

Kalita, S.J., Bose, S., Hosick, H.L., Bandyopadhyay, A. 2003. Development of controlled porosity polymer-ceramic composite scaffolds via fused deposition modeling. *Mater. Sci. Eng. C.* 23:611–620.

Kenzari, S., Bonina, D., Dubois, J.M., Fournée, V. 2012. Quasicrystal–polymer composites for selective laser sintering technology. *Mater. Des.* 35:691–695.

Khoda, A.K.M.B., Ozbolat, I.T., Koc, B. 2013. Spatially multi-functional porous tissue scaffold. *Procedia Eng.* 59:174–182.

Kieback, B., Neubrand, A., Riedel, H., 2003. Processing techniques for functionally graded materials. *Mater. Sci. Eng. C* 362:81–106.

Kim, J., Creasy, T.S. 2004. Selective laser sintering characteristics of nylon6/clay reinforced nanocomposite. *Polym. Test.* 23:629–636.

Klosterman, D.A., Chartoff, R.P., Osborne, N.R. et al. 1999. Development of a curved layer LOM process for monolithic ceramics and ceramic matrix composites. *Rapid. Proto. J.* 5(2):61–71.

Krishna, B.V., Bose, S., Bandyopadhyay, A. 2007. Low stiffness porous Ti structures for load bearing implants. *Acta Biomaterialia* 3(6):997–1006.

Krishna, B.V., Bose, S., Bandyopadhyay, A. 2009. Fabrication of porous NiTi shape memory alloy structures using laser engineered net shaping. *J. Biomed. Mater. Res. B: Appl. Biomater.* 89, B(2):481–490.

Krishna, B.V., Xue, W., Bose, S., Bandyopadhyay, A. 2008a. Engineered porous metals for implants. *JOM*. 60(5):45–48.

Krishna, B.V., Xue, W., Bose, S., Bandyopadhyay, A. 2008b. Functionally graded Co–Cr–Mo coating on Ti–6Al–4V alloy structures. *Acta Biomater.* 4: 697–706.

Kruth, J.-P., Levy, G., Klocke, F., Childs, T.H.C. 2007. Consolidation phenomena in laser and powder-bed based layered manufacturing, *CIRP Annals: Manufacturing Technology* 56(2):730–759,

Kumar, S. 2009. Manufacturing of WC–Co moulds using SLS machine. *J. Mater. Process. Technol.* 209(8):3840–3848.

Kumar, S., Kruth, J.-P. 2010. Composites by rapid prototyping technology. *Mater Des.* 31:850–856.

Kunce, I., Polanski, M., Bystrzycki, J. 2013. Structure and hydrogen storage properties of a high entropy ZrTiVCrFeNi alloy synthesized using Laser Engineered Net Shaping (LENS). *Int. J. Hydrog. Energy* 38:12180–12189.

Kusakabe, H., Sakamaki, T., Nihei, K., Oyama, Y., Yanagimoto, S., Ichimiya, M., Kimura, J., Toyama, Y. 2004. Osseointegration of a hydroxyapatite-coated multilayered mesh stem. *Biomaterials* 25(15):2957–2969.

Laoui, T., Froyen, L., Kruth, J.P. 2000. Effect of mechanical alloying on selective laser sintering of WC–9CO powder. *Powder Metal.* 42(3):203–205.

Leigh, S.J., Purssell, C.P., Bowen, J., Hutchins, D.A., Covington, J.A., Billson, D.R. 2011. A miniature flow sensor fabricated by micro-stereolithography employing a magnetite/acrylic nanocomposite resin. *Sensors and Actuat. A* 168:66–71.

Li, X.C., Stampfl, J., Prinz, F.B. 2000. Mechanical and thermal expansion behavior of laser deposited metal matrix composites of Invar and TiC. *Mater. Sci. Eng.: A.* 282:86–90.

Li, Y., Bai, P., Wang, Y., Hu, J., Guo, Z. 2009. Effect of TiC content on Ni/TiC composites by direct laser fabrication. *Mater. Des.* 30(4):1409–1412.

Liu, F.-H., Shen, Y.-K., Liao, Y.-S. 2011. Selective laser gelation of ceramic–matrix composites, *Composites: Part B* 42:57–61.

Liu, W., DuPont, J.N. 2003. Fabrication of functionally graded TiC/Ti composites by Laser Engineered Net Shaping. *Scr. Mater.* 48 (9):1337–1342.

Lopes, A.J., MacDonald, E., Wicker, R.B. 2012. Integrating stereolithography and direct print technologies for 3D structural electronics fabrication. *Rapid Prototyping J.* 18(2):129–143.

Lu, L., Fuh, J.Y.H., Chen, Z.D., Leong, C.C., Wong, Y.S. 2000. In-situ formation of TiC composite using selective laser melting. *Mater. Res. Bull.* 35:1555–1561.

Maeda, K., Childs, T.H.C. 2004. Laser sintering (SLS) of hard metal powders for abrasion resistant coatings. *J. Mater. Process Technol.* 149(1–3):609–615.

Martin, B., Chris, T., Ricky, W., Ian, A., Richard, H. 2016. Shape complexity and process energy consumption in electron beam melting: A case of something for nothing in additive manufacturing? *J. Ind. Ecol.* 21(S1): S157–S167.

Masood, S.H., Song, W.Q. 2004. Development of new metal/polymer materials for rapid tooling using fused deposition modeling. *Mater. Des.* 25:587–594.

Mazzoli, A.G., Moriconi, G., Pauri, M.G. 2007. Characterization of an aluminum-filled polyamide powder for applications in selective laser sintering. *Mater. Des.* 28:993–1000.

McCarthy, J.C., Bono, J.V., O'Donnel, P.J. 1997. Custom and modular components in primary total hip replacement. *Clin. Orthop.* 344:162–171.

Melchels, F.P.W., Domingos, M.A.N., Klein, T.J., Malda, J., Bartolo, P.J., Hutmacher, D.W. 2012. Additive manufacturing of tissues and organs. *Prog. Polym. Sci.* 37:1079–1104.

Melcher, R., Martins, S., Travitzky, N., Greil, P. 2006. Fabrication of Al_2O_3-based composites by indirect 3D-printing. *Mater. Lett.* 60:572–575

Merkt, S., Hinke, C., Schleifenbaum, H., Voswinckel, H. 2012. Geometric complexity analysis in an integrative technology evaluation model (ITEM) for selective laser melting (SLM). *S. Afr. J. Ind. Eng.* 23(2): 97–105.

Mohan, N., Senthil, P., Vinodh, S., Jayanth, N. 2017. A review on composite materials and process parameters optimisation for the fused deposition modelling process, *Virtual Phys Prototyp.* 12(1): 47–59. doi:10.1080/17452759.2016.1274490.

Moon, J., Caballero, A.C., Hozer, L., Chiang, Y-M., Cima, M.J. 2001. Fabrication of functionally graded reaction infiltrated SiC-Si composite by three-dimensional printing (3DP™) process, *Mat. Sci. Eng. A* 298:110–119.

Mortensen, A., Llorca, J. 2010. Metal matrix composites. *Annu. Rev. Mater. Res.* 40:243–270.

Mueller, B., Toeppel, T., Gebauer, M., Neugebauer, R. 2012. Innovative features in implants through beam melting: A new approach for additive manufacturing of endoprostheses. In *Innovative Developments in Virtual and Physical Prototyping*, Leiden, the Netherlands: CRC Press, 519–523.

Murr, L.E., Gaytan, S.M., Martinez, E., Medina, F., Wicker, R.B. 2012b. Next generation orthopaedic implants by additive manufacturing using electron beam melting. *Int. J. Polym. Mater.* 2012:245727.

Murr, L.E., Gaytan, S.M., Martinez, E., Medina, F.R., Wicker, R.B. 2012a. Fabricating functional Ti-alloy biomedical implants by additive manufacturing using electron beam melting. *J. Biotechnol. Biomaterial.* 2(3):1000131.

Nikzad, M., Masood, S.H., Sbarski, I. 2011. Thermo mechanical properties of a highly filled polymeric composites for fused deposition modeling. *Mater. Des.* 32:3448–3456.

Ovidiu, G.M., Mirela, T., Radu, P., Dinu, V. 2010. Influence of the lattice structures on the mechanical behavior of hip endoprostheses. *Advanced Technologies for Enhancing Quality of Life* 6–11.

Parthasarathy, J., Starly, B., Raman, S. 2011. A design for the additive manufacture of functionally graded porous structures with tailored mechanical properties for biomedical applications. *J. Manuf. Process* 13:160–170.

Postiglione, G., Natale, G., Griffini, G., Levi, M., Turri, S. 2015. Conductive 3D microstructures by direct 3D printing of polymer/carbon nanotube nano composites via liquid deposition modelling. *Composites Part A: Appl. Sci. Manufacturing.* 76:110–114.

Rehme, O., Emmelmann, C. 2009. Selective laser melting of honeycombs with negative Poisson's ratio. *J. Laser Micro Nanoen.* 4(2):128–134.

Ronca, A., Ambrosio, L., Grijpma, D.W. 2013. Preparation of designed poly(D,L-lactide)/nanosized hydroxyapatite composite structures by stereolithography. *Acta Biomaterialia* 9:5989–5996.

Rosen, D.W. 2007. Computer-aided design for additive manufacturing of cellular structures. *Comput. Aided. Des. Appl.* 4(5):585–594.

Rumpf, R.C., Pazos, J., Garcia, C.R., Ochoa, L., Wicker, R. 2013. 3D printed lattices with spatially variant self-collimation. *Prog. Electromagn. Res.* 139:1–14.

Ryder, M.A., Lados, D.A., Iannacchione, G.S., Peterson, A.M. 2018. Fabrication and properties of novel polymer-metal composites using fused deposition modeling, *Compos. Sci. Technol.* 158:43–50.

Sahasrabudhe, H., Bose, S., Bandyopadhyay, A. 2018. Laser processed calcium phosphate reinforced CoCrMo for load-bearing applications: Processing and wear induced damage evaluation, *Acta Biomaterialia.* 66:118–128.

Sahoo, S., Sinha, A., Balla, V.K., Das, M. 2018. Synthesis, characterization and bioactivity of $SrTiO_3$ incorporated titanium coating. *J. Mater. Res.* 33:2087–2095.

Samuel, S., Nag, S., Scharf, T.W., Banerjee, R. 2008. Wear resistance of laser-deposited boride reinforced Ti-Nb–Zr–Ta alloy composites for orthopedic implants. *Mater. Sci. Eng. C* 28:414–420.

Sharma, R., Singh, R., Penna, R., Fraternali, F. 2018. Investigations for mechanical properties of Hap, PVC and PP based 3D porous structures obtained through biocompatible FDM filaments, *Compos. Part B-Eng.*132:237–243.

Simchi, A., Godlinski, D. 2008. Effect of SiC particles on the laser sintering of Al–7Si–0.3Mg alloy, *Scripta Materialia.* 59(2):199–202.

Singh, R., Singh, S. 2014. Development of Nylon based FDM filament for rapid tooling application. *Journal of The Institution of Engineers (India): Series C.* doi:10.1007/s40032-014-0108-2.

Sommers, A., Wang, Q., Han, X., T'Joen, C., Park, Y., Jacobi, A. 2010. Ceramics and ceramic matrix composites for heat exchangers in advanced thermal systems – A review. Applied Thermal Engineering 30(11–12):1277–1291.

Song, B., Dong, S.J., Coddet, C. 2014. Rapid *in situ* fabrication of Fe/SiC bulk nanocomposites by selective laser melting directly from a mixed powder of microsized Fe and SiC. *Scripta Mater.* 75:90–93.

Stevinson, B., Bourell, D.L., Beaman, F.F. 2008. Over-infiltration mechanisms in selective laser sintered Si/SiC performs. *Rapid Prototyping J.* 14(3):149–154.

Stoica, A. 2009. Robotic, scaffolds for tissue engineering and organ growth. *Proceedings of Advanced Technologies for Enhanced Quality of Life* 47–51.

Sun, C.-N., Gupta, M.C. 2011. Effect of laser sintering on Ti–ZrB_2 mixtures. *J. Am. Ceram. Soc.* 94(10):3282–3285.

Tan, K.H., Chua, C.K., Leong, K.F., Cheah, C.M., Cheang, P., Abu Bakar, M.S. et al. 2003. Scaffold development using selective laser sintering of polyetheretherketone–hydroxyapatite biocomposite blends. *Biomaterials.* 24:3115–3123.

Tjong, S.C. 2007. Novel nanoparticle-reinforced metal matrix composites with enhanced mechanical properties. *Adv. Eng. Mater.* 9:639–653.

Tjong, S.C., Ma, Z.Y. 2000. Microstructural and mechanical characteristics of *in situ* metal matrix composites. *Mater. Sci. Eng. R* 29:49–113.

Travitzky, N.A., Shlayen, A. 1998. Microstructure and mechanical properties of Al_2O_3/Cu-O composites fabricated by pressureless infiltration technique. *Mater. Sci. Eng. A* 244:154–160.

Vaezi, M., Black, C., Gibbs, D.M.R., Oreffo, R.O.C., Brady, M., Moshrefi-Torbati, M., Yang, S. 2016. Characterization of new PEEK/HA composites with 3D HA network fabricated by extrusion freeforming, *Molecules.* 21:687.

Vaezi, M., Seitz, H., Yang, S. 2013. A review on 3D micro-additive manufacturing technologies. *Int. J. Adv. Manuf. Technol.* 67:1721–1754.

Vaezi, M., Srisit, C., Brian, M., Shoufeng, Y. 2013. Multiple material additive manufacturing—Part 1: A review. *Virtual. Phys. Prototyp.* 8(1):19–50.

VonWilmowsky, C., Vairaktaris, E., Pohle, D., Rechtenwald, T., Lutz, R., Münstedt, H., Koller, G., Schmidt, M., Neukam, F.W., Schlegel, K.A. et al. 2008. Effects of bioactive glass and β-TCP containing three-dimensional laser sintered polyetheretherketone composites on osteoblasts in vitro. *J. Biomed. Mater. Res. Part A.* 87A:896–902.

Wahab, M.S., Dalgarno, K.W., Cochrane, R.F., Hassan, S. 2009. Development of polymer nanocomposites for rapid prototyping process. *Proceedings of the World Congress on Engineering*, Vol. II. WCE, 2009.

Walther, M., Ortner, A., Meier, H., Löffelmann, U., Smith, P.J., Korvink, J.G. 2009. Terahertz metamaterials fabricated by inkjet printing. *Appl. Phys. Lett.* 95:251107.

Wang, F., Mei, J., Wu, X. 2007. Compositionally graded Ti6Al4V + TiC made by direct laser fabrication using powder and wire. *Mater Design.* 28(7):2040–2046.

Wang, F., Mei, J., Wu, X. 2008. Direct laser fabrication of Ti6Al4V/TiB. *J. Mater. Process. Technol.* 195:321–326.

Wang, X., Tuomi, J., Mäkitie, A.A., Paloheimo, K.-S., Partanen, J., Yliperttula, M. 2013. The integrations of biomaterials and rapid prototyping techniques for intelligent manufacturing of complex organs, in *Advances in Biomaterials Science and Biomedical Applications* 437–463, Prof. R. Pignatello (Ed.), InTech, doi:10.5772/53114.

Weisensel, L., Travitzky, N., Sieber, H., Greil, P. 2004. Laminated object manufacturing (LOM) of SiSiC composites. *Adv. Eng. Mater.* 6(11):899–903.

Wilson, J.M., Shin, Y.C. 2012. Microstructure and wear properties of laser-deposited functionally graded Inconel 690 reinforced with TiC. *Surf. Coat. Tech.* 207:517–522.

Wohlers, T. 2010. Additive manufacturing a new frontier for composites. *Composites Technology*, http://www.compositesworld.com/columns/additive-manufacturing-a-new-frontier-for-composites (Accessed May 22, 2014).

Xia, Y., Zhou, P., Cheng, X., Xie, Y., Liang, C., Li, C., Xu, S. 2013. Selective laser sintering fabrication of nano-hydroxyapatite/poly-ε-caprolactone scaffolds for bone tissue engineering applications. *Int. J. Nanomedicine.* 8:4197.

Xiong, Y., Smugeresky, J.E., Ajdelsztajn, L., Schoenung, J.M. 2008. Fabrication of WC–Co cermets by laser engineered net shaping. *Mater. Sci. Eng. A* 493(1–2):261–266.

Xue, W., Krishna, B.V., Bandyopadhyay, A., Bose, S. 2007. Processing and biocompatibility evaluation of laser processed porous titanium. *Acta Biomaterialia* 3(6):1007–1018.

Yakovlev, A., Trunova, E., Grevey, D., Pilloz, M., Smurov, I., 2005. Laser-assisted direct manufacturing of functionally graded 3D objects. *Surf. Coat. Tech.* 190:15–24.

Yan, C., Hao, L., Hussein, A., Bub, S.L., Young, P., Raymont, S. 2014b. Evaluation of light-weight AlSi10Mg periodic cellular lattice structures fabricated via direct metal laser sintering. *J. Mater. Process Technol.* 214:856–864.

Yan, C., Hao, L., Hussein, A., Raymont, D. 2012. Evaluations of cellular lattice structures manufactured using selective laser melting. *Int. J. Mach. Tool. Manu.* 62:32–38.

Yan, C., Hao, L., Hussein, A., Young, P., Raymont, D. 2014a. Advanced light weight 316L stainless steel cellular lattice structures fabricated via selective laser melting. *Mater Design.* 55:533–541.

Yan, C., Hao, L., Xu, L., Shi, Y. 2011. Preparation, characterisation and processing of carbon fibre/polyamide-12 composites for selective laser sintering. *Compos. Sci. Technol.* 71:1834–1841.

Yan, C.-Z., Shi, Y.-S., Yang, J.-S., Liu, J. 2009. Nanosilica/Nylon-12 composite powder for selective laser sintering, *J. Reinf. Plast. Comp.* 28/23:2889–2902.

Yan, M., Tian, X., Peng, G., Li, D., Zhang, X. 2018. High temperature rheological behavior and sintering kinetics of CF/PEEK composites during selective laser sintering. *Compos. Sci. Technol.* 165:140–147.

Yang, J.S., Shi, Y.S., Shen, Q.W., Yan, C.Z. 2009. Selective laser sintering of HIPS and investment casting technology. *J. Mater. Process Technol.* 209:1901–1908.

Yang, L., Cormier, D., West, H., Harrysson, O., Knowlson, K. 2012b. Non-stochastic Ti–6Al–4V foam structures with negative Poisson's ratio. *Mater. Sci. Eng. C* 558:579–585.

Yang, L., Harrysson, O., West, H., Cormier, D. 2012a. Compressive properties of Ti–6Al–4V auxetic mesh structures made by electron beam melting. *Acta Mater.* 60:3370–3379.

Yin, X.W., Travitzky, N., Greil, P. 2007. Three-dimensional printing of nanolaminated Ti_3AlC_2 toughened $TiAl_3$-Al_2O_3 composites. *J. Am. Ceram. Soc.* 90(7):2128–2134.

Yin, X.W., Travitzky, N., Melcher, R., Greil, P. 2006. Three-dimensional printing of $TiAl_3$/Al_2O_3 composites. *Int. J. Appl. Ceram. Technol.* 97(5), 492–498.

Zhang, W., Melcher, R., Travitzky, N., Bordia, R.K., Greil, P. 2009. Three-dimensional printing of complex shaped alumina/glass composites. *Adv. Eng. Mater.* 11(12):1039–1043.

Zhang, W., Travitzky, N., Greil, P. 2008. Formation of $NbAl_3$/Al_2O_3 composites by pressureless reactive infiltration. *J. Am. Ceram. Soc.* 91(9):3117–3120.

Zhang, Y., Hao, L., Savalani, M.M., Harris, R.A., Tanner, K.E. 2008. Characterization and dynamic mechanical analysis of selective laser sintered hydroxyapatite-filled polymeric composites. *J. Biomed. Mater. Res. Part A* 86A:607–616.

Zhang, Y., Wei, Z., Shi, L., Xi, M. 2008. Characterization of laser powder deposited Ti–TiC composites and functional gradient materials. *J. Mater. Process Technol.* 206:438–444.

Zhang, Y.Z., Sun, J.C., Vilar, R. 2011. Characterization of (TiB+TiC)/TC4 *in situ* titanium matrix composites prepared by laser direct deposition. *J. Mater. Process. Technol.* 211:597–601.

Zheng, B.L., Topping, T., Smugeresky, J.E., Zhou, Y., Biswas, A., Baker, D., Lavernia, E.J. 2010. The Influence of Ni-coated TiC on Laser-deposited IN625 metal matrix composites. *Metall. Mater. Trans. A* 41:568–573.

Zheng, H., Zhang, J., Lu, S., Wang, G., Xu, Z. 2006. Effect of core-shell composite particles on the sintering behaviour and properties of nano-Al_2O_3/polysterene composites prepared by SLS, *Mater. Lett.* 60 (9–10):1219–1223.

Zhong, W., Li, F., Zhang, Z., Song, L., Li, Z. 2001. Short fiber reinforced composites for fused deposition modeling. *Mater. Sci. Eng. A* 301(2):125–130.

13 Personalized Implants and Additive Manufacturing

Mukesh Kumar and Bryan Morrison

CONTENTS

13.1 INTRODUCTION

Additive manufacturing has opened a new manufacturing path to design and manufacture personalized products—this chapter focuses on one such segment of personalized products, namely, medical implants and associated instruments, with greater emphasis on implants for large joints, although great strides have been made and clinically implemented in the field of dentistry. The field of medical implants is not new—implants such as prostheses for joint replacements have been around for a long time. To some extent, some of these have already been personalized using traditional manufacturing methods. For example, the market can already boast of patient-matched implants, although limited to select patients with unusual boney defects or anomalies due to disease, functional deformity, or tumor resection. In these cases, a computed tomography (CT) or magnetic resonance imaging (MRI) scan of the patient is used to assess the dimensions of the bone defect, and a physical bone model of the defect is created. Using this physical model and the patient's CT or MRI scan, a personalized implant is designed and eventually machined using conventional methods. Normally, the machined implant and the physical bone model are shipped to the hospital where the surgeon can use the bone model to help determine the best placement of the implant. While clinically successful, this path to manufacture implant can be time consuming and involves a lot of waste of raw material. Moreover to some extent, the surgical outcome is dependent on the ability of the surgeon to place the implant at the correct location and with the correct orientation, using the provided plastic bone model as a preoperative visual and tactile aid. Additionally, if possible, but perhaps as important as the implant itself, making use of additive manufacturing could better facilitate the design and manufacture of patient-specific instruments and surgical guides that could help the surgeon position the implant accurately.

Recognizing that the field—and regulations—of additive manufacturing is still developing, the intention of this chapter is to expose the reader to the various nuances of medical implants, specifically those related to orthopedics and considerations of various aspects involved in the personalization and manufacture of such devices using additive manufacturing. The chapter closes with some thoughts on the future of additive manufacturing technology specifically as it relates to personalized medical implants and instruments.

13.2 PATH TO CLINICAL USE

Prior to clinical use, implantable devices must clear regulatory approval processes that vary by country. The following is a very simplified explanation of medical device regulations and how they could affect personalized medical products. The reader must realize that it takes time to gather information, understand implications of various aspects of new technology, and build consensus between researcher, corporations, and regulatory bodies—regulations almost always tend to follow technology. The reader is cautioned that medical device regulations are vast, complicated, and vary between countries. Additionally, the best place to obtain the latest on regulations, guidance documents, and standards on medical devices is the official Food and Drug Administration (FDA—United States), European Commission, and ISO (International Standardization Organization) websites.

In the United States, medical devices are classified according to the level of risk to patients. Generally, class I devices include tongue depressors; however, orthopedic implants fall under either class II or class III based on the level of risk and potential unknowns. Class II devices require the design holder to demonstrate that the proposed device under scrutiny is *substantially equivalent to* previously cleared device. If the proposed device is *substantially equivalent to a previously cleared device*, the characteristics of the device are relatively well known and thus the level of risk is known and risk mitigation procedures are well understood. At least for orthopedic implants that are not similar to previously cleared devices—and are thus associated with a higher level of risk—a separate classification is used. This class III category has a regulatory path that is extremely challenging and time consuming, sometimes running in years.

Among other things, regulations pose a challenge for the device design holder in adopting the use of additive manufacturing in making patient-specific instruments and devices—how to obtain regulatory approval for personalized implants for patients that generally perform the same function but are potentially shaped different to meet the anatomical and possibly the biomechanical requirements of a particular patient? This regulatory requirement becomes more complicated, as the quest for personalization involves greater complexity—it is conceivable that medical devices can be tailored to the patient's anatomy, bone density, cancellous bone pore structure, and possibly coating with specific antibiotic, peptides, and other personalized biomolecules. Advent of additive manufacturing technology definitely brings close the possibility of making implants that not only match the anatomy of the patient, but can also be used to manufacture implants with density that changes with location, thus effecting mechanical properties and thus the biomechanics of the implant reconstructed site. Implants could be manufactured with variations in porous ingrowth surfaces to achieve varied amounts of biological fixation. One can argue as to the need for such variations—but the fact remains that current state of additive manufacturing technology can deliver such implants. Unfortunately, at the time of writing this chapter, existing regulatory framework is not well tailored to accommodate the potential of this technology. Regulators realize the potential of additive manufacturing and its capability to create personalized implants. Robust and well-crafted regulations will eventually follow. However, a logical approach could be to lay down regulations for *low-risk products* and gradually expand regulations to include higher risk products. It is for this reason that personalized instruments and fixtures to help correctly place implants are currently available, whereas higher risk personalized implants are under development waiting for regulations. Until then, answering the question *what can additive manufacturing techniques make that can be safely used in the operating room (OR) with minimal risk* will help determine what products can be brought to the market with minimal risk of regulatory rejection. Of course, these devices will still have to meet all existing applicable standards and requirements of the FDA and foreign regulatory agencies to satisfy requirements for sale in those foreign countries.

13.3 SOFTWARE FOR TEMPLATES: IMPLANT SIZING

In the OR, other than the actual implant, the surgeon makes use of instruments. Prior to surgery, surgeons use a patient-specific X-ray to determine the correct implant size to be used in surgery. The use of an implant overlay on a patient-specific X-ray with implant shape and size information to determine the best fit for the patient is called templating. Templates are device specific and provided by the implant manufacturer, with the intent of helping decide the optimal size of the chosen implant. Additionally, templates help doctors as they determine surgical cut placement and where to position the implant. To some extent, templating is the first step in identifying a personalized product for each patient.

While almost a decade ago, film X-ray was commonly used, many hospitals have moved to digital X-ray systems. Current practice to determine the correct size of implant needed for a particular patient is based on analysis of patient X-ray and implant profile. This is done with proprietary software, currently available from numerous vendors who have gone through the 510 K regulatory clearance process as part of a picture archiving and communication system. An X-ray is presented on the screen and calibrated based on a known scaling object. This scaling object could vary in sophistication so much so that in some instances, even a quarter can be used as a marker. We will refer to this aspect later on when describing the various imaging modalities. Obviously, one universal scaling object would be optimal, but generally X-ray technicians use what they have. If no marker or scaling object is seen in the X-ray, some software tools assume that the X-ray was made at 115% scale. This allows for the template to be scaled and placed in the view and moved in to a proper surgical position. Through this procedure, a template is chosen from those available to provide the best fit, as determined by the operating surgeon. This specific template relates to an implant size for that patient.

13.4 DENTAL INDUSTRY: EXAMPLE OF MASS MANUFACTURED PERSONAL PRODUCTS

Perhaps the most visible advancement in personalized product is the use of digital manufacturing in dentistry. As recent as a decade ago, it was common to make use of bite impressions on a malleable or setting plastic to generate a negative model of the patient's mouth. This was then used to cast a hard model to generate a positive and true impression of the mouth and thus the missing/defective teeth. Using this information, the dentists would prescribe a restorative tooth that was manufactured by milling a stock material in a dental lab. This was sent back to the dentist to evaluate fit. This workflow required the patient to meet with a dentist a few times. Adding to this inconvenience, due to the inherent shrinkage of setting plastics and hardening plaster, errors in the final part were quite common. While a lot of development work ensued to find malleable setting materials that did not deform or shrink during curing, the field of dentistry saw a sea of change with the advent of a camera wand that could take images of the patient mouth and software that could be used to delineate the defect on the image of the tooth. To elaborate, using a camera system, multiple images of the patient's teeth and mouth are captured. Using specialized software, these images are stitched to create a three-dimensional rendition of the patient's mouth and teeth. On the captured digital image, the dentist demarcates the defect. The software then creates a three-dimensional volume of the defect, taking into account the shape and geometry of the mating teeth to ensure proper bite. The software generates a computer-aided design (CAD) file that is sent to an additive manufacturing machine that prints the implant while the patient waits. The surgeon then implants this additively manufactured personalized product. Comparing the present day workflow to the established workflow from just a decade ago shows the advantage of digital manufacturing—the patient goes home the same day and the quality of the fit of the implant is far superior as the usage of error causing impression materials has been completely eliminated.

13.5 ADDITIVE MANUFACTURED PATIENT-MATCHED SURGICAL GUIDES AND BONE MODELS

A recent push in orthopedics is to ensure that not only the correct implant size is obtained using the templating method described above, but there is growing awareness that placement of the device is critical to restore the biomechanical alignment. In the previous decades and to a large extent even today, the preparation of the surgical site involved the experience of a surgeon to visualize the bony anatomy and prepare the bone bed to achieve the proper orientation to place the implant based on experience and available two-dimensional X-ray films. To appreciate the complexity of this critical step, one must imagine looking at and identifying bony landmarks through a small surgical cut of the intervening soft tissue. To exacerbate the complexity, imagine doing this in a bloody and potentially bleeding environment. It was recognized that if there was a way to use the information on CT or MRI scans, to correctly identify a starting reference plane or line, it would be possible to determine the exact location of the various mechanical axes of the defective site. Further, the availability of a tool in the OR that held onto the bone and guided other tools could help the surgeon quickly and accurately make the necessary surgical cuts on the bony site to prepare for the implant. The recognition of this concept gave birth to the idea of using patient-specific guides to help in orthopedic implant placement.

To elaborate on how patient-specific guides are created, after digital data from CT or MRI of the damaged bone site and possibly other anatomical locations are obtained, the guide design engineering team works with the orthopedic surgeon to identify bony landmarks. Based on these landmarks, the biomechanical axis that the surgeon would want to restore and use in the correct placement of the implant is identified. The operating surgeon approves of this surgical plan, and using this information, a CAD file of a surgical guide is generated. To help the surgeon visualize the damaged bone site, a CAD file of the patient's existing bone structure is also created. In a production environment, where such guides are being manufactured for many patients, one possible method to ensure that there is no mix up of guides during manufacturing is to place unique identifiers on the CAD files, respecting patient privacy requirements. As the details of bony anatomy vary from patient to patient, it is impractical to write machine codes to turn and mill these guides from bar stock material. These CAD files are converted to stereolithography (STL) files that are used by additive manufacturing machines to make a patient-specific bone model and implant guide. After the making of these bone model and guides, the parts are removed from the machine, cleaned, and undergo quality checks. There can be many different kinds of quality checks but the most common is to ensure dimensional accuracy. The most common method employed is digital scanning where structured light is used to scan the surface of the additive manufactured part. This results in another CAD file that is compared to the original CAD file. These two CAD files must match within the specified tolerances for the guide to be considered to be acceptable for surgical use.

During surgery, the surgeon makes incisions to expose the bone site and uses the guide to make the cuts on the bone or determine the precise location to place standard instruments. If one has chosen a cut-through guide, once the cuts are made, the guide is removed and the implant is placed and secured per the normal surgical procedure. The use of the patient-specific surgical guide enables the surgeon better precision in implant placement. As the bone model and guide is patient specific, after the surgical procedure, these are disposed of as biological hazard materials. A few examples of commercial guides that are being used clinically are (1) Zimmer—Patient Specific Instruments, (2) Biomet—Signature, (3) Depuy—TruMatch, and (4) Conformis—iJig.

As the patient-specific guide is to be placed on the open surgical site, these patient-specific bone models and guides must meet certain regulatory requirements as listed below to ensure biological safety and effectiveness.

1. The material must be biocompatible for short-term exposure meeting the requirements of International Standard ISO–10993, "Biological Evaluation of Medical Devices Part 1: Evaluation and Testing."
2. Mechanical strength requirements as dictated by the surgical procedure to ensure that there is no mechanical failure of the guide during surgery. As these bone models and guides are generally not load bearing, this mechanical strength requirement maybe limited to demonstrating that the construct is strong enough to withstand forces encountered in normal surgical procedures.

3. And depending on the method of delivery to the OR, one or more of the below must be satisfied to ensure the product must meet shipping requirements and is sterile prior to clinical use.
   a. Packaging system integrity testing may include the following:
      i. Package integrity (ASTM F2096: Bubble Test)
      ii. Seal integrity (ASTM F1886: Visual Inspection, ASTM 1929: Dye Test)
      iii. Seal strength (ASTM F88: Peel Test, ASTM F1140: Burst Test)
      iv. Packaging system performance testing/distribution simulation (International Safe Transit Association [ISTA]) procedures
      v. ISO-11607 Packaging for terminally sterilized medical device
   b. Sterility
      i. ANSI/AAMI/ISO: 11137 (Sterilization of health care products—radiation)
      ii. ISO 17665 Steam sterilization for medical devices
      iii. ISO 11135 EtO sterilization for medical devices

Depending on the clinical use of the guides, there may be many other standards and specifications that must be met. The reader is advised to refer to the FDA and ISO websites for more up-to-date guidance or consult a regulatory specialist.

13.6 ADDITIVE MANUFACTURED GENERIC PRODUCT

To understand the impetus of additive manufacturing in the orthopedic industry, it is necessary to consider the evolution of bone ingrowth surfaces. Orthopedic implants make use of metallic systems (Ti6Al4V, CoCrMo to name a few common alloys) that allow for integration with surrounding bone (osseointegration) due to the roughened or porous structure on the surface of the metallic implant. Traditional manufacturing methods include sintering of beads or spraying powder particles to generate a porous structure. However, our understanding of the osseointegrating surface has advanced where there is push to make the osseointegrating layer more porous and possibly more biomimetic by making the structure more similar to cancellous bone. The thought being that the presence of large pore volume would provide ample bone integration space. There is a growing clinical need to achieve osseointegration even where bone loss is severe, bone quality is poor, and only focal contacts between implant and host bone are possible. This was evident by the introduction of and with the clinical success of Trabecular Metal™ (Zimmer, Indiana), Tritanium™ (Stryker, Michigan), and Regenerex™ (Biomet, Indiana), among others. The manufacturing methods for these modern porous structures include common processes in powder metallurgy and associated sintering, physical vapor deposition, and chemical vapor deposition. Although the initial clinical use of such porous structures is generally considered clinically successful, the resulting structures were not truly biomimetic. In the quest to make the porous structure biomimetic, additive manufacturing methods are being considered. Further, the advantage would be to make the bone integration layer of variable porosity, modulus, and pore structure to mimic the bone structure of the natural anatomy.

A generic production route may include the following:

1. Generation of the CAD files including that of the solid and porous region
2. Populating the porous region with the details of pore structure, making it as biomimetic as possible
3. Feeding this information to additive manufacturing machines to produce parts that, at least for Ti6Al4V alloy, meet the chemical and mechanical requirements listed in ASTM F2924-14, ASTM F3001-14, and FDA guidance documents (Guidance for industry and for FDA reviewers/staff—Guidance for industry on the testing of metallic plasma sprayed coatings on orthopedic implants to support reconsideration of postmarket surveillance requirements)
4. Machining mating surfaces to ensure that mating components such as ultrahigh molecular weight polyethylene (UHMWPE) fit precisely and the possibility of the inevitable micromotion between the mating parts do not generate polyethylene debris
5. Cleaning the additive manufactured parts of entrapped metal powder and machining additives such as coolant (ASTM F2847)
6. Passivation to meet ASTM A 967-13
7. Final packaging and terminal sterilization (meeting requirements listed in the Section 11.5)

As mentioned previously, substantial regulations dictate risk mitigation on the above workflow. However, at the time of writing this chapter, various underlying regulations are being discussed. The reader is advised to follow FDA-sponsored workshops to get a better perspective on existing and upcoming regulations. Additionally, the reader is advised to refer to the ISO for more up-to-date regulatory guidance.

13.7 ADDITIVE MANUFACTURED PATIENT-MATCHED IMPLANTS

The above workflow does not necessarily create a personalized product or patient-specific implant but can be adapted to create personalized implant utilizing additive manufacturing. To elaborate on the manufacturing route of a patient-specific implant and guide, let us consider the making of components needed for high tibial osteotomy (HTO) surgery. This is a corrective surgery normally used on patients to correct instability due to misalignment of the tibial plateau to the femoral condyles, without compromising or violating the cartilage or menisci of the knee. Patient-specific information is needed to determine the extent of malalignment—so CT or MRI data are required. Based on this information, a surgeon can plan on making a slot in the region inferior to the tibial plateau. This slotted region will receive the implant—thus pushing superior the tibial plateau to correct the misalignment. The design team in collaboration with the operating surgeon would then design the HTO guides that allow the surgeon to position, orient, and make the necessary cuts in the region of the tibia. At the same time, an implant wedge is designed such that the cortical wall of the wedge seamlessly mates with the cortical bone of the receiving bone tissue while correcting the malalignment. Based on the approved design of the wedge, a Ti6Al4V implant, nylon guide, and nylon trial are manufactured. As these components are patient specific, the CAD files could contain patient-specific code identifiers that indicate components are for a particular patient, thus avoiding mix-up with components for other patients. The implant is made per the workflow listing described above, and the guides and trials are made per the procedure described in the earlier section. During surgery, the guide is used to prepare the implant receiving bone bed. Finally, the implant is placed inside the wedgeshaped cavity following normal surgical protocol for HTO. As the guide and trials are patient specific, they are discarded following surgery.

Another example of a personalized product involves flanged acetabular components (i.e., triflange). Essentially this implant looks like an acetabular shell with three continuous flanges designed to mate with the ilium, pubis, and ischium of the pelvis. Surgically restoring function of the hip joint, where there is acetabular bone loss including pelvis discontinuities, presents a challenging situation for the operating surgeon. The problem of where to place the implant and achieve stability and restore functionality is compounded by the fact that there is little bone left in the pelvis to make use of a standard acetabular shell. Currently, a potential treatment option is the use of structural allografts, which may not always be available or may not be viable scaffold for bone integration. While allografts are generally considered safe, there is still an existing risk of disease transmission from the donor to the recipient. The surgeon often has little choice but to order a patient-specific triflange acetabular component. Using CT scan information of the patient, the sizes of the three flanges, their shapes, and orientation are designed to fit securely with the remaining bones. The orientation of the acetabular cup portion of the triflange implant is established and multiple screw holes are added into the design to accommodate retaining screws. Physical models of the patient's bone and the proposed implant are sent to the operating surgeon for approval. Upon surgeon's approval of the proposed implant, these are shipped back to the manufacturer and serve as a manufacturing tool to aid the manufacturing of the implant. As of now, most such triflanges are milled using traditional manufacturing methods. Subsequently, a porous surface is applied to allow for biologic fixation. Naturally, this process takes time. Additive manufacturing is poised to change this—as with traditional triflange workflow, the new method too will require surgeon's input in implant design and final approval of the implant. However, the manufacturing method will be streamlined where the body of the implant and porous structure are additively manufactured. This is expected to save a lot of wasted material and cut down on manufacturing time as the porous structure will be generated concurrently.

13.8 HARD TISSUE REPLACEMENT FOR CRANIAL RECONSTRUCTION

Perhaps the first patient-matched implant approved for clinical use and made by additive manufacturing technology is OsteoFab™ Patient Specific Cranial Device (Oxford Performance Material, CT). While the clinical device was not unique, the use of additive manufacturing technology made the work flow efficient. Additionally, this technology makes use of poly-ether ketone material that is alleged to be more biointeractive compared to the traditional material (poly methyl methacrylate, PMMA). The traditional method and the additive manufacturing method make use of patient CT data. In the older technology, the CT data are used to create a mold where PMMA beads are cast and chemically sintered. In the newer technology, the CT data are used to generate an STL file and an additively manufactured part is generated.

13.9 MANUFACTURING COST: IS ADDITIVE MANUFACTURING A VIABLE TECHNOLOGY?

In spite of the advantages apparent from the descriptions above, currently there are few major disadvantages that limit the wide scale adoption of additive manufacturing to make implants. The first is the prohibitive cost of additive manufacturing machines. At almost three to four times the expense of a generic mill, the capital cost of starting a manufacturing line comprised of additive manufacturing machines becomes steep. However, as with any new technology, the cost of machines is decreasing. There is substantial competition between machine manufacturers—this healthy competition and the rapid adoption in the aerospace, automotive, medical, and other industries are helping to make these machines feature-rich while driving the price down.

The second disadvantage is the speed of build. Additive manufacturing technology has come a long way in achieving better dimensional accuracy and surface finish. But even today it still takes a long time to make an average sized acetabular shell implant—with some machines, this is measured in hours. In comparison, the time to machine a similar sized acetabular shell is just about 30 minutes and requires machines that are half to a third as expensive. Naturally, there is a camp of manufacturing engineers that highlights the build speed as a deterrent in adopting this technology. However, the argument of build speed as the sole measure of cost of the implant is flawed. One must remember it is the sum of the cost and time of all operational steps that must be considered in estimating or calculating the cost of the implant. In additive manufacturing machines, the solid and the contiguous porous structure is printed concurrently. Therefore, the time to build the porous acetabular shell in additive manufacturing machines covers the time and therefore the cost of building the porous structure as well. In the build time for a generic porous shell via the traditional machining route, one must include the time necessary for post-machining operations such as masking, blasting, and applying a porous structure either via thermal spraying or sintering with beads. Further, as these operations are staged and progress in batches, there is inventory carrying cost as well. Additionally, there is the lead time to deliver the order to market that is quite long with manufacturing via traditional routes.

This difference in work flow for additive manufacturing and traditional machining is explained in detail in Table 13.1.

Recognizing that the cost structure of manufacturing facilities is different, it is imperative that a thorough cost calculation be conducted to ensure that the implant

TABLE 13.1
Difference in Work Flow for Additive Manufacturing and Traditional Machining

Traditional Manufacturing Route	Additive Manufacturing Route	Comments
Starting material ASTM F136 bar stock	Starting material ASTM F1584 powder	The cost of powder is generally much higher than the cost of bar stock. However, there is very little material waste in the additive manufacturing route.
Machine inner and outer diameter	Feed CAD files to additive manufacturing machine and build parts	Time to build via AM is longer but this time includes the concurrent building of the porous structure.
Blast and clean outer diameter to prepare surface for thermal spray	Machine ID	
Mask areas that cannot receive thermal spray		
Thermal spray		
Clean and passivate	Clean and passivate	Cleaning procedure must include steps to remove residual powder from porous structure.
Package	Package	

being considered for manufacture via additive manufacturing is economically viable. It is the authors' experience that a combination of design features not possible to achieve via traditional route and the judicious selection of implants that make economic sense usually draws a backing from decision makers.

Though not associated with cost of parts, nevertheless there is yet another deterrent to the adoption of the additive manufacturing technology. This is a perception among engineers that the printed material is inferior. For the uninitiated in additive manufacturing, this is perhaps natural—after all, the technology makes use of liquid or powder to make a solid part. This concept conjures the possibility of flaws in the parts. Interestingly, the two currently available ASTM standards on Ti6Al4V parts made by additive manufacturing technology specify the minimum strength requirements—these are about the same as that of wrought material. Therefore, processes that have been validated to the ASTM standards ensure strength and counter the belief that additive manufactured parts are not sufficiently strong. Another limitation, which perhaps is most crippling for now, is the lack of standards. While there are now two ASTM standards for Ti6Al4V alloys, there are no standards for CoCrMo alloy. At the time of writing this chapter, the ISO and ASTM committee have recognized these limitations and have signed a Partner Standards Developing Organization cooperative agreement to govern the ongoing collaborative efforts between ASTM

International Committee F42 on Additive Manufacturing Technologies and ISO Technical Committee 261 on Additive Manufacturing. The issue of perception of inherent weakness possibly results from earlier and perhaps overzealous attempts by machine manufacturers to sell their technology to implant original equipment manufacturers (OEMs). The earlier versions of the machines were not robust enough resulting in residual porosity or residual unmelted (or partially melted) powder particles. In the last few years, these issues have been corrected by better managing build speed and expectations. Moreover, there is ongoing work on thermal imagining of the build layers where the build layers are thermally imaged for residual porosity or residual unmelted (or partially melted) powder particles. These technologies are still in their infancy and may take time to mature. In the meantime, OEMs must adapt quality control test system similar to that employed by casting facilities. To this end, nondestructive testing methods such as X-ray imaging, ultrasound, and CT scanning would prove most beneficial.

13.10 ROLE OF IMAGING HUMAN ANATOMY

Custom implants and surgical guides require patient-specific anatomy. Without such data, the immense power of additive manufacturing is worthless as one of the critical inputs in making parts from such technology is STL data, which can only be generated from the 3D medical images. It is of value to note that the accuracy and resolution of existing 3D medical images are magnitudes less than what modern day additive manufacturing machines can deliver. To obtain this patient-specific anatomy, medical scanning devices are utilized. There are four primary modalities that are used to obtain these data. The most common 3D data sources are CT followed by MRI and the most recent ultrasound and X-ray. For CT and MRI, 2D stacked image slices are individually segmented or masked and combined to form a 3D shape. X-ray has been utilized as a 2D templating source and input to make custom sized standard implants. In Sections 13.11.3 and 13.11.4, new uses of ultrasound and X-ray will be discussed. These new uses are typically being utilized as alternatives to their more expensive medical scan counterparts in universities and other research and development environments.

13.11 MOST COMMON MODALITIES

13.11.1 Computed Tomography

CT is an X-ray-based technology for scanning the body in 3D. These scanners create images in Hounsfield units and are directly related to density of the scanned material. Air has a value of –1000 while bone can range from 700 to 3000, based on the quality of the patient's bone. These scans are relatively fast to generate and are the easiest to reconstruct for orthopedics. There is no soft tissue detail and only bone or calcified ligaments and tendons are visible. They do however expose the patient to radiation.

13.11.2 Magnetic Resonance Imaging

MRI is less dangerous in terms of radiation for patients than X-ray-based scans and provides details on soft tissues. These scanners create images in generic units that can only be compared to details in that current scan. There are methods to calibrate on a scan-by-scan basis but these are time intensive. The resolution of MRI scans is eight times less than CT and takes 10–30 times as long to capture and create the data set. This pushes the boundaries on clinically relevant scans and the patient's ability to sit still long enough to get a good scan. At least for orthopedics, MRIs do provide the most relevant data including that on cartilage and soft tissue. However, these data are much less homogenous and require much more human intervention to determine the 3D shape. There is clinical risk with MRI, but it is closely monitored by specific absorption rate, which is the amount of energy absorbed by the body. All scanners have fail safes to ensure the limit is not exceeded and this can lead to extending the scanning time even longer. A prohibitive aspect of using MRI is the associated high cost. The other modalities do not have issues of volumetric distortion in the 3D image—MRI is the only imaging technique that is encumbered with this issue. If MRI is being used to generate patient-specific CAD data for additive manufacturing, extreme care must be exercised to ensure that these distortions are minimized.

13.11.3 Ultrasound

Ultrasound can be combined with motion capture to create 3D shapes of bones in large point clouds. The 3D shapes can be very accurate, but may have gaps due to other bones or limitations to the motion capture system. This then requires multiple 3D shapes to be registered together in order to have a solid bone.

13.11.4 X-Ray

So far the data generated by the three previous modalities are actual patient three-dimensional data—X-ray image is an image on a single plane with no or little useful information on the third dimension. With X-ray, statistical shape modeling can be utilized to predict patient anatomy from one or more X-rays. Of course, the user must have access to such data. Additionally, this approach has a lot of dependencies on the underlying data that are driving the model. The dependencies are number of data sets, types of data sets, and quality of data sets. So the larger number of data sets one has the more accurate the 3D reconstruction. Having types of data sets that are specific to the anatomy being reconstructed is critical. This means if one wants to address hip orthopedic conditions, then they should have data of hips with similar clinical conditions.

Once the shape is available, it must be formatted to work with commercially available 3D printers. From these 3D point clouds or masks, an STL is typically generated and exported for use.

13.12 SEGMENTATION

This term refers to the act or art of generating 3D shapes from imaging data. This can be done in three different styles—fully manual method, semi-automated, or fully automated.

13.12.1 Manual

Manual segmentation utilizes a trained human to specify pixel by pixel which ones are to be included and which ones are to be excluded from the shape. This method is the most time consuming and can take from minutes on an easy CT with bones of high density to hours or even days on a complicated MRI. This method is still considered the gold standard when evaluating accuracy of segmentations of human anatomy.

13.12.2 Semi-automated

Semi-automated segmentations rely on algorithms to predict the shape based on guidance from the user. This method can save massive amounts of time on straightforward cases. As the anatomical shape and bone quality of the patient further deviate from normal anatomy, the complexity increases, the time savings decrease, and in extreme cases can actually take longer than manual segmentation.

13.12.3 Automated

Techniques where humans do not interact with the 3D reconstruction are known as automated. These techniques are the least accurate of the three types, but are the most repeatable.

13.12.4 Segmentation Accuracy

There are two components that go into segmentation accuracy—the first component is accuracy of the scanner and the second is the method of segmentation. CT scans are typically your most accurate scans with MRI and ultrasound, depending on how the data are collected from those two modalities coming in second. X-ray is typically the least accurate, due to the fact that it is a predicted model and not enough data sets are usually available to drive the model. This, however, does not mean that X-ray technologies are not clinically relevant.

When creating custom and semi-custom patient-specific products, one needs to look at the accuracy dependencies of the system. As a whole, the scan accuracy is the least accurate component. CT can be run at most facilities somewhere between 1 and 1.25 mm inter-slice distance and around 0.5×0.5 mm intra-slice pixel dimensions. This gives one a voxel of approximately 0.25 mm^3. MRI scanners produce 2 mm inter-slice distance and 0.8×0.8 mm intra-slice pixel dimensions for a voxel size of 1.28 mm^3. MRIs are typically enhanced by interpolation in all three dimensions to an approximate 0.16 mm^3 voxel. Even with this enhancement, one can see that the accuracy of the scanner is the limiting factor.

Besides the differences in accuracy, the other important difference when deciding on a modality is what tissues should be included. If the goal is a long bone with no bearing surface, CT is ideal as it is fast and accurate. If, however, one wants to build a guide for a bearing joint like the hip or knee, one should consider MRI as the mating surface will actually be cartilage which is not visualized in CT. MRI is less accurate, and it takes a lot longer to create a 3D model from the generated data, but includes tissues not seen in the other modalities.

13.13 SOFTWARE

There are numerous software vendors providing segmentation software or tool kits. For manual segmentation, some of the more popular options are as follows: Amira (FEI, Burlington, MA), ITK (Kitware, Clifton Park, New York), Mimics (Materialise, Leuven, Belgium), ScanIP (Simpleware, Exeter, United Kingdom), 3DSlicer (open source, Harvard, MA), ORS (Object Research Systems, Montreal, Quebec, Canada), or Osirix (Pixmeo, Bernex, Switzerland). Some of these are commercially available, while others are free for noncommercial use or even open source. Depending on the user needs and end goal, a thorough evaluation should be done. Two of the more common medically cleared semi-automated tools are TeraRecon (Foster City, CA) and VitalImages (Minnetonka, MN). They are very robust and handle most CT scans extremely well. With work they can output an STL. Beyond this there are some extremely skilled companies such as Imorphics (Manchester, United Kingdom), ImageIQ (Cleveland, OH), Qmetrics (Rochester, NY), and VirtualScopics (Rochester, NY) that have automated processes and typically work as processing houses or software as a service model (SaaS). Recently, cloud-based tools have been introduced by companies like 3DSystems (Rock Hill, SC)—Bespoke Modeling. All of these options export an STL or finite element analysis (FEA) format that is typically incompatible with many CAD packages.

13.14 STL TO CAD

There are CAD packages like SolidWorks (Waltham, MA) that can handle STL files and utilizing them for engineering design. However, if such software packages are unavailable, one is required to stay in the segmentation tool's proprietary design software or use a conversion package like 3D Systems' Geomagic Wrap. Autodesk's Maya (San Rafael, CA) and many other programming languages such as Mathworks MATLAB® (Natick, MA) have scripts available online to perform this conversion as well.

The discussion above is by no means the final word in software necessary in generating patient-specific CAD data, without which additive technology cannot be used to make patient-specific personalized product. Of course, software technology changes at a rapid pace—at the time of writing this chapter, CAD data based on CT scan are considered most prevalent.

13.15 MUCH NEEDED TECHNOLOGY

Additive manufacturing involves powder or liquid to be transformed into a physical object with requirements on dimensions and, perhaps more important, on structural integrity. A complex interplay of energy source, raw material quality, and build strategy contributes to the physical and chemical properties of the three-dimensional additive manufactured object. In some instances, finishing operations include thermal treatments as well. The need for devices that determine the dimensional accuracy and presence of structural flaws are apparent. Modern day computers with structured light scanning equipment are becoming indispensable in making additive manufacturing technology address the need for fast checks on dimension on production lines that are catering to personalized products being manufactured in large volumes. The technology has some limitations as it does require some surface preparation, that is, coating the surface with talc and the fact that the method is line of sight dependent so robotics may become necessary to ensure that the entire part is well exposed to the structured light source and camera. If it takes about 15 minutes to coat with talc, fixture the personalized product in the structured light system, make measurements, and compare with the specifications, each such quality control machine can process only 32 parts on an 8-hour work shift. An establishment engaged in mass production of personalized products needing to do 100% inspection on such parts will require multiple such structured light machines to be able to process parts through the quality control group. Suddenly, the cost of running a personalized product manufacturing facility jumps substantially. Newer machines should be able to scan parts regardless of how fixtured, must be faster and work without making use of ad hoc coatings such as talc.

To ensure that there are no structural flaws, common nondestructive tests can be used. The technology here is much more robust as it has had time to mature while addressing the needs of the casting, metal injection molding, and forging industries. However, on personalized products where there is a porous surface layer deliberately added for bone osseointegration, such traditional nondestructive tests may prove inadequate. Of course, process validations, routine monitoring of process parameters, and stringent adherence to preventive maintenance of additive manufacturing machines will help alleviate risk of introducing structural flaw in the additive manufactured part.

13.16 THE FUTURE

The future is wide open for mass customization of medical devices. As stated earlier, we will need to start simple, with low-risk medical devices and instruments, and with experience and data, gradually increasing complexity and risk. Regulations currently under development will soon be established and will guide the process in the near term. As the capabilities, creativities, and capacities grow, the then existing regulations will be challenged and hopefully expanded to accommodate this dynamic and prolific field of medical devices.

Some of the most significant advances will come as new materials and alloys are utilized in additive manufacturing machines. This is directly related to multi-material printing and dynamic material property printing. Well-known attributes of today's implants such as corrosion resistance and biocompatibility will still play an important role. However, design team will tailor make implants such that the biomechanical forces are better channeled to invoke Wolff's laws to ensure strengthening of surrounding remaining bone. Take the hip stem for instance—additive manufacturing a hip stem with strength directly proportional to the bone it is being inserted into and varying the same from the distal end to the proximal seat could potentially be a great advancement. Initially, this may be achieved by simply designing cavities in the body of the stem that do not act as stress riser but help in decreasing the stiffness of the metallic implant. Such design will eventually include organic-shaped cavities as our understanding of form and biomechanics advance. Additive manufactured multi-material will eventually follow but would most likely be limited to metal (or alloy) with ceramic, thus ensuring the elimination of galvanic corrosion. And perhaps well into the future, when collagen printing has advanced with associated and concurrent cross-linking, it is quite possible that metallic parts will eventually be replaced with well-designed composites of collage and calcium phosphates to help with the repair of diseased bone. In the remote future, such composites of collage and calcium phosphates will be seeded with cells obtained from the recipient patient making the implant truly biological and patient specific. There will be procedural controls in place to ensure patient privacy and process controls to ensure that the seeded cells remain biologically viable at the time of surgery. Moreover, quality control using DNA tests will ensure that the patient-specific implant is used with the correct patient. In the meantime, while metal or alloy-based implants are still the norm, additive manufacturing machines will become a part of traditional subtractive milling or grinding and polishing machines, essentially to obtain better surface finish. Such combination machines will compete with traditional additive manufacturing machine with downstream (electro) chemical-based surface polishing techniques.

The field of additive manufacturing, specially related to medical devices, will be very dynamic—it will enjoy heights of rapid adoption and experience downturns. There will be phenomenal successes and colossal failures. In the future, medical device industry will draw experts from cell and tissue architectural biology, materials scientists and bio-mechanics, automation, and software. The above portrayal of the future is simply the authors' view—the reader is encouraged to image the future and work to achieve and shape the future.

13.17 DISCLAIMER

The opinions in this chapter are solely of the authors and are not of their employer.

14 Teaching Additive Manufacturing Concepts to Students and Professionals

Amit Bandyopadhyay, Yanning Zhang, and Susmita Bose

CONTENTS

14.1 INTRODUCTION

The concept of additive manufacturing (AM) is to fabricate 3D objects layer-by-layer in one operation based on a computer-aided design (CAD) file. AM provides a new way of fabricating complex shapes and highly customized 3D objects that traditional manufacturing methods may or may not be able to produce. In addition, no part-specific tools are needed to produce these parts using AM, which means manufacturing cost per part remains almost constant for AM. This is significant for parts that are only needed in small production volume [1]. Nowadays, additive manufacturing has been utilized widely in many applications that we could never imagine before. The first 3D printed pedestrian bridge was accomplished in 2016 by a group from the Institute of Advanced Architecture of Catalonia, Barcelona (Figure 14.1) [2]. The bridge has a total length of 12 meters, a width of 1.75 meters, and is made by micro-reinforced concrete. The first 3D printed electric car, called "Strati," was developed by Local Motors and presented at the International Manufacturing Technology Show 2014 (Figure 14.2) [3]. This vehicle was printed using carbon fiber-reinforced acrylonitrile butadiene styrene (ABS) plastic (40 pounds per hour deposition rate) by a Big Area Additive Manufacturing machine and needed only 44 hours to print. Furthermore, a group of researchers has designed and successfully fabricated a 3D printed jet engine using both electron beam melting and selective laser melting methods, as shown in Figure 14.3 [4].

FIGURE 14.1 World's first 3D printed pedestrian bridge in Madrid. (The Institute for Advanced Architecture of Catalonia designs the first 3D printed bridge in the world, *Inst. Adv. Archit.* Catalonia. https://iaac.net/institute-advanced-architecture-catalonia-designs-first-3d-printed-bridge-world/.)

FIGURE 14.2 Strati, the first 3D printed electric car was developed by Local Motors. (From Robarts, S., "World's first" 3D printed car created and driven by Local Motors, New Atlas, 2014, https://newatlas.com/local-motors-strati-imts/33846/.)

There are still many amazing applications that are being pursued almost every day using different AM technologies. One of the key benefits of AM technologies is the ease of customization. We live in a world where customers are willing to pay premium prices for custom products. AM is already making a tremendous impact in that product market. For example, for complex bone fractures due to accidents, AM-based technologies can

FIGURE 14.3 3D printed jet engine. (From Smith, C.J. et al., *Mater. Today Commun.*, 16, 22–25, 2018.)

be used to manufacture patient-matched implants. For complex surgeries, AM-based technologies are being used to print plastic models for surgical planning. Before too long, AM can invade the food industry to help customers design their own birthday cake, for example, and print it while they shop. Your coffee mug, favorite bicycle, or everyday car can all be custom-designed and 3D printed even today, if there is a demand and the customer is willing to pay the extra cost. However, all of these products may seem unnecessary today for our everyday life. With time, more novel products will be designed and manufactured via AM technologies that can only be produced this way. It can be your additive kitchen tool to make your favorite pasta in unique shapes, toys that will be designed by children following simple instructions, or furniture that will have a unique design based on a customer's need and available space. The jewelry industry has already been at the forefront of customized products using AM technologies. Imagine your wedding ring having embedded information about your names, marriage date, and other information designed by you. That is the level of customization that we will get used to in the coming years, and AM technologies will help us accomplish that at very little added cost to manufacturers [5].

Apart from those fascinating ideas, AM is also being utilized to design products that are in need today. Micro-lattice structures have been reported to show great absorption of impact energy along with a significant reduction of weight [4]. It is almost impossible to fabricate such an object with complex structure through traditional manufacturing methods. With the innovation of AM technologies, researchers have successfully 3D printed and tested various micro-lattice structures (Figure 14.4) [6]. Based on this research, a new type of sports helmet has been designed with the potential to improve the safety performance (Figure 14.5) [7]. In addition, a micro-lattice structure also permits more air flow and provides better comfort than traditional foam materials. This is just one example to demonstrate how AM technologies will transform product design in the coming days. However, we can only do this if designers can begin thinking in terms of layer-by-layer manufacturing, as opposed to casting, forming, or machining operations that are typically used today.

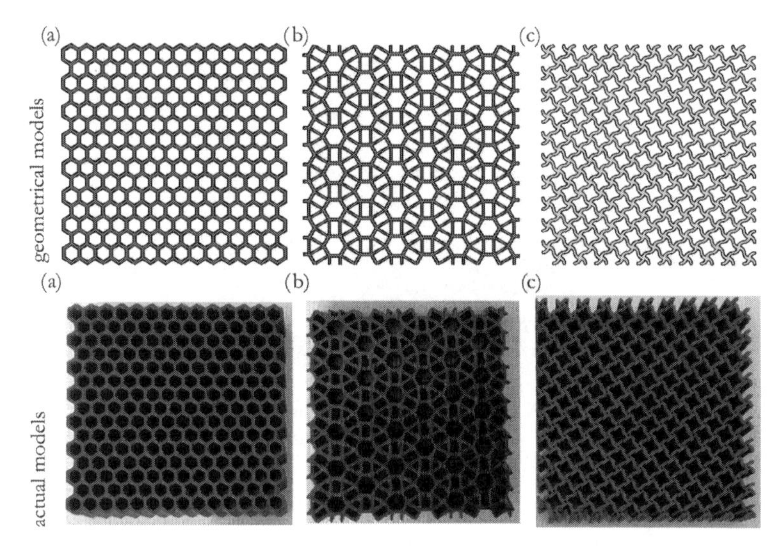

FIGURE 14.4 Top: CAD models and bottom: 3D printed structures: (a) honeycomb, (b) modified honeycomb, and (c) spiral. (From Kucewicz, M. et al., *Mater. Des.*, 142, 177–189, 2018.)

FIGURE 14.5 3D model of a bike helmet with a lattice structure. (From Müller, J., Process and Material Characterization in Additive Manufacturing, ETH Zürich/EDAC. http://www.edac.ethz.ch/Research/Process-Material-Characterization-in-AM.html.)

14.2 UNDERSTANDING ADDITIVE MANUFACTURING

The concept of AM can be taught in a classroom setting from various perspectives. A mechanical engineering faculty member can focus on design aspects and how different AM technologies work to integrate design and manufacturing, while a materials science faculty member can focus on how different materials can be additively manufactured and what machines are good for which material. However, a professor of fine arts may focus on the amazing creativity that can be unlocked using AM technologies. While they all are important, one common thing must be done in all cases—hands-on projects. Hands-on projects can help students see their designed product and feel the real sense of creativity. Even if some serious mistakes are made, the depth of learning using hands-on individual or group projects is extremely important to teaching AM. This approach can also help industry veterans, who have never used AM technologies, to learn how to fully exploit the future potentials of AM.

Since 2015, Washington State University began offering a graduate level additive manufacturing class. The goal of this class is to help a diverse group of students to understand the fundamental concepts of additive manufacturing, as well as its application to product design. As a part of this class, all students need to work on hands-on projects that involve design, AM with the concept model, followed by analysis, and design validation. A 2nd generation design is then produced and, finally, a 2nd generation model is printed. The idea behind this hands-on project is to understand how AM can be used to generate touch and feel models. Once students see the models and the building process (typically fused deposition modeling [FDM] is used to make these parts), ideas to further modify the model to improve its function evolve, which result in the 2nd generation model. In some cases, the differences between the 2nd generation and the 1st generation models are small, while in other cases they are quite significant. At the end, students give presentations about their design and 3D printing experience in front of the other students. Figure 14.6 shows a simple flowchart to explain this process.

14.2.1 Design for AM Processes

Before we start talking about a hands-on project, it is important to see some designs as examples that are unique to additive manufacturing, which are difficult, if not impossible, to fabricate by traditional manufacturing techniques, and shown in Figure 14.7. For example, the ball bearing was built in one operation using resorbable support material. Such a concept of fugitive material to support a structure during the build process is unique to AM and used widely for mostly polymeric materials. Similarly, the "look" piece with complex internal features is another common example of AM capability. These designs are common and can be printed using a variety of AM techniques.

14.2.2 Hands-On Project

One of the key objectives of an additive manufacturing class is to emphasize the unique capabilities of additive manufacturing, i.e., designing for AM or thinking of AM during the design. To demonstrate this point, hands-on projects can be assigned

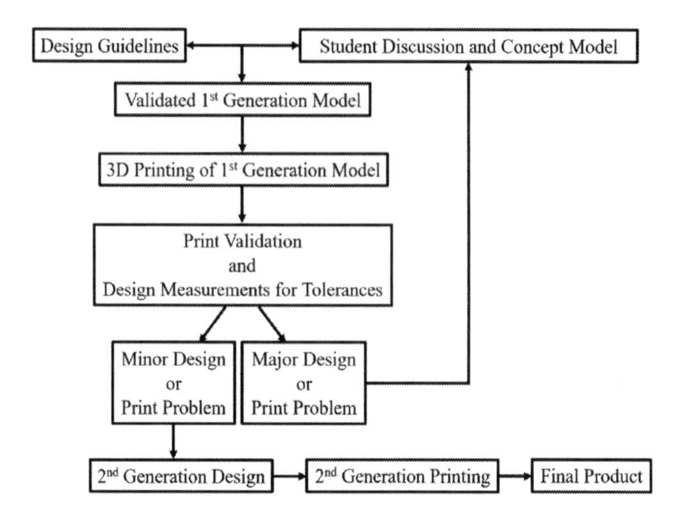

FIGURE 14.6 Design flowchart for 3D printing—from the concept model to the product design.

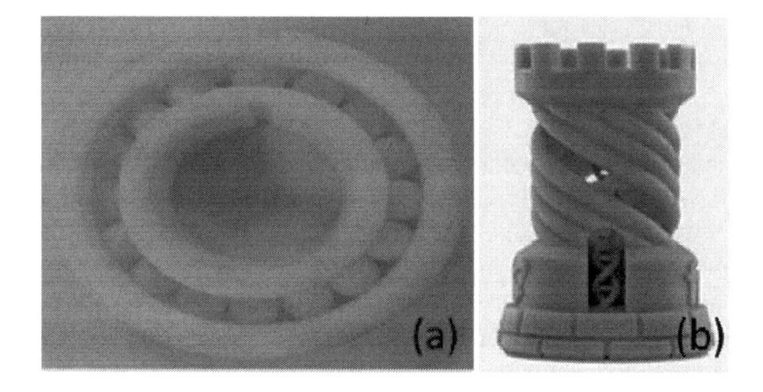

FIGURE 14.7 (a) 3D printed ball bearing. (From Lee, Y.J., et al., *Results Phys.*, 10, 721–726, 2018) (b) 3D printed rook. (Formlabs, The Ultimate Guide to Stereolithography (SLA) 3D Printing, Formlabs.com. https://formlabs.com/blog/ultimate-guide-to-stereolithography-sla-3d-printing/.)

to individual students or to a student group. Once the design is done, students should print at least one model, but it is better if two models can be printed in order to make any necessary changes to the design, as shown in Figure 14.6.

Project 1: Design a part with:

- A rotating component;
- Internal channels;
- Tapered walls;
- And must fit in a 2″ x 2″ x 2″ cube.

Such open-ended design guidelines can be used to promote creativity among students. To make this project more practical, students can also be asked to do a cost analysis based on a build time estimate from the AM machine and that can be reduced.

For this project, three different sets of designs are shown here in Figure 14.8 that three different groups produced. First, it is important to note how different these parts look, yet they all fulfill the design guidelines given in Project 1. That is perhaps the most important classroom learning experience among students when they see the other students design ideas. All parts were printed with FDM type printers using ABS polymers. While all of them have moving parts, the cylindrical structure with screw thread in Figure 14.8i has the simplest design. A round shape nut with six fan blades was screwed onto the cylindrical structure. The nut could move up and down by twisting or blowing air to the fan blades. In addition, the square shape base did not have a flat surface, but irregular curves. On top of the cylindrical structure, a small protrusion was designed to prevent nut detachment.

A red color ABS polymer was utilized to print eight fan blades for the 2nd design (Figure 14.8ii). The fan blades were installed inside a hollow cylinder. Specifically, six non-curved and two curved fan blades were designed. Additionally, an interlock system was designed to prevent detachment of the fan blades. The fan blades could be rotated by applying force or blowing air. Moreover, an internal channel was also designed along the wall. The air could be blown into the small tube on one side of the cylinder wall, which would exit from the hole on the other side.

For the 3rd design, a cubic structure with multiple features was made (Figure 14.8iii). This cubic structure had cross-section variation. Specifically, a spiral structure was designed and sandwiched in the middle. Additionally, in the top section, inversed "T" shape channels were designed in both the X and Y axes. These channels allowed two sliders to slide along each direction. The sliders were fastened with bolt shape structures and connected with a connector. Two holes were designed in this cubic structure, one at the bottom part, and another was located at the center of the cubic structure and were connected via an internal channel.

From a build time point of view, the part in Figure 14.8i took the least amount of time, while the part in Figure 14.8iii took the most time. All three teams could reduce build time during their 2nd iteration of printing by making hollow parts and adjusting proper build orientation.

Summary: Project 1 is a very simple project that can be the first of many hands-on projects a teacher can give to her/his students. It shows how AM technology can be used to make parts that are difficult to make by other operations. All of these parts were just one build, i.e., no assembly was allowed. Therefore, the teams had to learn how to keep the minimum gaps between the moving parts, how to stop them from going out of the part, how to minimize build time, and most importantly, how to design for AM that uses layer-by-layer manufacturing.

Project 2: Multiple parts with assembly

Instead of just a single part with unique features, hands-on projects can also be designed with multiple parts to test both the part design and assembly.

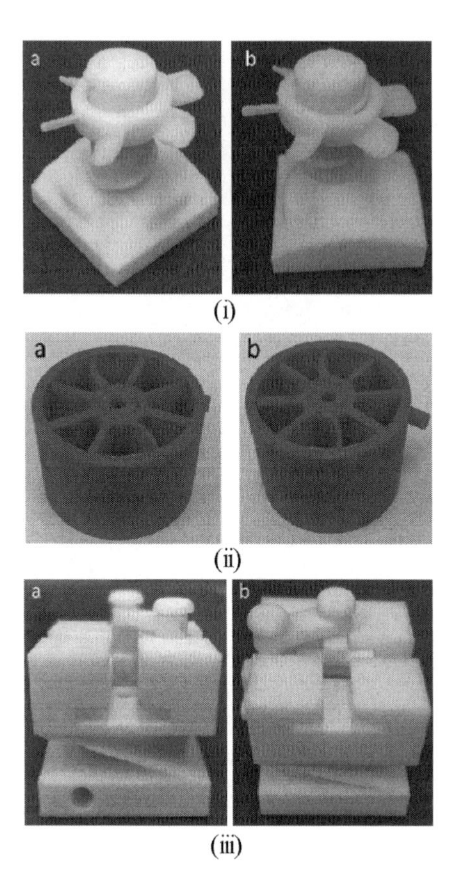

FIGURE 14.8 FDM processed parts based on the Project 1 guidelines: (i)–(iii).

In many instances, small mistakes in parts can prevent the assembly to make a product. One simple example is given in Figure 14.9. It is a gearbox design that consists of 12 different gears and was 3D printed with an ABS polymer using an FDM printer. Different numbers of tooth, shape, and size were designed to make the entire gearbox. In addition, each gear was fastened with a 3D printed screw. Although the entire gearbox was 3D printed in one run, each gear could be installed and uninstalled by screwing and unscrewing. The gearbox could be rotated by rotating the arm. Fortunately, many such design concepts can be found on the Internet, and instructors can make some modifications and design projects that are suitable for their class based on the class size and difficulty level.

These examples demonstrated the idea of hands-on projects in an additive manufacturing class. Through these projects, students could develop a deep appreciation for designing for AM and related challenges. From the design to the 3D printing, students fully participated in every single step. In addition, students could experience some of the limitations of additive manufacturing techniques as well. These limitations include processing defects (cracks and delamination),

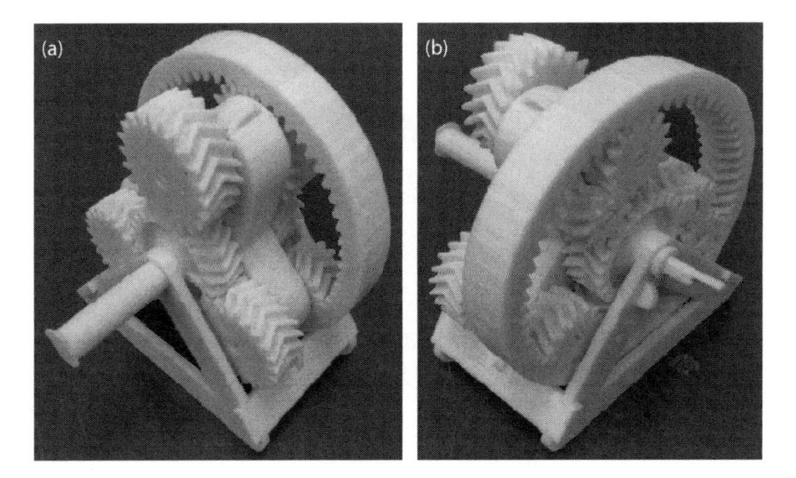

FIGURE 14.9 Designed gearbox: (a) front view and (b) rear view.

post-processing (surface finishing and support materials removing), and printing parameters optimization. Using different additive manufacturing methods, such as FDM and stereolithography (SLA), if available, to fabricate parts from the same digital file also help students compare the pros and the cons of each additive manufacturing method. Depending on the AM process and process parameter selection, the tolerances of AM processed parts could be different. The instructor may require students to measure the 3D printed parts and then compare them with the original design. Through research and classroom discussions, students could gain more experience that is related to the factors that can affect the part quality and tolerances.

14.3 SUMMARY

3D printing or additive manufacturing is becoming a very popular manufacturing tool for a large variety of parts from concept models to functional components. With the adaptation of this technology in the main stream manufacturing domain, there comes a critical need to educate and train our next generation of students on how to utilize and innovate using this technology. For the 1st generation of AM users, the justification of AM is usually shorter lead time or low volume production that is cheaper. However, most of those designs are done for other manufacturing operations, and AM is really an afterthought. The AM approach can actually do a lot more if the parts are designed specifically for layer-wise manufacturing. From that point of view, there is a critical need not only to train our younger students, but even to retrain our existing workforce, those who are generally familiar with subtractive and other forms of manufacturing processes. This chapter is designed to help instructors to use some hands-on projects with students to help them understand how additive manufacturing works. Our sincere hope is that the next generation of students trained in AM will design and manufacture more 3D printed products for our everyday life that wouldn't be possible to manufacture using any other techniques.

QUESTIONS

As a part of the question, here are a couple of simple ideas for student projects that can be used for hands-on learning using additive manufacturing.

Project 1: The idea is to design a single part with the help of AM that cannot be manufactured otherwise.

Pick a simple engineering part that you can think of (Examples—a carabineer or a gear). Define the basic function that this part is supposed to do. And then define parameters such as—

1. Maximum size of the part (Example: should fit in a 2″ x 2″ x 2″ cube)
2. The designer's initials must be present
3. Some unique AM features must be present (such as some part may be hollow)
4. Define a unique property (Example—The carabineer must withstand this much tensile strength).

Let students design the part and print their 1st generation concept model. If it meets all the requirements, then you can ask how they can re-design their part to reduce weight and build time, i.e., offer a 2nd generation design criteria and see the outcome.

Outcome: The main outcome is to see that the students can conceptualize a part, design it, and then improve its performance without sacrificing basic design criteria with the help of AM.

Project 2: Projects that deal with topology optimization.

Pick at least two or more parts that need to be connected to perform a job. It could be a base part and brackets; co-axial tubes for gas mixing; and so on. The idea is to come up with a new design that will eliminate the idea of multiple parts and make it one complex part with the help of additive technology. These types of projects are ideally suited for professional engineers or students who possess advanced knowledge of engineering systems.

These projects are unique in nature. First, students build a prototype that can connect all the parts as it is currently done. The question is—how can that design be improved utilizing topology optimization? Since no joining is involved in AM processed parts, through engineering stress analysis, better designs can be proposed that are only possible by AM. A 2nd generation model fabrication is desirable to appreciate the project.

REFERENCES

1. S. Bose, D. Ke, H. Sahasrabudhe, A. Bandyopadhyay, Additive manufacturing of biomaterials, *Progress in Materials Science*, 93, 45–111, 2018.
2. The Institute for Advanced Architecture of Catalonia designs the first 3D printed bridge in the world, Inst. Adv. Archit. Catalonia. https://iaac.net/institute-advanced-architecture-catalonia-designs-first-3d-printed-bridge-world/ (Accessed January 10, 2019).
3. S. Robarts, "World's first" 3D printed car created and driven by Local Motors, New Atlas. (2014). https://newatlas.com/local-motors-strati-imts/33846/ (Accessed January 10, 2019).

4. C.J. Smith, P.S. Mahoney, I. Todd, 3D printing a jet engine: An undergraduate project to exploit additive manufacturing now and in the future, *MaterialToday Communication* 16, 22–25, 2018.
5. Syed A.M. Tofail, E.P. Koumoulos, A. Bandyopadhyay, S. Bose, L. O'Donoghue, C. Charitidis, Additive manufacturing: Scientific and technological challenges, market uptake and opportunities, *Materials Today*, 21, 22–37, 2018.
6. M. Kucewicz, P. Baranowski, J. Małachowski, A. Popławski, P. Płatek, Modelling, and characterization of 3D printed cellular structures, *Material Design*, 142, 177–189, 2018.
7. J. Müller, Process and Material Characterization in Additive Manufacturing, ETH Zürich/EDAC. http://www.edac.ethz.ch/Research/Process-Material-Characterization-in-AM.html (Accessed January 10, 2019).
8. Y.J. Lee, K.H. Lee, C.H. Lee, Friction performance of 3D printed ball bearing: Feasibility study, *Results Physics* 10, 721–726, 2018.
9. Formlabs, The Ultimate Guide to Stereolithography (SLA) 3D Printing, Formlabs.com. https://formlabs.com/blog/ultimate-guide-to-stereolithography-sla-3d-printing/ (Accessed January 10, 2019).

15 Additive Manufacturing
The Future of Manufacturing in a Flat World

Susmita Bose and Amit Bandyopadhyay

CONTENTS

15.1 INTRODUCTION

Additive manufacturing (AM) or 3D printing is the most exciting news today in the world of manufacturing. Parts can be directly built without any part-specific tooling. Most of these parts are near net-shape and require only a small finishing operation before actual use. Unlike in the early 1990s, when most of these parts only mimicked the size and shape of the actual part, today's AM machines can produce functional parts that can perform regular operations and replace the existing parts. In the early stages of AM technology implementation, most companies were looking at low volume parts to see if those could be manufactured using AM-based techniques to reduce the inventory. The publishing industry has already gone through this transition. In today's world, many books are only printed when orders come in for them. The same concept is currently being implemented by many in the manufacturing industries using AM: manufacture on-demand. However, if those are multi-component or multi-materials parts, significant challenges still exist that are needed to be overcome to use AM techniques directly [1].

The rise of AM or 3D printing is making a significant impact on how products are designed and manufactured. The focus of this last chapter is to highlight some changes that are happening in our society and how those changes can influence our lives. Let us start with a simple example—buying a car. In the current practice, buyers will go to a dealership to look at various models, colors, and features. Buyers can also conduct a similar search over the Internet as well. Finally, when the buyer makes their decision based on the available options, the buyer purchases the car. However, we have already seen a functional 3D printed car. A large fused deposition modeling machine was developed to directly manufacture the entire car. Most of the parts for this car were printed with a carbon fiber-reinforced plastic material. Therefore, it can be envisioned that in the coming days, a buyer can go to a showroom, look at various options for new cars, and then design or modify her/his own car and get a manufacturer quote. The manufacturer will do the remaining engineering and make this a "customized car" that will be delivered in a few days. Such possibilities are no longer considered as "science fiction," but a reality that can happen anywhere in the world. The purpose of this example is to show how flexible manufacturing via AM can offer the customization of various products. Instead of a car, a similar concept can be applied to bicycles, skate-boards, or most other products where customization can be beneficial and important to the consumer.

Use of the AM approach can also have a significant effect on human health, where a physician can have access to a patient-specific or a defect-specific implant to treat a patient [2]. Depending on the clinical need, this technology can revolutionize today's treatment options and routine health care processes. Accepting the fact that more of these kinds of transformative changes will happen in our everyday life in the coming days, we like to ask a basic question—how will such changes impact the future of manufacturing industries?

15.2 FROM A 3D PRINTED CAR TO 3D PRINTER IN SPACE

The year 2014 marked two important events for additive manufacturing—(1) a 3D printed functional car and (2) a 3D printer in space. National Aeronautics and Space Administration (NASA) launched the first 3D printer to the International Space Station to experiment with printing parts in zero gravity. Made In Space built the fused deposition-based 3D printer that can operate in zero gravity to manufacture plastic parts than can be produced in space. Though the concept was envisioned by many for the past 20 years, its implementation marks an important new era for additive manufacturing. The race is now on to use AM technologies to produce functional parts in space using materials other than just polymers. More importantly, can *in situ* resources be utilized on the surface of the Moon or Mars to produce small and large structures for future human explorations? It is envisioned that in the coming years, AM in space will see a significant growth because that is the only way that space explorations can move forward. In 2010, the first direct fabrication of moon-rock regolith structures using a commercially available laser engineered net shaping (LENS) system was reported [3]. It was shown that direct laser melting of moon-rock regolith is possible due to the high silica contents and some simple shapes can be formed, as shown in Figure 15.1. Though the parts produced had low strength for tooling, this concept can be utilized to make parts for basic civil infrastructures. For example, it can be envisioned that solar powered 3D printers are busy at work on the Moon or Mars surface to make bricks that are being used for roads, small buildings, or launch pads. Based on recent advances in NASA and the European space agency's work with AM, such ideas are not far from reality. The concept of 3D printed building in outer space using *in situ* resources perhaps can be though of as mundane activities in the coming decade.

FIGURE 15.1 First demonstration of LENS-processed direct fabrication of moon-rock regolith simulants (JSC-1AC). (From Balla, V.K. et al., *Rapid Prototyp. J.*, 18, 451–457, 2012.)

15.3 FROM BIOPRINTING TO FLEXIBLE ELECTRONICS

Additive manufacturing offers significant potential toward solving long-standing challenges related to human health [2,4,5]. There are many facets to bioprinting, i.e., printing with living cells is still at the research and development stage. However, the basic concept of printing human bone or organs with the help of 3D printing is still fascinating. From plastic surgery to cancer treatment, from birth defects to amputees—all are looking for a potential breakthrough in the development of tissue engineering and regenerative medicine in order to harvest different body parts to enhance the quality of life or help patients live longer [2]. For example, 1st generation craniofacial metallic implants from computed tomography images of a fractured skull, shown in Figure 15.2, were manufactured in 2007, something that is impossible to accomplish using any other manufacturing techniques [6]. Similarly, bone tissue engineering using 3D printing in which defect-specific porous bioresorbable ceramic scaffolds are manufactured for bone healing are also becoming popular [7,8]. A scaffolds porous architecture can be tailored to match the bone density of the patient or specific defect location. Such implants can also be used for site-specific drug delivery to enhance healing. Moreover, bioprinting, the direct deposition of cells on a substrate using additive manufacturing, although exciting and feasible on a small scale, still needs to be developed further to ensure longer shelf-life and cell viability during and after processing. The coming years will be very exciting in this field, and many new products are poised to come to the market with the help of different 3D printing technologies.

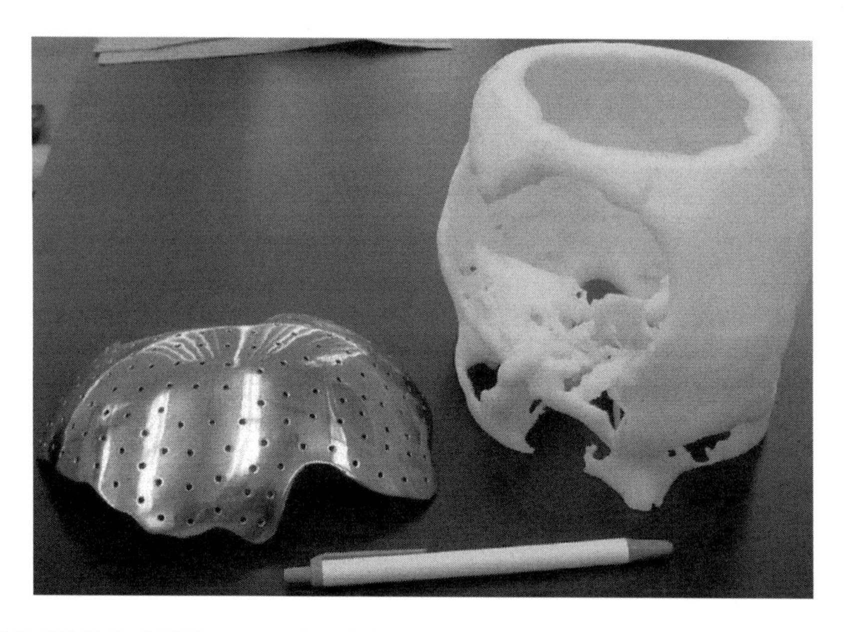

FIGURE 15.2 LENS-processed craniofacial Ti implant and FDM-processed polymer prototype of the skull with a large defect. (From Bandyopadhyay, A. et al., *MRS Bull.*, 40, 108–115, 2015.)

Flexible electronics is also an area of significant promise in which additive manufacturing has already started to play a critical role. As we move beyond 2D structures and enter the domain of 3D structures for various electronic devices, flexibility in manufacturing will be a key to success. Low cost and single use devices will also dominate various application areas in the world market. Faster design optimization/validation of these devices is the key to success toward commercialization. Different 3D printing technologies are already making a difference in these application areas, and many more applications are considering such transition.

15.4 INNOVATION IN MANUFACTURING USING AM—MULTI-MATERIALS STRUCTURES

Apart from building parts without any part-specific tools, additive manufacturing can also help to innovate different components that are difficult, if not impossible, to make using conventional manufacturing. These innovations may come from the design of novel topology optimized parts of single or multi-material structures. In conventional manufacturing, multi-material parts need to be joined by welding, brazing, or soldering. However, additive manufacturing processes can be used to directly fabricate parts with multiple significantly different compositions offering different functionality. The directed energy deposition-based approach was used to design and manufacture parts having bimetallic composition—from Ti6Al4V to SS 316 [9], Inconel 718 to Ti6Al4V [10], or Inconel 718 to Cu alloy [11]. Large thermal residual stresses prevented them from being bonded directly, therefore an intermediate layer may be needed in some cases to minimize the residual stresses and related cracking. Figure 15.3a and b show images depicting an AM-processed bimetallic structure using LENS. Figure 15.3 shows Inconel 718 to Ti6Al4V (Ti64) bimetallic structures processed via LENS [10]. Figure 15.3a shows as build structure. Figure 15.3b shows polished cross-section. Figure 15.3c shows polished and etched cross-section. Smooth interphase from one composition to another helped to form this bimetallic structure between otherwise incompatible alloys with the help of vanadium carbide ceramics. Various research groups have also started looking at different combinations of bimetallic/multi-materials structures using AM [1,12]. The main challenge in multi-material additive manufacturing lies with the design of these structures with built-in properties that are different in various locations because *.stl* files are still based on one material property. Multi-material structures can be used in a variety of applications. For example, structures where one end may experience a high temperature can be manufactured in one operation by using two different materials in which one is suitable for high temperature application such as ceramics, while the rest

FIGURE 15.3 LENS processed Ti6Al4V to Inconel 718 bimetallic structure with composition bond layer, (a) shows as build structure, (b) shows polished cross-section and (c) shows polished and etched cross-section. (From Onuike, B., and Bandyopadhyay, A., *Addit. Manuf.*, 22, 844–851, 2018.)

of the part is made of a different material, such as a metal [13]. Similar structures can also be envisioned for materials where corrosion resistance is a key factor, where surfaces that will be exposed to a corrosive environment will be made of a corrosion resistant materials, such as ceramics and the rest of the body is made of metals [14].

15.5 APPLICATION OF ADDITIVE MANUFACTURING IN REPAIR

In the coming decade, the application of AM towards repair and reconstruction of engineering parts will find significant uses. Many engineering parts need replacements not due to major failure, but mainly because of wear or damage in small areas of the part. For large parts, such replacements are quite expensive and sometimes difficult if they happen to be one of a kind parts. Additive manufacturing can be used to selectively repair a specific region of an engineering part based on the computer-aided design (CAD) design. In most cases, a final machining is needed. RPM Innovations in South Dakota (USA) specializes in such repair operations using laser-based additive manufacturing [15]. Due to multi-axial AM operations, a repair in a selected region with a specific alloy composition can be accomplished easily. Figure 15.4 shows one such example of the laser repair of a drag line swing shaft. The swing shaft is an ~5 meters long shaft that weighs 25,000 pounds and is used on a drag line of coal mine. Conventional repairs of a swing shaft cause failure problems due to heat input. A novel laser-based repair process was developed at RPMI for bearing surfaces of a 4340 Swing Shaft utilizing 420 stainless steel. Such composition modifications increase the wear resistance of surfaces, which extends the shaft life as well. A repair cost using laser-based AM technology is approximately 65% less than a new part. Moreover, due to compositional variations during repair, a swing shaft's service is extended as compared to a new one. Such an approach not only saves money, but can also save considerable time delay for critical parts. Since AM technologies are not design- or material- specific, the same machine can be used to repair a variety of parts with different shapes and compositions.

FIGURE 15.4 (a) A swing shaft worn bearing surface, (b) shaft during deposition, and (c) final machined shaft. (From Nick Wald, RPM Innovations (SD, USA), Unpublished data, 2014.)

15.6 ADDITIVE MANUFACTURING IN TISSUE ENGINEERING AND DRUG DELIVERY

Application of additive manufacturing has a significant potential in the areas of implants, implants, tissue engineering and drug delivery. In 2018, over 100,000 metal implants have been manufactured using metal additive manufacturing approach that are certified for human use, and this number is growing rapidly. Moreover, there will always be shortage of organ donors for patients with critical needs. The advent of tissue engineering has offered significant promise to improve human health over the past three decades. However, patient-matched organs are not yet feasible due to many engineering challenges. Additive manufacturing technologies are currently being explored to fabricate different scaffolds that can mimic the size and shape of a particular defect of a patient. The idea behind combining AM technologies and tissue engineering is simple—produce the patient-matched scaffolds using AM with bio-resorbable materials that can dissolve within the body after some time. Therefore, when the defect is being healed, the scaffold material will slowly dissolve without any harmful effect to the body. A simple example is bone tissue engineering and additive manufacturing. Bone defects due to cancer, osteoporosis, or ordinary fracture require implants that are mostly made of metals. In most cases, once placed, those metal implants remain in the body. However, instead of metals, if ceramic or polymeric implants can be placed in the body that are bio-resorbable and shaped based on that specific defect size, better healing can be accomplished. To improve healing, implants can be made porous to introduce biological fixation, where bone tissue can integrate with the implants better. The pore size and shape can also be tailored based on the patient's anatomy [16–19]. More importantly, specific dopants can be added to the scaffold to further enhance angiogenesis and osteogenesis during healing [20]. All of these ideas are possible now due to additive manufacturing and are being pursued by different research groups around the world. Figure 15.5 shows images of tricalcium phosphate (TCP, $Ca_3(PO_4)_2$) based bioresorbable scaffolds with different shapes and porosity produced via a powder bed-based 3D Printer (ExOne, PA). Figure 15.5a shows photomicrograph of 3d Printed (3DP) pure (i & ii) and Mg-Si doped TCP (iii & iv) scaffolds showing the development of new bone formation inside the interconnected macro pores of the 3DP scaffolds after 16 (i & iii) and 20 (ii & iv) weeks in a rat distal femur model. Hematoxylin and Eosin (H&E) staining of transverse section. BM = Bone marrow; Arrows indicate the interface between scaffold and host bone; Star (*) indicates acellular regions derived from the scaffold. Color description: Black = Bone marrow; Pink/Reddish = New/old bone; Yellowish = acellular regions derive from scaffold; (b) Histomorphometric analysis of bone area fraction (total newly formed bone area/total area, %) from 800 μm width and 800 μm height H&E stained tissue sections (**$p < 0.05$, *$p > 0.05$, n=8) [18]. It can be seen that addition of dopants can influence new bone formation. TCP is a bio-resorbable ceramic, which degrades in the body at a slow rate. The degradation kinetics of TCP ceramics can also be tailored by adding different metal ions such as Mg or Zn w Sr [21–25]. Moreover, the scaffolds degradation kinetics can be further modified through the introduction of porosity of various sizes, shapes, and volume fractions [26,27]. Similar to resorbable ceramics, many resorbable polymers are also

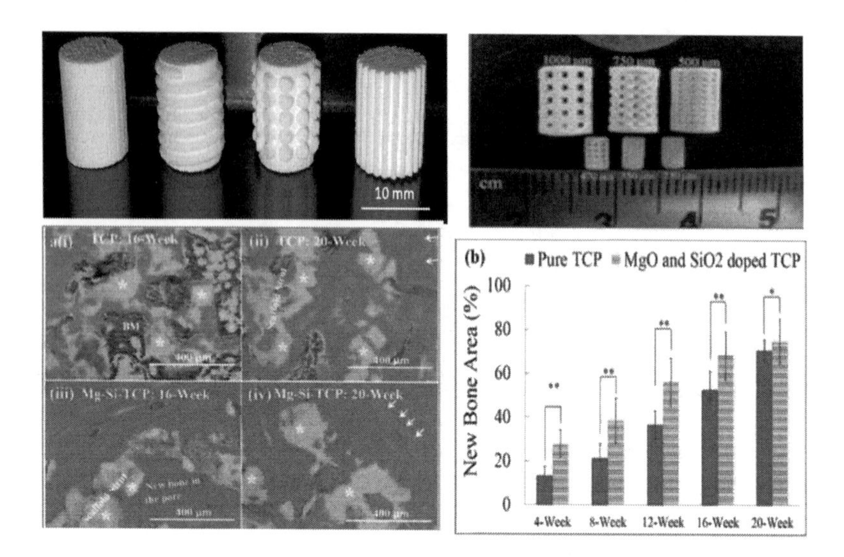

FIGURE 15.5 3D Printed TCP scaffolds and different structures with fine features printed in a powder bed based ExOne system. (a) Photomicrograph of 3DP pure (i and ii) and Mg-Si doped TCP (iii and iv) scaffolds showing the development of new bone formation inside the interconnected macro pores of the 3DP scaffolds after 16 (i and iii) and 20 (ii and iv) weeks in rat distal femur model. Hematoxylin and Eosin (H&E) staining of transverse section. BM = Bone marrow; Arrows indicate the interface between scaffold and host bone; Star (*) indicates acellular regions derived from the scaffold. Color description: black = bone marrow; pink/reddish = new/old bone; yellowish = acellular regions derive from scaffold; (b) Histomorphometric analysis of bone area fraction (total newly formed bone area/total area, %) from 800 μm width and 800 μm height H&E stained tissue sections (**$p < 0.05$, *$p > 0.05$, n=8). (From Bandyopadhyay, A. et al., *Journal of the American Ceramic Society*, 89 (9), 2675–2688, 2006.)

currently being used to fabricate scaffolds for various tissue engineering needs. It is expected that such developments will soon result in many commercial products to impact human health [28].

15.7 ON-SITE ON-DEMAND MANUFACTURING VS. MASS PRODUCTION

Today, there are many makerspaces available, where parts can be fabricated based on a small printer charge. Just like photocopiers that can be used to copy materials, customers can bring their own 3D drawings or create them and then 3D print it in a machine. Though the concept is simple, its implications can be far-fetched. Ordinary citizens now have access to inexpensive 3D printers to validate their creations—whether it is for fun, for a gift, or for a technical project. Such advances have the potential to change the culture of our society in the long run. Companies may start the same approach of distributed on-demand manufacturing houses at various locations, as opposed to bulk manufacturing in one location. Certainly cost is a big issue that generally drops significantly as a function of production volume. However, for many parts, prediction of future market trends can be tricky, and companies may deal with large unsold inventories because of the unpredictable market dynamics. On-demand manufacturing is not suitable for all products, but only for value-added products that are primarily made of one or two materials and smaller in size. These products can be for biomedical industries, space or aerospace industries, or applications in Department of Defense related areas, just to name a few. More importantly, innovative topologically optimized designs are also possible with the AM approach that can enhance efficiency of many parts and allow the incorporation of complex designs that are simply impossible to manufacture using conventional manufacturing approaches. A secondary advantage for the use of AM may come from the control over the core designs for a company, instead of sharing with many parties at various locations. On-demand manufacturing also offers the possibility of personalization that would be impossible to incorporate in a mass manufacturing platform, for example, special memorable photos such as marriages no longer need to be in 2D, but can be printed in 3D with color and embedded messages. Additionally, patients can see their body parts even before the planned plastic surgery and make changes with the use of AM. Due to the lower cost of the AM machines and a better understanding from the general population, the use of AM does not need to be restricted to only jewelry or dentistry, but it can spread to many other ordinary applications. That does not mean that global manufacturing centers will be obsolete in the coming days. However, more and more community centric manufacturing will certainly become a part of our everyday lives. Children can put their input into designing their own toys rather than just buy one. Cars can have the owner's name or other information printed on the body instead of just on the license plates. Physicians can order implants for their patients with special needs rather than try to retrofit what is commercially available. The possibilities are endless, and the only limiting factor will be our imagination and creativity. The applications of such technologies will have no geo-political boundaries. If there are raw materials and Internet access, additive manufacturing facilities can be installed in a few days, even in remote locations on earth, to enhance the quality of life for our generation, and the generations to come. The differences between developing countries and developed nations or the differences between manufacturing nations and predominantly consumer nations will decrease because of the use of additive manufacturing. A flat world will become even more uniform for the generations to come. Even 25 years ago, no one could have imagined such an impact that could come from the advent of additive manufacturing. We, the editors of this book, are fortunate to have worked with this technology for over two decades now and have had a chance to contribute to this technological revolution. Cheaper machines, reproducible parts, and reliable part quality will drive the AM-based industry to shape the future of main-stream manufacturing.

15.8 SUMMARY

A brief summary of the impact of additive manufacturing is discussed in this chapter. It is clear that additive manufacturing is changing the landscapes of current industrial practices. On-demand manufacturing using AM technologies is a new trend that will significantly influence many industries and product design protocols. It is envisioned that the differences between developing countries and developed nations or the differences between manufacturing nations and predominantly consumer nations will decrease due to the use of additive manufacturing. In combination with Internet-based technologies and the computer-aided designs, additive manufacturing will make a flat world even more uniform for the generations to come.

REFERENCES

1. Bandyopadhyay A. and Heer B., "Additive manufacturing of multi-material structures," *Materials Science and Engineering R*, Reports 129, 1–16 (2018).
2. Bose, S., Ke, D., Sahasrabudhe, H., and Bandyopadhyay, A., "Additive manufacturing of biomaterials," *Progress in Materials Science*, 93, 45–111 (2018).
3. Balla, V. K., Roberson, L. B., O'Connor, G. W., Trigwell, S., Bose, S., and Bandyopadhyay, A., "First demonstration on direct laser fabrication of lunar regolith parts," *Rapid Prototyping Journal*, 18 (6), 451–457 (2012).
4. Bose, S., Robertson, S. F., and Bandyopadhyay, A., "3D printing of bone implants and replacements," *American Scientist*, 106, 112–119 (2018).
5. Bose, S., Vahabzadeh, S., and Bandyopadhyay, A., "Bone tissue engineering using 3D printing," *Materials Today*, 16 (12), 496–504 (2013).
6. Bandyopadhyay, A., Bose, S., and Das, S., "3D printing of biomaterials," *MRS Bulletin*, 40 (02), 108–115 (2015).
7. Bose, S., Roy, M., and Bandyopadhyay, A., "Recent advances in bone tissue engineering scaffolds," *Trends in Biotechnology*, 30 (10), 546–554 (2012).
8. Bose, S., Fielding, G., Tarafder, S., and Bandyopadhyay, A., (2013) "Trace element doping in calcium phosphate ceramics to understand osteogenesis and angiogenesis," *Trends in Biotechnology*, 31 (10), 594–605 (2013).
9. Sahasrabudhe, H., Harrison, R., Carpenter, C., Bandyopadhyay, A., "Stainless steel to titanium bimetallic structure using LENS™," *Additive Manufacturing*, 5, 1–8 (2015).
10. Onuike, B., and Bandyopadhyay, A., "Additive manufacturing of Inconel 718–Ti6Al4V bimetallic structures," *Additive Manufacturing*, 22, 844–851 (2018).
11. Onuike, B., Heer, B., and Bandyopadhyay, A., "Additive manufacturing of inconel 718—Copper alloy bimetallic structure using laser engineered net shaping," *Additive Manufacturing*, 21, 133–140 (2018).
12. Bandyopadhyay, A., and Traxel, K., "Invited review article: Metal-additive manufacturing—Modeling strategies for application-optimized designs," *Additive Manufacturing*, 22, 758–774 (2018).
13. Gualtieri, T., and Bandyopadhyay, A., "Additive manufacturing of compositionally gradient metal-ceramic structures: Stainless steel to vanadium carbide," *Materials & Design*, 139, 419–428 (2018).
14. Stenberg, K., Dittrick, S., Bose, S., and Bandyopadhyay, A., "Influence of simultaneous addition of carbon nanotubes and calcium phosphate on wear resistance of 3D printed Ti6Al4V," *Journal of Materials Research*, 33 (14), 2077–2086 (2018).
15. Nick Wald, RPM Innovations (SD, USA), Unpublished data, 2014.
16. Darsell, J., Bose, S., Hosick, H., and Bandyopadhyay, A., "From CT scans to ceramic bone grafts," *Journal of the American Ceramic Society*, 86 (7), 1076–1080 (2003).
17. Bose, S., Darsell, J., Kintner M., Hosick, H., and Bandyopadhyay A., "Pore size and pore volume effects on calcium phosphate based ceramics," *Materials Science and Engineering C*, 23, 479–486 (2003).

18. Bose, S., Tarafder, S., and Bandyopadhyay, A., "Effect of chemistry on osteogenesis and angiogenesis towards bone tissue engineering using 3D printed scaffolds," *Annals of Biomedical Engineering*, 45 (1), 261–272 (2017).
19. Tarafder, S., Balla, V. K., Davies, N., Bandyopadhyay, A., and Bose, S., "Microwave sintered 3D printed tricalcium phosphate scaffolds for bone tissue engineering," *Journal of Tissue Engineering and Regenerative Medicine*, 7 (8), 631–641 (2013).
20. Fielding, G., and Bose, S., "SiO2 and ZnO dopants in 3D printed TCP scaffolds enhances osteogenesis and angiogenesis *in vivo*," *Acta Biomaterialia* 9 (11), 9137–9148 (2013).
21. Bandyopadhyay, A., Bernard, S., Xue, W., and Bose, S., "Feature article: Calcium phosphate based resorbable ceramics: Influence of MgO, ZnO and SiO_2 Dopants," *Journal of the American Ceramic Society*, 89 (9), 2675–2688 (2006).
22. Banerjee, S. S., Tarafder, S., Davies, N. M., Bandyopadhyay, A., and Bose, S., "Understanding the influence of MgO and SrO binary doping on the mechanical and biological properties of β-TCP ceramics," *Acta Biomaterialia*, 6, 4167–4174 (2010).
23. Fielding, G., Bandyopadhyay, A., and Bose, S., "Effects of SiO_2 and ZnO doping on mechanical and biological properties of 3D printed TCP scaffolds," *Dental Materials*, 28, 113–122 (2012).
24. Bose, S., Tarafder, S., Banerjee, S. S., and Bandyopadhyay, A., "Understanding in vivo response and mechanical property variation in MgO, SrO and SiO_2 doped β-TCP," *Bone*, 48 (6), 1282–1290 (2011).
25. Bandyopadhyay, A., Petersen, J., Fielding, G., Banerjee, S., and Bose, S., "ZnO, SiO_2 and SrO doping in resorbable tricalcium phosphates: Influence on strength degradation, mechanical properties and in vitro bone cell material interactions" *Journal of Biomedical Materials Research: Part B -Applied Biomaterials*, 100 (8), 2203–2212 (2012).
26. Tarafder, S., Dernell, W., Bandyopadhyay, A., and Bose, S., "SrO and MgO doped microwave sintered 3D printed tricalcium phosphate scaffolds: Mechanical properties and in vivo osteogenesis in a rabbit model," *Journal of Biomedical Materials Research-Applied Biomaterials*, doi:10.1002/jbm.b.33239 (2014).
27. Ke, Dongxu, Bandyopadhyay A., and Bose S., "Doped tricalcium phosphate scaffolds by thermal decomposition of naphthalene: Mechanical properties and in vivo osteogenesis in a rabbit femur model" *Journal of Biomedical Materials Research: Part B Applied Biomaterials*, 103 (8), 1549–1559 (2015).
28. Tofail, Syed A. M., Koumoulos, E. P., Bandyopadhyay, A., Bose, S., O'Donoghue, L., and Charitidis, C., "Additive manufacturing: Scientific and technological challenges, market uptake and opportunities," *Materials Today*, 21 (1), 22–37 (2018).

Index

Note: Page numbers in italic and bold refer to figures and tables, respectively.

T